Distributed Real-Time Architecture for Mixed-Criticality Systems

Distributed Real-Time Architecture for Mixed-Criticality Systems

Edited by
Hamidreza Ahmadian
Roman Obermaisser
Jon Perez

CRC Press
Taylor & Francis Group
Boca Raton London New York

CRC Press is an imprint of the
Taylor & Francis Group, an **informa** business

CRC Press
Taylor & Francis Group
6000 Broken Sound Parkway NW, Suite 300
Boca Raton, FL 33487-2742

First issued in paperback 2022

© 2019 by Taylor & Francis Group, LLC
CRC Press is an imprint of Taylor & Francis Group, an Informa business

No claim to original U.S. Government works

ISBN-13: 978-0-815-36064-3 (hbk)
ISBN-13: 978-1-03-233898-9 (pbk)
DOI: 1201/9781351117821

Visit the Taylor & Francis Web site at
http://www.taylorandfrancis.com

and the CRC Press Web site at
http://www.crcpress.com

Contents

Contributors   **xi**

1 Introduction   **1**
   R. Obermaisser
  1.1 Context in the Area of Mixed-Criticality Systems   1
  1.2 Scope of the Book   2
  1.3 Motivation and Objectives   4
  1.4 Structure of the Book   5

2 Architectural Style   **7**
  R. Obermaisser, M. Abuteir, H. Ahmadian, P. Balbastre, S. Barner, M. Coppola, J. Coronel,
  A. Crespo, P. Balbastre, G. Fohler, G. Gala, M. Grammatikakis, A. Larrucea Ortube, T.
  Koller, Z. Owda, and D. Weber
  2.1 System Model   9
      2.1.1 Physical Platform Structure   9
      2.1.2 Logical Application Architecture 10
      2.1.3 Mapping of Application to Platform 10
  2.2 Waistline Structure of Services 11
      2.2.1 DREAMS Core Services 12
      2.2.2 Architectural Building Blocks for Platform Services 14
      2.2.3 Communication Services: On-Chip 17
      2.2.4 Communication Services: Off-Chip 19
      2.2.5 Communication Services: Shared Memory 26
      2.2.6 Communication Services: IOMMU 27
      2.2.7 Communication Services: Security 29
      2.2.8 Global-Time Service 29
      2.2.9 Resource Management Services 40
      2.2.10 Execution Services: Software Architecture 46
      2.2.11 Execution Services: DREAMS Virtualization Layer 48
      2.2.12 Execution Services: Security 59
      2.2.13 Optional Service: Voting 60
  2.3 Model-Driven Engineering 64
      2.3.1 Model 65
      2.3.2 Metamodel 65
      2.3.3 Platform-Independent Model 67
      2.3.4 Platform-Specific Model 67
  2.4 DREAMS Certification Strategy 67
      2.4.1 Safety and Certification 68
      2.4.2 Modular Certification 69
      2.4.3 Mixed-Criticality Patterns 69
      2.4.4 Product Families 69
  2.5 Fault Assumptions 70

      2.5.1   Fault Containment Regions 70
      2.5.2   Failure Modes 72
      2.5.3   Threats 73
  2.6   DREAMS Harmonized Platform 77

3 State-of-the-Art and Challenges   79
H. Ahmadian, M. Coppola, M. Faugére, D. Gracia Pérez, M. Grammatikakis, and I. Martinez
  3.1   Avionics Domain 80
      3.1.1   State-of-the-Art: Integrated Modular Avionics 81
      3.1.2   Challenges 81
  3.2   Wind-Power Domain 83
      3.2.1   State-of-the-Art: Wind-Turbine Control and Supervision System 83
      3.2.2   State-of-the-Art: Safety Protection System 83
      3.2.3   Challenges 84
  3.3   Health-Care Domain 85
      3.3.1   State-of-the-Art Solutions 85
      3.3.2   Challenges: Platform Security and Functionality 86

4 Modeling and Development Process   87
S. Barner, F. Chauvel, A. Diewald, F. Eizaguirre, Ø. Haugen, J. Migge, and A. Vasilevskiy
  4.1   Introduction 89
      4.1.1   Mixed-Criticality System Modeling Viewpoints 89
      4.1.2   Fundamental Metamodels 91
  4.2   Architecture Design 92
      4.2.1   Logical Modeling Viewpoint 92
      4.2.2   Technical Modeling Viewpoint 95
      4.2.3   Platform Architecture Modeling Framework 95
      4.2.4   DREAMS Platform Metamodel 99
  4.3   Timing Requirements 114
      4.3.1   Temporal Modeling Viewpoint 114
      4.3.2   Generic Methodology Pattern 117
  4.4   Safety Management 120
      4.4.1   Safety Modeling Viewpoint 120
      4.4.2   Development Process of Safety Solution Design 123
      4.4.3   Verification of Safety Solution Design 124
  4.5   Deployment and Resource Allocation 131
      4.5.1   Deployment Modeling Viewpoint 131
      4.5.2   Resource Allocation Modeling Viewpoint 138
      4.5.3   Basic Deployment and Scheduling Workflow 141
      4.5.4   Adaptivity and Resource Management Workflow 144
  4.6   Service Configuration Generation 146
      4.6.1   Configuration Modeling Viewpoint 146
      4.6.2   Model-transformations and Configuration Synthesis 149
  4.7   Variability and Design Space Exploration 149
      4.7.1   Variability Modeling Viewpoint 150
      4.7.2   Variability Exploration Process 151
      4.7.3   Design-Space Exploration Process 155

5   Algorithms and Tools

5   Algorithms and Tools   **163**

*J. Migge, P. Balbastre, S. Barner, F. Chauvel, S. S. Craciunas, A. Diewald, G. Durrieu, G.
Fohler, Ø. Haugen, A. A. Jaffari Syed, C. Pagetti, R. Serna Oliver, and A. Vasilevskiy*

5.1   Variability and Design Space Exploration 166
    5.1.1   Variability Analysis and Testing Techniques 166
    5.1.2   Multi-Objective Design Space Exploration 170
5.2   Scheduling 184
    5.2.1   Timing Decomposition 184
    5.2.2   Partition Scheduling 192
    5.2.3   On-Chip Network Scheduling 194
    5.2.4   Off-Chip Network Scheduling 198
5.3   Adaptation Strategies 221
    5.3.1   Recovery Strategies 222
    5.3.2   Comprehensive Offline Schedules 230
    5.3.3   Flexibility 236
5.4   Timing Analysis 250
    5.4.1   Problem Definition 250
    5.4.2   On-Chip Network 251
    5.4.3   Off-Chip Network 251
    5.4.4   Task Timing Analysis 251
    5.4.5   Composition 252
    5.4.6   Tool: RTaW-Pegase/Timing 252
5.5   Generation of Configuration Files 252
5.6   Toolchain 254
    5.6.1   Use Case 1: Basic Scheduling Configuration 255
    5.6.2   Use Case 2: Scheduling Configuration with Resource Management 257
    5.6.3   Use Case 3: Variability and Design-Space Exploration 259

6   Execution Environment   **261**

A. Crespo, P. Balbastre, K. Chappuis, J. Coronel, J. Fanguède, P. Lucas, and J. Perez

6.1   Virtualization Technologies 262
    6.1.1   Introduction 262
    6.1.2   Basic Implementation Types: Bare-Metal and Hosted 264
    6.1.3   Provided Virtual Environment - Full Virtualization and Para-Virtualization
        Technology 265
6.2   Execution Architecture 267
    6.2.1   Hardware Layer 267
    6.2.2   Virtualization Layer 268
    6.2.3   Runtime Layer 270
    6.2.4   Application Layer 271
6.3   Bare-Metal Hypervisor: XtratuM Case 273
    6.3.1   Overview 274
    6.3.2   System and Partitions Operation 275
    6.3.3   Partitions 277
    6.3.4   Health Monitor 281
    6.3.5   Access to Devices 282
    6.3.6   Services 284
    6.3.7   Configuration 285
    6.3.8   Deployment 292
6.4   Operating System Hypervisor: Linux-KVM Case 294
    6.4.1   Overview 295

		6.4.2	Scheduling and Coordination for Linux-KVM	295
		6.4.3	'Memguard' to Boost KVM Guests on ARMv8	299
		6.4.4	Secure Monitor Firmware	304

7 Chip-Level Communication Services   317

M. Grammatikakis, H. Ahmadian, M. Coppola, S. Kavvadias, A. Mouzakitis, K.
Papadimitriou, A. Papagrigoriou, P. Petrakis, V. Piperaki, M. Soulié, and G. Tsamis

	7.1	Bandwidth Regulation Strategies in Linux	318
		7.1.1	Genuine MemGuard Principles and Extensions	319
		7.1.2	Genuine vs. Extended MemGuard (MemGuardXt)	319
		7.1.3	NetGuard Extension (NetGuardXt)	322
	7.2	Hardware MemGuard: Bandwidth Control at Target Devices	323
		7.2.1	Limitations of Hardware MemGuard	324
		7.2.2	Architecture of the Hardware MemGuard	325
		7.2.3	Synthetic Traffic Evaluation	325
		7.2.4	NoC-Based Evaluation	327
	7.3	Hardware Support at Network-on-Chip Level	330
		7.3.1	STNoC Implementation of Address Interleaving	330
		7.3.2	Evaluation Framework: Performance and Power Consumption . . .	332
	7.4	Mixed-Criticality Network-on-Chip	333
		7.4.1	Support for Mixed-Criticality	333
		7.4.2	Support for Heterogeneous Communication Paradigms	333
		7.4.3	Overall Architecture	333
		7.4.4	Network Interface	334
		7.4.5	Core Interface Using Ports	335
		7.4.6	Mixed-Criticality Unit	337
		7.4.7	Back-End	339

8 Cluster-Level Communication Services   341

T. Koller, M. Abuteir, A. Eckel, A. Geven, L. Kohútka, L. Rubio, and C. Zubia

	8.1	Off-Chip Network	342
		8.1.1	Time-Triggered Ethernet	342
		8.1.2	EtherCAT	354
	8.2	Security Services	360
		8.2.1	Risk Analysis	361
		8.2.2	Security at Multiple Levels	363
		8.2.3	Security Classification	364
		8.2.4	Cluster-Level Security	365
		8.2.5	Secure Time Synchronization	367
		8.2.6	Application-Level Security	369

9 Resource Management Services   377

G. Gala, G. Fohler, D. Gracia Pérez, and C. Pagetti

	9.1	Overview of DREAMS Resource Management	378
	9.2	Local Resource Monitor or MON	379
		9.2.1	MON for Core Failure	380
		9.2.2	MON for Deadline Overrun	381
		9.2.3	MON for Quality of Service	382
	9.3	Local Resource Scheduler or LRS	384
		9.3.1	General Approach	384
		9.3.2	Implementation	385

        9.3.3   Requirements on Applications 386
    9.4 Local Resource Manager or LRM 387
        9.4.1   Core Failure Management 387
        9.4.2   Deadline Overrun Management 388
        9.4.3   QoS Management 390
    9.5 Global Resource Manager or GRM 394
        9.5.1   Implementation 394
        9.5.2   Global Reconfiguration Graph 395
    9.6 Resource Management Communication 397
        9.6.1   Secure Resource Management Communication 400

10 Safety Certification of Mixed-Criticality Systems   403
I. Martinez, G. Bouwer, F. Chauvel, Ø. Haugen, R. Heinen, G. Klaes, A. Larrucea Ortube, C. F. Nicolas, P. Onaindia, K. Pankhania, J. Perez, and A. Vasilevskiy
    10.1 DREAMS Safety Certification Strategy 405
    10.2 Certification and Compliant Items 406
        10.2.1 Need and Importance of Certifications 407
        10.2.2 Accreditation 408
        10.2.3 EC Type-Examinations 408
        10.2.4 Certification Requirements According to IEC 61508 409
        10.2.5 Compliant Items According to IEC 61508 409
    10.3 Modular Safety Cases 409
        10.3.1 Modular Safety Case for Cluster-Level Mixed-Criticality Networks . . . 410
    10.4 Mixed-Criticality Patterns 414
        10.4.1 Hypervisors 414
        10.4.2 COTS Multi-Core Device 418
        10.4.3 Mixed-Criticality Network 423
    10.5 Functional Safety Management Process for DREAMS Architecture 425
        10.5.1 IEC 61508 Functional Safety Management 425
        10.5.2 Tools 427
    10.6 Certification of Mixed-Criticality Product Lines 428
        10.6.1 Families of Systems and Product Lines 428
        10.6.2 Piecewise Certification 431
        10.6.3 IEC 61508 Certification 435
    10.7 Method for Certifying Mixed-Criticality Product Lines 437
        10.7.1 Certification Support in DREAMS 437
        10.7.2 Certification Arguments 440
        10.7.3 Database of Argument Models: Mixed-Criticality System 440
        10.7.4 Arguments of Compliance to Safety-Standards 441
        10.7.5 Arguments Based on Verification, Validation and Testing 444
        10.7.6 Summary 445

11 Evaluation   447
J. Perez, M. Coppola, M. Faugère, D. Gracia Pérez, M. Grammatikakis, A. Larrucea Ortube, A. Mouzakitis, A. Papagrigoriou, P. Petrakis, V. Piperaki, I. Sarasola, and G. Tsamis
    11.1 Wind-Power Domain 448
        11.1.1 Introduction 448
        11.1.2 Demonstrator Description 448
        11.1.3 Results and Conclusions 452
    11.2 Safety-Critical Domain 453
        11.2.1 Mixed-Criticality and Multi-Cores 455

        11.2.2 Fault Management 458
    11.3 Healthcare Domain 459
        11.3.1 Out-of-Hospital Use-Case: Security-Performance Tradeoffs 460
        11.3.2 In-Hospital Use-Case: Hospital Media Gateway 469

References   **477**

Index   **505**

Contributors

Mohammed Abuteir
TTTech Computertechnik AG
Vienna, Austria

Hamidreza Ahmadian
Universität Siegen
Siegen, Germany

Patricia Balbastre
Universitat Politécnica de Valéncia
Valéncia, Spain

Simon Barner
fortiss GmbH
Munich, Germany

Gebhard Bouwer
TÜV Rheinland Industrie Service GmbH
Cologne, Germany

Kevin Chappuis
Virtual Open Systems
Grenoble, France

Franck Chauvel
SINTEF
Oslo, Norway

Marcello Coppola
STMicroelectronics
Grenoble, France

Javier Coronel
Fent Innovative Software Solutions, S.L.
Valéncia, Spain

Silviu S. Craciunas
TTTech Computertechnik AG
Vienna, Austria

Alfons Crespo
Universitat Politécnica de Valéncia
Valéncia, Spain

Alexander Diewald
fortiss GmbH
Munich, Germany

Guy Durrieu
ONERA
Toulouse, France

Andreas Eckel
TTTech Computertechnik AG
Vienna, Austria

Fernando Eizaguirre
IK4-Ikerlan
Mondragón, Spain

Jeremy Fanguède
Virtual Open Systems
Grenoble, France

Madeleine Faugère
THALES Research & Technology
Paris, France

Gerhard Fohler
Technische Universität Kaiserslautern
Kaiserslautern, Germany

Gautam Gala
Technische Universität Kaiserslautern
Kaiserslautern, Germany

Arjan Geven
TTTech Computertechnik AG
Vienna, Austria

Daniel Gracia Pérez
THALES Research & Technology
Paris, France

Miltos D. Grammatikakis
TEI of Crete
Crete, Greece

Øystein Haugen
SINTEF
Oslo, Norway

Robert Heinen
TÜV Rheinland Industrie Service GmbH
Cologne, Germany

Ali Abbas Jaffari Syed
Technische Universität Kaiserslautern
Kaiserslautern, Germany

Stamatis Kavvadias
TEI of Crete
Heraklion, Greece

Gernot Klaes
TÜV Rheinland Industrie Service GmbH
Cologne, Germany

Lukáš Kohútka
TTTech Computertechnik AG
Vienna, Austria

Thomas Koller
Universität Siegen
Siegen, Germany

Asier Larrucea Ortube
IK4-Ikerlan
Mondragón, Spain

Pierre Lucas
Virtual Open Systems
Grenoble, France

Imanol Martinez
IK4-Ikerlan
Mondragón, Spain

Jörn Migge
RealTime-At-Work
Nancy, France

Angelos Mouzakitis
Virtual Open Systems
Grenoble, France

Carlos Fernando Nicolas
IK4-Ikerlan
Mondragón, Spain

Roman Obermaisser
Universität Siegen
Siegen, Germany

Peio Onaindia
IK4-Ikerlan
Mondragón, Spain

Zaher Owda
Universität Siegen
Siegen, Germany

Claire Pagetti
ONERA
Toulouse, France

Kanti Pankhania
TÜV Rheinland Industrie Service GmbH
Cologne, Germany

Kyprianos Papadimitriou
TEI of Crete
Heraklion, Greece

Antonis Papagrigoriou
TEI of Crete
Heraklion, Greece

Jon Perez
IK4-Ikerlan
Mondragón, Spain

Polydoros Petrakis
TEI of Crete
Heraklion, Greece

Voula Piperaki
TEI of Crete
Heraklion, Greece

Leire Rubio
IK4-Ikerlan
Mondragón, Spain

Ibai Sarasola
IK4-Ikerlan
Mondragón, Spain

Ramon Serna Oliver
TTTech Computertechnik AG
Vienna, Austria

Michaël Soulié
STMicroelectronics
Grenoble, France

George Tsamis
TEI of Crete
Heraklion, Greece

Anatoly Vasilevskiy
SINTEF
Oslo, Norway

Donatus Weber
Universität Siegen
Siegen, Germany

Cristina Zubia
IK4-Ikerlan
Mondragón, Spain

1

Introduction to Distributed Real-Time Mixed-Criticality Systems

R. Obermaisser

Universität Siegen

1.1   Context in the Area of Mixed-Criticality Systems 1
1.2   Scope of the Book 2
1.3   Motivation and Objectives 4
1.4   Structure of the Book 5

This chapter provides an overview about the content of the book and gives an introduction about the concept of mixed-criticality. Section 1.1 elaborates on the context of mixed-criticality and introduces the major challenges in such systems. Section 1.2 describes the scope of the book and offers an abstract view of the content of the book to help the reader to find the needed content, taking into account the large size of the book. Furthermore, Section 1.3 elaborates on the most important motivation and objectives of the DREAMS project, with which the structure of the chapters are aligned. At the end, Section 1.4 concludes this chapter by introducing the structure of the book.

1.1 Context in the Area of Mixed-Criticality Systems

Embedded systems have led to tremendous improvements with respect to functionality, efficiency, safety and comfort in many application areas such as avionics, space, railway, process control and medical systems. At the same time, the number of electronic components and communication is steadily increasing. For example, a modern car contains up to 100 Electronic Control Units (ECUs). The high number of components and network links is becoming a challenge due to the hardware cost, the required weight and space. In addition, the high number of components and wiring increases the likelihood of hardware faults. In particular, connector faults have become a prevalent source of failures in embedded systems [1, 2].

A promising solution is the integration of electronic functions using fewer ECUs, thereby reducing the mentioned issues due to a high component number and cabling. This integration is in particular facilitated by advances in the semiconductor industry resulting in powerful multi-core processors. A multi-core processor can host and execute several electronic functions in parallel, while meeting stringent temporal requirements.

However, a fundamental challenge in this trend towards integrated architectures are safety and certification requirements. A typical safety-critical embedded system comprises subsystems with varying requirements concerning safety assurance levels. For example, the criticality of avionic functions according to DO-178C [3] ranges from level A with catastrophic consequences in case of failures (e.g., flight control) to level E with no safety effects (e.g., entertainment services). Likewise, automotive functions are classified according to the effect of failures using Automotive Safety Integrity Levels (ASILs) ranging from ASIL A to ASIL D [4].

The integration of functions with different criticality on shared platforms is only feasible if each function can be validated and certified according to the respective level of criticality. The establishment of such a compositional safety argument is known as *modular certification*, which combines safety arguments from different applications subsystems. In addition, *modular safety arguments* are introduced for the building blocks of the platform (e.g., operating system, network, processors).

A prerequisite for modular certification is *temporal and spatial partitioning* in order to ensure the independence of functions from each other in both the time and value domains. Spatial partitioning ensures the data integrity of each function's internal state and the integrity of information exchanged between functions. In the temporal domain, the timing of resources available to one function shall be independent from the behavior of other functions. Examples of temporal properties are the latencies, jitter and through out in networks, processors and input/output resources. Ideally, the timing of a resource as perceived by one resource is completely independent from other functions. This gold standard with respect to partitioning is ideal for the compositional validation of safety-critical systems and modular certification. Research platforms (e.g., COMPSOC, TTSoC [5]) support an invariant timing with respect to processor cycles upon the composition functions. However, today's COTS processors do not efficiently support this level of isolation. Performance depends on numerous shared resources, such as caches and interconnects, the disabling of which would result in unacceptable performance degradations.

As a consequence, the demand for independence between functions is often weakened to acclaim for a bounded effect between functions. While a function can indeed effect the timing of other functions, the increase of latencies, jitter and throughput degradation can be limited. Thus, worst-case execution and communication times can be determined assuming the worst-case interference between functions. This weakened level of isolation is sufficient for modular certification, despite complicating validation due to an increased state space for testing.

For many types of platform the determination of tight time bounds for delays induced by worst-case interference is a challenging research problem. As a consequence, research in the area of probabilistic worst-case execution times attempts to establish stochastic models of applications and platforms. An opposing approach is the research in the area of predictable platforms, facilitating a deterministic timing analysis.

1.2 Scope of the Book

This book describes a cross-domain architecture and design tools for networked complex systems, in which application subsystems of different criticality coexist and interact on networked multi-core chips. The presented results were established in the European research project DREAMS with 16 partners including many major European embedded system suppliers and OEMs encompassing a broad range of application domains, supported by leading research and academic organizations.

DREAMS resulted in an architectural style that guides the development of a mixed-criticality system by providing rules how to structure the application into components and how to specify the interfaces between the components. The architectural style also defines the services to be provided by the platform as the foundation and stable baseline for the development of applications. These services perform the virtualization of processors, networks, memory and inputs/outputs, while ensuring real-time support, safety, reliability and security.

Platforms in today's mixed-criticality systems are typically hierarchically structured. Platforms often comprise multi-core chips because of high performance requirements, the potential for higher integration of functions and the more efficient use of semiconductor resources compared to single-core processors [6].

This book presents an in-depth explanation of the virtualization technologies at the three integration levels including hypervisors, networks-on-a-chip, off-chip networks and memories. The management and virtualization of the cores are the purpose of operating systems and hypervisors. A hypervisor establishes partitions, which serve as protected execution environments for the execution of functions. Hypervisors virtualize the computational resources and permit the coexistence of different guest operating systems. In addition, multi-core hypervisors enable the virtualization of the cores and allow application functions to abstract from the actual processing hardware. Thereby, hypervisors also decouple the number of software functions from the number of cores.

The processor cores typically interact via a shared memory and can be grouped into tiles, which are interconnected by Networks-on-a-Chip (NoCs). NoCs play a strong role in the temporal and spatial partitioning at the hardware level, complementing the software-based segregation of the hypervisor. Partitioning in the NoC results from monitoring and controlling the message transmission based on a priori knowledge about the permitted behavior. Common techniques are central or distributed guardian functionalities using time-triggered or rate-constrained communication.

At the cluster-level, the system is decomposed into nodes interconnected by off-chip networks. Many mixed-criticality systems require resources exceeding the ones of a single chip. In addition, safety-critical fail-operational systems with the highest safety integrity levels can only be implemented with redundancy at cluster-level given the failure rates of today's chips.

A further focus of the book is the DREAMS adaptation strategies, which allow to react to changing environmental conditions and fluctuating resource availability. For instance, fault recovery strategies after the occurrence of faults enable a continued provision of correct services with less cost than active redundancy. DREAMS introduced services for a system-wide adaptivity of mixed-criticality applications consuming several resources via global integrated resource management. This approach is based on the separation of system-wide decisions to meet global constraints from the local execution on individual resources. Resources are monitored individually with abstract information provided to global resource management. Should significant changes demand adaptation, *global resource manager* takes decisions on a system-wide level, based on off-line computed configurations, with orders, such as bandwidth assignment, or scheduling parameters for all resources, which are controlled by *local resource manager*. Thus, system-wide constraints, such as end-to-end timing, reliability and energy integrity can be addressed without incurring the complexity and overhead of individual negotiations among resources directly.

Another focus of the book is the development methodology along with tooling. In DREAMS, model-driven engineering methods are applied in order to cope with the potential complexity and of the hierarchical systems comprised of networks of virtualized multi-cores. The first contribution beyond the state of the art is in the definition of models for the description of the DREAMS platform and the according adoption of domain-independent application models. The second contribution is in the area of design-space exploration and scheduling. In both cases, the novelty consists in the consideration of mixed-criticality systems and extra-functional properties. With its alignment to the V-model, which is widely used in the area of safety-critical systems, the DREAMS approach is therefore suited for the development of mixed-criticality systems under realistic assumptions.

The methodology also addresses certification and mixed-criticality product-lines. DREAMS provided fine-grained modular building blocks that can be combined in a safety case for certification and increase the re-use possibilities of available evidence. Furthermore, it avoids the lift-up effect in order to allow modularized certification and make it possible to limit certification efforts through identifying exactly what needs to be re-certified.

1.3   Motivation and Objectives of DREAMS

This book describes a flexible platform and associated design tools for embedded applications where subsystems of different criticality, executing on networked multi-core chips, can be integrated seamlessly. The objective is a unified view of the system through systematic abstraction of the underlying hierarchic network topology and related constraints. DREAMS offers pervasive support for design, modeling, verification, validation and certification up to the highest criticality levels in multiple domains (e.g., transportation, wind-power, health-care). The cross-domain architecture enables synergies between application domains and exploitation of the economies of scale.

- **Objective 1 – Architectural Style and Modeling Methods based on Waistline Structure of Platform Services:** DREAMS consolidated and extended architectural concepts from previous projects (e.g., GENESYS, ACROSS, RECOMP, ARAMIS) towards a new architectural style for a seamless virtualization of networked embedded platforms ranging from multi-core chips to the cluster-level with support for security, safety and real-time performance as well as data, energy and system integrity. The waistline architecture with domain-independent platform services can be successively refined and extended to construct more specialized platform services and application services. The platform services provide a stable foundation for the development of applications and enable the safe and secure composition of mixed-criticality systems out of components. Models of hierarchical platforms comprised of networked multi-core chips serve as the foundation for the model-driven development methodology.

- **Objective 2 – Virtualization Technologies to Achieve Security, Safety, Real-Time Performance as well as Data, Energy and System Integrity in Networked Multi-Core Chips:** Mixed-criticality systems typically consist of subsystems with different programming models (e.g., message passing vs. shared memory, time-triggered vs. event-triggered) and have different requirements for the underlying platform (e.g., trade-offs between predictability, certifiability and performance in processors cores, hypervisors, operating systems and networks). DREAMS established certifiable platform services for virtualization and segregation of resources at cluster and chip-level (e.g., I/O virtualization, message-based networks and memory architectures, dynamic resource management). Gateways enable the end-to-end segregation as a means for integration of mixed criticalities at chip-, network- and cluster-level. Resource managers achieve the virtualization for heterogeneous applications and platforms. The monitoring and dynamic configuration of virtualized resources is the foundation for integrated resource management.

- **Objective 3 – Adaptation Strategies for Mixed-Criticality Systems to Deal with Unpredictable Environment Situations, Resource Fluctuations and the Occurrence of Faults:** The integrated resource management of DREAMS addresses the requirements of mixed-criticality systems and offers monitoring, runtime control and virtualization extensions recognizing system-wide and high-level constraints, such as end-to-end deadlines and reliability. The integrated resource management combines off-line and on-line scheduling algorithms and provides segregation of activities of different criticalities in a flexible way.

- **Objective 4 – Development Methodology and Tools based on Model-Driven Engineering:** The development process of DREAMS ranges from modeling and design to validation of mixed-criticality systems. DREAMS targets a methodology and prototypes of tools for mapping mixed-criticality applications to heterogeneous networked platforms including algorithms for scheduling and allocation, analysis of timing, energy and reliability. The methodology also includes a test-bed for validation, verification and evaluation of extra-functional properties.

- **Objective 5 – Certification and Mixed-Criticality Product Lines:** DREAMS resulted in a modular safety-concept for mixed-criticality systems with safety arguments for different parts of the platform, such as networks, Commercial Off-The-Shelf (COTS) processors and hypervisors. DREAMS also offers architectural support for an eased definition of mixed-criticality product lines with certification support across product lines. Hence, variability in applications and platforms serves as another architectural dimension to handle different criticalities and domains.

1.4 Structure of the Book

Chapter 2 introduces the DREAMS architectural style with platform services for mixed-criticality systems. The platform services include time services, communication services, execution services and resource management with strict temporal and spatial partitioning.

Chapter 3 provides an overview of the state-of-the-art in the area of mixed-criticality systems and discusses research challenges with respect to system architectures, platform technologies and development methods. Transportation, wind-power and health-care systems serve as the application areas for the analysis of requirements and research gaps.

Chapter 4 is dedicated to the modeling of hierarchical mixed-criticality systems and the development process. The modeling considers different aspects including logical, physical, safety, timing, variability, resource-allocation and configuration viewpoints.

Chapter 5 focuses on algorithms and tools to support the design space exploration, timing analysis, variability management, scheduling of resources and the generation of configuration information for the platform. Algorithms and tools for adaptation strategies enable fault recovery, higher flexibility and an adaptive system behavior.

Chapter 6 addresses execution environments for mixed-criticality systems. This chapter explains different software layers ranging from the hardware-abstraction layer to the application layer. Generic virtualization technologies are introduced and mapped to a secure monitor firmware as well as the hypervisors XtratuM and KVM. The DREAMS abstraction layer (DREAMS Abstraction Layer) provides fundamental services for a mixed-criticality application (e.g., time services, communication services, health monitoring services), while abstracting from the implementation technology of the execution environment.

Chapter 7 introduces the chip-level communication services for the virtualization of different types of resources. Memory resources are managed using bandwidth regulation strategies such as MemGuard. On-chip communication resources are subject to traffic shaping for time-triggered and rate-constrained communication in order to establish fault isolation and temporal predictability.

Chapter 8 is dedicated to cluster-level communication services. Wirebound and wireless off-chip networks based on TTEthernet and IEEE 802.11 support the mixed-criticality requirements. Gateways are responsible for the information exchange between on-chip and off-chip networks, while performing the necessary property transformations and providing fault isolation.

Chapter 9 introduces the resource management services of DREAMS. The resource management architecture can be hierarchical encompassing a global resource manager as well as a hierarchy of local resource managers, local monitors and local resource schedulers. Thereby, global strategies are combined with local resource monitoring and local management schemes. Security communication services ensure confidentiality, integrity and authenticity for the interactions between local and global resource managers.

Chapter 10 deals with the certification and mixed-criticality product lines. The modular safety concept is the foundation for a *safe by construction* modular approach with arguments to demonstrate that safety properties are satisfied. It provides mechanisms for review by Reliability, Availability, Maintainability and Safety (RAMS) engineers and certification authorities. The presented safety concept is based on the Claim, Arguments and Evidence (CAE) notation and was successfully assessed with respect to functional safety (IEC 61508) by TÜV Rheinland. In a linking analysis, the generic safety arguments are mapped to concrete building blocks such as mixed-criticality networks, COTS processors and hypervisors.

Chapter 11 concludes the book with an evaluation of the presented DREAMS results with respect to the requirements and research challenges. The chapter analyzes key performance indicators from different application areas such as wind-power and health-care.

2

DREAMS Architectural Style

R. Obermaisser

Universität Siegen

M. Abuteir

TTTech Computertechnik AG

H. Ahmadian

Universität Siegen

P. Balbastre

Universitat Politécnica de Valéncia

S. Barner

fortiss GmbH

M. Coppola

STMicroelectronics

J. Coronel

Fent Innovative Software Solutions, S.L.

A. Crespo

Universitat Politécnica de Valéncia

P. Balbastre

Universitat Politécnica de Valéncia

G. Fohler

Technische Universität Kaiserslautern

G. Gala

Technische Universität Kaiserslautern

M. Grammatikakis

TEI of Crete

A. Larrucea Ortube

IK4-Ikerlan

T. Koller

Universität Siegen

Z. Owda

Universität Siegen

D. Weber

Universität Siegen

2.1	System Model		9
	2.1.1	Physical Platform Structure	9
	2.1.2	Logical Application Architecture	10
	2.1.3	Mapping of Application to Platform	10
		2.1.3.1 Mixed-Criticality Namespace	10
		2.1.3.2 Virtual Links	11
2.2	Waistline Structure of Services		11
	2.2.1	DREAMS Core Services	12
	2.2.2	Architectural Building Blocks for Platform Services	14
		2.2.2.1 Building Blocks for Core Services	14
		2.2.2.2 Building Blocks for Optional Services	15
		2.2.2.3 Building Blocks for Application Services	16
		2.2.2.4 Technology Independence of Architectural Style	16
	2.2.3	Communication Services: On-Chip	17
		2.2.3.1 Storing Messages at Ports	17
		2.2.3.2 Message Delivery	18

		2.2.3.3	Collision Handling	18
		2.2.3.4	Monitoring and Reconfiguration Services	18
	2.2.4	Communication Services: Off-Chip		19
		2.2.4.1	Off-Chip Network Interface	19
		2.2.4.2	Off-Chip Router	21
		2.2.4.3	Gateway	23
	2.2.5	Communication Services: Shared Memory		26
	2.2.6	Communication Services: IOMMU		27
		2.2.6.1	IOMMU and Page-Level Security	28
		2.2.6.2	NoC Firewall and Multi-Compartment Technology	28
	2.2.7	Communication Services: Security		29
	2.2.8	Global-Time Service		29
		2.2.8.1	On-Chip Clock Synchronization Services	32
		2.2.8.2	Off-Chip Clock Synchronization Services	36
	2.2.9	Resource Management Services		40
		2.2.9.1	Global Resource Manager	41
		2.2.9.2	Local Resource Management	42
		2.2.9.3	Resource Management Communication	44
		2.2.9.4	Resource Management Architecture	44
	2.2.10	Execution Services: Software Architecture		46
		2.2.10.1	Software Architecture	46
		2.2.10.2	Key Services for Certification	47
	2.2.11	Execution Services: DREAMS Virtualization Layer		48
		2.2.11.1	Interrupt Virtualization Services	50
		2.2.11.2	Processor Virtualization Services	50
		2.2.11.3	DREAMS Abstraction Layer Services	53
	2.2.12	Execution Services: Security		59
	2.2.13	Optional Service: Voting		60
2.3	Model-Driven Engineering			64
	2.3.1	Model		65
	2.3.2	Metamodel		65
	2.3.3	Platform-Independent Model		67
	2.3.4	Platform-Specific Model		67
2.4	DREAMS Certification Strategy			67
	2.4.1	Safety and Certification		68
	2.4.2	Modular Certification		69
	2.4.3	Mixed-Criticality Patterns		69
	2.4.4	Product Families		69
2.5	Fault Assumptions			70
	2.5.1	Fault Containment Regions		70
		2.5.1.1	Fault Containment Regions for Design Faults	70
		2.5.1.2	Fault Containment Regions for Physical Faults	72
	2.5.2	Failure Modes		72
		2.5.2.1	Failure Rates and Persistence	73
	2.5.3	Threats		73
		2.5.3.1	Threat Models	74
		2.5.3.2	Threat Analysis for Communication Services	74
		2.5.3.3	Threat Analysis for Global Time Services	75
		2.5.3.4	Threat Analysis for Resource Management Services	76
		2.5.3.5	Threat Analysis for Execution Services	76
2.6	DREAMS Harmonized Platform			77

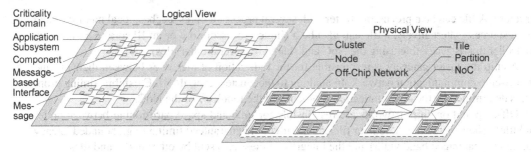

Figure 2.1: System Structure of Application (Logical View) and Structure of Platform (Physical View), © IEEE 2014, Reprinted with Permission from [7]

This chapter describes the architectural style of mixed-criticality systems and introduces the DREAMS waistline structure of services. It provides structuring rules according to several integration levels to support the integration of applications with different timing models and different safety assurance levels. Section 2.1 introduces the system model in mixed-criticality systems and describes the logical and physical system structures. In addition, it shows how those two system models can be mapped by a mixed-criticality name space. Section 2.2 elaborates on the DREAMS waistline structure of services and lists the core services as well as the optional services. In addition, the architectural building blocks and services of communication services, global time services, execution services as well as local and global resource management services are elaborated in this section. Section 2.3 gives an overview of the model-driven development based on the described building blocks. The DREAMS certification strategy along with the concept of modular certification and mixed-criticality patterns are introduced in Section 2.4. Fault assumptions is another topic which is covered in Section 2.5. The concept of fault containment regions is introduced to overcome the design faults as well as the physical faults. In addition, the failure modes, threats, threat models and threat analysis for the four major core services are elaborated in this chapter. At the end, the DREAMS harmonized platform as instantiation of the DREAMS architectural style is introduced in Section 2.6.

2.1 System Model of a Mixed-Criticality System

In the system model we distinguish between the physical and the logical system structure. The physical system structure comprises a platform with networked multi-core chips, while the logical system structure encompasses applications and a corresponding namespace. The end-to-end communication between application components is mapped to the networks of the platform and contention is resolved with support for different traffic types.

2.1.1 Physical Platform Structure

The overall system is physically structured into a set of *clusters*, where each cluster consists of nodes that are interconnected by a real-time communication network in a corresponding network topology (e.g., bus, star, redundant star, ring). Inter-cluster gateways serve as the connection between clusters.

Each *node* is a multi-core chip containing tiles that are interconnected by a Network On Chip (NoC). Each *tile* provides a Network Interface (NI) to the NoC and can have a complex internal structure. A tile can be a processor cluster with several processor cores, caches, local memories and I/O resources. Alternatively, a tile can also be a single processor core or an IP core (e.g., memory controller that is accessible using the NoC and shared by several other tiles).

A chip-to-cluster gateway is responsible for the redirection of messages between the NoC and the off-chip communication network. In analogy to the cluster-level, the NoC exhibits timing properties determined by the communication protocol and the topology of the NoC.

Off-chip and on-chip networks are responsible for time and space partitioning between nodes and tiles. They ensure that a node or tile cannot affect the guaranteed timing (e.g., bounded latency and jitter, guaranteed bandwidth) and the integrity of messages sent by other nodes and tiles.

The processor cores within a tile can run a hypervisor that establishes partitions, each of which executes a corresponding software component (or component for short). The hypervisor establishes time and space partitioning, thereby ensuring that a software component cannot affect the availability of computational resource in other partitions (e.g., time and duration of execution on processor core, integrity and timing of memory).

2.1.2   Logical Application Architecture

The overall system is logically structured into *criticality levels*. Several criticality levels are distinguished in different application domains such as Classes A to E in avionics [8], ASILA to D in automotive [4] and SIL1-4 in multiple domains according to IEC 61508 [9].

For each criticality level, there can be multiple *application subsystems*. In the automotive domain, steer-by-wire and brake-by-wire would be examples of subsystems belonging to the highest criticality level (ASILD). An application subsystem can be further subdivided into *components*, which interact by the exchange of messages via ports.

Each component provides services to its environment. The specification of a component's interface defines its services, which are the intended behavior as perceived by the transmission of messages as a response to inputs, state and the progression of time.

In order to provide the services, components require resources of the underlying platform as identified in the physical system structure. Each component must be assigned to a partition with suitable computational resources (e.g., CPU time, memory). Messages must be mapped to the communication networks with suitable timing and reliability properties.

Since components can be mapped to partitions residing on different nodes and even different clusters, messages must be transmitted over different on-chip and off-chip networks.

2.1.3   Mapping of Application to Platform

In order to provide the services, components require resources of the underlying platform as identified in the physical system structure. Each component must be assigned to a partition with suitable computational resources (e.g., CPU time, memory). Messages must be mapped to the communication networks with suitable timing and reliability properties. Since components can be mapped to partitions residing on different nodes and even different clusters, messages must be transmitted over different on-chip and off-chip networks.

2.1.3.1   Mixed-Criticality Namespace

Based on the logical and physical system structure, we introduce the following namespace:

$$\underbrace{Criticality.Subsystem.Component.Message}_{Logical\ Name} : \underbrace{Cluster.Node.Tile.Port}_{Physical\ Name} \qquad (2.1)$$

Components are only aware of logical names, whereas the platform requires physical names for the routing of messages. The conversion between logical and physical names can occur using a translation layer (in software or hardware) between the components and the communication system.

Alternatively, components and messages can be hard-bound to the platform by fixing the translation to the physical namespace at development time.

2.1.3.2 Virtual Links

Virtual Links (VLs) [10] are an abstraction over these networks and they hide the physical system structure of the platform from the components. A VL is an end-to-end multicast channel between one sender component and multiple receiver components. The timing and reliability of the VL is determined by the properties of the constituent physical networks.

We distinguish two types of VLs:

- **Time-triggered VL for periodic messages:** Periodic messages are supported by time-triggered VLs with communication at predefined points in time according to a communication schedule.

- **Rate-constrained VL for sporadic messages:** Rate-constrained communication establishes the transport of sporadic messages with minimum interarrival times. A rate-constrained VL has a priority that determines how contention with other rate-constrained VLs is resolved. Rate-constrained communication guarantees sufficient bandwidth allocation for each transmission with defined limits for delays and temporal deviations.

Aperiodic messages do not require VLs, but are subject to a connectionless transfer. Therefore, each aperiodic message must include naming information for routing through the network.

The decoupling between the application and the platform results in location transparency, reduces complexity, facilitates reusability of components and addresses the technology obsolescence of platforms. Presently, end-to-end channels are supported in distributed systems (e.g., based on AUTOSAR in the automotive domain [11]), whereas end-to-end channels in hierarchical systems with networked multi-core systems are not addressed.

2.2 DREAMS Waistline Structure of Services

In order to support cross-domain reusability and an independent development of platform services, the DREAMS architecture defines generic platform services (e.g., operations to send and receive a message) as a foundation for the development of applications. There are two types of platform services at any given level of integration: core services and optional services.

The *core services* are mandatory in every instantiation of the architecture. At any given integration level, the core services form a waist that should include exactly those features, which are required by all the targeted application domains. For instance, a message transport service is a core service.

Optional services build upon the core services and provide functionality to customize the architecture towards a particular application domain. However, these services need not to be present in every instantiation of the architecture. For instance an encryption service could be an optional service.

The platform services of the DREAMS architecture are structured in a waistline as shown in Figure 2.2.

This waistline structuring of services is inspired by the Internet, where the Intellectual Property (IP) provides the waist for different communication technologies and protocols. Towards the bottom, a variety of implementation choices is supported. IP can be implemented on Ethernet networks, ATM networks, different wireless protocols, etc. Towards the top, different refinements to higher protocols depending on the application requirements occur. IP can be refined into UDP or TCP, thereafter into HTTP, FTP, etc.

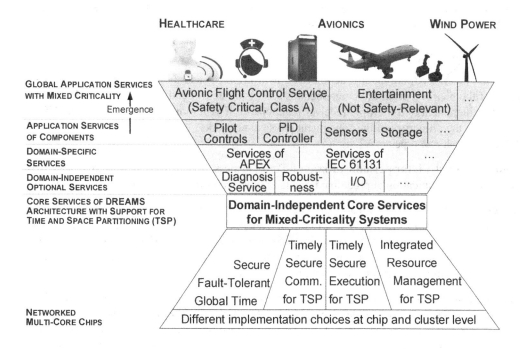

Figure 2.2: Waistline Structure of Services

In a similar way, the services of the DREAMS architecture are structured. The core services are a stable waist encapsulating all those capabilities that are required in all targeted application domains for the realization of mixed-criticality systems. These core services also lay the foundation for exploiting the economies of scale as they can be implemented in a space and energy efficient way in hardware for a multitude of application domains. The core services offer capabilities that are required as the foundation for the construction of higher platform services and application services.

Different underlying implementation options exist for each of the core services. For example, the core communication services can be realized using different protocols in NoCs or off-chip networks. DREAMS is not restricted to specific protocols (e.g., TTNoC and Spidergon NoC), but any protocol providing the core services is a suitable foundation for the DREAMS architecture. In analogy, variability increases towards the waistline's top where the application services are implemented. Platform services can be successively refined and extended to construct more specialized platform services.

2.2.1 DREAMS Core Services

Four core services are mandatory and part of any instantiation of the DREAMS architecture, since they represent capabilities that are universally important for mixed-criticality systems and all considered application domains. The core services are absolutely necessary to build higher services and to maintain the desired properties (e.g., Time and Space Partitioning (TSP)) of the architecture.

Secure and fault-tolerant global time base:

The global time service of DREAMS provides to each component a local clock, which is globally synchronized within the system of networked Multi-Processor Systems-on-a-Chip (MPSoCs) and within each MPSoC. The main rationale for the provision of a global time is the ability for the temporal coordination of activities, the establishment of a deterministic communication infrastructure and the ability for establishing a relationship between timestamps from different components.

Timely and secure communication services for time and space partitioning:

DREAMS provides services for the message-based real-time communication among components. The DREAMS communication services establish end-to-end channels over hierarchical, heterogeneous and mixed-criticality networks respecting mixed-criticality safety and security requirements. Based on an intelligent communication system with a priori knowledge about the allowed behavior of components in the value and time domain, DREAMS ensures TSP. The shared memory model is supported on top of message-based NoCs and message-based off-chip networks. Thereby, application subsystems are able to exploit programming models based on shared memory, while TSP of the message-based network infrastructure ensures segregation.

Timely and secure execution for time and space partitioning:

For the sharing of processor cores among mixed criticality applications, including safety-critical ones, partitioning OSes and hypervisors (e.g.,XtratuM and KVM) are used, which ensure TSP for the computational resources. The scheduling of computational resources (e.g., processor, memory) in DREAMS ensures that each task obtains not only a predefined portion of the computation power the processor core, but also that execution occurs at the right time and with a high level of temporal predictability. On one hand, DREAMS supports static scheduling, where an off-line tool creates a schedule with pre-computed scheduling decisions for each point in time. In addition, we support dynamic scheduling by employing a quota system in the scheduling of tasks in order to limit the consequences of faults. Safety-critical partitions establish execution environments that are amenable to certification and worst-case execution time analysis, whereas partitions for non safety-critical partitions provide more intricate execution environments (e.g., based on Linux). In addition, the separation between safety-critical and non safety-critical applications is supported using dedicated on-chip tiles with respective OSes.

Integrated resource management for time and space partitioning:

DREAMS provides services for system-wide adaptivity of mixed-criticality applications consuming several resources via global integrated resource management. The approach is based on the separation of system-wide decisions to meet global constraints from the local execution on individual resources: resources are monitored individually with abstract information provided to Global Resource Manager (GRM).

If significant changes should demand adaptation, the GRM takes decisions on a system-wide level, based on off-line computed configurations, with orders, such as bandwidth assignment, or scheduling parameters for all resources, which are controlled by Local Resource Manager (LRM). Thus, system-wide constraints, such as end-to-end timing, reliability, of energy integrity, can be addressed without incurring the complexity and overhead of individual negotiations among resources directly.

The distinction between mandatory core services and optional higher services allows to prevent deep service chains that would make real-time guarantees difficult and increase the level of uncertainty. The modular DREAMS architecture introduces a minimal set of services for safety-critical subsystems for ensuring the required properties of the DREAMS architecture. Application with less stringent timing and certification requirements can use optional services with increased functionality and flexibility.

The DREAMS architecture supports the information exchange between safety-critical and non-safety critical subsystems. While the information flow from safety-critical towards non safety-critical parts is supported with no restriction, the reverse direction requires restrictions, namely

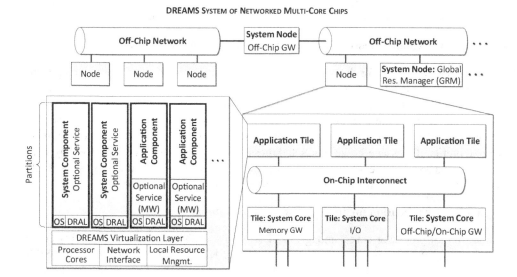

Figure 2.3: Realization of Platform Services in Networked Multi-Core Chips (Core Services in Yellow, Optional Services in Blue and Application Services in Red)

the separation of interactions by communication channels with temporal and spatial partitioning. The DREAMS technologies contribute hardware and software solutions for this constraint.

2.2.2 Architectural Building Blocks for the Provision of the Platform Services

The mapping of the DREAMS platform services of the waistline architecture to the networked multi-core chips is depicted in Figure 2.3.

2.2.2.1 Building Blocks for Core Services

The core communication services are realized at the chip-level by the (1) network interfaces, (2) the on-chip interconnect, (3) memory gateways and (4) on-chip/off-chip gateways. The network interface acts as the injection point for messages (and their constituting packets and flits) generated by a tile or core. The on-chip interconnect transports the messages between network interfaces inside one chip. The memory gateway establishes access to external memory (e.g., DRAM) and supports the shared memory paradigm on top of the message-based NoC. An on-chip/off-chip gateway relays selected messages from the NoC to an off-chip network and vice versa, while performing the necessary protocol transformations. At off-chip level, the (1) off-chip networks and (2) off-chip gateways belong to the core communication services. Each cluster has a corresponding off-chip network, where the networks of different clusters can be connected through an off-chip gateway.

The core execution services are realized by a virtualization layer inside a tile. Either each processor core runs its own hypervisor or the virtualization layer manages the entire tile including one or more processor cores. The virtualization layer establishes the partitions for the execution of components with guaranteed computational resources. Within each partition, an operating system and a DREAMS Abstraction Layer (DRAL) are deployed to provide software-support for utilizing the platform services from the application software (e.g., including communication drivers, drivers for time services, domain-specific APIs such as ARINC653).

The resource management services are realized by a Global Resource Manager (GRM) in combination with local building blocks for resource management. A DREAMS system contains a single

GRM, which can be realized by a single node or a set of nodes for improved fault-tolerance and scalability. The GRM performs global decisions with information from local resource monitors. It provides new configurations for the virtualization of resources (e.g., partition scheduling tables, resource budgets). The GRM configuration can include different pre-computed configurations of resources (e.g., time-triggered schedules) or parameter ranges (e.g., resource budgets). Alternatively, the GRM can dynamically compute new configurations.

Three local building blocks for resource management are distinguished: (1) Local Resource Managers (LRMs), (2) Local Resource Schedulers (LRSs) and (3) Resource Monitors (MONs). These local resource management building blocks are located at the individual resources at chip and cluster level. The LRS is responsible for controlling the access to particular resource based on a configuration that has been set by the LRM. Each resource has a corresponding built-in LRS such as the on-chip network interface, the hypervisor layer inside a tile, the memory gateway, the I/O gateway, the on-chip/off-chip gateway and the off-chip network interfaces. For example, the LRS in the on-chip network interface is responsible for dispatching time-triggered messages according to the schedule tables in the network interface and for traffic shaping of sporadic messages.

The Local Resource Scheduler (LRS) performs the runtime scheduling of resource requests (e.g., execution of tasks on processor, processing of queued memory and I/O requests). The LRS in DREAMS will support different scheduling policies (e.g., dispatching of time-triggered actions, priority-based scheduling). The LRMs adopt the configuration from the GRM at particular resources (e.g., processor core, memory, I/O). It is responsible for mapping global decisions to the local scheduling policy of the LRS. In some cases LRMs are able to take decisions for local reconfiguration.

The Resource Monitor (MON) monitors the resource availability (e.g., energy). Resource monitors also observe the timing of components (e.g., detection of deadline violations), check the application behavior (e.g., stability of control) and perform intrusion detection. Small changes will be handled locally, while significant changes will be reported to the GRM, who in turn can provide a different configuration at system-level.

2.2.2.2 Building Blocks for Optional Services

On top of the core services, the optional platform services establish higher-level capabilities for certain domains (e.g., control systems, multimedia). Optional services are capabilities that are not needed in all targeted applications, thus they can be integrated when needed or omitted if unnecessary to minimize resource consumption. In addition, using optional services we can support complex platform services for non safety-critical applications without affecting certification of safety-critical application subsystems.

Optional services are subject to partitioning and segregation performed by the core services. Hence, any fault of an optional service only affects the application and other optional services building on top of it. This fault containment is a key enabler for modular certification, because optional services not used by an application subsystem do not need to be considered in its certification.

One can distinguish three implementation choices for optional services:

1. System core: Optional services are implemented as self-contained IP cores with a message-based interface towards the NoC. The segregation is established by the NoC.

2. System component in a partition: An optional service is realized as a component within a partition. The optional service is provided to components in other partitions inside the tile using inter-partition communication mechanisms of the virtualization layer. In addition, the platform service can be made available using the NoC.

3. Middleware in a partition: The platform service is realized as middleware within a component and provides services to the application component within the same partition.

2.2.2.3   Building Blocks for Application Services

An application service is realized by an application component inside a partition. The application service provides its service to other components using the core communication services, where a partition with an application service is a communication end point.

2.2.2.4   Technology Independence of Architectural Style

The architectural style including the logical system structure, the physical structure and the architectural services is not restricted to a particular implementation technology. Different types of processors, on-chip networks, off-chip networks and operating systems can serve as the starting point for the establishment of the DREAMS architecture.

At the chip-level, we can distinguish the following categories of instantiations of the architectural style depending on the type of the underlying multi-core processor:

- **Shared memory-based chip architectures** are a special case of the architectural style with a multi-core chip containing only a single tile. This single tile contains multiple cores that interact via a shared memory. Instead of realizing the off-chip gateway via the NoC, there exists a dedicated (memory-mapped) I/O peripheral for the off-chip network interface. Instantiations of the DREAMS architectural style using PowerPC P4080 and x86 belong to this category.

- **NoC-based architectures** are another instantiation where the multicore architecture contains tiles each of which contains only a single core. The tiles are interconnected by a message-based NoC. Instantiations of the DREAMS architectural style using TTSoC belong to this category.

- **Full-scale architecture instantiations** provide multiple tiles with multiple cores per tile. The tiles are interconnected by an NoC. Each tile can contain a shared memory for the interaction within the tile. The interaction between the tiles is message-based, although a shared memory interaction can be realized on top of message passing based on a tile serving as a memory gateway (e.g., DDR controller).

Likewise, different scales can be distinguished at the cluster level including single-cluster and multi-cluster DREAMS mixed-criticality systems. The latter types of systems depend on off-chip gateways in-between the off-chip networks of different clusters. Based on the different integration levels, the architectural style supports different types of communication mechanisms:

1. The intrapartition communication between tasks within a partition is the responsibility of the application software or guest operating system within the partition and thus transparent to the DREAMS architecture.

2. Interpartition communication between partitions on the same tile is supported by the hypervisor and can be implemented using the local shared memory within the tile.

3. Interpartition communication between partitions on different tiles of the same chip occurs using the message-based NoC. Shared memory interactions are possible using a memory gateway and shared memory accesses on top of message passing.

4. Interpartition communication between partitions on different chips occurs using the on-chip/off-chip gateway. The gateway is accessed using either the NoC or via the tile's shared memory depending on whether a single-tile or multi-tile node is considered.

The application interface for interpartition communication is identical, regardless of which communication type (2 to 4) is used.

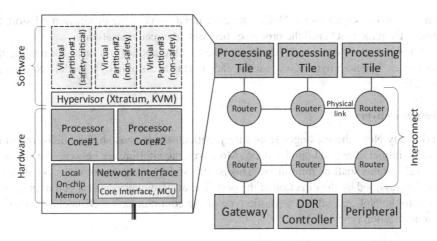

Figure 2.4: Chip-Level Communication Services, © IEEE 2016, Reprinted with Permission from [12]

2.2.3 Communication Services: On-Chip

The on-chip communication services close the gap between the execution services and the off-chip communication services. In addition to the execution services and off-chip networks, the on-chip communication services serve the global time base and the resource management services by conveying their messages.

The message-based on-chip communication services are realized mainly by the on-chip NI, the on-chip router and the on-chip physical link. The NI serves as an interface to the NoC for the tiles by exchanging the messages from the tile into the NoC as well as delivering the received messages from the NoC to the tile (cf. Figure 2.4). Routers on the other hand, are responsible to relay the flits from the sender's NI to the destination NIs. The number of input and output units at each router and the connection pattern of these units represent the topology of the on-chip network (e.g., star, ring, spidergon). The physical links act as a glue element among NIs and routers and realize the interconnection among them.

On-chip communication services can be grouped in the following three main categories:

1. **Core interfacing** establishes an interface to the interconnect for the execution services. For instance, sending and receiving the messages, reading the status values and performing the on-the-fly reconfiguration of the communication services.

2. **Mixed-criticality handling** establishes the mixed-criticality requirements by maintaining the temporal constraints as well as the priorities.

3. **NoC interfacing** performs the NoC-specific operations, such as creating the packets, the flits and embedding the path and destination information in the header.

   Below these services are elaborated.

2.2.3.1 Storing Messages at Ports

The NI acts as an interface between the tile and the NoC by providing *ports*. They are communication terminals of application components and decouple them from each other and from the NoC. Using the ports, the design of the NoC can abstract from the application components and the implementation technology of the tiles. From the point of view of the application layer, ports can be either *input* or *output*. Output ports are used at the source side and store the message from the core until the NI inject the message into the NoC. Similarly, input ports at the destination NI store the messages coming from the NoC until the processor performs the read operation.

If the tile hosts processor core, each port is accessed by the application layer for reading and writing the messages. In case a hypervisor is used, each port is accessible by a predefined partition from the core side.

2.2.3.2   Message Delivery

In a mixed-criticality NoC, the messages from safety-critical subsystems shall be delivered to the destination with minimal jitter and bounded delay. Moreover, the interference between the messages of low-critical subsystems shall be minimized. This is achieved by a number of services offered by the NI and can be classified by the direction of the port. The available messages at output ports need to be injected into the NoC, based on the traffic type of the communication channel by one of the following communication types:

1. **Time-Triggered Dispatching of Periodic Messages:** The service reads the periodic ports and feeds the data into the dedicated queue for periodic messages in the serialization layer at the ·defined instant given by the time-triggered schedule.

2. **Traffic Shaping of Sporadic Messages:** The service reads the sporadic messages from the ports and enqueues the respective buffers at the serialization layer. The sporadic messages will be read from the port only if the minimum interarrival time is already elapsed. This parameter is available in the port configuration.

3. **Relaying of Aperiodic Messages:** Since the aperiodic messages have no timing constraints on successive message instances and no guarantees with respect to the delivery and the incurred delays, the service only forwards them once there is new message available at the respective port. Afterwards, the serialization layer will send the aperiodic messages only if there is bandwidth available which has not been used by the periodic and sporadic messages.

In case of input pots, the NI acts as a bridge between the NoC and the cores to support a bidirectional communication by supporting the communication from the NoC towards the cores. This bridging is done via packet classification and dispatching the messages to the destination ports. The one-to-one mapping between the VLs and the ports enables this service to classify the incoming messages and write them to the respective port. Thereafter the respective tile will be able to read the message.

2.2.3.3   Collision Handling

In order to handle the collisions between the messages of different criticalities, a mechanism is needed. Collisions between the time-triggered and event-triggered messages can be avoided by timely block or shuffling, whereas to resolve the collision among sporadic messages and also between sporadic and aperiodic messages, only shuffling is employed.

The timely block mechanism guarantees no collision between two messages by blocking the bandwidth during a guarding window for a message of higher priority. Shuffling in contrast, requires no guarding window. In this method, priorities are used for the arbitration of using a shared resource.

2.2.3.4   Monitoring and Reconfiguration Services

The resource management services are built on top of the communication services. As a result, ports provide the interface to the resource management service for the application layer. More precisely, ports offer the following two interfaces:

1. **Monitoring interface** provides the read-only information which can be read by the application

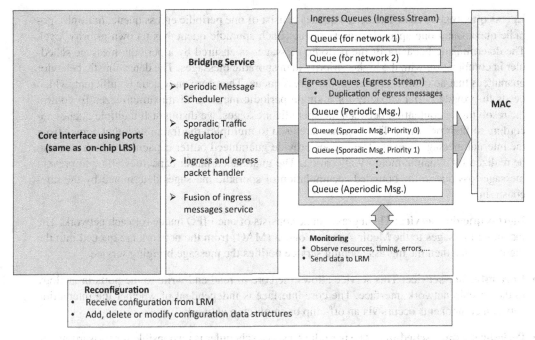

Figure 2.5: Services of Off-Chip Network Interface

layer and is composed of two types of information. STATistical information (STAT) includes information which can be used by the resource manager or in some cases, by the application layer (e.g., number of available messages at a port, if a port is empty or full). DIAGnostic information (DIAG) will be generated, only if any unexpected behavior is spotted at the resource (reading an empty port, writing into a full port). This information will aid the resource manager to detect the faulty software or hardware element.

2. **Reconfiguration interface** provides the adaptability based on the application needs (e.g., change to the fail-operational mode), environment changes (e.g., significant change in the temperature) and platform feedbacks (e.g., low battery). This provides energy efficiency (e.g., by shutting down selected entities) or fault-tolerance (e.g., by blocking the communication channels of the faulty entities).

2.2.4   Communication Services: Off-Chip

The off-chip communication services comprise the off-chip network interface, the off-chip communication routers and the gateway services.

2.2.4.1   Off-Chip Network Interface

The off-chip network interface provides the interface of a node to an off-chip network with a suitable communication protocol (e.g., TTEthernet). The off-chip network interface acts as the injection point for messages generated by a node for the off-chip network. Likewise, the network interface is a sink for messages from an off-chip network destined to the respective node. The connection between network interfaces occurs using one or more off-chip routers in a given topology (e.g., star, ring).

The off-chip network interface provides the following services as depicted in Figure 2.5:

- **Egress queuing service:** The egress queues consist of one periodic egress queue, multiple sporadic queues and one aperiodic egress queue. Each sporadic queue has its own priority level. The deterministic behavior of the periodic messages is ensured by a periodic message scheduler in combination with the higher priority than sporadic messages. The deterministic behavior guarantees that no conflict appears at the egress queue. Therefore one queue is sufficient, which needs to provide buffer capacity for a single periodic message of maximum size. To control the resolving of contention between the sporadic messages, we distinguish multiple queues according to their priorities. These queues are used to multiplex the frame flow that comes from the internal message queues. The queues provide guaranteed buffer capacities, which can also be realized by dynamic memory allocation. The guaranteed buffer capacities allow to prevent message loss due to the bounded accumulation of sporadic messages determined by the rate-constraints.

- **Ingress queuing service:** The ingress queue consists of one FIFO queue for each network. The incoming messages to the Media Access Control (MAC) from the network are queued into the ingress queue, then the ingress queuing service notifies the message bridging service.

- **Core interface service:** This service allows the core to read and write to the ports in analogy to the on-chip network interface. The core interface is independent of whether the interaction between components occurs via an off-chip or an on-chip network.

- **Periodic message scheduler:** The periodic message scheduler is responsible for forwarding the periodic messages from a corresponding virtual-link to the egress queue at the time specified in the static communication schedule. The periodic message schedule uses the port configuration parameter to determine the point in time when the periodic message needs to be forwarded with respect to the global time base.

- **Sporadic traffic regulator:** The sporadic traffic regulator guarantees the minimum interarrival time between two consecutive instances of sporadic messages on the respective virtual link. If this timing constraint is satisfied, then the sporadic traffic regulator relays these sporadic messages from its queue to one of the sporadic queues at the egress queue according to the message priority.

- **Ingress and egress packet handler:** The packet handler is responsible for redirecting the incoming aperiodic messages from the off-chip network to the respective ports. In addition, the packet handler polls the aperiodic ports and redirects the respective messages to lowest priority egress queue.

- **Fusion of ingress messages:** This service performs message deduplication using different mechanisms according to the traffic type. In order to hide the paths and different latencies for periodic messages in different networks, the fusion of ingress messages service requires a priori knowledge about the time-triggered schedule. This schedule includes information about the receiving time, the sending time and the corresponding buffer identification. The fusion of ingress messages service checks the corresponding virtual-link buffer before the sending time and takes the decision to send one of the redundant periodic messages accordingly. Moreover, the fusion of ingress messages service establishes deterministic arrival times of these messages.

  For each incoming sporadic message the fusion of ingress messages service checks the sequence number and compares it with the sequence number that is listed in the configuration parameters. The "First Valid Wins" policy is used to take the decision on the forwarding of redundant messages. Upon the transmission of a message, the fusion of ingress messages service updates the sequence number in the configuration parameters.

- **Reconfiguration and monitoring services:** The off-chip network interface has two building blocks (i.e., reconfiguration and monitoring) that are responsible for rewriting the configuration parameters and for the observation of the communication resources, the message timing and for retrieving error detection information. The monitoring block monitors the time of the message arrival and transmission and compares it with its configuration parameters (i.e., period and phase of sporadic messages with tolerance windows, minimum interarrival times of sporadic messages). In addition, the monitor is responsible for monitoring the application behavior (e.g., monitoring deadlines, overload detection).

The resource management services of DREAMS can check and analyze the monitored behavior and send a new configuration to the reconfiguration building block. The reconfiguration adopts the modified configuration parameters of the ports. Supported configuration parameters are the timing configuration of ports, address information of ports, guaranteed buffer capacities and redundancy degrees for replication/fusion.

2.2.4.2 Off-Chip Router

The model of the off-chip router is illustrated in Figure 2.6. The off-chip router architecture includes several building blocks to segregate messages from subsystems of different criticality, to ensure the deterministic behavior of the periodic messages and the bounded end-to-end delay of sporadic messages. The off-chip router provides multiple physical links and a bridge layer. Each physical link contains a physical layer and a MAC layer. The bridge layer is responsible for handling ingress messages and forwarding them to the egress ports depending on the traffic type (i.e., periodic, sporadic and aperiodic).

The router provides several services to redirect the messages and guarantee spatial and temporal partition of the critical traffic.

- **Internal message queuing:** Each periodic VL has one periodic VL buffer which provides buffer space for exactly one message. In case this buffer is full and another message arrives with the same VLID, the newer message replaces the old one. Likewise, each sporadic VL has one queue. It is possible to store several messages of the respective VL in this queue. All aperiodic messages are stored in one queue since aperiodic messages have no timing constraints on successive message instances and no guarantees.

- **Egress queuing service:** The egress queues consist of one periodic egress queue, multiple sporadic queues and one aperiodic egress queue. Each sporadic queue has its own priority level. The deterministic behavior of the periodic messages is ensured by the *periodic message scheduler* in combination with the higher priority than sporadic messages. The deterministic behavior guarantees that no conflict appears at the egress queue. Therefore one queue is sufficient, which needs to provide buffer capacity for a single periodic message of maximum size. To control the resolving of contention between the sporadic messages, we distinguish multiple queues according to their priorities. These queues are used to multiplex the frame flow that comes from the internal message queues. The queues provide guaranteed buffer capacities, which can also be realized by dynamic memory allocation. The guaranteed buffer capacities allow to prevent message loss due to the bounded accumulation of sporadic messages determined by the rate-constraints.

- **Packet classification service:** The packet classification service distinguishes between traffic types based on connection-oriented and connectionless communication. The connection-oriented communication is used for the periodic and sporadic messages. Aperiodic messages use the connectionless communication.

We regard a message as a tuple with the following elements:

- <type, VLID, data> for messages in connection-oriented communication

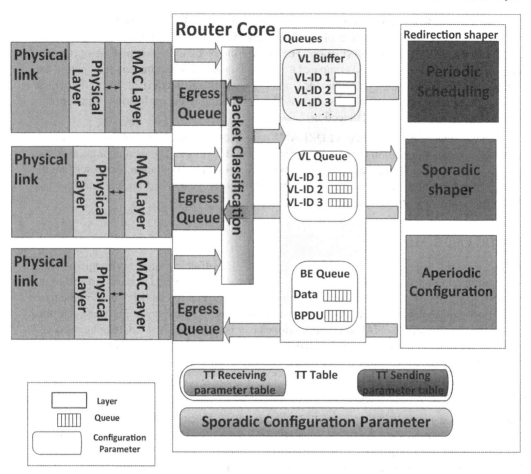

Figure 2.6: System Model of Off-Chip Router

  – <type, destination address, data> for messages in connectionless communication

- **Periodic scheduling service:** The periodic scheduling service is responsible for relaying the periodic message from the virtual-link buffer to the queue for periodic messages at the correct egress port according to a TT table. The TT table also determines the point in time when the periodic message is relayed, thereby ensuring the deterministic communication behavior.

- **Sporadic shaper service:** The sporadic shaper realizes the traffic policy for the sporadic messages by implementing an algorithm known as token bucket [10]. This service checks the time interval between consecutive frames on the same virtual link and moves sporadic messages from the virtual-link queue to one of the sporadic egress queues according to the message priority.

- **Aperiodic self-configuration service:** For aperiodic message the spanning tree protocol is used to establish a loop-free topology for communication of aperiodic messages [13]. The supported aperiodic messages include Bridge Protocol Data Units (BPDU) and aperiodic data messages. BPDU messages are exchanged between off-chip routers to determine the network topology, e.g., after a topology change has been observed.

- **Configuration parameters and reconfiguration service:** The time-triggered communication is based on a predefined schedule where there are two groups of parameters for each periodic message: a time-triggered receiving parameter table and a time-triggered sending parameter table providing the message period and phase with respect to a global time base [14].

The sporadic communication is based on configuration parameters that define a minimum interarrival time and jitter for each virtual link. The minimum interarrival time is defined as the time interval between two consecutive messages that are transmitted on the same virtual link. The jitter is the maximum timing variability that can be introduced by multiplexing the virtual links into shared egress queues. A message that arrives within the jitter is considered as timely, otherwise a new minimum interarrival time is started.

The global resource management can switch time-triggered tables in case the system has multiple scenarios of the periodic messages. In addition, the global resource management can rewrite or modify the configuration for one or several virtual links for the periodic and sporadic communication. Moreover, the global resource management may modify the non-active time-triggered tables and later switch to this new table instead of the current one.

- **Monitoring services:** The off-chip network offers a number of monitoring features that can provide the basis for reconfiguration decisions. These monitoring features cover the behavior of switches themselves as well as the communication that is transferred by the switch. Switch-level monitoring is supported for each off-chip network router (e.g., invalid router configuration, not enough router memory). In addition, the off-chip monitoring services support network traffic monitoring on the level of each individual virtual link. For example, timing errors are detected in case a frame is received outside of the expected window. A length error is reported when a frame exceeds the configured maximum length for the specific VL.

- **Serialization service:** The serialization service forwards the messages from the egress queues to the MAC layer according to the priority. The highest priority is assigned to periodic messages, whereas aperiodic messages have the lowest priority. Also the serialization service uses one of the following mechanisms to solve the collision between different traffic types, the shuffling or timely block mechanisms. The timely block mechanism disables the sending of other messages in the router-core during a guarding window prior to the transmission of a periodic message. For the shuffling mechanism, no guarding window is needed. In the worst case, the router-core delays a periodic message for the duration of maximum size message. In addition, the message serialization supports timely block and shuffling service.

- **MAC interfacing:** The MAC interfacing sends and receives the message from the off-chip network by encapsulating the message in the frame or decapsulating the message from the frame. In case of incoming messages, the MAC layer filters messages that are not destined to this node based on the MAC address.

2.2.4.3 Gateway

Off-chip/on-chip gateways are used to establish the end-to-end communication over heterogeneous and mixed-criticality networks. The connection between off-chip and on-chip networks is established through gateways as illustrated in Figure 2.7 [15, 16]. The gateway consists of the gateway core functionality, network interfacing and MACs.

The gateway core is responsible for processing incoming messages based on timely redirection, protocol conversion, monitoring and configuration services. The network interfacing provides the interface between the MAC and the gateway core. Furthermore, classification and serialization of the packets is performed in the network interfacing. In order to realize fault-tolerance, the gateway can include multiple network MACs. Each network MAC connects the gateway to either an off-chip network or an on-chip network. In case of network redundancy, multiple network MACs are required. Thus, the network interfacing is responsible for merging identical incoming messages

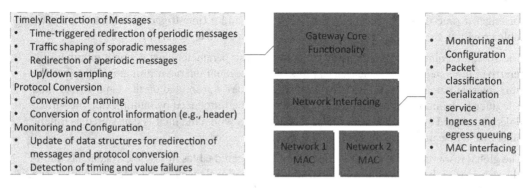

Figure 2.7: System Model of a Gateway

and duplicating outgoing messages to be sent to different MACs. The architecture is illustrated in Figure 2.8.

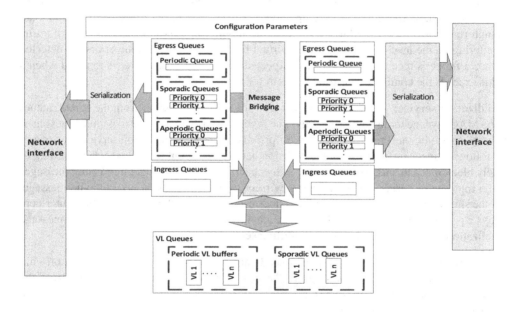

Figure 2.8: Architecture of the Off-Chip/On-Chip Gateway

The services of the gateway are as follows:

- **Message-Classification Service:** The message classification is based on the concept of VLs. The message-classification service is responsible for classifying the incoming messages from the network interface in order to decide on the corresponding buffer (i.e., VL queues and egress queues) according to the message type and the configuration parameters. Additionally, the message-classification service checks the message format and its control information, e.g. the Virtual Link IDentifier (VLID). In case the message has an invalid message format, it is discarded. Moreover, the message-classification service uses the configuration parameters to check the integrity and validity of the periodic and sporadic messages. This includes the verification of the message size and checking whether messages arrive with correct VLID. In addition, the gateway checks whether the periodic messages arrive within the specified reception windows of the VL.

- **Message-Scheduling Service:** This service guarantees the determinism of the redirection of periodic messages within the on-chip/off-chip gateway. Each periodic message has predefined timing parameters such as a period and a phase. According to the predefined configuration for the message scheduling, this service determines the points in time when the periodic messages are relayed.

- **Traffic-Shaping Service:** This service is responsible for guaranteeing the minimum interarrival time between two consecutive sporadic messages on the respective VL. The minimum interarrival time is part of the configuration parameters for each VL.

- **Relaying of Aperiodic Messages:** This service is responsible for relaying the aperiodic messages between ingress and egress queues based on the respective data direction and the destination addresses.

- **Down Sampling:** This service provides the message exchange between networks with different periods of periodic messages or different rate-constraints of sporadic messages. Down sampling is also required to resolve the differences in the bandwidths of off-chip and on-chip networks. The gateway has to redirect a subset of the incoming messages to satisfy the timing requirements of the target network. In addition, the redirection needs to be synchronized between networks to ensure the forwarding of consistent data. In the down sampling service, the gateway will send the most recent periodic message that arrived before the next transmission instant. In case of sporadic messages, the traffic shaper will drop all messages that arrive within the minimum interarrival time.

- **Protocol Conversion:** The protocol conversion service is responsible for the encapsulation and decapsulation of incoming and outgoing messages. The gateway adapts the message format and the control information according to the used communication protocols (e.g., headers with addresses, flow-control information, CRC). In addition, the gateway needs to establish for each incoming message a new address for the destination network. This computation is performed using the address information of the incoming message and differs according to the traffic and network types.

  In case of periodic and sporadic traffic, the new addressing information is either a VLID or a routing path towards the final destination or towards another gateway. The routing path is required for source-based routing, which is common in many NoCs. The VLID or the routing path can be acquired by a lookup of the incoming address information in the gateway's configuration. In case of aperiodic traffic, the new addressing information is either a destination address or a dynamically computed routing path. The gateway can use the spanning tree protocol [13] to establish the destination address in a dynamic way.

- **Egress-Queuing Service:** The egress queues consist of one periodic egress queue, multiple sporadic queues and one aperiodic egress queue. Each sporadic queue has its own priority level. The deterministic behavior of the periodic messages is ensured by the message scheduling service in combination with a higher priority than sporadic messages. The deterministic behavior guarantees that no conflict appears at the egress queue. Therefore one queue is sufficient, which needs to provide buffer capacity for a single periodic message of maximum size.

  To control the resolving of contention between the sporadic messages, we distinguish multiple queues according to their priorities. These queues are used to multiplex the frame flow that comes from the internal message queues. The queues provide guaranteed buffer capacities, which can also be realized by dynamic memory allocation. The guaranteed buffer capacities allow to prevent message loss due to the bounded accumulation of sporadic messages determined by the rate-constraints.

- **Ingress Queuing Service:** The ingress queue consists of one FIFO queue for each network. The incoming messages from the network are queued into the ingress queue, then the ingress queuing service notifies the message classification service.

- **Virtual-Link Queuing Service:** The VL queues belong to two groups: one for the periodic messages and the other one for the sporadic messages. Each periodic VL has one periodic VL buffer, which provides buffer space for exactly one message. In case this buffer is full and another message arrives with the same VLID, the newer message replaces the old one. Each sporadic VL has one queue. It is possible to store several messages of the respective VL in this queue.

- **Serialization Service:** The serialization service forwards the messages from the egress queues to the network (off-chip or on-chip) according to the priority. The highest priority is assigned to periodic messages, whereas aperiodic messages have the lowest priority. Also, the serialization service uses either shuffling or timely blocking to resolve contention between different traffic types. The timely block mechanism disables the sending of other messages in the egress queues during a guarding window prior to the transmission of a periodic message. For the shuffling mechanism, no guarding window is needed. In the worst-case, the gateway delays a periodic message for the duration of a sporadic or aperiodic message of maximum size.

- **Configuration Parameters:** The configuration parameters of the gateway are as follows:

  - **Guaranteed buffer capacity**: Each ingress queue, egress queue and VL queue is associated with a corresponding guaranteed minimum buffer capacity. The buffer capacity is determined by the maximum message size and the message timing. This buffer capacity can avoid message omissions of sporadic and periodic messages based on rate-constraints and message periods.

  - **Address information of ports**: The VL associated with a port and the data direction (from the off-chip network or to the off-chip network) are defined.

  - **Message type**: The message type is defined such as periodic, sporadic or aperiodic.

  - **Timing parameters**: In case of periodic messages, the parameters include the period and phase. For sporadic messages, the interarrival time, the jitter and the priority are specified. In case of aperiodic messages, no timing parameters are required.

2.2.5 Communication Services: Shared Memory

The shared memory model is supported on top of message-based networks. A shared address space is established for external memories and input/output devices. Thereby, application subsystems can exploit programming models based on shared memory in addition to message-based interactions, while exploiting the temporal and spatial partitioning of the message-based network infrastructure.

Shared memory communication allows efficient data exchanges between multiple programs running on the same processor of a tile. This can be further extended to the communication of multiple threads within a single program.

From the hardware perspective, shared memory consists in a typically large amount of RAM that can be accessed by means of read/write instructions issued by several CPUs in a multiprocessor computer system. This requires that all CPUs implicitly share a common application memory space, which classically consists of several on-chip DDR memory dies accessed through on-chip DDR memory controllers.

From the software perspective, two processes communicating through shared memory are using the same physical memory location as their regular working memory. This requires that the two processes are located on the same machine (running a given OS/hypervisor). While being very fast (the communication between the processes happens with a data rate in the order of a memory access), specific care must be taken with respect to memory inconsistency when the communicating processes are executed on two different CPUs. An underlying cache coherent architecture is necessary in this case. Cache coherency might be guaranteed using cache controllers coupled to OS services. In this case, part of the shared memory traffic will be constituted by read/write accesses generated by the cache controllers upon cache refill, cache miss (reads) or cache flush, clean (writes) events.

The following services as distinguished:

- **Write access service:** Write operations (also known as store operations) are used to write data in a shared memory location. Then any other process implied in a communication exchange may observe the data at the same memory mapped address. Writes are classical operations from the processor's instruction set for which IP protocols and underlying on-chip communication layers such as bus or NoC offer full support. Note that write operations might be used not only for writing data structure (such as strings, tables...) but also to access memory location that can be considered as flags. Furthermore, write operations reaching the shared memory might not always be generated from the processor itself but from an intermediate communication stage, such as a cache controller executing a flush/invalidate/write back operation.

- **Read access service:** Read operations (also known as Load operations) are used to read data in a shared memory location. Reads are classical operations from the processors instruction set for which IP protocols and underlying on-chip communication layers such as bus or NoC offer full support. Note that read operations might be used not only for accessing data structure (such as strings, tables...) but also to access memory location that can be considered as flags. Furthermore, read operations reaching the shared memory might not always be generated from the processor itself but from an intermediate communication stage, such as a cache controller executing a speculative fectch/cache refill operation.

- **Shared memory coherency service:** To avoid data inconsistency in shared memory, when in multiprocessor context, special care must be taken with respect to memory coherency. Furthermore, modern generations of processors embed L1/L2 caches memory which speeds up access to memory at the cost of a higher processing for maintaining the cache/shared memory coherency. Coupling cache controllers is important for modern multiprocessor real-time OS. Cache coherency and shared memory consistency is a well-defined problem with standard solutions. We propose to rely on existing offered SW services for this aspect in DREAMS.

- **Monitoring and configuration services:** For these resource management services, a shared memory controller must be assumed. Depending on the model chosen, different criteria might be monitored (internal queues status, number of page misses/hits) or reconfigured (internal queues allocation, power-off of part of the memory area, etc.).

2.2.6   Communication Services: I/O Memory Management Unit and NoC Firewall

On-chip performance and security can be enhanced via I/O Memory Management Unit (IOMMU) and NoC Firewall services. These system services are directly related to global shared memory services and are further developed in DREAMS (and beyond) compliant with ARM v7 processor architecture and related virtualization extensions.

2.2.6.1 IOMMU and Page-Level Security

The IOMMU is a system module designed to translate addresses from the virtual space of a guest device to globally shared physical address space, thereby allowing DMA requests originating from a device to efficiently access external shared memory. This translation is similar to a processor's Memory Management Unit (MMU), except that the IOMMU translates memory accesses of fully virtualized devices rather than the CPU, as the MMU does.

However, IOMMU functionality is not limited to virtual address translation from device DMA addresses to physical addresses. The IOMMU also supports secure memory access services by isolating device accesses using page-level granularity. For instance, in a virtualization-aware environment, the hypervisor can configure (or remap) the I/O page table to safely map a device to a particular guest OS without risking integrity of other guests, i.e. a guest cannot break out of its address space with rogue DMA traffic. Similarly, the IOMMU is able to provide an increased amount of security for system processes in scenarios without virtualization. In addition, the OS is able to protect itself from corrupt device drivers by limiting a device's memory accesses and managing the permissions of all peripheral devices. Typically, upon an address translation request from a device, the IOMMU consults the I/O page table to compute the physical address. If a device tries to access memory without a valid entry in its I/O page table, then the IOMMU will access a default translation context and inform the hypervisor through an interrupt, rejecting the access if configured to do so; notice that different types of system exceptions can occur, such as address translation requests from a device with uninitialized context, and even more critical security-related events, such as request access violations arising from malevolent or corrupt devices. Besides address translation, IOMMU can also provide monitoring services at page-level access granularity through a specialized hardware monitoring unit (HMU) that logs events related to: internal IOMMU activity (counter statistics and error logs) and interface transactions (AMBA AXI bus). These events can be used to perform access pattern analysis, estimate key performance metrics, e.g. latency, throughput and resource utilization, and optimize the architecture by introducing novel decision control mechanisms, including system-wide services for

- dynamic address space management, such as I/O remapping,

- performance-oriented system adaptation, including pre-fetching and/or pinning of certain pages (e.g. this in particular is related to hard real-time processing at process- or VM-level), and

- fault tolerant services, such as dynamic reconfiguration of the page entries in order to recover from hardware faults.

In particular, within DREAMS, TEI has demonstrated using RTL simulation that a dedicated IOMMU (or MMU) module for mixed criticality able to pin the most critical pages can help improve performance of critical healthcare applications.

2.2.6.2 NoC Firewall and Multi-Compartment Technology

In addition to IOMMU services, a virtualization-aware hardware NoC Firewall unit implemented at the on/off-chip network interface can support VM/process isolation services throughout the multicore SoC by tagging NoC transactions, establishing access rules for virtual components on physical address regions and ensuring that rules are obeyed at each network interface. An initial solution envisioned in DREAMS targeted fine grain rule-checking at memory page-level by invoking a rules table walk to an external memory which stores the rules defined with page-level granularity. However, due to the higher complexity of this architecture, an alternative segment-level of the NoC Firewall was implemented (see Chapter 11).

The NoC Firewall concept supports a multi-compartment philosophy by specifying firewall rules that relate not only to a separate BRAM (NoC output port), but also to a separate input port depending on the path of the access request. The technique extends existing protection mechanisms available in virtualization-aware technologies, such as ARM v7 Trustzone architecture (and related IOMMU support). More specifically, ARM v7 Trustzone architecture defines only two security domains (secure and non-secure) identified using an NS bit available within the memory page descriptors. Notice that alike our rules, the NS bit can be statically set (e.g. for a secure or non-secure peripheral), or dynamically modified either at boot time or by a system security thread. Further information on the implementation of the NoC Firewall infrastructure, the corresponding Linux driver hierarchy, and its use to protect privacy of healthcare data is provided in Chapter 11.

2.2.7 Communication Services: Security

Security services are required to protect the system against passive and active attacks. Passive attacks do not influence the system behavior, but they lead to information disclosure that may be not intended. Performing active attacks, the system is manipulated, e.g., an attacker can send false messages or shut down services. This may result in an unintended system behavior and threatens the integrity of the system.

The security services in DREAMS protect the system against passive and active attacks. They provide the basic security services confidentiality, integrity, and authentication. Access control is an additional capability that requires authentication and prohibits the access of unauthorized users or applications, e.g., an attacker or a malicious application.

The secure communication services extend the communication services of DREAMS to achieve the protection. They are divided into two levels: security services for the cluster-level communication and security services of the application-level. The security services for the cluster-level communication provide a basis protection for all data sent through the off-chip network. This service operates on OSI layer 2 [17], the data link layer, and hence, it is transparent for the higher layers, e.g., the applications on the application layer. The security services for the application-level provide a secure end-to-end connection between the communication partners. This services operates on OSI layer 7, the application layer, and they have to be used explicitly by the applications.

The other core services of the DREAMS architecture, i.e., the global time services, the resource management services and the execution services make use of the secure communication services. To provide a secure global time base, the time synchronization mechanism uses the secure communication services of the cluster-level. The resource management services and the execution services use the secure communication services of the application-level.

2.2.8 Global-Time Service

From a generic point of view the clock synchronization problem appears in systems that consist of entities connected to each other by using a network, where some, or all of, the entities are equipped with local clocks. In such systems, the aim of the clock synchronization services is then to establish a concept of "global time" between those entities that have local clocks.

Global time is established, when the distributed local clocks in a system, which are usually implemented as counters (e.g., as SW variables or HW registers) have "about the same value" at "about the same points in real-time".

Clock synchronization services establish exactly that. Consequently, as the presented definition of a global time is vague it makes sense to discuss the clock synchronization services in a generic sense first and specialize them hand in hand with concertizing the definition of a global time, as we will do as an example for off-chip and on-chip networks. In this section, we give a general overview of the clock synchronization problem.

In the following we will use Figure 2.9 in the discussion of the generic clock synchronization services. The diagram plots real time on the x-axis vs. computer time on the y-axis, where the computer time is the simulation of real time by the local clocks. The diagram depicts the traces of three local clocks, a slow clock, a fast clock, and a clock that perfectly resembles the progress of

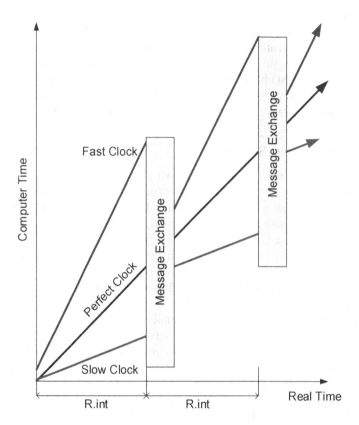

Figure 2.9: Computer Time vs. Real Time [18]

real time – the perfect clock. At any point (or small interval) in real time, the difference in computer time of the local clocks in the system (in this case the system of the three clocks) has an upper bound. The key challenge in clock synchronization is to design specialized services that ensure that a given upper bound can be guaranteed in system-specific settings.

We will use the following quality aspects of synchronization in the discussion of the synchronization services:

- precision: worst-case difference of any two non-faulty clocks in the system

- accuracy: worst-case difference of the clocks in the system to an external time reference

- startup time: worst-case time after startup of the time sources until the system is synchronized (with given precision and/or accuracy)

- integration time: worst-case time for a non-synchronized component in the system to become synchronized

- changeover time: worst-case time for the components in the system to change from one time source to another one (e.g., in the case that the original time source fails)

- recovery time: worst-case time for the synchronized timebase to recover after global synchronization loss

Before discussing the generic services in detail, we should note that there is frequently ambiguity by what "time" actually means and we thus provide the following differentiations:

- Phase synchronization vs. TAI synchronization:

  - Phase synchronization refers to clock synchronization in a way that the time represented by the local clocks is a circular counter, e.g., starting with 0, and counting up to a maximum value (usually referred to as the "epoch" of time), once the epoch is reached the counter wraps around and starts counting at 0 again. Definition 1 holds as it stands above.

  - TAI synchronization in contrast to phase synchronization means that the local clocks are not only synchronized to each other in conformance to Definition 1, but TAI synchronization also requires the local clocks to represent Time Atomique International (TAI time).

- State synchronization vs. rate synchronization

  - State synchronization refers to the process of the distributed local clocks instantaneously changing the current value of the counters that are used to implement the clocks.

  - Rate synchronization refers to the process of the distributed local clocks changing the rate according which the counters are updated. Rate correction can be done post-factum or into the future, or both: post-factum means that a local clock that found that it is currently deviating from other local clocks in the system applies rate correction for some time to gradually reach alignment with the other local clocks again, while into the future means that the local clock updates its rate with the aim not to generate a deviation to other local clocks in the first place.

Figure 2.9 depicts phase and state synchronization.

Time Representation

The representation of the global time base within the DREAMS architecture is based on a uniform time format for all configurations, which has been standardized by the IEEE Standard 1588 [19]. A digital time format can be characterized by three parameters: granularity, horizon and epoch. The granularity determines the minimum interval between two adjacent ticks of a clock, i.e., the smallest interval that can be measured with this time format. The reasonable granularity can be derived from the achieved precision of the clock synchronization. The horizon determines the instant when the time will wrap around. The epoch determines the instant when the measuring of the time starts. The unified time format (see Figure 2.10) is a binary time-format that is based on the physical second and nanoseconds. According to this time format, the highest possible granularity of the global time base is in nanoseconds.

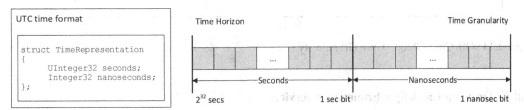

Figure 2.10: IEEE 1588 Time Format [19]

The range of the absolute value of the nanoseconds member shall be restricted to:

$$0 \leq | \, nanoseconds \, | < 10^9$$

The sign of the nanoseconds member shall be interpreted as the sign of the entire representation and a negative time stamp shall indicate time prior to the epoch.

Note that the time horizon of the off-chip network and the on-chip network may differ and require conversion. In particular, the off-chip network provides global time only on a granularity of major and minor cycles of the communication as shown in Figure 2.11.

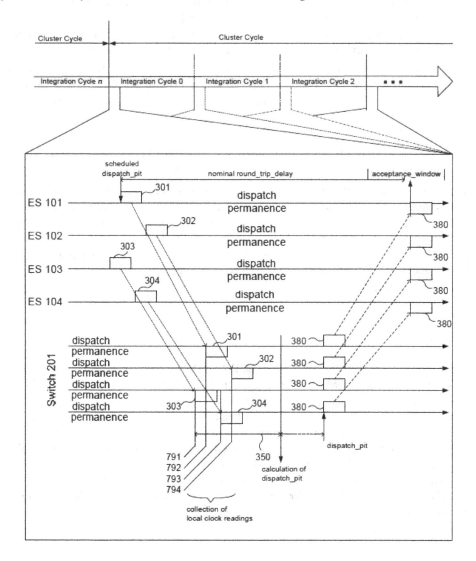

Figure 2.11: Off-Chip Global Time Granularity [20]

2.2.8.1 On-Chip Clock Synchronization Services

In general, a multi-core chip cannot be assumed to provide a single clock signal for the entire chip. The reasons why designers introduce multiple clock domains include the handling of clock skew, the clocking down of individual IP blocks as part of power management, or the support for heterogeneous IP blocks with different speeds (e.g., high-clocked special purpose hardware and a slower general purpose CPU).

Despite the existence of multiple clock domains, the DREAMS architecture will support a global time base at chip-level that is also externally synchronized with respect to a chip-external reference time (i.e., the cluster-level global time base). Figure 2.12 shows the global time at chip-level and the provision of multiple clock domains by providing different clocks to different components.

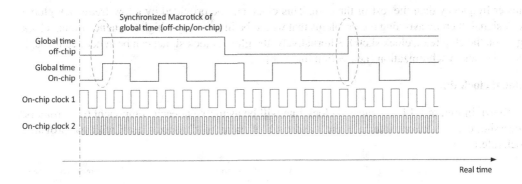

Figure 2.12: Example of Different Clock Domains in the DREAMS Architecture

Different On-Chip Clock Domains

The DREAMS architecture supports different clock domains by design. As shown in Figure 2.13, different parts of the system can operate at different clock speeds and components can include an arbitrary number of local clock domains, which are not visible outside of the tiles. For instance, a tile can be assembled by processor cores, memories, and network interface, which operate at their own frequencies.

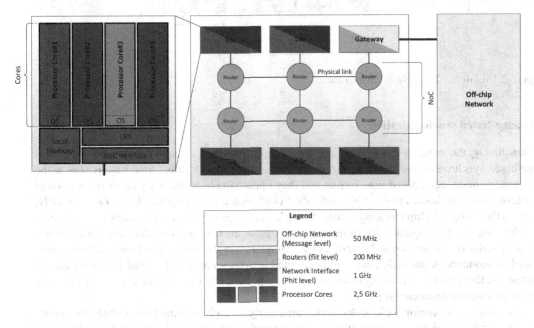

Figure 2.13: Example of Different Clock Speeds at Different Parts of the System

On the other hand, the aim of DREAMS is to introduce an architecture which provides a system-wide synchronized global time base. This global time base allows the temporal coordination of actions on the distributed components (e.g., avoidance of contention at resources based on TDMA). In addition, timestamps assigned at different components can be related to each other. Timestamps become also meaningful outside the component where the event has been observed.

The global time base at chip-level embodies an independent clock domain, which typically has a lower frequency than the rest of the chip. This clock can be provided by a low-frequency global clock signal, thereby avoiding the problems that would be incurred by a high frequency global clock signal on the chip (e.g., clock skew). Alternatively, the global clock signal can be generated through internal clock synchronization (i.e., within the chip).

Global clock line

As shown in Figure 2.14, a dedicated clock line will be available at each component (e.g., routers, processing cores, network interface, etc.) and each of them synchronizes itself with the provided clock reference.

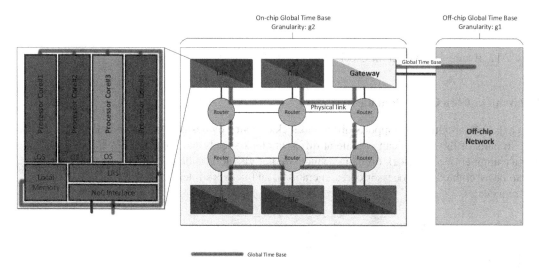

Figure 2.14: Global Time Base Clock Line

Message-based synchronization

Alternatively, the value of the global time base can be provided to each component via a message based synchronization protocol. In this approach, the value of global time base is sent to the components in defined unified time format and they update the local clock by either of mentioned synchronization methods. For example, individual clock domains can operate in the range of GHz, whereas the global on-chip clock signal can have a lower frequency by several orders of magnitude.

The choice of the frequency determines the precision of the temporal coordination and the meaningful granularity of timestamps. In particular, the frequency of the global time base determines how densely a sequence of mutually exclusive distributed actions with time-triggered execution can be packed together while still avoiding collisions at the respective resources. (An example is given later for the on-chip communication.)

The existence of multiple clock domains, particularly of a global time base, entails the decoupling of synchronization of actions within the system and the operation of local entities. The global time base is allowed to maintain a relatively slow clock domain compared to the remainder of the system and the frequency associated with this clock domain determines the global granularity, to which actions in the system are synchronized. More precisely, the activities are not driven by the global time base, but they are synchronized by the global time base.

For instance, the on-chip communication of flits and phits can take place at a frequency that is higher than the rate of the global time base while operating in a synchronized manner with the global time base. The frequency at which the LRS at on-chip NI operates, is higher than the frequency of the global time base (as shown in Figure 2.12), but fully synchronized with it.

In the example in Figure 2.12, after every 16 clock cycles of the LRS, there must be a single clock cycle of the global time base. This synchronization is necessary for the transmission of periodic messages. The global time is used at the LRS to align the start of the transmission of a periodic messages with other NIs, in order to guarantee bounded delay and minimum jitter for periodic messages (cf. Figure 2.15). In contrast, the global time base will not be necessary for sporadic and aperiodic transmission of messages.

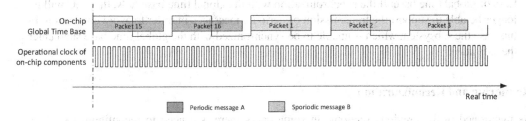

Figure 2.15: Global Time Base vs. Transmission of Packets and Flits

On-Chip Synchronization

The synchronization between the on-chip global time base and the off-chip global time base is based on rate correction in combination with overflow time intervals. Figure 2.16 shows an example, where the on-chip global time base is four times faster than the off-chip global time base, but supposed to be synchronized, in a sense that each fourth rising edge of the on-chip global time is associated with a rising edge of the off-chip global time base. However, the on-chip global time base runs faster and as shown in the figure, after the fourth occurrence, the next rising edge waits until the rising edge of the reference clock, i.e., the off-chip global time base. The reflow interval determines the tolerable deviation between the rates of the off-chip and on-chip global time base.

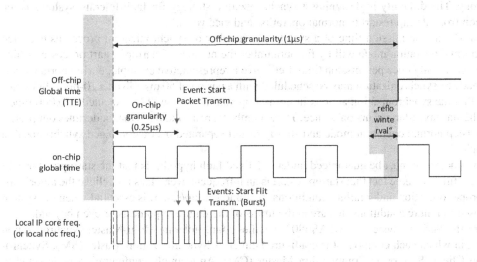

Figure 2.16: State Synchronization for On-Chip Global Time Base

In addition, one can adjust the rate of the on-chip global time base in a way that in coming cycles the drift becomes smaller.

Loss of synchronization

We can consider a system as clock synchronization perspective in one of the following statuses:

- System wide synchronization: in this case, the synchronization between multiple clock domains is operating without any problem and all entities are well synchronized.

- Loss of off-chip synchronization (on-chip only): in case of a loss of off-chip clock synchronization, the on-chip transmission of periodic messages is still possible, since the NoC is still able to correct the on-chip clock with the global time base.

- Loss of global time base: if the synchronization with the global time base fails, the NoC will no longer be able to support the transmission of periodic messages in order to avoid contention. In this case the subsystem which is unable to be synchronized with the global time base shall enter the safe state.

Monitoring and Reconfiguration

As mentioned in the previous sections, in some cases there is a need to reconfigure the clock system. For instance, in case of loss of the global clock line, the monitoring interface shall report the failure to the LRM in order to provide the new configuration. Furthermore, local modifications, for instance tuning frequencies in components and the communication subsystem clock parameters (e.g., horizon, epoch, etc.) can be established using the reconfiguration services.

2.2.8.2   Off-Chip Clock Synchronization Services

We assume that distributed local clocks are being driven by independent oscillators. In addition, there shall be non-negligible transport delays in the communication of the local clock values between nodes.

There are two different modes of operation in an off-chip network: normal operation and startup/restart. During normal operation, the synchronization strategy assumes initial synchronization is established and maintains this synchrony. It is the task of the startup/restart to establish initial synchrony. The difficulty in designing a synchronization strategy for fault-tolerant systems is the transition from startup/restart to normal operation and vice versa.

Considering the mission time of a system, the number of synchronization processes executed under normal operation mode will by far outnumber the number of startup/restart processes which ideally occurs only once per mission time. Let's give a representative example: during normal operation mode re-synchronization may be scheduled with a period of 50 ms. Given a 10-hour flight, this means that the synchronization actions in normal operation mode will be executed 720,000 times, while the startup/restart occurs only once. These numbers are a solid basis that underlines our preference to keep normal operation mode and startup/restart separated over a combined synchronization approach.

Nevertheless, it must be guaranteed under a defined fault hypothesis that the startup/restart will be successful. The mere fact that startup/restart is an infrequent event does not relieve the algorithms from proper operation under failure conditions. A sound startup/restart is essential when the system is exposed to failure conditions that are at the limits of the failure hypothesis or even beyond.

For safety-critical systems SAE AS6802 specifies a fault-tolerant Multi-Master synchronization strategy, in which each component is configured either as Synchronization Master (SM), Synchronization Client (SC), or as Compression Master (CM). An example configuration is depicted in Figure 2.17. Typically, the end systems would be configured as SM, while the central role of the CM suggests its realization in the switch in the computer network, though this is not mandatory. All other components in the network are configured as SCs and only react passively to the synchronization strategy. The synchronization information is exchanged in Protocol Control Frames (PCFs). There are three types of PCFs: integration (IN) frames are communicated in normal operation mode, coldstart (CS) and coldstart acknowledgment (CA) frames are communicated during startup/restart.

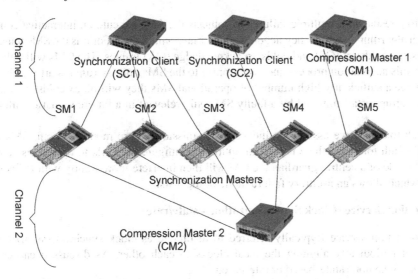

Figure 2.17: Example Configuration of the Synchronization Services for an Off-Chip Network [20]

Time-Preserving Transmission Service

As discussed, in general entities use a network to exchange the current values of their local clocks. In order to allow synchronization at all, the network must provide a time-preserving transmission service with known timing error. For example if the local clock values are exchanged by using a message-based transmission service, the transmission latency and transmission jitter need to be predictable. The quality of the transmission latency and jitter of the service typically also directly influence the quality of the synchronization, i.e., the smaller the latency and jitter the better the local clocks can be synchronized to each other.

The off-chip network implements a one-step transparent clock mechanism – a mechanism implemented in the nodes and switches in the off-chip network to measure the delay of Ethernet frames used for the synchronization services. In particular the transparent clock mechanism operates as follows: Ethernet frames used for the synchronization services, called Protocol Control Frames (PCFs) contain a field in their payload called "transparent clock".

The off-chip nodes and switches modify this transparent clock field in the following way:

- Nodes will measure the duration it takes from the internal trigger to send a PCF until the first bit of the PCF will be transmitted on the Ethernet network and add this delay into the transparent clock field.

  – Switches will measure the duration it takes from reception of a PCF until the forwarding of the PCF and add this delay into the transparent clock field.

  – Additionally the nodes and switches may add delays to the transparent clock field that reflect the transmission delays imposed by the wiring itself.

  – A receiver of a PCF will thus be able to learn from the value of the transparent clock field inside the PCF, for how long the PCF has been in transmission.

Synchronization Startup Service

The synchronization startup service refers to the process of initially synchronizing the local clocks to each other, e.g., after initial power up of the system.

During startup/restart in a multiple-failures hypothesis the SMs execute an interactive consistency agreement algorithm in which they negotiate the initial point in time. For this the SMs transmit dedicated PCFs, the coldstart (CS) and coldstart acknowledge (CA) frames. The CMs will only interfere minor in this negotiation process and synchronize to the SMs once startup/restart is finished. Once, the CMs see a sufficiently high number of operational SMs they will block coldstart frames and so prevent startup/restart initiated by a faulty SM (only relevant in a failure scenario with two faulty SMs).

For two-fault tolerance we assume an inconsistent omission failure mode for both, SMs and CMs. For single-fault tolerance there is also an option to configure the CMs to operate as central guardians. In the role of a central guardian the CM will then interfere more tightly with PCFs sent from the SMs which allows an arbitrary failure mode of the SMs.

Resynchronization Service (Clock Synchronization – state/rate)

The resynchronization service (typically referred to in literate as clock synchronization) refers to the process of periodically aligning the local clocks to each other. As discussed earlier resynchronization can be done state-based or rate-based.

The local clocks in the off-chip network are resynchronized in two steps. In the first step, the SMs send PCFs to the CMs. The CMs extract from the arrival points in time of the PCFs the current state of their local clocks and execute a first convergence function, the so-called compression function. The result of the convergence function is then delivered to the SMs in form of new PCFs (the compressed PCFs). In the second step the SMs/SCs collect the compressed PCFs from the CMs and execute a second convergence function.

Integration Service

The integration service refers to the process of entities joining an already synchronized system, e.g., in case the entity is powered-on late or after a transient failure of the entity.

The nodes and switches in the off-chip network use information in the payload (the membership field) of the IN frames for the integration service:

- The membership field in the PCFs is a bit vector that statically associates each bit with a specific SM in the network.

- When the CMs generate the compressed PCFs, they will set the bit of a respective SM in the membership vector of the compressed PCF if the SM has provided a PCF and clear the bit otherwise.

- Thus, the compressed PCFs carry in the membership field a current view on how many (and also which) SMs are currently supporting a given time line.

- A node and/or switch that is powered-up (or re-integrates) waits for at least one synchronization interval to receive an IN frame.

- If the number of bits set in the membership field of a received IN frame are equal or higher than an offline configured threshold, then the node/switch will adopt its local clock to the time associated with the received IN frame.

Clique Detection Service

The clique detection service refers to the process of detecting global synchronization failure, in particular the identification of situations in which several subsets of local clocks have been formed where the local clocks within the subsets are synchronized to each other but not over subset boundaries.

The nodes and switches use the membership information carried in IN frames also for clique detection. In particular the devices execute three clique detection mechanisms: synchronous clique detection, asynchronous clique detection, and relative clique detection.

Clique Resolution Service

The clique resolution service refers to the process of resolving clique scenarios once they have been formed. Clique resolution typically follows clique detection as discussed above. Once a device has detected a clique scenario it will resolve it by either executing the Synchronization Startup Service or the Integration Service.

Synchronization Restart Service

The synchronization restart service refers to the process of globally restarting the global time within a system. Synchronization restart can be a means for clique resolution. See Synchronization Startup Service.

External Clock Synchronization Service

The external clock synchronization service refers to the process of synchronizing the local clocks to a system-external time source. Such a system-external time source may be for example a GPS receiver.

The nodes in the off-chip network can be configured to apply a configuration-specific value in addition to the value as calculated by the Resynchronization Service when resetting their local clocks. This mechanism can be used to synchronize the nodes to an external time source.

Time-Hierarchy Service (Up, Down)

The time-hierarchy service refers to the processes of translating global time between the different layers in the hierarchy of networks in the DREAMS architecture.

Using the Clock Synchronization Services

At the application level these services should be transparent. To achieve this transparency, the Operating system or the virtualization layer should take into account the mechanisms proposed and offer the time services in a transparent way.

In DREAMS, the main component to provide the time services is the virtualization layer which should support the selected mechanisms. The main property for clock management in real-time applications is to deal with a monotonic increasing clock and timers based on it.

As stated above, the clock synchronization at node level can introduce some problems:

1. If the local clock is faster than the global clock, at synchronization time, the local clock can be set "before". This situation implies that the clock is not monotonic increasing.

2. If the local clock is slower than the global clock, at synchronization time, the local clock can be set "after". This situation can generate that some timers set on this clock can be past at time synchronization. The property is preserved but a jump in the clock can generate that several timers can expire at the same time. It can generate punctual overloads that should have to be considered in the real-time schedulability analysis.

The virtualization (operating system) layer has to provide methods and techniques to deal with the possible problems and to guarantee the correct system behavior. One of the solutions is to use the local clock for all the application services and to identify synchronization points that should not affect the applications.

Synchronization time points (interval)

The virtualization layer or operating system has to define secure synchronization points to perform the clock synchronization. In order to consider the adjustment instead a point, it is considered an interval.

This secure interval has to fulfill two basic requirements:

- It shall set the local clock before or later without affecting the application behavior.

- No application timers shall be pendent.

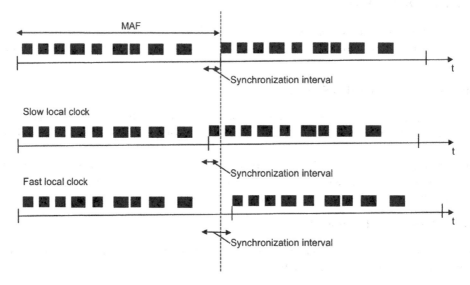

Figure 2.18: Synchronization Interval at the End of the MAF [21]

This activity at virtualization layer or operating system has to be scheduled properly to guarantee the requirements. In the case of a virtualization layer with cyclic schedule, the secure interval to perform this synchronization is at the end of the Major Active Frame (MAF) that corresponds to the hyperperiod. At the end of the MAF all periodic activities have been completed and no pending application timers should be set. Figure 2.18 shows the scheme for the MAF synchronization. It shows how the clock synchronization is achieved. Next to the end of the MAF, the virtualization layer waits for the synchronization periodic message from the network. When the synchronization message arrives, the local clock is updated and the next MAF is executed.

2.2.9   Integrated Resource Management Services

The resource management services provide services for system-wide adaptivity of mixed-criticality applications. The approach is based on the separation of system-wide decisions to meet global constraints from the local execution on individual resources: resources are monitored individually by the local resource management with abstract information provided to global resource management. If significant changes should demand adaptation, the global resource management takes decisions on a system-wide level, based on offline computed configurations, with orders, such as bandwidth assignment, or scheduling parameters for all resources, which are controlled by local resource management. Thus, system-wide constraints, such as end-to-end timing, reliability, can be addressed without incurring the complexity and overhead of individual negotiations among resources directly.

The resource management services support the requirements for resource management in networked multi-core chips based on a global resource manager in coordination with local resource management.

The following sections give an overview of the DREAMS resource management. It is explained further in detail in Chapter 9.

2.2.9.1 Global Resource Manager

The resource management services are realized by a Global Resource Manager (GRM) in combination with local building blocks for resource management. A DREAMS system contains a single GRM, which can be realized by a single node or a set of nodes for improved fault-tolerance and scalability. The GRM performs global decisions with information from local resource monitors. It provides new configurations for the virtualization of resources. The GRM configuration can include different pre-computed configurations of resources (e.g., time-triggered schedules) or parameter ranges (e.g., resource budgets). Alternatively, the GRM can dynamically compute new configurations.

The GRM provides the following main services (Figure 2.19):

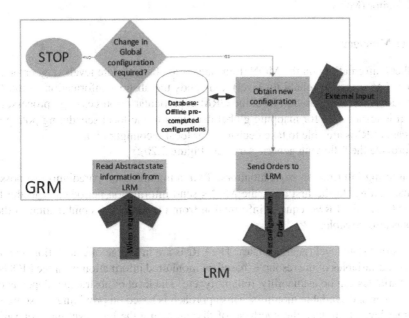

Figure 2.19: Overview of GRM

- Gather status from LRM: GRM receives regular updates from the LRM informing the GRM about the current configuration of the LRM, and failed cores on the corresponding nodes.

- Obtain configuration: The GRM is in charge of the data base of all off-line pre-computed configurations.

- Global reconfiguration decision: The GRM will analyze the update sent by the LRMs, and take reconfiguration decisions that allow the system to adapt to different faults. The GRM takes into account updates from all LRMs to make a decision.

- Send orders to LRM: Once a reconfiguration decision has been taken, the GRM will communicate it to the LRMs involved in the reconfiguration, via network and middleware. For example, if a change in the scheduling plans of an application tile is required, the GRM will provide a new mode for the corresponding LRMs, as well as for the network interfaces of the application tiles involved, because reconfiguration of the network is expected.

- Manage external input: The GRM can manage an external input that can trigger a global reconfiguration.

2.2.9.2   Local Resource Management

These local resource management building blocks are located at the individual resources at chip and/or cluster level. The local building blocks for resource management can be distinguished as:

- Local Resource Manager (LRM)

- Local Resource Scheduler (LRS)

- Resource Monitor (MON)

Local Resource Manager

The LRM gathers information via the MON, translates it to abstract state levels (e.g. error counts may be associated to a certain reliability level) and sends the abstract information to the GRM. The LRM also adopts the configuration from the GRM at particular resources (e.g., processor core, memory, I/O). It is responsible for mapping global decisions to the local scheduling policy of the LRS. In some cases LRMs are able to take decisions for local reconfiguration.

The LRM provide the following generic services (Figure 2.20):

- Gather monitoring information from monitors: Two communication paradigms are possible - interrupt and polling. In the first case, the MONs send information periodically to the LRM. In the second case, the LRM requests information from the MONs. A combination of the two approaches is also possible.

- Calculate abstract state level (generic state): The LRMs are in charge of calculating an abstract level of the state variables of the resource, based on monitored information from the MONs. Abstract state variables can be availability, reliability, etc. The level of abstraction depends on the specific resources and available monitors. This approach is based on providing a resource view on an abstract level, to reduce the overhead of disseminating the low-level monitor variables and only provide information requiring a system-wide reconfiguration.

- Send information: Each LRM will transmit the abstract state of the resources in its domain, via the network and middleware, to the next LRM in the hierarchy, or to the GRM if it stands at the second to last level.

- Receive orders from GRM/Supervisor LRM: The LRMs can only receive orders from Supervisor LRMs or the GRM, never from application components, other system components or LRMs at same or lower level in hierarchy.

- Translate orders and configure LRS: After receiving an order from the GRM, the LRM maps it to the local scheduling policies of the LRS of the resource. For excample, LRSs that implement online scheduling of the resource, the LRM can provide the scheduling parameters to the corresponding LRS. In the case of table-based scheduling policies, the LRM can provide the table itself, or a reference to it. This approach is based on the conceptual separation between implementation details of the scheduler of a resource, and the abstract view of the component that is kept by the GRM.

- Trigger local reconfiguration: Small changes in the state of a resource can be handled locally by the LRM locally. For example, suspension of a low-criticality partition that only communicates with other partitions in its own node, can be performed by the LRM of that node, as no reconfiguration of resources outside of the node is necessary. In that case, the LRM will report the new state of the resource (the node) to the GRM to maintain coherence in the state of the system.

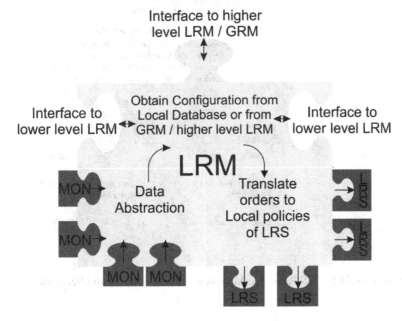

Figure 2.20: Overview of LRM

Local Resource Monitor

The Resource Monitor (MON) monitors the resource (e.g., availability, failures, etc). MONs may also observe the timing of components (e.g., detection of deadline violations), check the application behavior (e.g., stability of control) and perform intrusion detection. Small changes are handled locally by the LRM, while significant changes will be reported to the GRM, who in turn can provide a different configuration at system-level.

Local Resource Scheduler

The Local Resource Scheduler (LRS) performs the runtime scheduling of resource requests (e.g., execution of tasks on processor, processing of queued memory and I/O requests). The LRS can support different scheduling policies (e.g., dispatching of time-triggered actions, priority-based scheduling). It is also responsible for controlling the access to particular resource based on a configuration that has been set by the LRM.

For example, the LRS in the hypervisor layer is responsible for dispatching application according to the offline scheduling tables. There can be multiple offline scheduling tables present in the system. The LRS selects a table based on input from the LRM.

Figure 2.21 gives an overview of the resource management components and services.

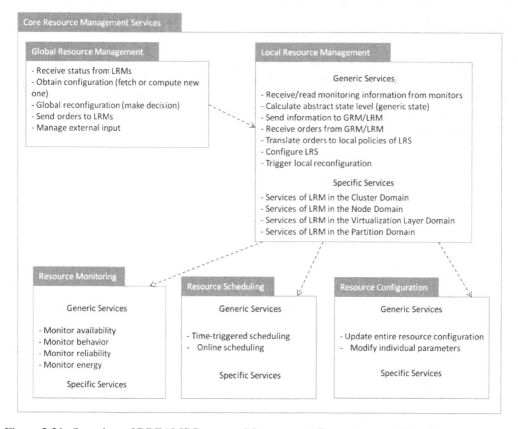

Figure 2.21: Overview of DREAMS Resource Management Components and Services

2.2.9.3   Resource Management Communication

In order to guarantee that the resource management components have a correct view of the system, these services are not intended to be used at the application level, or by any component that is alien to the resource management architecture. This means that only resource management building blocks can communicate with each other.

The resource management is a promising target for a passive as well as an active attacker since it deals with critical information of the system. Furthermore, the fact of having the authority to actively take decisions on resource allocation makes it an interesting target. Therefore, security mechanisms are required to ensure an adequate protection of the system's resource management communication.

2.2.9.4   Resource Management Architecture

The four types of Resource Management (RM) building blocks (GRM, LRM, MON, LRS) can be arranged across the DREAMS platform in many different configurations. We take a look at two main type of architectures:

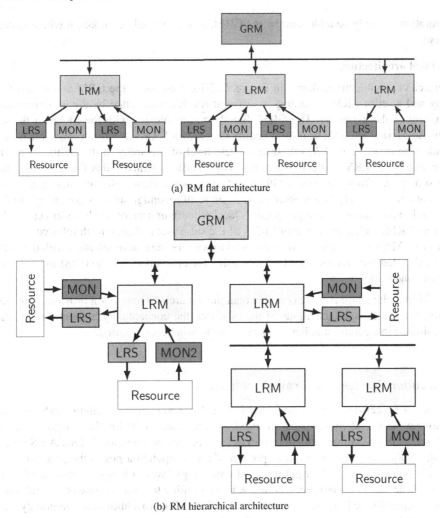

(a) RM flat architecture

(b) RM hierarchical architecture

Figure 2.22: Resource Management Architectures

1. **Flat architecture**

The flat architecture, shown in Figure 2.22(a), consists of the GRM at the top of the hierarchy, which directly supervises and controls a set of LRMs and has a complete view of the system. All LRMs stand all at the same level. Each of them manages one resource, together with MONs and LRSs. The resources can be processor cores, memories, I/O components, hardware accelerators, etc. In this scheme, each LRM directly communicates with the GRM, with disregard for where the resource is located in the system, i.e., inside which node or off-chip cluster.

A Flat architecture for resource management is a simple solution. However, this approach only provided limited scalability, because having the LRMs of each resource communicate with the GRM will become infeasible soon, as the number of resources increases. Another important disadvantage of this structure is that it cannot cope with granularity issues, especially from a timing perspective. Different resources realize their activities at considerably different speeds. When all LRMs are treated equally by the GRM, it is not possible to take that fact into account. Furthermore, there can be faults that require a reconfiguration of only a subset of resources, e.g. all resources inside a single node (chip). In this architecture, such faults and the subsequent reconfiguration can only be addressed by the GRM, as it is the only component with a system-wide view.

2. **Hierarchical architecture**

The hierarchical architecture, shown in Figure 2.22(b), consists of the GRM at the top of the hierarchy and a set of LRMs, standing at different levels (represented by the horizontal lines in the figure) of the hierarchy. The GRM directly communicates with the LRMs at the second to highest level, while those in-turn communicate with LRMs at a lower level. Each LRM communicating to another LRM introduces a new level of hierarchy in the architecture. This structure allows the LRMs to act as a granularity interface, which hides fine-grained activities of a sub-system from the view of GRM, and only communicate relevant information when global reconfiguration may be necessary, e.g. when local reconfiguration is not enough to deal with the fault. From the temporal perspective, local reconfiguration of a sub-system can be initiated by an LRM, which in-charge of LRMs of the sub-system (hence forth referred to as the Supervisor LRM), much sooner without the need to wait for the communication with the GRM. Faults could be temporarily mitigated while waiting for instructions by the GRM to implement a more permanent solution.

A hierarchical architecture is more complex than the flat architecture, but it provides a lot more flexibility. For some implementations of the platform, the conceptually hierarchical structure could be limited to a certain number of specific levels, whenever required.

2.2.10   Execution Services: Software Architecture

For the sharing of processor cores among mixed criticality applications, including safety-critical ones, partitioning OSes and hypervisors are used, which ensure TSP for the computational resources. The scheduling of computational resources (e.g., processor, memory) in DREAMS ensures that each task obtains not only a predefined portion of the computation power the processor core, but also that execution occurs at the right time and with a high level of temporal predictability. On one hand, DREAMS supports static scheduling, where an offline tool creates one or several schedules with pre-computed scheduling decisions for each point in time. In addition, we support dynamic scheduling by employing a quota system in the scheduling of tasks in order to limit the consequences of faults. Safety-critical partitions establish execution environments that are amenable to certification and worst-case execution time analysis, whereas partitions for non safety-critical partitions provide more intricate execution environments. In addition, the separation between safety-critical and non safety-critical applications is supported using dedicated on-chip tiles with respective OSes.

2.2.10.1   Software Architecture

An important part of the DREAMS architectural core services are the execution services that provide basic operations to run the system.

This section describes the software architecture supported by DREAMS, that involve several applications with different levels of criticality. It details the execution environment and the services provided to support the application execution.

The software architecture is built on top of a DREAMS node. The services provided by the software architecture are realized by a virtualization layer inside a tile. Either each processor core runs its own hypervisor or the virtualization layer manages the entire tile including one or more processor cores. The virtualization layer establishes the partitions for the execution of components with guaranteed computational resources. Within each partition, an operating system and a DREAMS Abstraction Layer (DRAL) are deployed to provide software-support for utilizing the platform services from the application software.

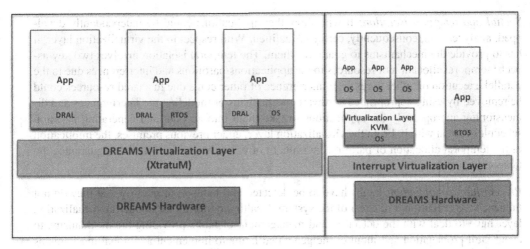

Figure 2.23: Software Architectures

In order to support mixed-criticality applications, the DREAMS software architecture is composed by:

- *Virtualization layer:* It is a software layer that provides hardware virtualization to the applications. Two different approaches are considered in DREAMS depending on the application constraints.

    - Partitioning kernel: It provides virtualization of the hardware resources by defining a set of services that are used by the partitions to access the virtualized resources. The partitioning kernel provides spatial and temporal isolation to the partitions (see Section 2.2.11).

    - Interrupt Virtualization layer: This layer virtualizes the Host OS interrupts and is only introduced when KVM hypervisor is used. The main objective is to take hardware interrupts control away from Host OS and handle them in a thin layer, so as to preserve timing guarantees for the RTOS. Thus, an interrupt virtualization layer (ADEOS or similar) is introduced below the Host OS and real-time partition to prioritize the RTOS (see Section 2.2.11.1).

- *Partitions:* A partition is the execution unit in the DREAMS architecture. It provides the basic infrastructure to execute an application. Different partitions are supported in the DREAMS architecture (see Section 2.2.11.2).

    - Basic single-thread application to be executed near a native hardware.

    - Multi-thread real-time applications to be executed on top of a real-time operating system.

    - Multi process applications to be executed on top of a full featured operating system.

    - Multi-partition applications to be executed on top of an operating system that provides the ability to build virtualized multiple process applications.

Figure 2.23 sketches the two proposed software architectures.

2.2.10.2 Key Services for Certification

From the point of view of certification/qualification the next properties are considered key elements:

1. *Spatial and temporal isolation:* It will allow that applications could be independently developed, analyzed and, consequently, certified/qualified. With respect to the virtualization layer, it has to provide the mechanisms to guarantee them. The temporal isolation involves two key aspects: temporal allocation of resources to the applications/partitions and interferences due to the parallel execution on other cores. The interference of other cores due to shared resources could be removed by using appropriated hardware mechanisms or modeling the interferences and dimensioning appropriately the applications to take them into account and generating partition schedules to deal with it. From the virtualization layer, under previous premises, the implication of the temporal allocation of partitions permits to offer the basic mechanisms to guarantee the temporal isolation.

2. *Prevent fault propagation:* Faults have to be detected and handled in the way that they do not influence the execution of the rest of the system. Health monitoring techniques at virtualization layer have to deal with the detection and management of faults providing the mechanisms to avoid fault propagation and monitor the generated faults to implement additional mechanisms for global fault management.

3. *Static resource allocation:* The system architect is responsible of the system definition and resource allocation. This system definition is detailed in the configuration file of the system specifying all system resources, namely, number of CPUs, memory layout, peripherals, partitions, the execution plan of each CPU, etc. Each partition has to specify the memory regions, communication ports, temporal requirements and other resources that are needed to execute the partition code. Static resource allocation is the basis of predictability and security of the system. The hypervisor has to guarantee that a partition can access the allocated resources and deny the requests to other not allocated resources.

2.2.11   Execution Services: DREAMS Virtualization Layer

The virtualization layer is the software layer that abstracts the underlying hardware and provides virtualization of the CPUs. This virtualization layer is a hypervisor (XtratuM hypervisor [22]) that permits to execute multiple isolated virtual machines. Each virtual machine is a partition.

As the virtualization layer is a common layer for all the partitions, in order to support mixed criticality applications, it has to achieve the highest level of criticality in the system.

The basic properties that the virtualization layer shall accomplish are:

- *Spatial isolation:* A partition is completely allocated in a unique address space (code, data, stack). This address space is not accessible by other partitions. The hypervisor has to guarantee the spatial isolation of the partitions. The system architect can relax this property by defining specific shared memory areas between partitions.

- *Temporal isolation:* A partition is executed independently of the execution of other partitions. In other words, the execution of a partition cannot be disturbed by the execution of other partitions. It influences directly on the scheduling policies at hypervisor level. The hypervisor has to schedule partitions under a scheduling policy that guarantees the partition execution.

- *Fault isolation and management:* A fundamental issue in critical systems is the fault management. Faults, when occur, have to be detected and handled properly in order to isolate them and avoid the propagation. A fault model to deal with the different types of errors is to be designed. The hypervisor has to implement the fault management model and permits to the partitions to manage those errors that involve the partition execution.

- *Predictability:* A partition with real-time constraints has to execute its code in a predictable way. It can be influenced by the underlying layers of software (guest-OS and hypervisor) and by the hardware. From the hypervisor point of view, the predictability applies to the provided services, the operations involved in the partition execution and the interruption management of the partitions.

- *Security:* All the information in a system (partitioned system) has to be protected against access and modification from unauthorized partitions or unplanned actions. Security implies the definition of a set of elements and mechanisms that permit to establish the system security functions. This property is strongly related with the static resource allocation and a fault model to identify and confine the vulnerabilities of the system.

- *Confidentiality:* Partitions cannot access the space of other partitions neither to see how the system is working. From its point of view, they only can see its own partition. This property can be relaxed to some specific partitions in order to see the status of other partitions or control their execution.

Partition schedule

The virtualization layer schedules partitions in a fixed, cyclic basis (ARINC-653 scheduling policy). This policy ensures that one partition cannot use the processor for longer than scheduled to the detriment of the other partitions. The set of time slots allocated to each partition is defined in the configuration file during the design phase by means of a cyclic plan in a temporal interval referred as MAF.

Each partition is scheduled for a time slot defined as a start time and a duration. If there are several concurrent activities in the partition, the partition shall implement its own scheduling algorithm. This two-level scheduling scheme is known as hierarchical scheduling.

Multi-core schedule

The virtualization layer provides different policies that can be attached to any of the CPU [23]. Two basic policies are defined:

- *Cyclic scheduling:* Pairs ¡partition, vcpu¿ are scheduled in a fixed, cyclic basis (ARINC-653 scheduling policy). This policy ensures that one partition cannot use the processor for longer than scheduled to the detriment of the other partitions. The set of time slots allocated to each ¡partition, vcpu¿ is defined in the configuration file. Each ¡partition, vcpu¿ is scheduled for a time slot defined as a start time and a duration. Within a time slot, the virtualization layer allocates the system resources to the partition and virtual CPU specified.

- *Priority scheduling:* Under this scheduling policy, pairs ¡partition, vcpu¿ are scheduled based on the partition priority. The partition priority is specified in the configuration file. Priority 0 corresponds to the highest priority. All pairs ¡partition, vcpu¿ in normal state (ready) allocated in the configuration file to a processor attached to this policy are executed taking into account its priority.

Multiple scheduling plans

The system can define several scheduling plans or modes. A system partition can request the change from one plan to another. Once the change is accepted, it is effective at the end of the MAF.

2.2.11.1   Interrupt Virtualization Services

KVM converts Host (Linux) Processes into virtual machines, and re-uses most of the common features provided by Host OS such as Process Scheduling, Memory Management, Interrupt Handling etc. In order to support a hard real-time partition, we can either introduce a thin interrupt virtualization layer below the Host kernel or modify most of the Host kernel sub-systems. The former approach is considered a better option, such as using ADEOS (Adaptive Domain Environment for Operating Systems) or a similar one than modifications to the Host kernel, thanks to its smaller TCB (Trusted Computing Base). For example, ADEOS "nanokernel" is composed of a few KLOC for ARM processors as opposed to a fully featured Host OS such as Linux, which has a very large TCB. Thus, an interrupt virtualization layer along with the KVM hypervisor is necessary for realizing the RTOS-GPOS co-existence use-case.

Previous real-time efforts for the KVM hypervisor [24, 25] have focused on either semi-automatic virtual machine prioritization/shielding techniques or modification of guest system to realize a paravirtual interface. All of these techniques have failed to produce a hard real-time virtualization solution, so we consider them inadequate for DREAMS project. Moreover, maintaining a paravirtualized solution is difficult as it requires modifications to the guest operating systems.

The interrupt virtualization layer schedules multiple operating system instances running above it, and allows for the co-existence of multiple prioritized domains (real-time and non real-time). This layer implements an interrupt management scheme, which allocates specialized interrupt handlers for the Host OS and RTOS. The RTOS-specific interrupts are given higher priority to ensure real-time behavior. KVM will run on the Host OS (within the non-RT domain) and create multiple virtual machines.

2.2.11.2   Processor Virtualization Services by Partitions as Execution Units

The execution unit in partitioning is a partition. A partition is basically a program in a single application environment. It can comprise: the application code, the partition runtime and the configuration file. A partition runtime can have a minimal layer to facilitate the application execution and a guest Operating System adapted to be executed on top of the virtualization layer. The software that resides in a DREAMS partition can be:

- *Application software:* It refers to the code designed to deal with the specific application requirements

- *Runtime support:* It provides the services to execute the application code.

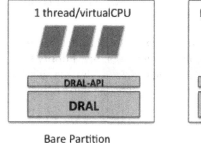
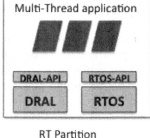
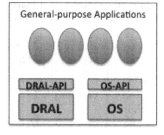

Bare Partition	RT Partition	General Purpose Partition

Figure 2.24: Partition Classes

Different types of partitions can be built (see Figure 2.24):

- *Bare Partitions:* Partitions that are executed as they were on top of the hardware. The application code can be a single thread executed in one core or several single-threads.

- *Real-Time Partitions:* These partitions shall contain a real-time operating system adapted to be executed on top of the DREAMS virtualization layer. Additionally, it can include the DRAL layer that complements the RTOS services with specific services for partitioning. The partition boot is managed by the Real-Time Operating System (RTOS).

  - *Real-Time Partitions (XtratuM Case):* The real-time partitions for XtratuM will be similar to non real-time partitions, as XtratuM is a baremetal hypervisor and can fully control scheduling of these partitions.
  - *Real-Time Partition (KVM Case):* The real-time partition for KVM will be based on a minimal interrupt virtualization layer, in order to ensure hard real-time behavior for a given RTOS. This design change is necessary as KVM uses Linux kernel for scheduling its virtual machines, which is soft real-time at best.

  Figure 2.25 shows the two types of RT partitions.

- *General purpose Partitions:* These partitions shall contain a full featured operating system (e.g. Linux) that offers the OS services to the partitions. Additionally, it can include the DRAL layer that complements the OS services with specific services for partitioning. The partition boot is managed by the OS.

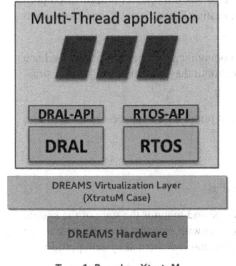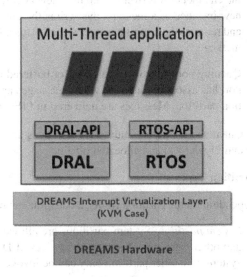

Figure 2.25: RT-Partition Types

A partition is seen by the virtualization layer as a piece of code with an entry point and a set of access points (communication ports) that allow it to communicate with other partitions. Figure 2.26 sketches the partition view.

Input and output ports permit a partition to send/receive messages to/from other partitions. A message is a variable block of data that is sent from a source partition to one or more destination partitions. The data of a message is transparent to the message passing system.

The message transport mechanism is a communication channel that is the logical path between

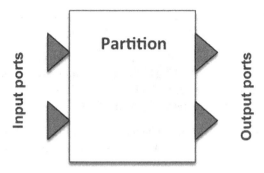

Figure 2.26: Partition View

one source and one or more destinations. Partitions send/receive messages through ports. The virtualization layer is responsible of the message transport from the memory area of a source partition to a memory area of the destination(s) partition(s).

Two basic inter-partition communication ports are supported: sampling and queuing.

- Sampling port: It provides support for broadcast, multicast and unicast messages. No queuing is supported in this mode. A message remains in the source port until it is transmitted through the channel or it is overwritten by a new occurrence of the message, whatever occurs first. Each new instance of a message overwrites the current message when it reaches a destination port, and remains there until it is overwritten. This allows the destination partitions to access the latest message.

- Queuing port: It provides support for buffered unicast communication between partitions. Each port has associated a queue where messages are buffered until they are delivered to the destination partition. Messages are delivered in FIFO order.

Channels, ports, maximum message sizes and maximum number of messages (queuing ports) are entirely defined and allocated off-line.

Partition Types

Depending on the partition rights, a partition can be defined as:

- *System partition:* System partitions are allowed to manage and monitor the state of the system and other partitions. A subset of services of DRAL dealing with the change of the state of the system or another partition only can be invoked if the partition is defined as system partition.

- *Real-time system partition:* A real-time system partition only exists when an interrupt virtualization layer is used (KVM case). This partition is similar to a system partition, except that it is dedicated for real-time OS and will have a unique instance on a given DREAMS chip.

- *Normal partition:* It corresponds to the partitions that have not the system attributes.

Considering the virtual cores that a partition uses, it can be classified as follows:

- *Mono-core partition:* This partition only uses a virtual core.

- *Multi-core partition:* This partition is associated with several virtual CPUs. The virtualization layer only boots the virtual CPU0. It is the responsibility of the partition, to boot the rest of the virtual CPUs.

Partition states and transitions

The virtualization layer is not aware about the nature of a partition. Partitions can be based on bare applications or OS dependent applications. From the virtualization layer point of view, a partition has the states and transitions as shown in Figure 2.27.

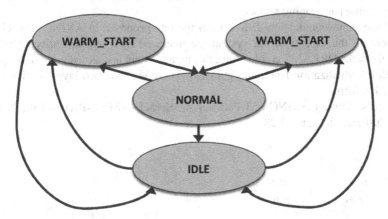

Figure 2.27: Partition States

After the virtualization layer initialization, each partition is loaded in memory and ready to be booted. When the resources are allocated to a partition, it boots (that is, initializes a correct stack and sets up the virtual processor control registers). From the virtualization layer, the partition is in NORMAL state.

From the virtualization layer point of view, there is no difference between the BOOT state and the NORMAL state. The NORMAL state is subdivided in three sub-states:

- *Ready:* The partition is ready to execute code, but it is not scheduled because it is not in its time slot.

- *Running:* The partition is being executed by the processor.

- *Idle:* If the partition does not need to use the processor during its allocated time slot, it can yield the processor and wait for an interrupt or for the next time slot.

A partition can halt itself or be halted by a system partition. In the HALT state, the partition is not selected by the scheduler and the time slot allocated to it is left idle (it is not allocated to other partitions). All resources allocated to the partition are released. It is not possible to return to normal state.

In SUSPENDED state, a partition will not be scheduled and interrupts are not delivered. Interrupts raised while in suspended state are left pending. If the partition returns to NORMAL state, then pending interrupts are delivered to the partition.

2.2.11.3   DREAMS Abstraction Layer Services

The DReams Abstraction Layer (DRAL) defines the common interface to the applications with the aim of standardizing the use of partitioning systems in the DREAMS Project. The result is an Application Programming Interface (API) that should be used standalone or in cooperation with an OS.

In order to define the API of DRAL, an analysis of the standard for partitioned systems ARINC-653 [26] has been performed. In ARINC-653-P1 and P2 [27,28], the partitioning kernel manages all services including the OS services. In DREAMS, the partitioning kernel is the hypervisor and the OS is isolated in a partition. The DREAMS approach allows to use different OSs at partition level while in ARINC-653, the OS is integrated in the partitioning kernel. The ARINC-653 approach presents difficulties to the partitioning kernel certification due to the complexity of the partitioning kernel. Recently, ARINC-653 published a subset of the standard (ARINC-653-P4) by removing OS services and maintaining partitioning services.

The DREAMS system organization differs from the one proposed in ARINC-653. Unlike in ARINC-653, where all the aspects of the system are managed by a unique entity (ARINC-653 partitioning kernel), in the DREAMS architecture, this management is handled by two independent, and distinct entities: Virtualization layer, and guestOS. There are thus two layers in the so-called System Executive Platform.

These differences between ARINC-653 Partitioning Kernel (ARINC-PK) and the DREAMS architecture are illustrated in Figure 2.28.

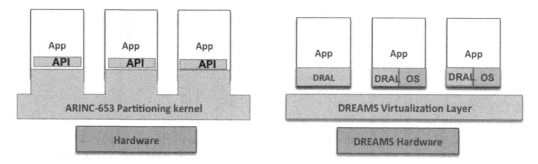

Figure 2.28: ARINC-653-PK versus DREAMS Platform Virtualization

The DRAL specification is based on ARINC-653 Part4 which provides a reduced subset of the ARINC-653 Part1 required services aimed at certifiable systems. However, the following precision has to be made: process management and intra-partition communication as defined in ARINC-653 have not been included in the DRAL and are left to the guest OS.

Table 2.1: Functional Services Provider

Service Class	ARINC-653-P1	ARINC-653-P4	DRAL
System Management	-	-	DRAL
Partition Management	ARINC-PK	ARINC-PK	DRAL
Process Management	ARINC-PK	-	GuestOS
Time Management	ARINC-PK	ARINC-PK	DRAL/GuestOS
Inter-Partition Communication	ARINC-PK	ARINC-PK	DRAL
Intra-Partition Communication	ARINC-PK	ARINC-PK	GuestOS
IRQ Management	-	-	DRAL
Health Monitoring	ARINC-PK	ARINC-PK	DRAL
Multiple Scheduling Plans	ARINC-PK	-	DRAL

Table 2.1 shows a comparison between ARINC-653 P1 and P4 and DRAL. The DRAL services are grouped in the categories summarized below:

System Management: Services related to the underlaying layer are not provided by ARINC-653 that does not offer the possibility to act on the global system directly by partitions. However, it is relevant for specific partitions to get the state or act on the global system. These services should be invoked by partitions with appropriated rights.

Partition Management: ARINC-653 defines a set of services to deal with the partition itself. However, partitions with the appropriate rights should be able to act or get the status of other partitions.

Process Management: DRAL shall not offer any service related to processes or tasks in the partition. These services shall be provided by the guestOS if any.

Time Management: DRAL shall offer the same services as ARINC-653 for time management.

Inter-Partition Communication (IPC): DRAL shall offer the same services as ARINC-653 for IPC.

Intra-Partition Communication: DRAL shall not offer any service related to processes or tasks in the partition. These services shall be provided by the guestOS if any.

IRQ Management: DRAL shall offer a new set of services for interrupt management. These services are not defined in ARINC-653.

Health Monitoring: DRAL shall offer the same services as ARINC-653 for HM.

Scheduling: DRAL shall offer a set of services related to scheduling.

A detailed description of the services provided by DRAL is presented in this section.

System Management Services

System Management Services refer to the services that a partition can invoke to get the status of the system or virtualization layer and perform actions. The services are as follows:

DRAL_GET_SYSTEM_STATUS: It returns the status of the virtualization layer. The result is a data structure that provides information related to the system status.

DRAL_SET_SYSTEM_MODE: It provides to a partition the ability to change the status of the hypervisor. Actions to be invoked are:

- Perform a cold reset of the system. As result of this invokation, the system is reset and boots. A counter informs about the number of consecutive warm resets that have been performed. This counter is set to zero when the cold reset is invoked.

- Perform a warm reset on the system. As result of this invokation, the system is reset and boots. The reset counter is increased.

- Perform a system halt. As result of this invokation, the system is halted. A physical reset is required to restart the system.

All system services are restricted to partitions with "system" attributes. A partition without the "system" rights will fail when invoking these services.

Partition Management Services

Partition Management Services refer to the services that a partition can invoke to get its own status or other partition status or perform actions on them. The services are as follows:

DRAL_GET_PARTITION_ID: It permits to get the partition identifier.

DRAL_GET_A_PARTITION_ID: It permits to get the partition identifier of another partition identified by its name.

DRAL_GET_PARTITION_STATUS: It returns the status of the partition. The result is a data structure that provides information related to the current partition status.

DRAL_GET_A_PARTITION_STATUS: It returns the status of another partition. The result is a data structure that provides information related to the referenced partition status. This service only can succeed if the invoker has the system rights.

DRAL_SET_PARTITION_MODE: It provides to a partition the ability to change its own status. The partition invokes an action to perform a HALT, COLD_RESET, WARM_RESET, NORMAL, SUSPEND or RESUME.

DRAL_SET_A_PARTITION_MODE: It provides to a partition the ability to change the status of another partition.

Previous services that involve a second partition (e.g. read the status of another partition) require "system" rights. The possible actions to be performed on a partition are:

- Perform a cold reset on a partition. As a result of this invokation, the partition is reset and boots. A counter informs about the number of consecutive warm resets. This counter is zeroed when the cold reset is invoked.

- Perform a warm reset on a partition. As a result of this invokation, the partition is reset and boots. The reset counter is increased.

- Perform a partition halt or idle. As a result of this invokation, the partition is idled.

- Perform a partition suspend. As a result of this invokation, the partition is suspended.

- Perform a partition resume. As a result of this invokation, the partition is resumed and its state is NORMAL.

- Perform a partition resume immediately. As a result of this invokation, the partition is resumed immediately and its state is NORMAL.

Table 2.2 shows the status observation of a partition when the status is read from the own partition or another partition.

Table 2.2: Status of a Partition

STATE	GET_PARTITION_STATUS	GET_A_PARTITION_STATUS
HALTED	HALTED	HALTED
COLD_START	COLD_START	NORMAL
WARM_START	WARM_START	NORMAL
NORMAL	NORMAL	NORMAL
SUSPENDED	SUSPENDED	SUSPENDED

Table 2.3 shows the actions that a partition can do on its own current and final state. However, if a partition performs a change on other partition, the final states are shown in Table 2.4.

Time Management Services

Time Management Services refer to the services that a partition can invoke to get time information or set timers.

Table 2.3: Transitions of the Partition under Internal Actions

INITIAL STATE	SET_PARTITION_MODE	FINAL STATE
any	HALT	HALTED
any	COLD_RESET	COLD_START
any	WARM_RESET	WARM_START
CLD_ST. | WARM_ST. | NORMAL	NORMAL	NORMAL
CLD_ST. | WARM_ST. | NORMAL	SUSPEND	SUSPENDED

Table 2.4: Transitions of a Partition under External Actions

INITIAL STATE	SET_A_PARTITION_MODE	FINAL STATE
any	HALT	HALTED
any	COLD_RESET	NORMAL
any	WARM_RESET	NORMAL
any	NORMAL	No effects
NORMAL	SUSPEND	SUSPENDED
NORMAL | SUSPEND	RESUME	NORMAL

Time can be global or local. Global time refers to a monotonic clock of the system (HW_CLOCK). Local time refers to a partition clock that runs when the partition is executed (EXEC_CLOCK). Timers can be set taking as a reference the global or the local time.

The services are as follows:

DRAL_GET_TIME: It permits to get the current time (global or local).

DRAL_SET_TIMER: It permits to set a timer referred to the global or local clock.

Inter-Partition Communication Services

A partition can send/receive messages to/from other partitions using sampling or queuing ports. The services are:

DRAL_CREATE_SAMPLING_PORT: Creates a sampling port.

DRAL_WRITE_SAMPLING_MESSAGE: Writes a message in a sampling port.

DRAL_READ_SAMPLING_MESSAGE: Reads a message from a sampling port.

DRAL_READ_SAMPLING_MESSAGE_CONDITIONAL: Reads a message from a sampling port only if it has been updated since a given reference time.

DRAL_READ_UPDATED_SAMPLING_MESSAGE: Reads a message from a sampling port only if it has been updated since the last request.

DRAL_GET_SAMPLING_PORT_ID: Gets a sampling port identifier by specifying the sampling port name.

DRAL_GET_SAMPLING_PORT_CURRENT_STATUS: Gets the current status of a sampling port.

DRAL_GET_SAMPLING_PORT_STATUS: Gets the attributes of a sampling port.

DRAL_CREATE_QUEUING_PORT: Creates a sampling port.

DRAL_SEND_QUEUING_MESSAGE: Sends a message in a queuing port.

DRAL_RECEIVE_QUEUING_MESSAGE: Receives a message in a queuing port.

DRAL_GET_QUEUING_PORT_STATUS: Gets the attributes of a queuing port.

DRAL_GET_QUEUING_PORT_ID: Gets a queuing port identifier by specifying the queuing port name.

DRAL_CLEAR_QUEUING_PORT: Removes all messages in a queuing port.

Health Monitor Services

A partition can raise Health Monitor (HM) events to the virtualization layer. These HM events are detected and generated by the application or the partition runtime. The events that the partition can raise are:

- APPLICATION_ERROR: An error in the application.

- DEADLINE_MISSED: A deadline miss has been detected.

- NUMERIC_ERROR: The application has detected a numeric error.

- STACK_OVERFLOW: The partition detects a stack overflow.

- MEMORY_VIOLATION: The partition detects an illegal memory access.

    The services are as follows:

DRAL_GET_ERROR_STATUS: Permits to the partition to access the reported errors.

DRAL_RAISE_APPLICATION_ERROR: The partition raises an HM event that will be handled by the virtualization layer.

Scheduling Services

A partition is scheduled under the virtualization layer policy. It is relevant for the partition to get the information related to its own schedule. The services are as follows:

DRAL_GET_PARTITION_SCHEDULE_ID: Gets the identifier of the schedule plan.

DRAL_GET_MODULE_SCHEDULE_STATUS: Gets the current schedule plan status.

DRAL_GET_SCHEDULE_INFO: Gets the current slot information.

DRAL_SET_MODULE_SCHEDULE: Requests for a schedule plan change.

DRAL_SET_IMMEDIATE_MODULE_SCHEDULE: Performs a schedule plan change.

Interrupt Management Services

Interrupt management services have been included in DRAL to allow a partition to enable/disable extended interrupts and install interrupt handlers. DRAL defines a set of symbols to deal with extended interrupts:

HW_TIMER: Interrupt generated by timers that use the system clock as reference

EXEC_TIMER: Interrupt generated by timers that use the partition clock as reference

START_SLOT_IRQ: Interrupt generated by the schedule of a new slot

SAMPLING_MESSAGE_IRQ: Interrupt generated by the arrival of a new sampling message

QUEUING_MESSAGE_IRQ: Interrupt generated by the arrival of a new queuing message

The interrupt management services are as follows:

DRAL_CPU_INSTALL_IRQ_HANDLER: Installs an interrupt handler for a specified interrupt.

DRAL_CPU_ENABLE_IRQS: Enables all IRQs.

DRAL_CPU_ENABLE_IRQ: Enables a specified extended IRQ.

DRAL_CPU_DISABLE_IRQS: Disables all IRQs.

DRAL_CPU_DISABLE_IRQ: Disables a specified extended IRQ.

2.2.12 Execution Services: Security

In this section the security services for the end-to-end communication on application level are described. Hence, there is a secure communication from one application to another application. The secure communication from one application to another application includes all parts in the communication between the application like on-chip communication as well as off-chip communication.

Encryption Service

The encryption service encrypts data with a given cryptographic key. It transforms a plaintext into a cipher text so that the un-intended recipients cannot understand the messages exchanged between two legitimate communication partners. The encryption service for end-to-end communication is used for a confidential communication between two applications. Even the system components between the two applications, e.g., gateways and routers, cannot interpret the content of the communication.

Decryption Service

The decryption service decrypts data with a given cryptographic key. It transforms a cipher text into plain text, if the key is correct and there was no transmission error. The decryption service for end-to-end communication is used for a confidential communication between two applications. The adversaries and the unintended recipients, such as the gateways and the routers cannot interpret the exchanged messages because they do not possess the key to decrypt the exchanged messages. Only the legitimate communication partners, owning the cryptographic key, can decrypt the exchanged data.

Integrity Service

The integrity service generates a cryptographic hash (or secure checksum) for a message, which is transmitted together with the message. With this checksum, any modifications in the message are detectable. The integrity service for end-to-end communication ensures that all changes are noticeable and that not only the changes during the off-chip communication are detectable. For example, this service can be used by the monitoring and resource scheduling components (GRM, LRM, LRS and MON) to ensure the integrity of the communication.

Integrity Check Service

The integrity check service verifies the integrity of a message by re-calculating the cryptographic hash (or secure checksum) on the received message and comparing it with the received checksum. With this checksum, even a single bit modification is detectable. The integrity check service for end-to-end communication ensures that all changes are noticeable and that not only the changes during the off-chip communication are detectable. For example, this service can be used by the monitoring and resource scheduling components (GRM, LRM, LRS and MON) to check the integrity of the communication.

Authentication Code Generation Service

The authentication code generation service generates a message authentication code (MAC) tag or digital signatures for ensuring the data origin respectively to verify the communication partner. This service generates the MAC tag or the digital signatures on the application layer. This implies that the service can be used by the monitoring and resource scheduling components (GRM, LRM, LRS and MON) to ensure the authenticity of the communication.

Authentication Code Verification Service

The authentication code verification service verifies the data origin or the communication partner by verifying the message authentication code (MAC) tag or the digital signatures received with the message. This service verifies the authentication tag or the digital signatures on the application layer. This implies that this service can be used by the monitoring and resource scheduling components (GRM, LRM, LRS and MON) to verify the authenticity of the communication.

Access Control Service

The access control service verifies if a system resource is allowed to access the requested object. For end-to-end communication it checks the permission on application layer for access to secure memory. Either the access control service or secure storage service (or both of them together) will ensure the concept of secure memory storage.

2.2.13   Optional Service: Voting

The DREAMS services that have been presented so far already provide assurances to prevent or contain the effects of perturbations that might result into a malfunctioning of the system. On the one hand, the architecture foresees security services at different levels such as the secure communication service (see Section 2.2.7), the secured resource management services (see Section 2.2.9) and the execution security services (see Section 2.2.12) that prevent or confine the impact of *malicious perturbations* that are caused by malign actions such as denial-of-service attacks, manipulation of data integrity, or side-effects of intrusion attempts. On the other hand, several services such as the off-chip and on-chip communication services (see Sections 2.2.3 and 2.2.4, respectively) or the services for the virtualization of I/O (see Section 2.2.6), interrupt (see Section 2.2.11.1) or processor resources (see Section 2.2.11.2) provide fault-containment regions in order to confine the impact of *accidental perturbations* (e.g., caused by internal or external physical faults, or design faults) covered by the DREAMS fault hypothesis (see Section 2.5).

In the following, we describe a voting service that employs some of the aforementioned guarantees provided by the DREAMS architecture to enhance the robustness of a system (i.e., its capability to withstand certain perturbations). The resulting *software-based voting service* is a building block for fault-tolerant systems (i.e., systems that continue to operate while faults have occurred) that builds upon on the containment property for accidental faults that is ensured by the communication and execution services referred to above. Software-fault tolerance mechanisms typically require the application of adjudicators, i.e., decision mechanisms that employ redundancy to determine if the result computed by an application component is correct. A voter is an adjudicator that compares the input of two or more replicas (possibly variants) of an application component and decides the correct result, if it exists.

Figure 2.29 illustrates the architecture pattern of the voting service that combines a number of (non-fault-tolerant) replicas of an application component, a voter component that implements a certain *voting strategy* to determine the correct result based on the received inputs, and the communication services into a Fault-Tolerant Unit (FTU). Regarding algorithms underlying the voting strategy, we refer to a large body of work on voting algorithms that are suitable for software-based voting [29, 30], such as the majority voter [24, 31], the consensus voter [32], the median voter [30, 33], or the mean voter [34].

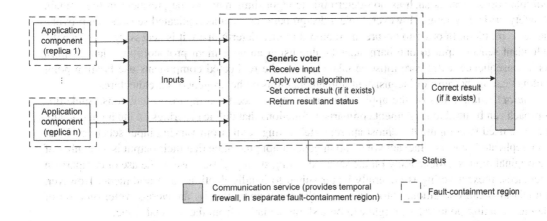

Figure 2.29: Generic Voter Architecture Used to Provide a Fault-Tolerant Unit

The voting service relies on the assumptions discussed below and depends on the guarantees provided by other services of the DREAMS architecture:

- In order to prevent a single fault to compromise an arbitrary number of replicas (and possibly the voter itself), each replica and the voter must reside in *isolated fault-containment regions* (FCRs).

- FTUs exclusively based on voters can only be used to handle errors in the value domain of the expected service of the underlying application component. Since for real-time systems also the temporal correctness of the FTU's output must be guaranteed (i.e., its observable behavior), voters here also have to cope with input that is erroneous in the time domain (e.g., early, belated or too frequent messages). In this case, the *temporal firewall* mechanism provided by the off-chip and on-chip communication services prevents the delivery of input that violates its "timing contract", which makes it appear as missing input from the voter's point-of-view.

- Missing input might also occur because of a fault (e.g., design fault, or lockup due to a physical fault), or a deliberate decision at originating components (e.g., failed admittance test on inputs, failed acceptance test on computed results). Hence, the DREAMS voting service implements the dynamic versions of the particular voting strategies that are also defined on partial input (in contrast to static voting that assumes that all inputs have been received).

Comparison function

The comparison of two inputs is the basic building of the voting service, where two approaches can be distinguished [30]:

Exact voting refers to voting strategies that are based on the procedure of performing a bit-by-bit comparison of the inputs. The advantages of this approach are that it is an efficient, scalable, and generic method. It maintains a strict separation of concerns between the voter and the application components that enables the universal usability of the comparison function. Since the bit-wise comparison induces a number of equivalence classes on the input data, it is typically used in the majority or the consensus voting strategies. Exact voting is based on the assumption of replica determinism where – in the absence of faults – all replicated components are guaranteed to produce exactly the same output messages with a bounded temporal deviation. Exact voting is not compatible with FTUs where multiple correct results of the different replica are possible. Some causes for multiple correct results such as non-deterministic algorithms may not be prevalent in the domain of safety-critical systems. However, typical design patterns such as replicated sensors can provoke the same problem. In order to ensure the required replica determinism, it is necessary to reduce the redundant sensor input to one harmonized value using an agreement protocol (e.g., data fusion). As a consequence, extra care must be taken in case the replicated components use floating-point arithmetic, e.g., because of inconsistent implementations on heterogeneous architectures.

Inexact voting relies on the application or data-type specific comparison functions. Since this approach can be used to implement comparison functions that define orderings on the input domain, it can be used to compute the "most appropriate" voting result from varying input sets (e.g., input from replicated sensors). Inexact voters do in general not guarantee that their output is a member of the original input set (depending on the selected voting strategy). Because of the use of comparison tolerances, inexact voting is generally better suited to handle floating-point arithmetic. However, it might also cause additional problems. e.g., the result computed by an average voter on a set of identical floating-point values might exhibit a slight deviation from the original value.

Voting strategy

Since the DREAMS architecture ensures replica determinism for safety-critical subsystems, the voting service implements the dynamic exact majority voting (DEMV) strategy [29]. DEMV masks a fault if and only if a majority exists among the non-faulty inputs with respect to the size of the entire input vector. If the voter detects an agreement of all inputs, it returns the correct result and indicates a successful return status. Otherwise, i.e., if a (correct) majority exists but there is at least one deviating input source, the voter returns the correct result, and determines which of the originating application component replicas are erroneous. If no majority exists, DEMV has detected faults in the application component replicas whose detailed origin cannot be determined. Here, it will not produce a return value and assume that a potential error in all of the originating application component replicas exists. In particular, this case also occurs when less than the half of the input vector has been received on time. DEMV is defeated if a "tainted" majority exists, in which case it will deliver an incorrect result. The dimensioning of the system, i.e. the degree of redundancy and the deployment of the individual components of the FTU to separate FCRs, must ensure that this case can only occur due to a rare fault that is not covered by the system's fault hypothesis.

The modular architecture of the voting service enables the integration of further voting strategies, such as dynamic consensus voter (relaxation of majority voter that does not require an absolute majority), and inexact voters such as the dynamic average or the dynamic median voter.

In the presence of faults in application component replica(s), the situations described below can arise:

- If the FTU contains "enough redundancy" to tolerate all current simultaneous faults, it is able to mask the fault(s) in the value domain, and therefore exhibits fail-operational behavior. Additionally, it is possible to identify the erroneous replica(s) and to possibly initiate appropriate counter-measures (normal fault).

• The FTU is able to detect the presence of value errors in the application component replicas, but it is not able to decide the correct result (and therefore also not to identify erroneous replicas). In this case, the FTU will not output any result at all in order to contain (value) errors. In combination with aforementioned temporal firewalls, voters can be used to guarantee fail-silent behavior in this case.

• If the voter is not able to detect/correct the presence of faults in one or more replicas, it is "defeated" and will forward an erroneous result. The design and dimensioning of the system must guarantee that the probability of this case is sufficiently low (rare fault).

• In the case of a fault in the voter itself, the voter might either be defeated due to a computation error in the voter logic or it might not return any value at all (e.g., lockup because of a corruption of the program counter). In the latter case, the voter is also likely to fail to provide information about its status.

In the following, we describe typical usage scenarios of the voting service in terms of (informal) design patterns:

Inexact voting as sensor agreement protocol

As pointed out above, it is possible to use inexact voters such as the average or median voter as

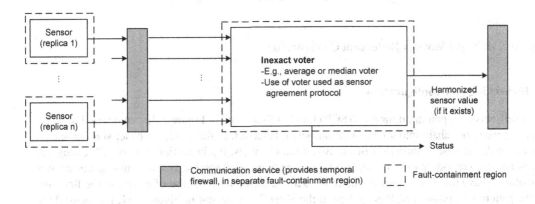

Figure 2.30: Use of Voting Service as Sensor Agreement Protocol

simple sensor value agreement protocol (see Figure 2.30). This is useful in order to obtain the replica determinism that is required to apply exact voting. It should be noted that in this configuration the inexact voter (and the subsequent processing chain) is a single point of failure in the system. However, this limitation can be removed by applying the Triple Modular Redundancy (TMR) pattern presented in the next section.

Triple modular redundancy configuration

In the configuration depicted in Figure 2.31, a processing chain is transformed into a chain of FTUs by replicating it into three redundant channels (triple modular redundancy – TMR [35]). The original non-fault-tolerant design that corresponds to the example consisted of merely a sending and receiving component (now represented by *Replica 1.x* and *Replica 2.x*). It should be noted that also the voters of the intermediate stages of the processing chain (here *Voter 1.x*) have been replicated. As it has been pointed out during the discussion of the voting strategy, this approach can be used to obtain fail-operational behavior in case at most a single fault occurs in each of the stages of the FTU, or fail-silent behavior in case of at most two faults in each of the stages of the FTU. However, it should be noted, that *Voter 2* is still a single point of failure in the above design. It is possible to generalize the above configuration to more than $N > 3$ input channels (NMR configuration).

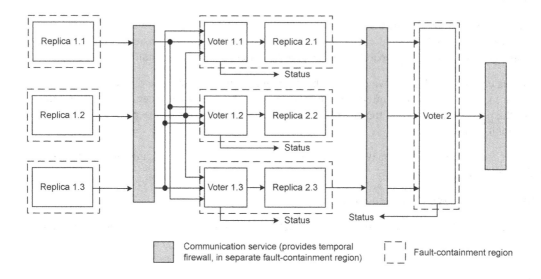

Figure 2.31: Triple Modular Redundant Configuration

Self-checking pair configuration

The self-checking pair configuration (SCP) [36, 37] illustrated in Figure 2.32 can be used to transform a processing chain into a FTU with fail-silent behavior. In this case, the exact voter collapses into a single bit-wise comparison of the two input channels. By instantiating two SCP configurations, and corresponding runtime support (e.g., communication service, resource management), it is possible to have the second pair maintain the service of the FTU in case of a fault in the first one. If the pattern of a (single) SCP is applied to the *Voter 2* component in Figure 2.31, it is possible to transfer the single point of failure in the output path from the more complex "genuine" voter (e.g., majority voter) to the simple comparison function that is sufficient in this case.

2.3   Model-Driven Engineering

DREAMS promotes the use of a Model-Driven Engineering (MDE) approach [38] in order to provide the different stakeholders in the system development process with appropriate means to provide a unified description of the system under design that is used for all activities in the development process, including design and analysis [39]. Consequently, MDE provides models for as many development artifacts as possible, which provide different views onto the same system ("everything is a model" [40]).

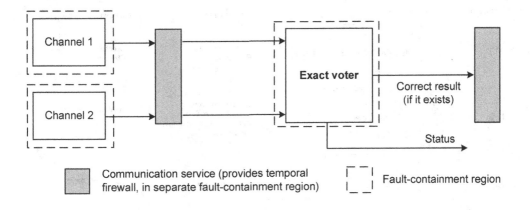

Figure 2.32: Voter in Self-Checking Pair Configuration

2.3.1 Model

Kopetz defines a model as "a deliberate simplification of reality with the objective of explaining a chosen property of reality that is relevant for a particular purpose" [41]. Two general types of models can be distinguished:

- *Descriptive models* are used in the engineering and scientific disciplines to describe the corresponding *system under study* [42]. These *analysis models* are used as abstraction of those properties of the system that are relevant for the intended use [41, 42]. A descriptive model is considered *correct* in case all statements implied by the model are satisfied by the system [42].

- *Prescriptive models* are used as a specification for a *system under design* [42], which can be given at different levels of abstraction. On the one hand, a model could provide a specification of the requirements on the system. On the other hand, a more detailed model could contain a description of different aspects of that system such as its architecture or the desired behavior (see below). For prescriptive models, Seidewitz defines a system to be *valid* with respect to its (specification) model in case all statements of the model are true for the respective system [42].

In DREAMS, like in most other development methodologies, both types of models are used. A prescriptive model of the system under design is expressed based on a metamodel for mixed-criticality systems (see Section 4.1.1). Descriptive models are provided on the one hand by means of requirements and constraint models expressed using the Mixed-Criticality System (MCS) metamodel, (e.g. timing requirements models, safety integrity levels of functions and derived constraints such as separation constraints). On the other hand, also the algorithms described in Chapter 5 implicitly provide a descriptive model of the system (e.g., schedulers and timing analysis tools encode scheduling policies of platform resources).

2.3.2 Metamodel

While the concepts used in object-technology and MDE are almost identical at first glance [40], there are some subtle differences that will be discussed in the following. Object-oriented languages are based on the two relations *instanceOf* (between objects and classes) and *inherits* (between a class and the super-class it refines). As shown in Figure 2.33, MDE can also be described by two relations, where "a particular view (or aspect) of a system can be captured by a model" [40] (relation

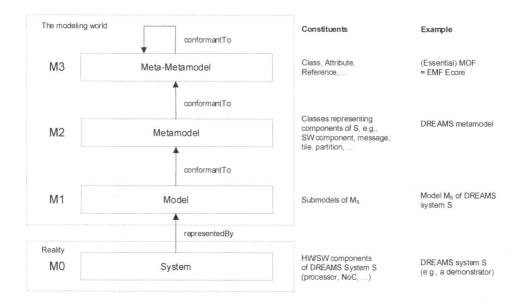

The modeling world

		Constituents	Example
M3	Meta-Metamodel (conformantTo)	Class, Attribute, Reference, ...	(Essential) MOF ≈ EMF Ecore
	conformantTo		
M2	Metamodel	Classes representing components of S, e.g., SW component, message, tile, partition, ...	DREAMS metamodel
	conformantTo		
M1	Model	Submodels of M_S	Model M_S of DREAMS system S
	representedBy		

Reality

M0	System	HW/SW components of DREAMS System S (processor, NoC, ...)	DREAMS system S (e.g., a demonstrator)

Figure 2.33: 4-Level Model Architecture (based on [40])

representedBy) and "that each model is written in the language of its metamodel" [40] (relation *conformantTo*). While MDE's relation *representedBy* looks very similar to the relation *instanceOf* from object-technology, it is important to note that there are arbitrarily many models of some (real-world) system, which individually provide suitable abstractions required for the specific purpose for which the model has been created. In contrast to that, the *instanceOf* is always a 1:1 relation between one particular object (instance) and its type (class). Accordingly, *conformantTo* means that a model is expressed in terms of its metamodel, which can therefore can be interpreted as the abstract syntax of a language that is used to formalize a given model.

The bottom part of Figure 2.33 illustrates the relationship between system, model and meta-model discussed before (also referred to as levels M0, M1 and M2 [40]). The metamodel is expressed in the "language" provided by the meta-metamodel (at level M3), that provides basic concepts like classes, references, attributes, atomic types. This principle could be continued infinitely, i.e. the meta-metamodel at level M3 is linguistically expressed in terms of a "meta-meta-metamodel". However, the figure illustrates a different approach where the meta-metamodel at level M3 conforms to itself. This has the advantage that the modeling architecture is built upon a "self-contained" meta-metamodeling kernel where it is possible to reason about M3 (in terms of M3 concepts) in a similar fashion like M3 concepts can be used to reflect M2. Since there is obviously no unique choice of the meta-metamodel, numerous suggestions have been made (see the overview by Kern *et al.* [43]). However, the efforts of the Model-Driven Architecture (MDA) initiative to establish standards for MDE, resulted in the definition of the Meta Object Facility (MOF) [44], a meta-metamodel that is basically a simplified version of the class modeling capabilities provided by UML 2.0 [45, 46]. The MOF 2 core specification [44] provides the basis for further standards in the area of model serialization and exchange, model life-cycle management, transformation languages and constraint specifications. MOF defines two compliance levels for implementations: Essential MOF (EMOF) and complete MOF (CMOF) [44].

For the DREAMS metamodel, mainly the EMOF compliance level is relevant because the Eclipse Modeling Framework (EMF), which is the basis for the implementation of the metamodel described in Chapter 4, provides the Ecore meta-metamodel which – to a large extent – is compliant to EMOF (mainly naming differences) [47].

2.3.3   Platform-Independent Model

A Platform-Independent Model (PIM) is a prescriptive model that is used to provide a specification of the system under development that is independent from the implementation technology [48]. This definition applies to the following metamodels underlying the model-driven engineering process for MCS presented in Chapter 4:

- The metamodels defined in the logical (see Section 4.2.1), temporal (see Section 4.3.1), and safety (see Section 4.4.1) viewpoint provide a platform-independent specification of mixed-critical applications that serve as the input to the MCS development process.

- The viewpoints described in Section 4.5 provide metamodels for a description of the output artifacts of the MCS development process that is independent from both the method or tools used to produce the result (e.g., offline scheduler), and the technological building blocks of the HW/SW platform that is ultimately configured based on these results (e.g., hypervisor configuration). Here, the deployment viewpoint (see Section 4.5.1) provides a mapping and a virtual link metamodel, whereas the resource allocation viewpoint (see 4.5.2) contributes a generic hierarchical schedule metamodel as well as a reconfiguration metamodel.

- The PIM is complemented by a representation of the platform that is independent from the implementation of the DREAMS HW/SW architecture. The design-principles of the technical architecture model presented in Section 4.2.2 mirror the approach of the waist-tail architecture described in Section 2.2. The technical architecture model provides metamodels for the different architectural levels of the architectures that built upon a platform model framework (see Section 4.2.3) that reflects the DREAMS core services.

2.3.4   Platform-Specific Model

In contrast to the PIM discussed in the previous section, a Platform-Specific Model (PSM) is bound to a specific technology [48].

The service configuration viewpoint (see Section 4.6.1) contributes metamodels to describe configurations of building blocks of the DREAMS platform. It consists of a configuration infrastructure that defines the interface to the configuration generators (see Section 5.5), as well as configuration metamodels for the components of the virtual and physical DREAMS platform that contains all required parameters in a representation that is tailored to the generation of configuration artifacts.

Figure 2.34 illustrates the role of the PSM in the MCS development process described in Chapter 4, where the PSM is used as juncture between the implementation-independent resource allocation tools, and the device- or service-specific configuration generators. A PSM is obtained from a resource allocation model using an automated model transformation that performs a conversion to the expected implementation-specific format and that sets default values for additional parameters (for which manual adaptations by the system integrator might be required).

2.4   DREAMS Certification Strategy

The DREAMS certification strategy sets out a modular and variable product line certification strategy, which is based on cross-domain mixed-criticality patterns. It supports modularity, thus allowing

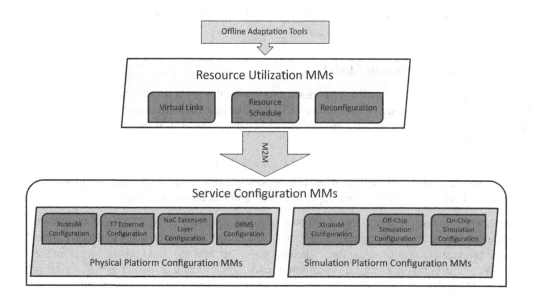

Figure 2.34: Integration of PSM into Mixed-Criticality System Development Process

independent component development and certification. It considers variation points from standards (such as safety standards and security standards) and requirements. And it also defines reusable cross-domain mixed-criticality patterns, which add solutions (e.g., diagnosis techniques, servers) applicable to partitioned and networked multi-core mixed-criticality systems.

2.4.1 Safety and Certification

Certification is a third-party attestation related to products, processes, systems or persons [49]. An attestation is the issue of a statement, based on a decision following review, which fulfillment of specified requirements has been demonstrated.

IEC 61508 [9] is a generic international safety standard that defines the () as a discrete level corresponding to a range of safety integrity values where 4 is the highest and 1 is the lowest. There are multiple domain specific standards relating to functional safety that consider IEC 61508 as a generic reference safety standard: e.g., IEC 62061 and ISO 13849 for safety of machinery, ISO 26262 for automotive, EN 5012X for railway, IEC 60880 for nuclear power plants, IEC 61511 for safety instrumented systems.

The integration of applications of different criticality (safety, security, real-time and non-real time) reducing the number of Electronic Control Units (ECUs) is referred to as mixed-criticality system. Multi-core and virtualization technology, independently or in combination, can support the development of integrated architectures in mixed-criticality platforms by means of software partition, or partition for short. However, safety certification according to industrial standards such as IEC 61508 becomes a challenge because sufficient evidence must be provided to demonstrate that the resulting integrated system is safe for its purpose [50].

Traditional safety certification relies on the concept of the whole system, where if a requirement of the system changes, the entire system shall be re-assessed. The implicit cost attached to the whole system certification can be reduced using a modular methodology approach. Modularity enables dividing the systems into modules which may be independently generated and certified with different criticality levels. Furthermore, this method improves the re-usability and the scalability of the overall system and allows reducing the complexity and cost of certification in the development of mixed-criticality systems. For example:

- Compliant item: Modularity is considered by several safety standards where the modules are called compliant items in IEC 61508, Safety Element out of Context (SEooC) in ISO 26262 and generic products in EN 5012X.

- Modular Safety Case (MSC): A safety case is a documented body of arguments and evidences, intended to justify that a system is acceptably safe for a given set of constraints (e.g., application, operating environment, hypothesis of usage). A safety case can be developed as a composition of MSCs. This modular approach can be used to limit the impact of changes to specific modules of the system, enable re-usability and reduce the complexity of the system (simplification strategy).

2.4.2 Modular Certification

Traditional certification relies on the concept of the whole system, where if a safety/security/real-time requirement of the system changes, the entire system shall be re-certified. Modularity methodology can be applied to tackle traditional certification process. Modularity refers to the division of a system in smaller parts, also called modules, which can be independently developed and certified according to a standard. This method is integrated into the DREAMS certification strategy, thus allowing a module (component) based design and certification.

2.4.3 Mixed-Criticality Patterns

Nowadays mixed-criticality system architectures define partitioned and networked multi-core solutions which are prompt to spatial and temporal interferences. Those interferences challenge the assessment of spatial and temporal independence and therefore, challenge to certification. For instance, according to the IEC 61508 safety standard (IEC 61508-3 Annex F), the common causes of interferences include shared use of random access memory, shared use of peripheral devices, shared use of processor time, communication between the elements necessary to achieve the overall design and fault-propagation.

Cross-domain patterns are widely used and universal approach, that guides and supports engineers, to describe and document recurring solutions for design problems of mixed-criticality systems (from design to verification and validation). Those patterns are used in research project DREAMS to define a set of reusable and generic solutions to solve and mitigate the issues on networked and partitions multi-core mixed-criticality systems.

2.4.4 Product Families

A product family is a range of similar products (product samples) which are developed and sold by the same company with different features. Product samples of a product line may differ from the features and the standards that they accomplish. That means, that whether a requirement of a product sample changes, it shall be re-certified. Furthermore, the complexity of such product samples of product families gives rise to unacceptable development and certification cost.

The certification strategy defined in DREAMS considers product families and sets out a modular and extensible safety argumentation scheme that supports variation points from standards and requirements.

2.5   Fault Assumptions

The fault hypothesis specifies assumptions about the types of faults, the rate at which components fail and how components may fail [51]. The fault hypothesis is a central part in any safety-relevant system and provides the foundation for the design, implementation and test of the fault-tolerance mechanisms [52]. The consideration of security mechanisms for the DREAMS architectural style requires a clear definition of threats. Section 2.5.3 provides this information.

2.5.1   Fault Containment Regions

A Fault Containment Region (FCR) is a subsystem that operates correctly regardless of any arbitrary logical or electrical fault outside the region [51]. A FCR is a set of subsystems that share one or more common resources that one single fault may affect. Based on the distinction between design faults and physical faults, one can distinguish corresponding FCRs (see Table 2.5).

An FCR restricts the immediate impact of a fault, but fault effects manifested as erroneous data can propagate across FCR boundaries. For this reason the system must also provide error containment [51] to avoid error propagation by the flow of erroneous messages. An Error Containment Region (ECR) is a subsystem of the mixed-criticality system that is encapsulated by error-detection interfaces such that there is a high probability that the consequences of an error that occurs within this subsystem will not propagate outside this subsystem without being detected and/or masked [41]. The error detection mechanisms must be part of different FCRs than the message sender. Otherwise, the error detection mechanism may be impacted by the same fault that caused the message failure.

2.5.1.1   Fault Containment Regions for Design Faults

Design faults include hardware and software faults that are introduced during the development of the platform and the application. For design faults, we can distinguish between the faults affecting the DREAMS platform (e.g., system tiles, DREAMS virtualization layer, communication networks) and the application software within the partitions.

For design faults affecting the application software and guest operating systems, we regard a partition as a FCR. Mechanisms for temporal and spatial partitioning of the DREAMS virtualization layer provide design fault containment between partitions. If a software component is replicated along multiple partitions (possibly located on multiple tiles or nodes) as part of a fault-tolerance concept, the FCR includes all partitions with distributed replicas of the software component. Replicated software components cannot be assumed to fail independently, since all replicas of a software component are based on the same programs and use the same input data.

The role of software components as design FCRs holds also in case of software diversity. When design diversity is applied for addressing common mode failures, replicas are necessarily different and ideally employ different specifications in addition to separate implementations. Consequently, we denote these diverse replicas as separate software components. Nevertheless, the decision of regarding the respective partitions with these software components as different design FCRs depends on the independence of the diverse software versions. Practical analyses of software diversity have demonstrated that diverse implementations often exhibit correlation with respect to design faults.

Since all partitions hosted on a tile depend on the correct behavior of the DREAMS virtualization layer, the partitions cannot be assumed to be unaffected by a fault affecting the virtualization layer. Therefore, *all tiles on which a particular virtualization layer is deployed* represent a common FCR for design faults affecting the virtualization layer. The virtualization layer is thus a critical resource in the mixed-criticality system. It is thus necessary to ensure the absence of software faults in the virtualization layer. In particular, the system software needs to be designed for validation and

Table 2.5: Fault Containment Regions

Fault		Fault Containment Region	Containment Coverage (Correlated Failures per Hour)
Design Fault	Design fault of the application component in the partition or the guest OS of the partition	Partition	$< 10^{-9}$
	Replicated design fault in copies of a component or the same guest OS	Multiple partitions containing the same application component or the same guest OS	$< 10^{-9}$
	Design fault of virtualization layer in an application tile	Application tiles with the virtualization layer	$< 10^{-9}$
	Design fault of a system tile (e.g., I/O gateway, memory GW)	System tile	$< 10^{-9}$
	Design fault of on-chip network (including network interfaces and on-chip routers)	Nodes with the on-chip network	$< 10^{-9}$
	Design fault of off-chip network	Cluster with the off-chip network	$< 10^{-9}$
	Design fault of global resource manager	Dynamically reconfigurable part of the platform and respective application subsystems	$< 10^{-9}$
Physical Fault	Affected physical resource of a node (e.g., power supply, clock source, on-chip network)	Node	$< 10^{-9}$
	Affected physical resource of the off-chip network (e.g., short circuit of physical link, clock source of router)	Off-Chip router with corresponding physical links	$< 10^{-9}$
	Affected physical resource only required for a tile (e.g., local memory)	Tile	10^{-5} to 10^{-6}

kept simple in order to permit a thorough validation (e.g., including formal verification). Moving functionality from the virtualization layer into the partitions is a viable strategy to achieve this goal, which is similar to the well known concept of micro-kernels in operating system design.

A *system tile* is an FCR for a design fault of the respective higher platform service. An example is a design fault of an input/output gateway, which affects the corresponding higher platform service provided on top of the core platform services of DREAMS. The entire *node* is an FCR for design faults for shared resources that are required for the correct operation of the node. For example, the on-chip network is a critical resource for the entire node where a design fault of a network interface or router has the potential to disrupt the timely communication of any tile.

In case of design faults affecting an off-chip communication network, the respective *cluster* is a FCR. Faults of the global resource manager can affect the *dynamically reconfigurable parts* of the platform and the respective applications, whereas static subsystems remain unaffected.

2.5.1.2 Fault Containment Regions for Physical Faults

A physical fault affects physical resources, such as mechanical or electronic parts. Physical faults typically originate from conditions that occur during operation. Examples are physical deterioration (i.e., wear-out) and external interference through physical phenomena (e.g., lightning stroke). Early and premature wear-out failures are caused by the displacement of the mean and variability due to manufacturing, assembly, handling, and misapplication.

To form a fault containment boundary around a collection of hardware elements, one must provide independent power and clock sources and additionally electrical isolation and spatial separation. These requirements make it impractical to provide more than one FCR within a *node* at a safety-critical rigor (at a containment coverage with a probability of correlated failures of $< 10^{-9}$ failures per hour).

We also regard each *off-chip router* with the corresponding physical links to the nodes as a FCR. For example, a central guardian of a time-triggered network (e.g., TTEthernet switch) serves as an FCR [53].

For physical faults, the hardware approach can provide certain containment coverage by providing spatial separation of the tiles and cores and multiple clock domains and pin-out (e.g., grounding) on the chip layout (e.g., for SEEs [54]). These on-chip FCRs for physical faults (i.e., *tiles*) work only at single chip failure probabilities (e.g., around $< 10^{-5}$ to $< 10^{-6}$ correlated failures per hour [55]).

Physical fault containment and design fault containment are orthogonal properties. Physical fault containment does not assure design fault containment and vice-versa. For instance, one may use two separated chip processors (two FCRs for physical faults) to implement a function but both can fail simultaneously due to a single design fault on the software. In the same way, a hypervisor can assure design fault containment for two independent operating systems within the same chip and a single physical fault can make both fail.

2.5.2 Failure Modes

The assumed failure modes include those identified by IEC 61508-2, according to which transmission errors, deletion, corruption, delay, repetitions, masquerading and insertion need to be addressed [56]. Furthermore, additional critical failure modes for mixed-criticality systems are introduced.

The following failure modes are distinguished:

1. **Babbling idiot failure:** This failure occurs when an application core or an off-chip router starts sending untimely messages (e.g., insertions according to IEC 61508-2), possibly generating a high traffic load by generating more messages than specified.

2. **Delay:** Faulty core or off-chip router can delay the transmission of messages.

3. **Masquerading:** A masquerading failure is an erroneous core that assumes the identity of another core. In case of periodic and sporadic communication, a faulty core sends messages with the incorrect virtual link identification. For aperiodic messages, the core will send messages with an incorrect logical namespace.

4. **Component crash:** The crash failure occurs when the DREAMS chip or the off-chip router exhibits a permanent fault and produces no outputs.

5. **Link failures:** The link failure occurs when the link exhibits a permanent or transient failure and fails to redirect a message. In combination with the component crash, this failure corresponds to the transmission error according to IEC 61508-2.

6. **Omission:** An omission failure is a transmission failure where a sender is not able to generate a message and/or a receiver is not able to receive a message. This failure corresponds to the deletion according to IEC 61508-2.

7. **Slightly-off-Specification (SOS):** Slightly-off-specification failures can occur at the interface between the analog and the digital world in the value and time domain. For example, consider the case that the specification requires every correct node to accept an analog input signal if it is within a specified receive window of a parameter (e.g., timing, frequency, or voltage). Every individual node will have a wider actual receive window than the one specified in order to ensure that even if there are slight variations in manufacturing it can accept all input signals as required by the specification. These actual receive windows will be slightly different for the individual nodes. If an erroneous FCR produces an output signal (in time or value) slightly outside the specified window, some nodes will correctly receive this signal, while others might fail to receive this signal [53].

2.5.2.1 Failure Rates and Persistence

Part of the fault hypothesis is a specification of the failure rate of FCRs. In general, a differentiation of failure rate with respect to different failure modes and the failure persistence is necessary.

For example, fault injection experiments [57] have shown that restrictive failure modes, such as omission failures, are more frequent by a factor of 50 compared to arbitrary failures.

Related to the failure rates in industrial communication the residual error rate needs to be calculated according to IEC 61784-3. The residual error rate needs to stay below 1% of the PFH of the target Safety Integrity Level (SIL) according to IEC 61508.

Also, failure persistence is an important factor in the differentiation of failure rates. In the temporal domain a fault can be transient or permanent. Whereas physical faults can be transient or permanent, design faults (e.g., software errors) are always permanent. While transient failures disappear without an explicit repair action, permanent failures prevail until removed by a maintenance engineer (e.g., software update in case of a software fault, replacement or repair of hardware in case of a hardware fault).

The permanent failure rate of a FCR with respect to hardware faults is typically considered to be in the order of 100 FIT, i.e., about 1000 years [53]. Motivated by literature on SER we assume that the transient failure rate of a FCR with respect to hardware faults is in the order of 10.000-100.000 FIT [58].

2.5.3 Threats

The DREAMS architecture defines four different core services as shown in Figure 2.2. These core services have different security requirements and there exist different potential attacks and threats which are described in this section. Threat models as well as threats and attacks are related to the cluster-level, e.g., communication services, global time services and resource management services.

2.5.3.1   Threat Models

A threat model describes and analyzes the security risks associated with the system. It identifies potential threats to the system as well as the vulnerabilities in the system which can be exploited. There are four important questions, which have to be considered while creating a threat model [59]:

1. Who is the attacker?
    There are two general types of attacker, a user and an application. Each one of them could be authorized or unauthorized to access a certain component. It is not always necessary to distinguish the attackers as users and/or applications. Considering attacks on the network layer (OSI Layer 3), the attacks are independent of the application layer (OSI layer 7). Hence, in the threat model for communication services, only the "internal" and "external" attackers are considered.

2. What is attacked? A system has different parts which could be attacked. These parts of the system are components and applications.

3. Where is the attacker? An attacker can attack a system from different locations. The attacker could be inside the system or he can attack the system from outside.

4. How is the attack performed? The attacker has different capabilities to perform an attack. Depending on the questions "Who is the attacker?", "What is attacked?" and "Where is the attacker?", the attacker has various options to realize an attack.

2.5.3.2   Threat Analysis for Communication Services

There are different types of communication services in the DREAMS architecture: the on-chip communication and the off-chip communication separated by the on-chip/off-chip gateway (refer to Figure 2.35). As described in Section 2.1, there is a physical and a logical view of the communication system. This section focuses on the physical view. Since there is a distinction between on-chip and off-chip communication, the security aspects can also be divided into on-chip security and off-chip security with different threats which are discussed as follows.

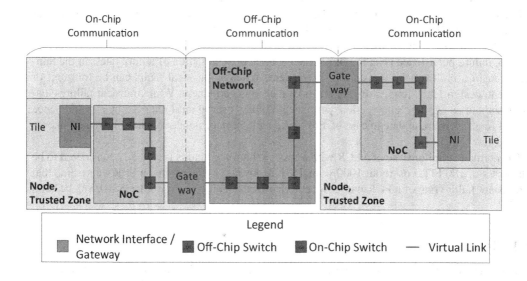

Figure 2.35: On-Chip/Off-Chip Communication

The distinction between on-chip and off-chip security allows the division of attacks on the on-chip and the off-chip communication. This leads to the distinction between internal and external attackers which is based on [60]. The main difference between internal and external attackers is the access point to the system and the knowledge about secret information. The access point of an internal attacker is inside of a trusted part of the system. He has access to the cryptographic keys on the network layer including access to other secret information. Hence, he can generate valid messages and can act as a legal part of the system. In contrast to an internal attacker, an external attacker has no access to the trusted part of the system and does not know the cryptographic keys. Thus, an external attacker can intercept and replay existing messages but cannot generate new ones.

In DREAMS, the communications take place at the NoC and at the off-chip network, respectively. These two types of communications correspond to the internal and external attackers. An attacker who has access to the on-chip communication is an internal attacker and an attacker who has only access to the off-chip communication is an external attacker. Therefore, it is assumed that the SoC, including the NoC and the gateway, is a trusted zone and is inaccessible to an external attacker.

Hence, an internal attacker has access to the NoC and to the other parts of the SoC, e.g., the CPU-cores and the memory. If the cryptographic keys are stored in the memory which is accessible to all components connected to the NoC, then the attacker also has access to these keys.

An external attacker has only access to the off-chip network. He can intercept and replay previously sent messages, but cannot read encrypted messages. Also he cannot generate new legal messages.

The gateway between the on-chip and the off-chip network forms the border among the two network types. Therefore, all communication leaving the gateway towards the off-chip network leaves the trusted zone of the DREAMS architecture. Hence, the gateway separates an internal attacker from an external attacker (Figure 2.3).

There are several types of attacks which can be performed on the communication services of the DREAMS architecture. An attacker can perform sniffing attacks, denial-of-service attacks, spoofing attacks, man-in-the-middle attacks, packet injection attacks and replay attacks. External and internal attackers have different opportunities performing an attack.

2.5.3.3 Threat Analysis for Global Time Services

The global time services should ensure that every local clock in the system has "about the same value" at "about the same points in real-time" (refer to Section 2.2.8). There are two main attack targets on the global time services. On the one hand there are attacks against the clocks or the time values in the components itself, on the other hand there are attacks against the time synchronization.

The attacks on the time synchronization are covered in the threat analysis for communication services. Authorized users as well as unauthorized users could perform attacks on the synchronization process. An attack could aim on a single target with the result that one component gets a false time value or it could aim on the entire synchronization process with the result that no component gets the proper time value.

A single target can be attacked with man-in-the-middle, packet injection and replay attacks. In a man-in-the-middle attack, the attacker can change the time value of the synchronization message before sending it to the receiver. In a packet injection attack, the attacker inserts new synchronization messages with false time values. The receiver of the new messages synchronizes to the false time value. In a replay attack, the attacker sends an old message again to the receiver and the receiver uses the old time value. Man-in-the-middle and packet injection attacks are only possible for authenticated users having access to keys needed to generate new valid messages. An unauthorized user can only perform replay attacks because he cannot generate new valid messages.

The entire synchronization process can be attacked by performing a denial-of-service attack on the master clock. Spoofing attacks can attack both a single target and the entire synchronization process. Denial-of-service attacks are possible for authenticated and unauthenticated users. A spoofing attack is only possible for an authenticated user if they have access to the needed keys masquerading as another user.

The impact of an attack against the clocks in the components itself is similar to the attacks on the time synchronization. However, the communication process for the time synchronization is not the objective of this type of attack. Attacking a clock in a component acting as a slave in the synchronization process only affects the behavior of this component, e.g., the component sends untimely messages or causes untimely actions. If an attacker changes the master clock all clocks in the system synchronizing with the master clock get the false time value. This might lead to measurements taken at the false point in time or to incorrect behavior of the system relating to real-time. Changing the clock values needs additional access privileges and can be performed only by an authorized user or an attacker which can masquerade as an authorized user.

2.5.3.4 Threat Analysis for Resource Management Services

In the DREAMS architecture the resource management services are realized by a Global Resource Manager (GRM). In addition to the GRM, there are Local Resource Managers (LRMs), Local Resource Schedulers (LRSs) and Resource Monitors (MONs) located in the different Tiles. The GRM performs global decisions by selecting configurations. These decisions are based on the information received from the LRM. Decisions for new configurations are sent back to the LRM. The LRM gets information from the MON and maps the global decisions from the GRM to the LRS.

There are several attacks on the resource management services. On the one hand there are attacks against the resource management components. An attacker could masquerade as one of the GRM, LRMs, LRSs or MONs. Acting as a trustworthy GRM or LRM, an attacker can apply wrong or invalid global or local configurations. If an attacker acts as an LRS, he can select other scheduling tables or he can use invalid scheduling parameters. The MON provides monitoring services. Hence, an attacker could send wrong availability, energy or error information to the LRM.

In addition, there are pre-computed configurations. If an attacker can change these offline-computed configurations, a genuine resource management component selects wrong configurations. This could lead to wrong configurations of resources, e.g., false partition scheduling tables or false resource budgets. These attacks can only be performed by an authenticated user who is inside of the system. An unauthenticated user has no access to the components.

On the other hand there are potential attacks on the communication process of the resource management services. An attacker could perform sniffing attacks providing him more information about the behavior of the system. He could perform denial-of-service attacks suppressing the availability of a resource management component. Man-in-the-middle, spoofing and packet injection attacks could lead to wrong configurations and a wrong scheduling. The same risk applies for a replay attack, but at least the configuration or scheduling was valid before. Nevertheless, the system or a part of the system will not operate as intended.

2.5.3.5 Threat Analysis for Execution Services

The execution services provide basic operations to run the system. The service includes the virtualization layer as the software layer that abstracts the underlying hardware and provides virtualization of the CPUs.

The virtualization layer provides the following properties that ensure protection against many attacks related to security.

- **Spatial isolation:** The address space of a partition is not accessible to other partitions. No application of one partition can access the data from another partition. Hence, no unauthorized as well as authorized user or application from one partition can attack another partition. There could only be an attacker inside of the partition. Therefore he can only be an application running

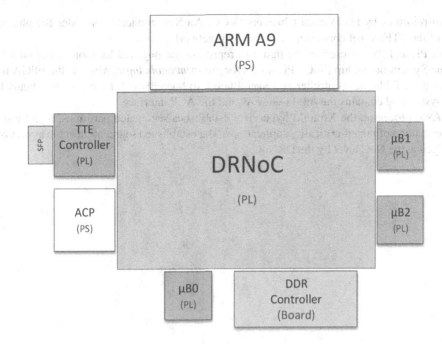

Figure 2.36: Overview of the Implemented Hardware Platform (PL stands for "Programmable Logic", PS for "Processing System" and µB for MicroBlaze)

in the partition or an authorized attacker who can access the partition. But the system architect can define specific shared memory areas between partitions. In these areas, no confidential information should be stored.

- **Temporal isolation:** The temporal isolation ensures that the execution of a partition is independent of the execution of other partitions. Hence, an attacker in one partition cannot prohibit the execution of another partition by performing attacks like sleep deprivation, where an attacker is keeping a partition active, so that he can prevent the calculation of other partitions than the active one from the attacker. Since no unauthorized user can access a partition, only authorized users or applications being inside of the partition can perform such attacks.

2.6   DHP an Instantiation of the Architectural Style

The DREAMS Harmonized Platform (DHP) is as an FPGA-based hardware platform, which has been identified as a domain-independent platform for instantiation of the DREAMS architectural style. This platform is called harmonized in the sense that it is used in the evaluation prototype of the project industrial partners in three use-cases, i.e., avionics, wind-power and health-care. The DHP is implemented by a Xilinx ZC706 ZYNQ-7000 evaluation board and can be used stand-alone or in conjunction with any other board via Ethernet, EtherCAT or TTEthernet.

Figure 2.36 represents the overall architecture of the platform. The hardware is composed of one dual-core ARM Cortex A9 processor, three MicroBlaze processors, TTEthernet gateway and a DDR controller interconnected by the DREAMS Network-on-a-Chip (DRNoC), which supports mixed-criticality communication. The central part of the NoC is STMicroelectronics NoC (STNoC), which supports two priorities by two Virtual Channels (VCs). An SFP connector provides the physical connection of the TTEthernet controller to the off-chip network.

The term PL and PS are used in the figure to represent the physical location of the building blocks on the System-on-a-Chip (SoC). PL stands for *programmable logic*, which is the FPGA part and contains the TTEthernet controller and MicroBlaze 0 to MicroBlaze 2. Likewise, PS stands for *processing system* and contains the ARM Cortex A9 and the ACP interface.

On the ARM processor, the XtratuM hypervisor establishes segregated partitions, thereby supporting the execution of mixed-criticality applications. The established segregation at the hypervisor layer is bridged at the NoC level by the DRNoC.

3

State-of-the-Art and Challenges

H. Ahmadian

Universität Siegen

M. Coppola

STMicroelectronics

M. Faugére

THALES Research & Technology

D. Gracia Pérez

THALES Research & Technology

M. Grammatikakis

TEI of Crete

I. Martinez

IK4-Ikerlan

3.1	Avionics Domain	80
	3.1.1 State-of-the-Art: Integrated Modular Avionics	81
	3.1.2 Challenges	81
	3.1.2.1 Support for More Performance	81
	3.1.2.2 Support for Mixed-Criticality	82
	3.1.2.3 Cyber-Security	82
3.2	Wind-Power Domain	83
	3.2.1 State-of-the-Art: Wind-Turbine Control and Supervision System	83
	3.2.2 State-of-the-Art: Safety Protection System	83
	3.2.3 Challenges	84
	3.2.3.1 Integration and Flexibility (Cost Reduction)	84
	3.2.3.2 Mixed-Criticality	85
	3.2.3.3 Safe Communication	85
	3.2.3.4 Safety Certification	85
3.3	Health-Care Domain	85
	3.3.1 State-of-the-Art Solutions	85
	3.3.2 Challenges: Platform Security and Functionality	86

This chapter provides an overview about the state-of-the-art solutions in three mixed-criticality industrial use-cases. It discusses the research challenges of three use-cases with respect to system architectures, platform technologies and development methods. Section 3.1 is dedicated to the avionics domain and introduces the modular architecture as the state-of-the-art solution in this domain. The challenges are introduced, as the performance and safety requirements of the domain are increasing. It is described how the demanded performance and support for mixed-criticality domains are achieved by the concept of partitioning and at the end the concept of security is discussed. Wind-power is the next industrial domain that is described in Section 3.2. In this section, the state-of-the-art solutions for control and supervision systems of wind-turbines are elaborated. Integration and flexibility, support for mixed-criticality, safe communication and safety certification are identified as the main challenges for such systems. At the end, the challenges of the health-care domain are elaborated in Section 3.3. Wearable intelligent devices are identified as the state-of-the-art solutions in the smart hospitals. In such solutions, security and functionality are identified as the key research and technological challenges.

3.1   Avionics Domain

The avionics domain is one of the industrial domains, in which safety has been the main driver for the development of new solutions, particularly in the civil domain. Both governmental and industrial bodies have defined regulations and standards to develop the exploitation of avionics systems. Regulations are legal documents and are the basis to define the certification of the system, if the system can operate or not. Regulation bodies are typically state agencies, like the Federal Aviation Administration (FAA) in the United States or the European Aviation Safety Agency (EASA) in the European Union. Standards are typically industry practices accepted by regulation bodies as means to develop products that follow the regulations, and in some cases as solutions to ensure the interoperability of products components.

From the technical perspective, two standards are used as basis for the development of aircraft computation systems: DO-178C [3] and RTCA DO-254/EUROCAE ED-80 [61]. These standards define the Development Assurance Level (DAL) and categorize the effects of a failure condition in the safety of the aircraft. Based on the DAL, these standards define the properties and development process that a system must follow. Table 3.1 illustrates the defined DALs by the standards, their associated failure condition in case of occurrence and the maximum accepted occurrence rate.

Table 3.1: Development Assurance Level (DAL), the Associated Failure Condition in Case of Occurrence and the Maximum Accepted Occurrence Rate

Level	Failure condition	Failure rate
A	Catastrophic	10^{-9}/hour
B	Hazardous	10^{-7}/hour
C	Major	10^{-5}/hour
D	Minor	10^{-3}/hour
E	No effect	n/a

The RTCA DO-178/EUROCAE ED-12 standard defines two properties that need to be ensured in a safety-critical system (i.e., with DALs from A to D): spatial partitioning and temporal partitioning. Spatial partitioning refers to the capability of a system to ensure that an application data is isolated in the system, i.e., it cannot be read or modified by external entities and the application cannot access other data than its own. In modern systems, temporal partitioning is often established by configuring the processor Memory Management Unit (MMU) to ensure the spatial partitioning requirements, typically by the operating system software.

Temporal partitioning refers to the capability of a system to allocate time windows to an application, during which the application is executed without being interrupted, and the application will not be executed outside the allotted time windows. Typically, temporal partitioning is established by the operating system or a hypervisor by implementing cyclic scheduling with time windows allocation in the scheduling cycle.

Avionic products must comply with the regulation of the legal bodies of the countries, where they will be used. Following the above standards during the development process allows the product developers to ensure that their products can be certified, commercialized and used. It must be noted that standards provide sufficient guidelines to ensure that the development of a product complies with the regulations, however, product developers might use other approaches than those proposed in the standards and still certify their products. The certification of products that are developed using practices other than those proposed by the standards are typically longer to certify, as the developer has to provide additional proofs that the product complies with the regulations, while in case of using the standard guidelines, the proofs are already provided by the standard.

3.1.1 State-of-the-Art: Integrated Modular Avionics

Similar to automotive domain, the electronic solutions have been introduced in the avionics domain to improve the safety of the aircraft and to reduce their weight by removing mechanical components. Initially, each functionality (or application) was typically delivered by a different single computing unit (or Local Resource Scheduler (LRS)). As a consequence, in case communications between different functions were required, safe networks, like AFDX, were deployed. This approach is known as *Federated Avionics Architecture* or *Federated Avionics System* (*Federated Architecture* or *Federated System* for short).

While the federated design has been successfully used for the development of several generations of aircraft products, the increase in usage of computing solutions for implementing the avionics functions has become a challenge in terms of *Size, Weight, Power and Cost (SWAP-C)*.

Integrated Modular Avionics (IMA) [62] was introduced as a solution to combine functionalities in a computing unit and to reduce the number of computing units in an aircraft. Basically, this solution involves an enhanced operating system or hypervisor, which ensures disjoint time windows (defined by the system integrator) in the cyclic schedule for each of the functionalities combined in the system. Furthermore certain IMA approaches enable the application of incremental certification: for a certified IMA system that combines functions A and B, the addition of a new function (e.g., C) does not require a new certification of the entire system, but only the added function. The IMA is standardized in the RTCA DO-297/EUROCAE ED-124 standard [62] and also considers the modularity of the hardware parts of an embedded system.

3.1.2 Challenges

The demanded performance and safety requirements in avionics are continuously increasing. The SESAR [63] project has the objective to to increase its capacity of the European air traffic by modernizing it, through the development and implementation of new procedures and technologies impact both ground control and the aircraft themselves. The Clean Sky [64] project develops technologies to reduce CO^2, gas emissions and noise levels produced by aircrafts. The solutions proposed by those programs require embedded processing systems with increased performance to support their implementations. Furthermore, the ongoing trend towards the usage of avionics solutions in urbanized areas, like UBER with their on-demand urban air transportation project Elevate [65], the Dubai flying taxi trials [66,67] or the Airbus Vahana project [68] to provide urban air mobility solutions, introduces new and major challenges. To target this new market, embedded systems do not only need to provide more performance, but also more integration, i.e., able to combine multiple applications in a single device.

With the new performance and safety requirements introduced by these projects, the expectation on the embedded systems capabilities is increased, forcing the industry to move away from the single-core solutions. Evidently, single-cores performance has stalled and new processing solutions need to be studied to create the new products and services these projects target. Furthermore, apart from what the technology and the industry will deploy, the safety requirements (e.g., space and time partitioning) will also be required to be implemented, and security requirements have to be addressed to satisfy the openness and integration needs.

3.1.2.1 Support for More Performance

In order to satisfy the previously described performance requirements, alternative approaches need to be explored. Multi-core processors are the most promising alternative to the single-core architectures used in the past. Theoretically, multi-core would satisfy the new performance requirements. In addition, they are widely applicable as the current software applications can be reused.

However, in order to use multi-core processors in safety-critical solutions, a temporal and spatial partitioning between the deployed application(s) need to be ensured. Hardware solutions, like the

MMU can still be used to ensure space partitioning, but due to the interferences time partitioning is difficult to ensure, when using shared resources on multi cores. These interferences may come from the usage of processor peripherals or accelerators (e.g., Direct Memory Access (DMA)), or from the caches, buses and memories that are shared between multiple cores. Studies as [69, 70] proved that applications suffer from a slowdown in the execution time proportional to the number of processors cores due to the interferences in the usage of the shared resources (caches, buses and memory). These studies effectively render the usage of multi-cores inefficient on safety-critical solutions, unless mitigation solutions are introduced.

Hardware [71–74] and/or software [75–82] approaches have been proposed to address time partitioning on multi-core processors. However, the following requirements of safety-critical industrial products [83] demand for additional consideration to be taken into account: support for legacy applications, efficiency of the proposed software/hardware solution, robust partitioning assessment complexity, integration into an industrial process, easiness to adapt existing applications, and complexity and certifiability of the solution.

3.1.2.2 Support for Mixed-Criticality

With the advent of multi-core processors, new challenges were introduced for the integration of applications of different criticality levels. IMA solutions have been used for the integration of applications on single-core processors. In multi-core processors, new approaches need to be yet defined to ensure that for safety-critical applications time and space partitioning are respected and at the same time, low-critical applications employ the highest possible performance.

Fisher [84] introduced a commercial solution that enabled the combination of critical and non-critical applications by disabling the execution of non-critical applications when a critical application was executed in the system. In this solution the execution of non-critical applications on all available cores is disabled, only when the critical applications (executing in a single core of the processor) are not scheduled. This effectively reduces the utilization of the system, as the cores remain unused when the critical application executes.

Multiple projects (including DREAMS) and studies [85–88] have addressed (or are addressing) this topic, but few have addressed this topic while considering solutions for integrating multiple critical applications in a multi-core processor (see previous subsection in Section 3.1.2.1). Moreover, performance is not the only aspect that multi-core processors introduce, when combining critical and non-critical applications in a system. This integration brings new challenges and solutions, when considering hardware faults. For example, when a core fault occurs, it does not mean that the whole processor needs to be disregarded. However, fault-tolerance mechanisms need to be introduced in the development and execution of such systems to ensure that the safety-critical functionalities of the system are maintained in such events.

3.1.2.3 Cyber-Security

Safety-critical systems have been mostly autonomous and disconnected from non-critical systems. However with the introduction of electronics on those systems, there is now a trend to connect and furthermore integrate them, e.g., federated systems, IMA. With the new requirements added for solutions to further enhance the aircraft safety and provide new exploitation services, these previously completely autonomous systems require now to communicate to the external world. As a result, to ensure the safety of those systems, security becomes essential, for instance, the security support virtualization for safety-critical systems [89–92].

3.2   Wind-Power Domain

In the wind-power domain, there is a tremendous market push towards off-shore operation. The road to off-shore introduces new technological challenges, stringent safety requirements and new standards to comply with. In this section, the state-of-the-art solutions in the wind-power domain are discussed.

3.2.1   State-of-the-Art: Wind-Turbine Control and Supervision System

Commercial wind-turbines are governed by a control and supervision platform, which implements the following two groups of functionalities:

- Control and supervision

- Human-machine interface and communications with the Supervisory Control and Data Acquisition (SCADA)

Figure 3.1: Galileo Platform Version 5.0

Figure 3.1 represents the latest version of the Galileo platform that is currently used in the state-of-the-art wind-turbines. Galileo is a commercial hardware (industrial PC APC 910) based on an x86 dual core processor that customized at operating system and software levels. Though the Galileo platform is used for the control and supervision of the wind-turbines, it may support other real-time applications such as wind farm control. The Galileo platform requires several inputs and outputs that are connected through an EtherCAT field bus.

3.2.2   State-of-the-Art: Safety Protection System

The protection system is in charge of maintaining the wind-turbine in a safe state, by assuring that the design limits of the wind-turbine are not exceeded. The protection functions are activated as a result of a failure of the control function (running in the supervisory system) or of the effects of an internal or external failure or dangerous event. It should be activated in cases such as:

- Over-speed

- Generator overload or fault

- Excessive vibration

- Abnormal cable twist (due to nacelle rotation by yawing)

The state-of-the-art protection system is an external module integrated in the EtherCAT ring. This is the only module that checks safety data in order to decide over the safety chain. The protection system works independently from the Galileo platform, just sharing the EtherCAT bus, as shown in Figure 3.2.

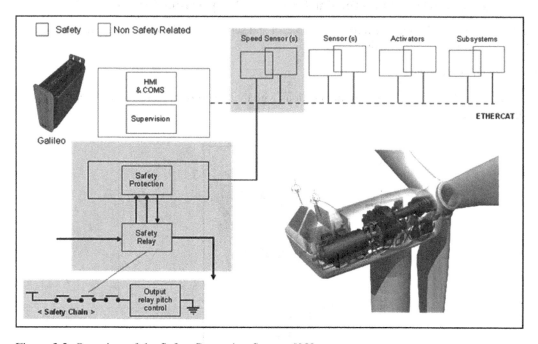

Figure 3.2: Overview of the Safety Protection System [93]

The act of checking safety data and deciding over the safety line is done in a non-redundant way. Thus, the current solution is a zero Hardware Fault Tolerance (HFT) system, which would mean that a failure in the safety protection system causes the loss of the safety function.

3.2.3 Challenges

The proposed state-of-the-art solutions for the control and supervision system and the protection system face several challenges:

3.2.3.1 Integration and Flexibility (Cost Reduction)

The protection system is implemented in an external module that is integrated in the EtherCAT ring and it lacks the flexibility to implement complex logics, as it is able to only handle digital inputs and outputs and it is mainly a commercial hardware based system.

Furthermore, there are other limitations, such as a reduced reliability and availability due to interconnection of subsystems (low integration), improvable maintainability due to the high number of functionalities being executed in different devices, operating systems not suitable for the purpose of all hosted applications, limited scalability and lack of composability (introducing more functionalities in such a complex and static architecture is not an easy task).

3.2.3.2 Mixed-Criticality

Galileo control and supervision platform is home to different kinds of processes and tasks. High criticality processes such as wind power control algorithms are run together with lower criticality processes such as external communication management. An overload in low criticality tasks could affect the performance of other processes such as the control algorithms.

3.2.3.3 Safe Communication

Galileo platform uses redundant EtherCAT communication in a ring topology to communicate with all subsystems in the wind turbine. Most of these subsystems are not safety-related and therefore the EtherCAT solution has been configured to maximize availability, but without safety concerns. Normally, Fail Safe over EtherCAT (FSoE) is used to add safety traffic over this EtherCAT network, but this protocol is not open and has some limitations imposed by the manufacturer.

Safety data is sent by an EtherCAT node to the safety protection system which is connected to the EtherCAT ring. The safety protection unit then evaluates this data and decides whether the safety line should be opened or not. Data integrity is not guaranteed in this communication channel. Therefore, decisions over the safety line can be taken based on corrupted data.

3.2.3.4 Safety Certification

The current solution is a zero HFT system and achieving SIL 3 integrity level is difficult.

3.3 Health-Care Domain

In the health-care domain there is an increasing trend towards mobile and smart solutions, to achieve potential improved care, cost savings and reduction of errors. Such solutions allow physicians to monitor patients remotely in real-time, ultimately avoiding medical errors, providing patient comfort, while also reducing treatment costs. This section offers an overview of the challenges and state-of-the-art solutions in the health-care domain.

3.3.1 State-of-the-Art Solutions

Miniaturized medical sensors detect biological signals and extract health-care data, assisting to a proliferation of services based on wearable intelligent devices [94–96]. In particular, heart-related disorders account for one of three deaths in the US in 2017 and include *atrial fibrillation* related to stroke risk and *ventricular arrhythmia* associated to cardiac arrest [97]. These disorders occur sporadically and are now treated with off-line ECG analysis based on 72-hour ambulatory Holter that has a low diagnosis rate.

Prolonged real-time ECG monitoring is needed for improved detection. However, most industrial products, e.g., AliveCor [98, 99], BodyGuardian [100], LifeMonitor [101], NowCardio [102], and PhysioMem [103], capture, process and transmit ECG data to a server for off-line analysis by specialists. In addition, research on wearable monitoring and arrhythmia diagnosis concentrates on detection capability than real-time, cf. Android [104] or iOS [99]. Android smart-phone is often used to capture, analyze and visualize ECG for alerting the patient in real time [105, 106]. Transmission of annotated ECG signals to a remote monitoring center for *off-line processing* by medical personnel is considered in [107–109], whereas ECG data packetization and TCP parameters can be considered prior to network transmission [110].

Two years ago, a lightweight mobile wearable cardiac pulse sensor (called BodyGateway, or

BGW) was developed with several micro-electromechanical sensors and an ARM Cortex M3 micro-controller supporting real-time OS. The BGW is an open version of BodyGuardian, which is attached to one of three places on the patient's chest (similar to a bandage) and allows physicians and care providers to monitor important biometric data (ECG signal, heart rate, respiration rate and physical activity level and body position) in real-time. The patch can be programmed to either stream or store and periodically transmit vital physiological data via Bluetooth over a continuous 30-day period to a host device.

The BGW can be used for remote monitoring for in- and out-hospital use cases, such as *rhythm monitoring* to understand the cardiac role of rare unexplained symptoms, *vitals monitoring* to study cardiac rhythm respiration and activity, or *long-term treatment effectiveness monitoring* to evaluate arrhythmia medication therapy. In all cases, we avoid expensive clinical trials, whether it is for daily checkups, or after heart attacks, surgery, or implants, while reducing patient concerns and improving engagement with care plans.

Recently the BGW functionality was integrated in a BodyGuardian Heart product to provide a discreet, pocket-sized wearable monitor. This solution accommodates patient mobility, enhances compliance, and streamlines data collection via wireless. In addition, a *Remote Monitoring Center* allows physicians to monitor patients' experience for individualized monitoring and care plan support. In this context, BodyGuardian Heart supports the following functionalities:

- Remote monitoring of ambulatory ECG and average heart rate for arrhythmias, including Atrial Fibrillation, Tachycardia, Bradycardia, Pause and others

- Recording and wireless transmission of periodic ECG at specified intervals for static analysis, so that physicians can access their patients' data set and review anytime, anywhere periodic cardiac event notifications (e.g., maximum and minimum heart-rate) via the web in a secure way through the PatientView and PatientFlow portals or a connected electronic medical record system

3.3.2 Challenges: Platform Security and Functionality

Digital medical information, whether stored on a computer or transmitted via wireless is vulnerable to hackers and fraudsters. In fact, there is a much higher rate of cyber attacks, e.g., as identity theft, on health-care providers and insurers. Despite rigorous security precautions and governmental rules and civil penalties, medical data is not secure, as demonstrated by recent hacks, e.g., Verizon's data breach affected 14 million patients.

The Health-care Insurance Portability and Accountability Act (HIPAA) establishes US standards for the protection of health information data, including technological safeguards for enforcing compliance. The BodyGuardian heart monitoring system complies with HIPAA Privacy and Security Rules, by supporting device-, network- and application-level security services.

A major challenge in DREAMS focuses on extending the state-of-the-art in biometric signal processing by realizing a health-care demonstrator, which exploits the advent of real-time and time-triggered technologies. More specifically, required support for a real-time diagnostic ECG arrhythmia detection application was considered, which can periodically capture and communicate in real-time asymptomatic events based on patient's vital signs to physicians, even in the presence of mixed-criticality traffic, such as infotainment. Settings of the ECG analysis algorithms can be customized to other non-fatal arrhythmias with high predictability.

In Chapter 7, key hardware/software architectural components are addressed. Furthermore, Chapter 11 focuses on the preliminary evaluation of the DREAMS health-care demonstrator solution in both in- and out-of-hospital scenarios. In addition, the health-care demonstrator supports hard real-time communication components (XtratuM hypervisor, TTEthernet router/driver).

4

Modeling and Development Process

S. Barner

fortiss GmbH

F. Chauvel

SINTEF

A. Diewald

fortiss GmbH

F. Eizaguirre

IK4-Ikerlan

Ø. Haugen

SINTEF

J. Migge

RealTime-at-Work

A. Vasilevskiy

SINTEF

4.1	Introduction	89
	4.1.1 Mixed-Criticality System Modeling Viewpoints	89
	4.1.2 Fundamental Metamodels	91
4.2	Architecture Design	92
	4.2.1 Logical Modeling Viewpoint	92
	4.2.1.1 Logical Component Architecture Metamodel	92
	4.2.1.2 Logical Architecture Annotations	93
	4.2.1.3 Logical Architecture Example Instance	94
	4.2.2 Technical Modeling Viewpoint	95
	4.2.3 Platform Architecture Modeling Framework	95
	4.2.4 DREAMS Platform Metamodel	99
	4.2.4.1 Cluster Domain	99
	4.2.4.2 Node Domain	100
	4.2.4.3 Tile Domain	102
	4.2.4.4 NoC Domain	104
	4.2.4.5 Processor Domain	105
	4.2.4.6 Hypervisor Domain	107
	4.2.4.7 Technical Architecture Annotations	110
4.3	Timing Requirements	114
	4.3.1 Temporal Modeling Viewpoint	114
	4.3.2 Generic Methodology Pattern	117
	4.3.2.1 Create Solution	117
	4.3.2.2 Attach Timing Requirements to Solution	118
	4.3.2.3 Create Timing Model	118
	4.3.2.4 Analyze Timing Model	119
	4.3.2.5 Verify Solution against Timing Requirements	119
	4.3.2.6 Specify and Validate Timing Requirements	119
	4.3.2.7 DREAMS Timing Lifecycle Specification	119
4.4	Safety Management	120

	4.4.1	Safety Modeling Viewpoint	120
		4.4.1.1 E/E/PE System Safety Requirements Specification	121
		4.4.1.2 Safety Functions	121
		4.4.1.3 Safety Manual	122
		4.4.1.4 Faults Management	122
		4.4.1.5 Usage Constraints	122
		4.4.1.6 Safety Cases	123
	4.4.2	Development Process of Safety Solution Design	123
		4.4.2.1 Safety Solution Design Development Steps	123
	4.4.3	Verification of Safety Solution Design	124
		4.4.3.1 Verification Safety Rules	125
		4.4.3.2 Safety Functions Architectural Style Verification	125
		4.4.3.3 Checking Whether SIL Claimed by Safety Compliant Items Is Achievable	126
		4.4.3.4 System-Subsystem Composition and Usage Constraints	127
		4.4.3.5 Verification Safety Constraints for Deployment	127
		4.4.3.6 Safety Case Generation	128
		4.4.3.7 Relationships to Certification and Mixed-Criticality Product Lines	130
4.5		Deployment and Resource Allocation	131
	4.5.1	Deployment Modeling Viewpoint	131
		4.5.1.1 Deployment Metamodel	132
		4.5.1.2 Virtual Link Metamodel	135
	4.5.2	Resource Allocation Modeling Viewpoint	138
		4.5.2.1 Schedule Metamodel	138
		4.5.2.2 Reconfiguration Metamodel	140
	4.5.3	Basic Deployment and Scheduling Workflow	141
	4.5.4	Adaptivity and Resource Management Workflow	144
4.6		Service Configuration Generation	146
	4.6.1	Configuration Modeling Viewpoint	146
		4.6.1.1 Configuration Infrastructure Metamodel	147
		4.6.1.2 Example: Physical On-Chip Network Interface Configuration Metamodel	147
	4.6.2	Model-transformations and Configuration Synthesis	149
4.7		Variability and Design Space Exploration	149
	4.7.1	Variability Modeling Viewpoint	150
		4.7.1.1 Modeling Variability	150
		4.7.1.2 Exploiting Variability	151
	4.7.2	Variability Exploration Process	151
		4.7.2.1 The Complexity of Product Lines	151
		4.7.2.2 Variability in Mixed-Criticality Systems	152
		4.7.2.3 Exploring Business-level Variations	154
	4.7.3	Design-Space Exploration Process	155
		4.7.3.1 Architectures of Safety Critical Functions	156
		4.7.3.2 Input Specification	157
		4.7.3.3 Synthesized Artifacts	157
		4.7.3.4 Technical Variability Exploration	159

This chapter introduces the DREAMS metamodel and a model-driven development process ranging from variability exploration to configuration synthesis. The metamodel is described in Section 4.1

and is organized into a set of viewpoints, each of which represents one system aspect. The logical viewpoint is introduced in Section 4.2 and allows for the platform-independent description of applications. The technical viewpoint enables the hierarchical description of the architecture and services of the platform. The timing viewpoint is introduced in Section 4.3 to model timing requirements that must be satisfied in order to guarantee a correct and safe operation of the system. Safety management is another topic that is covered in Section 4.4 and is supported by a safety modeling viewpoint. The deployment and resource allocation viewpoints are addressed in Section 4.5 and link the application model with the platform model. Section 4.6 describes a configuration viewpoint and defines a model-driven process to generate deployable configuration artifacts for HW/SW target platform. Lastly, in Section 4.7 a variability viewpoint constitutes the basis of a product-line exploration process.

4.1 Introduction

The availability of the DREAMS platform whose properties and guarantees have been described in Chapter 2, and whose implementation will be described in depth in the remainder of this book, is an important building block to solve the mixed-criticality integration problem. However, due to the size and complexity of the considered systems (both regarding the application sub-systems and the considered instances of the DREAMS platform), the platform is only one side of the coin: deploying mixed-critical applications to shared resources typically requires design-time configurations (e.g., to ensure real-time constraints or separation constraints mandated by safety regulations). These configurations are the outcome of complex optimization problems that are intractable in a manual process that also hardly can guarantee the consistency of all deployable artifacts nor their traceability to the requirements.

4.1.1 Mixed-Criticality System Modeling Viewpoints

In this chapter, we introduce the DREAMS model-driven development process for Mixed-Criticality Systems (MCSs) [38]. It is based on a mixed-criticality metamodel that is organized into a set of complementing viewpoints. Figure 4.1 provides an overview of the selected MCS viewpoints that represent different aspects of the system under development and reflect the views of different stakeholders.

Each viewpoint consists of a set of metamodels, and contributes to a modular MCS metamodel that separates different concerns. As an orthogonal abstraction perspective, the MCS metamodel provides a granularity dimension that enables to describe the recursive decomposition of (parts of) the system into atomic building blocks. In the following, we provide an overview of the different viewpoints that will be introduced in this chapter:

- The *logical viewpoint* described in Section 4.2.1 provides a metamodel for the platform-independent description of applications in terms of architecture, behavior and non-functional requirements.

- The *technical viewpoint* (see Section 4.2.2) allows for the hierarchical description of the architecture and services of MCS platforms, extending the concepts introduced in [111–113].

- In Section 4.3.1, we describe the *temporal viewpoint* that provides a metamodel to capture timing requirements that must be satisfied in order to guarantee a correct and safe functioning of the implemented system based on concepts from TADL2 [114].

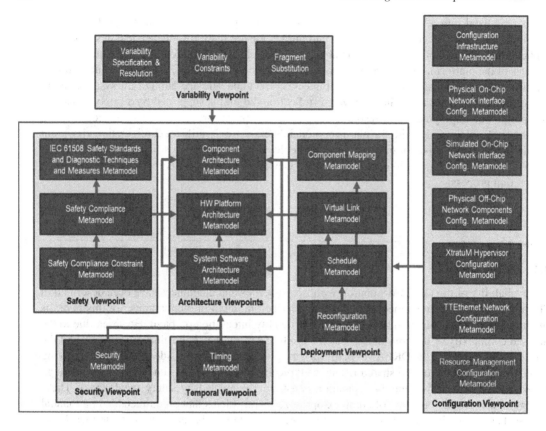

Figure 4.1: Viewpoints of Mixed-Criticality System Metamodel

- The metamodels of the *deployment viewpoint* (see Section 4.5) link the application model from the logical viewpoint with the platform model defined in the technical viewpoint (e.g., mapping of components to execution units). Further, it contains metamodels to describe *resource allocations* (e.g., time-triggered schedules) and system *reconfigurations* to mitigate hardware faults.

- The *safety viewpoint* presented in Section 4.4.1 augments the metamodels from the logical and technical viewpoint to support the early detection of safety-related errors during the realization phase of the system.

- The *configuration viewpoint* (see Section 4.6.1) provides metamodels representing the configurations of DREAMS platform services. Configuration models are usually derived from deployment and resource allocation models using automated model-transformations, and serve as input to platform configuration generators (c.f. Chapter 5).

- The metamodels of the *variability viewpoint* allow to specify variation points of the aforementioned domain models. This enables to define MCS product-lines based on variability models and models of all reusable assets [115].

4.1.2 Fundamental Metamodels

The kernel metamodel and the hierarchic element metamodel described in this section serve as a common basis for the metamodels introduced in this chapter. Table 4.1 depicts the kernel that provides fundamental modeling entities shared between the majority of metamodels introduced in this section. It defines root metaclasses to depict model elements and to describe their basic properties such as names, comments, and unique ids, and introduces the notion of modeling projects that serve as containers for model artifacts.

Table 4.1: Modeling Kernel Metamodel

Metaclass	Description
IIdLabeled	Model elements implementing this interface have a unique identifier.
INamedElement	Model elements implementing this interface have a unique id and a name.
INamedCommentedElement	Model elements implementing this interface have a unique id, a name and a comment.
IElementWithURI	Elements that can be referenced using a URI.
IProjectRootElement	Super class of all root elements contained in the project.

The hierarchic element metamodel depicted in Table 4.2 extends the aforementioned kernel metamodel to enable the hierarchic decomposition of model entities (parent-child relationship). Further, it provides base metaclasses to describe the interface of model elements in terms of connectors, and to define the relationship of model elements using connections. Lastly, the hierarchic element metamodel introduces the notion of specifications that can be used to flexibly extend the metamodel with additional attributes.

Table 4.2: Hierarchic Element Metamodel

Metaclass	Description
IModelElement	Super class of first class model elements. Its specifications attribute defines the list of model element specifications providing additional model element properties.
IModelElementSpecification	Super class of model element specifications that provide additional properties to an IModelElement.
IHiddenSpecification	Super class of hidden model element specifications (for internal use).
IAnnotatedSpecification	Super class of model element specifications that represent annotations (i.e., IModelElementSpecifications that are guaranteed to exist exactly once for the respective IModelElements for which the annotation has been registered).
IDerivedAnnotation	Interface for IAnnotationSpecifications whose value is derived / computed from the state of other annotations and/or model elements.

`IConnector`	Super class of connectors. Connectors reference incoming and outgoing `IConnections` of model elements.
`IConnection`	Super class of connections. Connections are aggregated in an `IHierarchicModelElement` (see below) and reference two connectors from that element or any direct sub-element.
`IHierarchicElementContainer`	Super class of all hierarchic model elements that defines their parent-children containment relation.
`IHierarchicElement`	Super class of hierarchic model elements that defines the `IConnections` between them.

4.2 Architecture Design

In this section, we describe the *architectural viewpoints* used to support the model-driven engineering process for MCSs described in this chapter. These viewpoints allow describing structural aspects of the system, i.e., the applications (see Section 4.2.1) and the underlying platform (see Section 4.2.2).

4.2.1 Logical Modeling Viewpoint

4.2.1.1 Logical Component Architecture Metamodel

The (logical) component architecture metamodel is used to describe logical and functional aspects of mixed-criticality applications. It is derived from the metamodel presented in [111] and is based on the hierarchic element metamodel (see Section 4.1.2).

The logical architecture consists of a hierarchical network of components whose data interface is described using typed logical input and output ports. To define data-dependencies between components, logical channels may be used to connect compatible output and input ports. Ports may also remain unconnected to model external inputs or outputs of the system, i.e., from sensors or to actuators.

Table 4.3: Logical Architecture Metamodel

Metaclass	**Description**
`ComponentArchitecture`	Root element of logical architectures that contains components.
`Component`	Describes a logical/functional block and may be a logical container for sub-`Components`.

Port	Defines the input or output of the `Component` to which it is attached, and allows `Components` to interact with their environment. `Ports` are transparent to the sub-elements of the Component to which they are attached.
`InputPort`	Represents a `Port` that receives input data of a `Component`.
`OutputPort`	Represents a `Port` that emits data from a `Component`.
`Channel`	Connects `Ports` and is used to define the data flow of a modeled application. Since `Channels` are directed, they describe the data-dependencies of `Components`.

Table 4.3 gives an overview of the metaclasses defining the logical component architecture meta-model.

4.2.1.2 Logical Architecture Annotations

Since the DREAMS Model-Driven Engineering (MDE) process focuses on deployment, resource allocation and configuration generation, the behavior of logical components is not explicitly modeled. In the following, we present a number of annotations for logical components and ports that depict the relevant extra-functional properties, and refer the reader to [111] for adequate behavioral models. Table 4.4 depicts attributes of logical components and annotations that have been defined in terms of specifications and annotations (see hierarchic element metamodel defined in Section 4.1.2).

Table 4.4: Annotations of Logical Components

Metaclass	Description
`SafetyIntegrityLevel`	This annotation allows defining the required safety integrity level for a `Component`. The annotation for the top level `Component` (which is associated with the `ComponentArchitecture`, is used to select the safety standard that defines the available levels (considered standards DO-178C, IEC 61508 and ISO 26262). The information provided by the `SafetyIntegrityLevel` annotation is mainly intended to support the architecture design and deployment phase of the development process. The safety viewpoint (Section 4.4.1) provides additional concepts that are used to support verification and validation activities in the development process. It therefore provides the safety standard metamodel that is used for safety consistency checks and report generation.
`MemoryRequirement`	Defines the amount of memory required by a `Component`.
`EventTriggerAnnotation`	This annotation allows defining the trigger of a `Component` based on the `EventTrigger` defined in the timing viewpoint (see Section 4.3.1).

Table 4.5 summarizes the available attributes for logical ports.

Table 4.5: Annotations of Logical Ports

Metaclass	Description
`PortSpecification`	Defines the type of data that is emitted or received by the annotated `Port`. It also defines the initial value of the associated `Port`.
`MessageSize`	Size of the raw data that is sent via the annotated `OutputPort`. It is given in bits and calculated via the data type (see `PortSpecification` introduced above) that is defined for the annotated `OutputPort`.
`InputEventAnnotation`	This annotation allows defining the event triggering at an `InputPort` based on the `InputEvent` defined in the timing viewpoint (see Section 4.3.1).
`OutputEventAnnotation`	This annotation allows defining the event triggering at an `OutputPort` based on the `OutputEvent` defined in the timing viewpoint (see Section 4.3.1).

4.2.1.3 Logical Architecture Example Instance

As an example, a model of the logical architecture of a navigation application is illustrated in Figure 4.2. The model has been created in AutoFOCUS 3 (AF3) [116] that provides the reference implementation of the metamodel introduced in this chapter.

The example application consists of the `Components` *SensorAcquisition* (reads and preprocesses sensor data), *Controller* (algorithm performing the navigation), *MapProcessing* (provides access to a stored map), *HMI* (user input / display), and *ActuatorControl* (controls a motor or similar). The exemplary application model is centered on the *Controller* `Component` that receives refined sensor data from the `Component` *SensorAcquisition* and performs the actual navigation using additional information from a map. The results from the *Controller* are output to the *Actuator-Control* `Component` to transform these results into physical actions and to the *HMI* `Component` that displays the results and forwards commands issued by the user to the *Controller*. Each of these `Components` has attached `InputPorts` (white circles) and `OutputPorts` (black circles) that are used to connect `Components` via `Channels` (black arrows). Disconnected `Ports` are used to model in- and outputs from or to the environment of the logical architecture, like data from sensors (e.g.,, the *GPS* `Port` at the `Component` *SensorAcquisition*) or sending commands to actuators (e.g., via the *Actuator* `Port` or the *ActuatorCtrl* `Component`).

Figure 4.3 depicts the annotated properties of the navigation application's `Component`. Since the `Components` related to the path planning are safety-relevant due to their direct impact on the maneuvers of the vehicle, their annotated Automotive Safety Integrity Level (ASIL) value is high (ASIL D), whereas the visualization function (HMI) is considered as uncritical (QM). Furthermore, the figure illustrates that further non-functional properties can be annotated to logical components (here: memory consumption). As it will be pointed out in Section 4.5.1, parameters that depend on the mapping of a logical `Component` to an `ExecutionUnit` provided by the platform (see Section 4.2.2) are described by the `Deployment` metamodel.

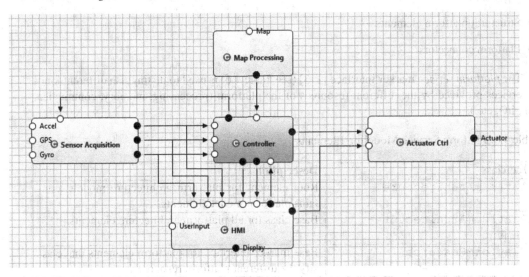

Figure 4.2: Model of Automotive Navigation Application

Model Element	Comment	Memory	Safety Level
Navigation App		0	ISO26262
Actuator Ctrl		1024	ASILD
Controller		20480	ASILD
HMI		524288	QM
Map Processing		15728640	ASILD
Sensor Acquisition		4096	ASILD

Figure 4.3: Annotations of Navigation Application Components

4.2.2 Technical Modeling Viewpoint

4.2.3 Platform Architecture Modeling Framework

The *technical viewpoint* is based on a platform architecture modeling framework that extends the hardware model introduced in [111] to enable the hierarchical description of the architecture and services of MCS platforms, incorporating the platform taxonomy introduced in [113]. It is composed of two complementing sub-viewpoints for the *physical platform*, and the *system software* abstracting the underlying hardware and providing services such as Time and Space Partitioning (TSP). Both sub-viewpoints base on a *generic platform modeling framework* that provides a number of base classes and marker interfaces serving as basic language primitives to define metamodels for particular platforms.

The modeling framework is structured into four groups of elements, which we describe in the following:

- Platform service marker interface

- Platform element type marker interface

- Structural platform elements

- Platform connectors

The *platform service* marker interface [113] (see Table 4.6) is used to distinguish different kinds of services provided by the different resources of the platform (processing, memory, communication, I/O, etc.).

Table 4.6: Platform Service Modeling Elements

Metaclass	Description
PlatformArchitecture	Root element for platform architecture models, contains platform model elements.
IPlatformArchitecture-Element	Base class for all platform architecture elements.
IPlatformResource	Base marker interface for platform elements that classify the different platform resources.
IPlatformCommunication-Resource	Interface to mark communication resources (i.e., resources that move data in system).
IPlatformProcessing-Resource	Interface to mark processing resources (i.e., resources that support the execution of software).
IPlatformMemoryResource	Interface to mark memory resources (i.e., resources that support the storage of data).
IPlatformIOResource	Interface to mark I/O resources that interface the platform to its environment.

The *platform element type* marker interface is described in Table 4.7. It enables to distinguish the element type of derived meta-classes such as *physical platform* architecture elements (with refinements *box*, *board*, *chip* and *IP-core*), *software platform* architecture elements (with refinements *middleware*, *virtualization platform* and *operating system*) and *logical* platform architecture elements (for grouping).

Table 4.7: Platform Element Type Modeling Elements

Metaclass	Description
PlatformArchitecture-ElementType	Marker interface to specify the type of platform architecture elements.
ILogicalPlatform-ArchitectureElement	Model element is a logical grouping.
IPhysicalPlatform-ArchitectureElement	Base marker interface for platform elements implemented in hardware.
IBoxPlatform-ArchitectureElement	Marker interfaces for "boxes", i.e., electronic devices hosting one or more computer systems (ECUs).
IBoardPlatform-ArchitectureElement	Marker interface for electronic circuit boards (hosting multiple chips).
IChipPlatform-ArchitectureElement	Marker interface for electronic chips that can host multiple hardware IP components in a single package.
IIpCorePlatform-ArchitectureElement	Marker interface for hardware IP component (may contain chip IP components).
ISoftwarePlatform-ArchitectureElement	Marker interface for platform architecture elements implemented in software.

`IVirtualization- Platform-ArchitectureElement`	Marker interface for software platform architecture elements that provide a virtualization layer of the underlying hardware.
`IOperatingSystem-Platform-ArchitectureElement`	Marker interface for operating systems and their subcomponents.
`IMiddlewarePlatform-ArchitectureElement`	Marker interface for middleware components (i.e., platform architecture elements implemented in software that belong into neither the virtualization nor the operating system layer).
`PlatformArchitecture-ElementGroup`	Logical group of platform architecture elements.

Structural elements in derived platform metamodels build on the base classes described in Table 4.8, which are based on the hierarchical element metamodel introduced in Section 4.1.2.

Table 4.8: Platform Structural Element Modeling Elements

Metaclass	Description
`ExecutionUnit`	Base class for execution units, i.e., platform elements that allow the execution of software.
`TransmissionUnit`	Base class for transmission units, i.e., communication platform elements that allow the transmission of data (e.g., busses, networks, etc.).
`GatewayUnit`	Base class for gateways units, i.e., dedicated communication platform elements that move data between transmission units residing at different levels of the platform architecture.
`MemoryUnit`	Base class for memory units (e.g., RAM, ROM resources).
`GenericPlatformUnit`	Placeholder for generic platform elements (e.g., custom IP blocks) that are not described by any of more specific base classes.

Platform connectors (see Table 4.9) are attached to the above structural elements, and are specialized into *transmitters*, *receivers*, and *transceivers* (depending on the direction of the information flow). They can be decorated with different marker interfaces, such as a *communication role* marker interface (depicts if the port is a bus *master* or *slave*), or connector type (*port* or *interface*). A *connection* between two platform connectors links two structural elements.

Table 4.9: Platform Connector Modeling Elements

Metaclass	Description
`PlatformConnectorUnit`	Base class for connectors of platform architecture elements.
`Transmitter`	Platform connector that supports only outbound traffic.
`Receiver`	Platform connector that supports only inbound traffic.
`Transceiver`	Platform connector that supports both inbound and outbound traffic.

`TransmissionConnection`	Connection between platform connector units of two platform architecture elements. It should be noted that the `TransmissionConnection` is a purely logical link that is used to model the connection of any platform architecture elements. All required attributes are described in the corresponding platform architecture elements and platform connector units. If not noted otherwise, `TransmissionConnections` are undirected (despite the fact that they inherit the source and target attributes from the `IConnection` interface).
`ICommunicationRole`	Marker interface to specify which role a platform element takes in the communication.
`ICommunicationMaster`	Marker interface to specify that platform element is a communication master that actively initiates the communication.
`ICommunicationSlave`	Marker interface to specify that platform element is a communication slave that can accept communication requests from communication masters.
`IPlatformConnectorType`	Marker interface to further classify the type of platform connector units.
`IPlatformPort`	Platform connector unit is a port that can be connected to / that can implement a given platform interface.
`IPlatformInterface`	Platform connector unit is an interface that can be implemented by platform ports.
`IPlatformExport`	Platform connector unit exports services for use at the parent level.
`IGenericPlatform-SourceConnector`	Generic platform (source) connector used to connect platform elements where the interconnect has no special semantics.
`IGenericPlatform-TargetConnector`	Generic platform (target) connector used to connect platform elements where the interconnect has no special semantics.

To derive a metamodel for a concrete platform, dedicated meta-classes are defined by inheriting from the base classes for structural elements or platform ports and the marker interfaces introduced above. For example, a meta-class for a processor core extends *execution unit*, and inherits the corresponding interfaces to mark it as a *physical platform architecture element*, *processing resource*, *communication master*, and *IP core*. The taxonomy induced by these abstract modeling concepts is useful to provide generic implementations of algorithms (e.g., a safety analysis), and is suitable to build a library of meta-classes for elements of platform architectures.

However, a finite meta-type system cannot be used to define the composition rules for a possibly infinitely large set of architectural styles that define rules of how basic platform elements can be combined (e.g., DREAMS architectural style, see Chapter 2). While this limited expressiveness of type systems has motivated the introduction of constraint languages such as OCL [117], our approach foresees a dedicated *platform domain* marker interface. Table 4.10 depicts the generic part of the platform domain interface that is specialized for each architectural style.

Table 4.10: Platform Domain Base Model Elements

Metaclass	Description
IArchitectureDomain	Marker interface to specify a platform architecture domain of hierarchical platforms. Platforms / platform element libraries must provide concrete domains (and derive its platform elements from these domains), as well as a programmatic implementation of the compositions rules that define the composability of the different domains.
IPlatformDomain	IArchitectureDomain depicting the PlatformArchitecture itself.

For the DREAMS platform, the domains *cluster, node, tile, off-chip network, network-on-chip* and *hypervisor* have been defined and will be introduced in Section 4.2.4. The actual composition rules are encoded as program logic in the modeling tool (based on the compositor framework provided by the AF3 tool [116] that is used for the prototype implementation).

4.2.4 DREAMS Platform Metamodel

4.2.4.1 Cluster Domain

As defined in Chapter 2, the purpose of the elements at the cluster domain is to provide a logical grouping of physically distributed computer systems (which are modeled at the node domain, see Section 4.2.4.2). Hence, the cluster domain elements are modeled as logical elements using the base marker interface ILogicalPlatformArchitectureElement (see Section 4.2.3).

Table 4.11 provides an overview of the model elements defined to describe the cluster level of DREAMS platforms.

Table 4.11: DREAMS Cluster Metamodel

Metaclass	Description
IClusterDomain	IArchitectureDomain identifying model elements of the cluster domain.
ClusterDomainElement	Base class for structural elements of the cluster domain.
Cluster	A DREAMS cluster, i.e., a (logical) group of nodes that are connected via an off-chip network (see Section 4.2.4.2).
OffChipNetworkGateway	GatewayUnit providing a bridge between the OffChipNetworks of connected Clusters.

Clusters are modeled as ExecutionUnits, and hence they (or, model elements in their offspring, respectively) are deployment targets (see Section 4.2.4.7) for software, which is described using logical components (see Section 4.2.1.1).

The communication between the Clusters and the OffChipNetwork- Gateways is realized by OffChipNetworkInterfaces and OffChip- NetworkPorts (see Section 4.12). The ports from the node metamodel are reused in the cluster metamodel since it provides only a logical grouping.

The mode of communication is modeled as bidirectional (base class Transceiver of OffChipNetworkPort and OffChipNetwork- Interface) with masters actively initiating the communication (marker interface ICommunciationMaster).

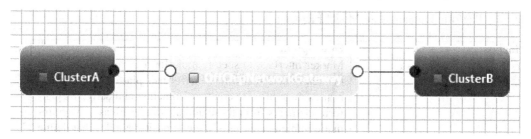

Figure 4.4: Example Model of the Cluster Level of a DREAMS Platform

In Figure 4.4, a simple model consisting of two `Clusters` can be seen. Both `Clusters` are connected via an `OffChipNetworkGateway`. The `Clusters` have attached `OffChipNetworkPorts` (connectors represented by black circles) which are each connected to an `OffChipNetworkInterface` of the contained `OffChipNetwork` and to the `OffChipNetworkInterface` of the `OffChipNetworkGateway` element (see Section 4.2.4.2). Thus, a connection between the internal `OffChipNetworks` of *Cluster_A* and *Cluster_B* is modeled.

4.2.4.2   Node Domain

The node metamodel is used to model the DREAMS node level, i.e., to define the internals of a single DREAMS cluster. Hence, a model at the node level describes the structural elements and the topology of a physically distributed computer system.

Table 4.12 provides an overview of the metaclasses defined for the node metamodel.

Table 4.12: DREAMS Node Metamodel

Metaclass	Description
INodeDomain	IArchitectureDomain identifying model elements of the node domain.
NodeDomainElement	Base class for structural elements of the node domain.
NodeDomainConnector	Base class for node domain IPlatform-ConnectorUnits.
Node	A DREAMS node, i.e., electronic control unit (or computer) hosting a multi-core chip containing tiles connected by a network-on-chip.
OffChipNetwork	An off-chip network to interconnect multiple nodes.
OffChipNetworkPort	Off-chip communication port of structural elements at the node level (Nodes, OffChipClusterGateways).
OffChipNetworkInterface	Communication interface of an OffChipNetwork.
PowerSupply	Model element of an individual (independent) power supply.
PowerOut	NodeDomainConnector attached to PowerSupply for connecting Nodes.
PowerIn	NodeDomainConnector allowing to connect power supplies to Nodes.

The node metamodel uses the following concepts from the platform modeling framework introduced in Section 4.2.3:

- The base marker interface `IBoxPlatformArchitectureElement` of `NodeDomainElement` and `NodeDomainConnector` indicates that the system entities modeled by the node domain are electronic devices that provide a dedicated housing.

- The structural elements `Node`, `OffChipNetwork` and `OffChip- ClusterGateway` are hierarchic model elements (see Table 4.2).

- Nodes are modeled as `ExecutionUnits`, and hence they (or, model elements in their offspring, respectively) are deployment targets (see Section 4.2.4.7) for software, which is described using logical components (see Section 4.2.1.1).

- Likewise, `OffChipNetworks` (modeled as `TransmissionUnits`), and `OffChipClusterGateways` (modeled as `GatewayUnits`), are part of the communication facilities of a DREAMS system.

- In the node metamodel, communication is modeled as bidirectional (base class `Transceiver` of `OffChipNetworkPort` and `OffChip- NetworkInterface`) with masters actively initiating the communication (marker interface `ICommunciationMaster`). Here, `OffChipNetworkPorts` constitute the interface of `Nodes` and `OffChipClusterGateways` to the `OffChipNetwork` (whose interface is modeled by `OffChipNetworkInterfaces`).

An exemplary model from the `Node` domain is shown in Figure 4.5 that illustrates the internal structure of a `Cluster`. The example consists of two `Nodes`, one `OffChipNetwork`, and two `PowerSupplys`. The `OffChipNetwork`, which represents e.g., a TTEthernet (see Section 8.1.1) network or an EtherCAT (see Section 8.1.2) network, has three attached `OffChipNetworkInterfaces`. Three `OffChipNetworkPorts` (represented by black connectors) at the `Nodes` and at the right-hand side of the `OffChipClusterGateway` are connected to these `OffChipNetworkInterfaces`. The `NetworkInterface` located at the left side of the `OffChipNetwork` in the example is connected to an `OffChipNetworkPort` of the containing `Cluster`.

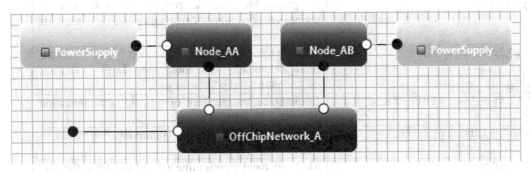

Figure 4.5: Example Model of the Node Level of a DREAMS Platform

As pointed out in Section 4.2.4.1, the `Cluster`'s `OffChipNetworkPort` can be connected to the `OffChipNetworkInterface` of an `OffChipNetworkGateway`. Since `Clusters` only represent a logical grouping of platform elements, the `OffChipNetworkGateway` (that is used to describe the connection of the off-chip networks of two different clusters) resides at the cluster-domain. In contrast to that, `OnChipOffChipGateways` (see Section 4.2.4.3) for an example and `NetworkInterfaces` (see Section 4.2.4.5 for an example) are used to route communication from different levels of the architecture. Hence, `OnChipOffChipGateways` and

`NetworkInterfaces` reside at the lower of the two architecture levels that are connected by them (tile-domain, and processor-domain, respectively) and their interface to the containing architecture level is expressed using specializations of `IPlatformExport` (`OnChipOffChipExport` and `OnChipNetworkExport`, respectively).

Each of the two `Nodes` in the example is connected to an independent `PowerSupply`. The connection is established via `PowerOuts` at the `PowerSupplys` and `PowerIns` attached to the `Nodes`. The information about the power supply of `Nodes` can be considered during safety analysis (e.g., shared vs. separated power supply).

4.2.4.3   Tile Domain

The metamodel described in this section is used to model the DREAMS tile level, i.e., to model the internals of a single DREAMS node. Hence, a model at the tile level describes the structural elements of a multi-processor system on-chip whose elements are interconnected by an on-chip network.

Table 4.13 shows the metaclasses that have been defined at the tile level.

Table 4.13: DREAMS Tile Metamodel

Metaclass	Description
`ITileDomain`	The `IArchitectureDomain` identifying model elements of the tile domain.
`TileDomainElement`	Base class for structural elements of the tile domain.
`TileDomainConnector`	Base class for tile domain `IPlatform-ConnectorUnits`.
`Tile`	A DREAMS tile, i.e., a multi-core or single-core processing unit that is connected to the `OnChipNetwork` via its `OnChipNetworkPort`.
`OnChipNetwork`	An on-chip network to connect multiple `Tiles`.
`OnChipOffChipGateway`	A gateway from the on-chip to the off-chip level.
`IpCore`	A placeholder for a generic IP core that is connected to the `OnChipNetwork` via its `OnChipNetworkPort`.
`OnChipNetworkPort`	On-chip communication port of structural elements at the tile level (`Tiles`, `OnChipOffChipGateways`).
`OnChipNetworkInterface`	Communication interface of on-chip communication network.
`OnChipOffChipExport`	It is required to model the communication routes to other nodes.
`WatchDog`	Model element representing a watchdog timer that can trigger a reset of connected elements that fail to reset the watchdog timer in time and hence are considered to be in a "failed" state.
`WatchDogIn`	`IConnector` to be attached to elements which shall be monitored by a `WatchDog`.
`WatchDogOut`	`IConnector` at the `WatchDog` to which monitored elements can be connected.
`Clock`	Model element that represents clock sources.
`ClockIn`	`IConnector` of the model element to which a clock signal shall be provided.

ClockOut	IConnector at the clock source from which a clock signal is emitted.
GeneralPurposeInput	IConnector representing a digital input port of the respective model element.
GeneralPurposeOutput	IConnector representing a digital output port of the respective model element.

The tile metamodel uses the following concepts from in the platform modeling framework introduced in Section 4.2.3:

- The base marker interface IIpCorePlatformArchitectureElement of TileDomainElement and TileDomainConnector indicates that the system entities modeled by the tile domain are IP cores that possibly are contained in the same package.

- The structural elements Tile, IpCore, OnChipNetwork and OnChipOffChipGateway are hierarchic model elements (see Section 4.1.2).

- Tiles are modeled as ExecutionUnits, and hence they (or, model elements in their offspring, respectively) are deployment targets (see Section 4.2.4.7) for software, which is described using logical components (see Section 4.2.1.1).

- Likewise, OnChipNetworks (modeled as TransmissionUnits), and OnChipOffChipGateways (being modeled as GatewayUnits are part of the communication facilities of a DREAMS system.

- The mode of communication is modeled as bidirectional (base class Transceiver of OnChipNetworkPort and OnChipNetwork- Interface) with masters actively initiating the communication (marker interface ICommunciationMaster). Here, OnChipNetworkPorts constitute the interface of Tiles, IpCores and OnChipOffChipGateways to the OnChipNetwork (interface modeled by OnChipNetworkInterface). As mentioned above, in addition to OnChipNetworkPorts, also OnChip- OffChipExports can be attached to OnChipOffChipGateway. Then, the route to the off-chip communication can be described using a link from the OnChipOffChipExport to the OffChipNetworkPort owned by the Node that contains the respective OnChipOffChipGateway.

- WatchDogs and Clocks can be connected to Tiles to model different clock domains and the monitoring of tiles, which is especially relevant for safety analysis. Note that each WatchDog must be connected to a Clock since a clock signal is required to implement the watchdog timer.

Figure 4.6 shows an exemplary model at the Tile-level, i.e., the internal structure of a Node. There are two Tiles, one OnChipNetwork, and an OnChipOffChipGateway. The white connectors attached to the OnChipNetwork represent the OnChipNetworkInterfaces. Likewise, the connectors attached to the Tiles and at the right-hand side of the OnChipOffChipGateway represent OnChipNetworkPorts that are connected to the corresponding OnChipNetworkInterfaces of the OnChipNetwork. The OnChipOffChipGateway depicts the gateway of the OnChipNetwork to the OffChipNetwork at the Node layer. The connector in the very bottom left of the figure represents the OffChipNetworkPort of the Node that contains the model shown in Figure 4.6. It is connected to an OnChipOffChipExport (left connector of OnChipOffChipNetworkGateway) that is used to model the connection to the containing

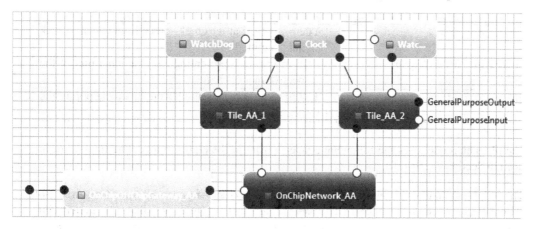

Figure 4.6: Example Model of the Tile Level of a DREAMS Platform

Table 4.14: DREAMS NoC Metamodel

Metaclass	Description
INocDomain	The `IArchitectureDomain` identifying model elements of the NoC domain.
NocDomainElement	Base class for structural elements of the NoC domain.
NocDomainConnector	Base class for `IPlatformConnectorUnits` of the NoC domain.
NocRouter	A router of the `OnChipNetwork`.
NocInputUnit	An input unit of a `NocRouter` of the `OnChipNetwork`.
NocOutputUnit	An output unit of a `NocRouter` of the `OnChipNetwork`.

Node's `OffChipNetworkPort`. In the example, a common `Clock` source provides a clock signal via `ClockOuts` to the connected `Tiles` that receive the signal via attached `ClockIns`. Furthermore, `WatchDogs` are connected to the two present `Tiles` via `WatchDogOuts` (at the `WatchDogs`) and `WatchDogIns` (at the `Tiles`). Since `WatchDogs` are essentially timers, they require a clock signal and, hence, they are connected to the `Clock` that provides the signal to the `Tiles`. The `Tile` *Tile_AA_2* additionally has an attached `GeneralPurposeOutput` port and a `GeneralPurposeInput` port modeling the generic GPIOs of processors or boards.

4.2.4.4 NoC Domain

The NoC metamodel (see Table 4.14) is used to describe the internals of OnChipNetworks (see Section 4.2.4.3).

The NoC metamodel uses the following concepts from in the platform modeling framework introduced in Section 4.2.3:

- The base marker interface `IIpCorePlatformArchitectureElement` of `NodeDomainElement` and `NodeDomainConnector` indicates that the system entities modeled by the node domain are IP cores that possibly are contained in the same package.

- `NocRouters` are modeled as `TransmissionUnits` and constitute the most fine-grained level in the metamodel of the DREAMS communication facilities.

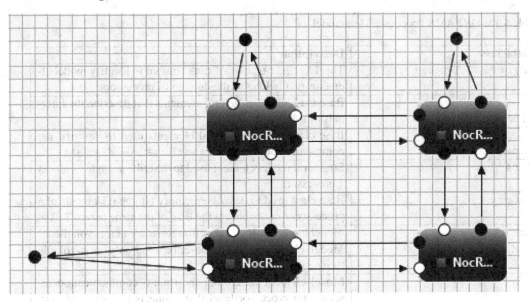

Figure 4.7: Example Model of the NoC Level of a DREAMS Platform

- The internal structure of an on-chip network is modeled using directed `TransmissionConnections` between the `OnChipNetworkInterfaces` of the `OnChipNetwork` and the `InputUnits` / the `OutputUnits` of the `NocRouters` contained by the `OnChipNetwork`.

Figure 4.7 shows an exemplary model at the NoC-Domain, i.e., the internal structure of an `OnChipNetwork`. The black connectors at the left and at the upper side of the figure represent the `OnChipNetworkInterfaces` of the containing `OnChipNetwork`. The example contains four `NocRouters` whose communication interfaces are represented by `Input`- and `OutputUnits`. The `InputUnits` are represented by the white connectors, while the black connectors attached to `NocRouters` are `OutputUnits`, respectively. Note the internal structure of the `OnChipNetwork`, i.e., the interconnection between the different `NocRouters` is modeled using directed connections (arcs) which allows to model complex communication topologies for `OnChipNetworks`. For instance, this can be used to segregate the communication of the platform components (e.g., `Tiles`) connected to the corresponding `OnChipNetworkInterfaces` into different classes. The topology depicted in the simple example in Figure 4.7 does not impose any restrictions onto the communication flow between Tiles connected to the corresponding `OnChipNetworkInterfaces`, but provides redundant communication routes. It should be noted that on all other levels of the platform metamodel, communication links are modeled as undirected connections (edges). Hence – unlike `InputUnits` and `OutputUnits` – `OnChipNetworkInterfaces` are modeled as bidirectional communication elements.

4.2.4.5 Processor Domain

The processor metamodel is used to model the DREAMS processor level, i.e., the internals of a DREAMS tile. Thus, the system elements of this package include busses, cores, memories and network interfaces. Hence, it is possible to describe multicore processors whose cores are connected via a bus and are able to access the `OnChipNetwork` using a `NetworkInterface` that is connected to the bus. Table 4.15 depicts the metaclasses defined in the processor metamodel:

Table 4.15: DREAMS Processor Metamodel

Metaclass	Description
IProcessorDomain	The IArchitectureDomain used to identify model elements that belong to the domain of processors.
ProcessorDomain-Element	Base class of the structural elements that describe the internals of a Tile.
ProcessorDomain-Connector	Base class of the structural elements used to describe the communication of ProcessorDomain- Elements.
Core	Structural model element used to describe a single Core of a processor.
Memory	Base class used to describe memory (storage) elements of a processor. Accessible via the Bus of the same parent Tile.
RAM	Model element used to describe Memory that is volatile.
ROM	Model element used to describe read-only, non-volatile Memory.
Bus	Model element describing the (main) communication resource on processor level that connects Cores, Memory, and NetworkInterfaces.
NetworkInterface	Model element that connects the processor elements to the OnChipNetwork via a BusOnChipNetworkExport to which this Tile is connected.
BusMasterInterface	Model element to describe interfaces of a processor Bus that is capable of handling bus master arbiters.
BusSlaveInterface	Model element to describe interfaces of a processor Bus that is only capable of serving slave devices.
BusMasterPort	The port of a ProcessorDomainElement that is connected to a BusMasterInterface of a processor Bus. The ProcessorDomainElement must be capable of fulfilling the role of a Bus master.
BusSlavePort	The port of a ProcessorDomainElement that is connected to a BusSlaveInterface of a processor Bus. The ProcessorDomain- Element cannot take over the master role at this Bus.

Application of concepts from platform modeling framework (see Section 4.2.3):

- The base marker interface IIpCorePlatformArchitectureElement of ProcessorDomainElement and ProcessorDomainConnector indicates that the system entities modeled by the tile domain are IP cores that possibly are contained in the same package.

- The structural elements Core, Memory, Bus, and NetworkInterface are hierarchic model elements (see Section 4.1.2).

- Cores are modeled as ExecutionUnits, and hence they are possible deployment targets (see Section 4.2.4.7) for software, which is described using logical components (see Section 4.2.1.1).

  Nevertheless, the typical lowest deployment granularity within a DREAMS architecture will consider Partitions (see Section 4.2.4.6) as deployment targets. Those will be executed on top of Cores and within Hypervisors providing the middleware between both model elements.

- Likewise, `Buses` are modeled as `TransmissionUnits`, and `Network- Interfaces` are modeled as `GatewayUnits`, which both are part of the communication resources of a DREAMS system.

- Furthermore, `Memory`, which appears at this level in the form of RAM and ROM, is modeled as a `MemoryUnit`.

- The mode of communication is modeled as bidirectional (base class `Transceiver` of `ProcessorDomainConnector`) with masters actively initiating the communication (marker interface `ICommunciationMaster`). Here, `BusMaster-` and `BusSlavePorts` constitute the interface of `Cores`, `Memorys` and `NetworkInterfaces` to the `Bus` whose interfaces are modeled as `BusMaster-` and `BusSlaveInterfaces`. As mentioned above, `BusOnChipNetworkExports` can be attached to `NetworkInterfaces` in addition to `BusMaster-` and `BusSlavePorts`. Then, the route to the off-chip communication can be described using a link from the `OnChipOffChipExport` to the `OffChipNetworkPort` owned by the `Node` that contains the respective `OnChipOffChipGateway`.

- The communication role (master or slave) at this architecture level is especially important considering the Bus architecture where one device must have absolute control over the communication. Otherwise, interfering access would render any information on the `Bus` unusable.

Figure 4.8 contains an exemplary model at the processor domain, i.e., the internals of a `Tile`. The model contains two `Cores`, one RAM and one ROM `Memory`, a Bus, and a `NetworkInterface` (NI). All mentioned elements are connected via the `Bus`. The ports (black connectors) attached to the `Cores` and to the `NetworkInterface` are `MasterPorts` since they need to be able to initiate communication via the `Bus`. These `MasterPorts` are connected to `BusMasterInterfaces`, and thus, the model elements mentioned above are able to communicate. In contrast, the `Memory` elements are connected via `BusSlavePorts` to the `BusSlaveInterfaces` of the Bus, as these elements do not initiate any communication (passive elements). The left-hand side of the `NetworkInterface` is a model of the gateway to the `OnChipNetwork` at containing layer (i.e., to the `Node` layer). The left-most black connector is the `OnChipNetworkPort` of the `Tile` that contains the discussed example model. This port is connected to the `NetworkInterface`'s `OnChipNetworkExport` (left connector of `NetworkInterface` component) that depicts the interface of the processor domain to the on-chip-network. As a result (and also considering the metamodels of the other levels of the DREAMS architecture discussed in the previous sections), the model describes that there is a possible communication route from the two `Cores` shown in Figure 4.8 to resources located in other `Tiles` (via the `OnChipNetwork`) or `Nodes` (via `OnChipNetworks` and `OffChipNetworks`). Likewise, the model contains the relevant information to determine routes to the `Cores` and the `Memorys` from Figure 4.8 from remote resources.

4.2.4.6 Hypervisor Domain

The metamodel described in this section is summarized in Table 4.16. It is used to model hypervisors within system software layer of a DREAMS system. The model of the system software layer is instantiated in a separate `PlatformArchitecture` that is linked to the model of the physical platform layer using the `ResourceLink` annotation (see Section 4.2.4.7).

Table 4.16: DREAMS Hypervisor Metamodel

Metaclass	Description
`IHypervisorDomain`	The `IArchitectureDomain` to identify model elements belonging to the domain of hypervisors.

`IVirtualizationLayer-Domain`	The `IArchitectureDomain` to identify model elements providing virtualization services.
`Hypervisor`	Class representing a hypervisor, i.e., a system software layer module that virtualizes `ExecutionUnits` of the physical platform (e.g., a processor (`Tile`)). The virtualized physical resources are designated by the `ResourceLink` annotation (see Section 4.2.4.7).
`HypervisorDomain- Element`	Base class for structural elements that are attached to `Hypervisors` or that are sub-elements of `Hypervisors`.
`HypervisorDomain-Connector`	Base class for describing communication structure of `HypervisorDomain- Elements`.
`Partition`	Isolated and virtualized execution environment for software components provided by a `Hypervisor`. Using the `ResourceLink` annotation, it is linked to `ExecutionUnits` of the physical platform resource to which its containing `Hypervisor` is linked (e.g., `Cores` of the corresponding `Tile`).
`OnChipNetworkDriver`	Model element representing a system partition of a `Hypervisor` that has access to the `OnChipNetwork` resource of the physical platform layer (referenced using the `ResourceLink` annotation).
`InterPartitionCom`	Class to express communication facility provided by the `Hypervisor` that provides message exchange between `Partitions`.
`InterPartitionComPort`	Communication port of virtual structural elements, i.e., `Partitions`.
`InterPartitionCom-Interface`	Communication interface located at `InterPartitionCom` that provides the inter-partition communication service.
`MemoryArea`	Model element used to represent memory areas assigned to `Partitions` or to `Hypervisors`. A `Partition` can have one or more assigned `MemoryAreas`, and a `MemoryArea` can be shared by multiple partitions. `MemoryAreas` assigned to `Hypervisors` have a 1:1 relation. Each `MemoryArea` is linked to a `MemoryUnit` of the underlying physical platform using the `ResourceLink` annotation.
`MemoryRequirement`	`HypervisorDomainConnector` that is attached to `Partitions` or `Hypervisors` to model their need of and the connection to an allocated `MemoryArea`.
`MemoryConnector`	`HypervisorDomainConnector` that provides access to MemoryAreas.

HealthMonitor-Configuration	Model element of the health status self-monitoring capabilities of `Hypervisors`. It can be connected to `Hypervisors` and parametrized by annotations to model the configuration of a health monitor.

The hypervisor metamodel applies the following concepts from the platform modeling framework described in Section 4.2.3:

- The base marker interface `IVirtualizationPlatformArchitecture-` Element of `Hypervisor`, `HypervisorDomainElement`, and `HypervisorDomainConnector` indicates that these system elements are part of the virtualization layer of the DREAMS system.

- Additionally, `Hypervisor` inherits from the base marker interface `ILogicalPlatformArchitectureElement` that indicates that this system element is a logical entity, i.e., it has realization in hardware.

- The structural elements `Partition`, `OnChipNetworkDriver`, and `InterPartitionCom` are hierarchic model elements (see Section 4.1.2).

- `Partitions` and `Hypervisors` are modeled as `ExecutionUnits`, and hence they are (possible) deployment targets (see Section 4.2.4.7) for software which is described using logical components (see Section 4.2.1.1).

- Likewise, `OnChipNetworkDrivers` are modeled as `GatewayUnits`, and `InterPartitionComs` are modeled as `TransmissionUnits`, both being part of the communication facilities of a DREAMS system.

- The communication within the `IHypervisorDomain` is modeled being bidirectional (base class `Transceiver` of `HypervisorDomainConnector`) with masters actively initiating the communication (marker interface `ICommunciationMaster`). Here, `InterPartitionComPorts` constitute the interface of `Partitions` and `OnChipNetworkDrivers` to the `InterPartitionCom` that is provided by the `Hypervisors`. As mentioned above, `InterPartitionOnChipNetworkExports` can be attached to `OnChipNetworkDrivers` in addition to `InterPartitionComPorts`. Then, the route to the off-chip communication can be described using a link from the `OnChipOffChipExport` to the `OffChipNetworkPort` owned by the `Node` that contains the respective `OnChipOffChipGateway`.

Hypervisor virtualize a processor, which in our metamodel is represented by an additional `PlatformArchitecture` that hosts the `Hypervisor` model elements that are linked to the corresponding `Tiles` of the physical platform.

An exemplary system software `PlatformArchitecture` that illustrates an instance of the Hypervisor metamodel is shown in Figure 4.9. In the example, two `Hypervisors` have been deployed to each of the `Tiles` contained by *Node_AA*. The mapping of a `Hypervisor` to the corresponding `Tile` is represented by a `ResourceLink` annotation that is bound to the `Hypervisor` instance (see Section 4.2.4.7). The structure above the `Hypervisors` reflects the structure of the referenced physical platform architecture, i.e., the node and cluster level is mirrored by corresponding logical `PlatformArchitectureElementGroup` model elements (represented by gray model elements). The internal model of the `Hypervisor` includes `Partitions`, `MemoryAreas`, `InterPartitionComs` and `OnChipNetworkDrivers`. As pointed out above, `Partitions` are modeled as `ExecutionUnits` of the system software layer that are linked to `Cores` that are contained in the `Tile` to which the `Hypervisor` is linked.

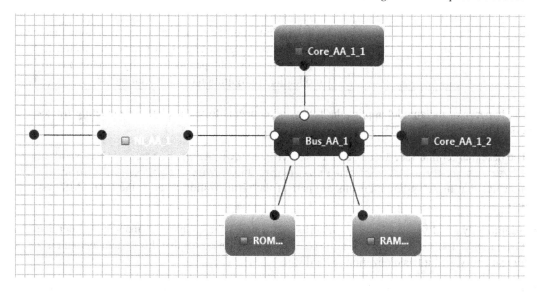

Figure 4.8: Example Model of the Processor Level of a DREAMS Platform

The `Hypervisor`'s partition-to-partition communication facility that enables the exchange of messages between the partitions hosted by the same hypervisor instance is represented by the `InterPartitionCom` model element. The connection is established by connecting `InterPartitionComPorts` (black connector) of `Partitions` with `InterPartitionComInterfaces` (white connectors) attached to `InterPartitionCom` model elements. Furthermore, system partitions such as the `OnChipNetworkDrivers` can be connected to `InterPartitionComs` by which the access of the `Hypervisor` to the `OnChipNetwork` is modeled. The resource mapping of these system partitions is again described using `ResourceLink` annotations (i.e., the `OnChipNetwork` hosted by the `Tile` to which the given `Hypervisor` is linked).

Finally, the access of partitions to physical memory resources is described using `MemoryAreas` that can be assigned to one or more `Partitions`. The `MemoryAreas` from the example are linked to the `RAM` resource hosted by the `Tile_AA2` using `ResourceLink` annotations.

4.2.4.7 Technical Architecture Annotations

Table 4.17 lists the annotations that are used to define extra-functional properties and additional parameters for the respective elements of a platform model.

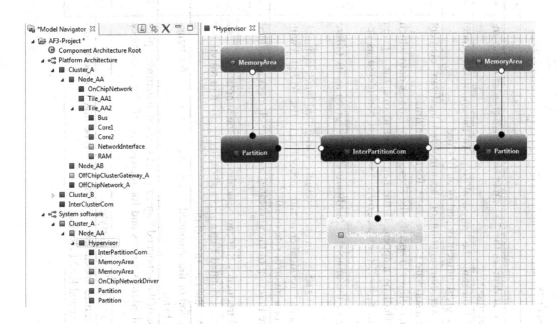

Figure 4.9: Example Model of the Hypervisor/Partition Level of a DREAMS Platform

Table 4.17: DREAMS Platform Architecture Annotations

Metaclass	Description	Model Elements
ArchitectureDomainLabel	Returns a label that denotes the IArchitectureDomain of the annotated model element.	all
PlatformArchitectureElementTypeLabel	Returns a label indicating the "physical" type of the annotated model element, like a logical element or part of an IP Core.	all
ExecutionUnitPower	The average power consumption of the annotated hardware element when executing a software Component for a given time.	ExecutionUnits
DeploymentGranularity	Boolean flag that allows to specify the ExecutionUnits onto which Components shall be mapped. If the flag is set to true for a given ExecutionUnit, its ExecutionUnits are not considered as deployment targets.	ExecutionUnits
FailureRate	The failure rate of the annotated hardware element given as its failure probability. In case a TransmissionUnit represents a so-called black channel, this parameter is the "residual error rate" according to IEC 61784-3 with the assumption of a bit error rate of 10^{-2}.	ExecutionUnits, TransmissionUnits
SafeFailureFraction	The Safe Failure Fraction of the annotated hardware element as defined in IEC 61508.	ExecutionUnits, TransmissionUnits
ProcessorSpeed	Maximum CPU frequency that can be achieved by the annotated Core.	Cores
TransmissionUnitBandwidth	Bandwidth of the annotated TransmissionUnit given in Mbyte per second. Describes the raw throughput.	TransmissionUnits
TransmissionUnitPower	Power consumption of the annotated TransmissionUnit for transmitting a single byte.	TransmissionUnits
MemoryAddress	The start address of the annotated MemoryUnit. Used in hardware platforms for global address spaces and for segregation of virtual memory allocations.	MemoryUnits
MemorySize	Capacity of a MemoryUnit in Bytes.	MemoryUnits
RamType	Allows a fine-grained specification of the RAM type that is used to implement the annotated RAM element.	RAMs

`ResourceLink`	Resource requirements (1:n relationship) between platform elements in different layers of the platform, e.g., from elements of the system software layer to elements of the physical platform.	`Tiles`, `Partitions`, `MemoryAreas`
`PartitionFlags`	Flags to be set when configuring the annotated `Partitions` (e.g., system or application partition).	`Partitions`
`HealthMonitor-Configuration`	Actions that shall be triggered by the health monitor of the connected Hypervisor if the defined event occurs (e.g., fault behavior is detected).	`HealthMonitor-Configurations`

4.3   Timing Requirements

The consideration of timing constraints and their verification is often neglected but mandatory in case of safety critical systems. If control orders do not arrive in time at the actuators or are not updated sufficiently often, then the system may get "out of control". The consequences may be damage to the system or harm to people and thus, without considering timing requirements during the development process, it is not possible to design a safe system. However, timing constraints and their verification through prediction techniques such as worst-case analysis or simulation are based on the assumption that in the real system all parts behave as supposed. If at some time an application actually sends its data over a network much more often than foreseen, it may hinder control orders of other applications to arrive in time, which could lead to failures. This is where safety considerations must come into play, in order to establish acceptable failure probabilities, which are then achieved through an appropriate safety design. In this sense, the safety and timing approaches are complementary.

The goal of the TIMMO / TIMMO-2-USE projects [118] was to elaborate a metamodel and a methodology for modeling and verifying timing constraints in the automotive domain. The main results, Timing-Augmented Description Language (TADL2) [114] and the Generic Methodology Pattern (GMP) [119] for timing, are however general enough to allow their application to other domains. For this reason, we decided to integrate into the temporal viewpoint of the DREAMS metamodel many of the TADL2 concepts (see Section 4.3.1). Furthermore, we describe in Section 4.3.2 the GMP and how it is applied in DREAMS.

4.3.1   Temporal Modeling Viewpoint

The metamodel developed in the TIMMO-2-USE project for the description of timing requirements [114], was designed to allow the extension of different systems views (of existing metamodels) with timing related information. This has been applied during the TIMMO-2-USE project to the different system views of the EAST-ADL and AUTOSAR metamodels. The same approach has been applied for DREAMS to extend the logical, technical, and mapping view of the DREAMS metamodel. Different kinds of timing requirements have been defined in TIMMO-2-USE. In the following, we explain those relevant for DREAMS.

The core concept of the timing metamodel is the so-called `TimingEvent`. It denotes a distinct form of state change in a running system, taking place at distinct points in time called occurrence of the event and to which the timing metamodel allows to attach `TimingConstraints` directly, or indirectly through `TimingChains`. Table 4.18 defines the `TimingEvents` needed for the DREAMS project.

Table 4.18: DREAMS Timing Metamodel: Events

Metaclass	Description
`Event`	It is a sequence of times indicating the times that each event occurrence is predicted to occur.
`InputEvent`	Links the timing model elements to component `InputPort`. Attributes: • `ref`: `InputPortAnnotation` of an `InputPort` from the logical component architecture.

OutputEvent	Links the timing model elements to component OutputPort from the logical component architecture. Attributes: • ref: OutputPortAnnotation of an OutputPort from the logical component architecture.
EventTrigger	Links the timing model elements to a Component. Attributes: • ref: ComponentAnnotation from the logical component architecture.

A TimingChain is a container for a pair of TimingEvents that are causally related (see Table 4.19). The "stimulus" event is supposed to trigger actions that lead to the "response" event. It is not necessary to know these actions, just that the "stimulus" event does lead to the "response" event. A TimingChain can be hierarchically decomposed into "segments", which are TimingChains themselves. This allows refining a TimingChain along with the refinement of the system description in the same or a different system view.

Table 4.19: DREAMS Timing Metamodel: Timing Chains

Metaclass	Description
EventChain	An EventChain is a container for a pair of events that must be causally related. Attributes: • id: a string identifier for event chain traceability. • description: a string description. • stimulus: Reference to the Event that stimulates the steps to be taken to respond to this event. • response: Reference to the Event that is a response to a stimulus that occurred before. • segment: Ordered list of references to EventChains in sequence.

Table 4.20 lists the meta-classes that can be used to define temporal constraints on events and timing chains.

Table 4.20: DREAMS Timing Metamodel: Constraints

Metaclass	Description
`TimingConstraint`	`TimingConstraint` is an abstract element. It is not a design constraint but either a requirement or the result of a validation. `TimingConstraint` offers several means to constrain the time occurrences of events. Attributes: • `id`: a string identifier for constraint traceability. • `description`: a string description of the constraint.
`PeriodicConstraint`	A `PeriodicConstraint` describes an event that occurs periodically. Attributes: • `period`: The effective ideal separation between two successive occurrences of the event without jitter. • `jitter`: Describes the local deviation from the strictly sporadic pattern. • `event`: Reference to the event for this constraint.
`SporadicConstraint`	A `SporadicConstraint` describes an event that occurs with a minimum inter-arrival time in between successive occurrences. Attributes: • `minimumDistance`: The effective minimum distance between any two occurrences of the event. • `jitter`: Describes the local deviation from the strictly sporadic pattern. • `event`: Reference to the event for this constraint.
`ReactionConstraint`	A `ReactionConstraint` defines how long after the occurrence of a stimulus a corresponding response must occur. Attributes: • `minimum`: Minimum value of the `ReactionConstraint`. • `maximum`: Maximum value of the `ReactionConstraint`. • `scope`: Reference to the `TimingChain` for this constraint.
`AgeConstraint`	An age constraint defines how long before each response a corresponding stimulus must have occurred. It applies to a `TimingChain`. Attributes: • `minimum`: Minimum value of the `AgeConstraint`. • `maximum`: Maximum value of the `AgeConstraint`. • `scope`: Reference to the `TimingChain` for this constraint.

Figure 4.10 shows an example where the timing viewpoint is instantiated for the expression of the requirements of a braking system. In this example, the following timing requirements are described:

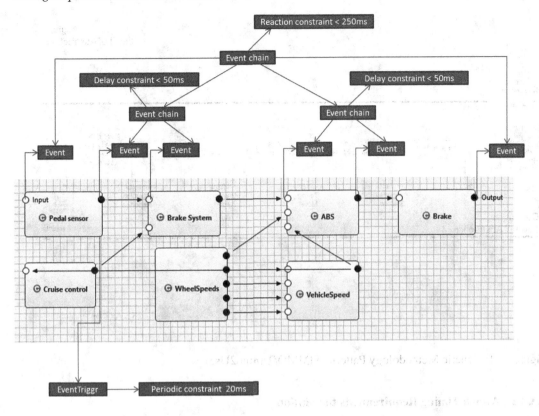

Figure 4.10: Braking System (Timing Viewpoint Illustration)

- End-to-end delay: The vehicle must start decelerating within the driver's reaction time (250ms) after the driver has indicated his wish to do so.

- Timing decomposition: This end-to-end delay is further decomposed into segments allowing time budget allocation between `InputEvent` and `OutputEvent` on `Components`.

- Event-trigger: The `EventTrigger` on the pedal sensor allows the specification of the brake pedal sensing period.

4.3.2   Generic Methodology Pattern

The Generic Methodology Pattern (GMP) is a set of process steps that identify design tasks that are relevant for considering timing constraints and their validation during the development of electronic systems. The GMP consists in a generic sequence of tasks that can be executed at every abstraction level of the development process (see Figure 4.11). These tasks are described in more detail in the following sections.

The natural flow of execution is from higher abstraction levels to low abstraction levels, but the GMP can be performed in top-down or bottom-up directions.

4.3.2.1   Create Solution

This task describes the definition of a solution architecture without any timing information. In the GMP, it is a placeholder for all other design activities at the current design level and system view.

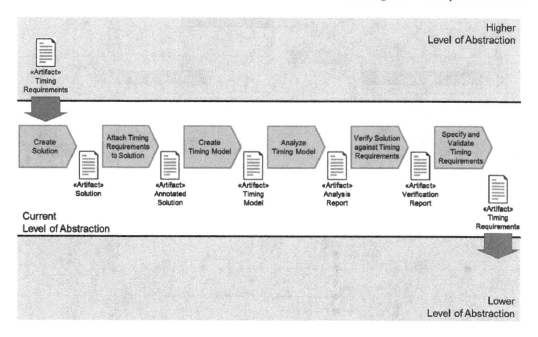

Figure 4.11: Generic Methodology Pattern (TIMMO/Timmo2Use)

4.3.2.2   Attach Timing Requirements to Solution

This task describes the formulation of timing requirements in terms of the current design level and system view. This means that for the input timing-requirements from the previous level, the corresponding entities need to be (re)defined in the current view:

- Definition of `TimingEvents`, attached to the structural entities of the current system view

- Redefinition of `TimingChains` in terms of the translated `TimingEvents`

- Redefinition of `TimingConstraints` with references to the translated `TimingEvents` and `TimingChains`

Let us consider the example of latency constraints in the technical view. As explained in Section 3.4, a latency constraint is based on a "stimulus" and a "response" event. In the technical view, these `TimingEvents` would be those related to the reception and the production of data in component ports.

4.3.2.3   Create Timing Model

This task describes the definition of a formalized model for the calculation of specific timing characteristics based on properties of the current design level and system view.

The goal is to define a model (or several models) that will serve as input for the timing analysis to be performed in the next task. This may mean to define additional information, like the decomposition of a `TimingChain` into segments.

Let us return to the latency constraint example, but this time at mapping level. The work of this task consists in decomposing the end-to-end timing chain into segments that span each a different execution or communication perimeter: execution on a tile, communication over NoC, communication over off-chip networks. This decomposition is generally needed so that worst-case analysis or simulation can be applied in order to evaluate the timing.

4.3.2.4 Analyze Timing Model

This task describes the actual execution and evaluation of all necessary calculations according to the timing model. This task consists of feeding the analysis tool(s) with the timing models built in the previous step, before running the analysis and finally retrieving the results for verification.

4.3.2.5 Verify Solution against Timing Requirements

This task describes the comparison of the obtained analysis results with the specified timing requirements. The simple part of this task consists of comparing analysis results with requirements.

In our latency constraint example, the verification would consist of simply checking that the upper bound on (end-to-end) delays is smaller than the latency constraint.

Notice however that if several kinds of analyses have been performed, the results might first have to be merged, before being able to perform the simple comparison. If two algorithms produce upper bounds on Worst Case Traversal Time (WCTT), and if algorithm 1 provides a bound larger than the constraint, then it can only be concluded that algorithm 1 cannot prove that the WCTT is below the constraint. But if for the same WCTT algorithm 2 provides a bound below the constraint then it can be concluded that the constraint is met. The problem with algorithm 1 is simply that the provided upper bound on WCTT is too pessimistic to allow a positive conclusion. The merging of the analysis results would consist of taking the minimum of the bounds computed by the different algorithms.

Notice also that the main objective of this design task is to decide whether the numbers are good enough for progressing or whether those numbers have to be revised. If not, it is necessary to return back to an earlier development step or level (i.e., iteration). The numbers might not be good enough if slack is needed for future extensions or in order to compensate for lack of precision in the estimation of timing characteristics.

4.3.2.6 Specify and Validate Timing Requirements

This task describes the identification of mandatory timing characteristics and their promotion to timing requirements for the next development phase. It is about stating which timing constraints are the inputs for the next design phase and whose satisfaction implies the satisfaction of the original input requirements of the current step. For example, in the case of latency constraints, one can either decide to only keep the end-to-end constraint or to impose sub-latency constraints (=sub-time budgets), for the different perimeters (e.g., processing, on-chip-, off-chip networks) that are covered by the end-to-end constraint. In the second case, the solution space is reduced, but the global problem is divided into sub-problems with lower complexity.

4.3.2.7 DREAMS Timing Lifecycle Specification

For the consideration of timing requirements throughout the development process, GMP can be mapped to the V-Model based on the safety-related development process of DREAMS.

The GMP can be instantiated at and in between design steps, where entities are considered that consume inputs and/or produce outputs. At a higher level one typically considers a functional architecture and at a lower level, a software architecture, consisting of software modules. Notice that timing requirements cannot apply to software alone, because software needs execution and communication resources. Thus, timing related activities generally concern the combination of software and hardware, unless a function is implemented only in hardware.

Timing requirements can be seen as constraints, imposed in a top-down manner or as properties of existing entities (bottom-up). When working in a top-down manner, one intends to design entities that satisfy the constraints coming from the previous higher level. The timing properties of the resulting entities are also expressed as timing requirements, since they play the role of constraints for (a) the design of the entities at the next lower level and in (b) the tests of the actual implementations in the corresponding timing related activities in the right branch of the V-Model. Only during the verification activities in the left branch of the V-Model, timing requirements belong to a lower level seen as properties that must be satisfied by higher level constraints.

4.4   Safety Management

In this section, we describe the DREAMS *Safety Model* and *Safety Management Tool* that supports the *model-driven engineering process* for MCSs. This model, and its corresponding tool, drives the system's designers towards valid designs from the safety point of view.

The next subsections are organized as follows: First, the *Safety Model* is described in detail. Next, the *Development Process and Safety Solution design* is explained, showing the typical steps followed to define a safety model. Finally, *Verification Rules and Safety Case generation process* is explained, showing how the tool helps designers to discover safety design potential errors and how the tool generates safety checking evidences to aid in the certification process.

4.4.1   Safety Modeling Viewpoint

The DREAMS Safety Compliance Model is associated to the Logical Component Architecture Model and the Platform Architecture Model. It aims at the early detection of errors during the *realization* phase in the V-Life-Cycle of IEC 61508-1 (see Section 10.5.1.1). The DREAMS safety-compliance model mainly addresses the following aspects:

- The architectural specification of a system or family of systems (e.g., family of wind power turbines).

- The process of choosing the specific safety functions architecture of a specific system by resolving variability.

- The deployment of software components into software partitions, software hypervisors, hardware tiles and cores.

The Safety Compliance Model (or Safety Model, for short) is attached to the Logical Component Architecture Model and the Platform Architecture Model of a given product, or product line, and is used as allows:

1. To support the *checking of safety consistency rules* that can help designers to reduce the risk of late discovery of safety related expensive design pitfalls that would prevent a certification.

2. To *help in the certification process by providing some evidences of safety aspects* that have taken into account during the realization phase of the product instance.

The scope of the safety model is mainly the architectural specification phase within the system realization because of the following reasons:

- The analysis, planning, installation and operation of the system do not usually consider the internal implementation details of the system (e.g., multicore partitioning technology); this is only relevant in the realization phase. Therefore, it is interesting to focus the scope on discovering early errors in the realization phase of multicore partitioning scenarios.

- Within the realization phase, the system architectural specification needs to deal with the non-trivial integration of mixed-criticality applications; multiple partitions mapped to multicore platform(s). This is the phase in which the definition of rules for safety consistency checking can provide higher benefits; improving productivity and reducing the risk of late discovery of safety related design pitfalls.

To summarize the scope of the safety metamodel is to support discovering errors during the realization phase of the hardware architecture, with a basic support to discover errors in the integration of software partitions, hypervisors and deployment of components in a mixed criticality multicore scenario. The focus of the approach is on IEC 61508-2 and IEC 61508-3 (but not going in depth in IEC 61508-3).

Although validation tests are a key aspect of the final certification process, they are not covered by the safety compliance model presented here, which scopes mainly the architectural specification for early detection of design errors. However, a final safety case generation for a certification process will have to be completed with validation tests.

The following subsections introduce the most relevant entities defined by the safety model. These entities are used during the DREAMS Development Process of Safety Solution Design (described in Section 4.4.2) to define and verify the safety model for a given product (or family of products).

4.4.1.1 E/E/PE System Safety Requirements Specification

The set of E/E/PE System Safety Requirements (defined during *IEC 61508 Phase 9*) defines safety requirements in terms of safety functions and safety integrity requirements to achieve the required level of functional safety. In the DREAMS safety model they are defined as entities with the following attributes:

- `Name` - Unique name identifying the requirement (for example SSR47).

- `Description requirement` - Textual description of the requirement, as for example, *When Speed Sensor Value is greater than 100% then activate Safety Relays in less than 100 ms*.

- `Operation mode` - Low demand mode, high demand mode or continuous demand mode.

- `Response time` - Maximum time to fulfill the requirement when demanded.

- `Safety Integrity Level (SIL) level` - SIL1, SIL2, SIL3, SIL4 level required/-claimed.

- `Systematic capability level` - Systematic capability level (i.e., SC1, SC2, SC3, SC4) required/claimed.

4.4.1.2 Safety Functions

These are the functions to be implemented by the system to achieve the safe state of the system under hazardous events. Note that the scope of the function, in the case of DREAMS is the electronic system itself. Key attributes of the system are:

- `Name` - Unique Name identifying the safety function.

- `Description` - Textual description of the function, as for example, *Function activating Safety Relays when Excessive Speed, Vibration or Voltage is detected in less than 100ms - SIL3*

- `Operation Mode` - Low Demand Mode/High Demand Mode or Continuous Demand Mode.

- `Safety Requirements` - List of safety requirements (defined in section above) met by this function.

- `Response Time` - Maximum time taken by the function to activate its output when demanded.

- `SIL level` - SIL1/SIL2/SIL3/SIL4 level required.

- `SC level` - Systematic Capability SC1, SC2, SC3, SC4 level required.

4.4.1.3 Safety Manual

Describes key aspects of safety management about the system owning the Safety Manual. The most important attributes are:

- `Functional Safety Management` where in the case of DREAMS, the IEC 61508 standard has been used.

- `Safety Functions` implemented by the system.

- `Faults Management` techniques implemented by software safety functions and hardware.

- `Usage Constraints` under which the system must be operated to be safe.

- `Safety Case` consisting in a structured argumentation justifying that the system's safety functions meet safety levels defined under the specific usage constraints.

4.4.1.4 Faults Management

This entity defines the fault management according to IEC 61508. Key aspects to be defined are:

- `Hardware Fault Tolerance` claimed by the system/subsystem (in the case of hardware elements).

- `Diagnostic Coverage` claimed by the system/subsystem. Defines the function of the effectiveness of the failure detection measures. Allowed values are Low, Medium, High.

- `IEC 61508-2 Random Failure Control Techniques & Measures`, it is, a set of random failure control techniques & measures from IEC 61508-2 tables A.2 to A.14 implemented by the system/subsystem.

- `IEC 61508-2 Systematic Failure Control Techniques & Measure`, it is, a set of systematic failure control techniques & measures from IEC 61508-2 tables A.15 to A.17 implemented by the system/subsystem.

- `IEC 61508-2 Systematic Failure Avoidance Techniques & Measure`, it is, a set of systematic failure avoidance techniques & measures from IEC 61508-2 tables B.1 to B.6 implemented by the system/subsystem.

- `IEC 61508-3 Systematic Failure Avoidance Techniques & Measure`, it is, a set of systematic failure avoidance techniques & measures from IEC 61508-3 tables A.1 to A.10 and B.1 to B.9 implemented by the system/subsystem.

4.4.1.5 Usage Constraints

Define the constraints and/or assumptions under which the system's safety model design has been defined. In the DREAMS safety model constraints and/or assumptions are defined as constraints on the allowed values of parameters. For example, a parameter named `WORKING_TEMPERATURE` can be constrained to be equal, greater, lower or within a given range of temperature. Given a system S with a `Usage Constraint` C, all subsystems of S must satisfy the constraint C.

4.4.1.6 Safety Cases

A safety case defines a structured argumentation justifying that the system's safety functions meet the required safety levels. The argumentation takes into account the elements defined in the `Functional Safety Management`, `Faults Management` and `Usage Constraint` entities. The subsection on the Safety Case Generation provides a detailed description about how a safety case is generated by the tools.

4.4.2 Development Process of Safety Solution Design

4.4.2.1 Safety Solution Design Development Steps

The safety solution design in DREAMS begins by defining:

- The System Safety Requirements Specification (corresponding to phase 9 *E/E/PE System Safety Safety Requirements Specification* of the IEC 61508 overall safety life cycle).

- The *Safety Functions Software Safety Requirements* (corresponding to IEC6108-3 Phase 10.1 & 10.3).

The tools do not consider all previous phases such as Concept, Overall Scope Definition, Hazard & Risk Analysis, etc. Therefore, the development process of the safety solution begins by defining a model like the one shown in Figure 4.12. This figure shows a model that is defined with the DREAMS Safety Compliance Model Editor to define Safety Requirements and Software Safety Functions.

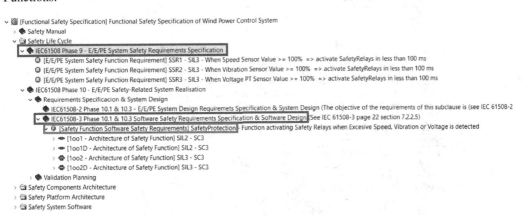

Figure 4.12: E/E/PE System Safety Safety Requirements Specification and Software Safety Functions

As shown in Figure 4.12, for each Software Safety Function its architectural style including the channels and software components per channel must be defined.

In addition to requirements and safety functions, the following Safety Compliant Items (SCI) must be defined:

- The list of software safety components.

- The list of hardware safety nodes, hardware safety tiles/cores.

- The list of software hypervisors and software partitions must be defined.

For each of these elements (as shown in Figure 4.13) safety attributed must be defined. The most important ones are listed in the following:

- SIL and SC levels claimed

- Hardware Fault Tolerance (HFT) in case of hardware elements

- IEC 61508-2/ IEC 61508-3 random/systematic control/avoidance techniques & measures that the SCI item implements.

With all this information, as it will be explained in Section 4.4.3 a DREAMS tool called the Safety Constraints and Rules Checker (SCRC) is able to infer if SIL claimed by SCI items is achievable, and therefore if the SIL claimed by Safety Functions can be met.

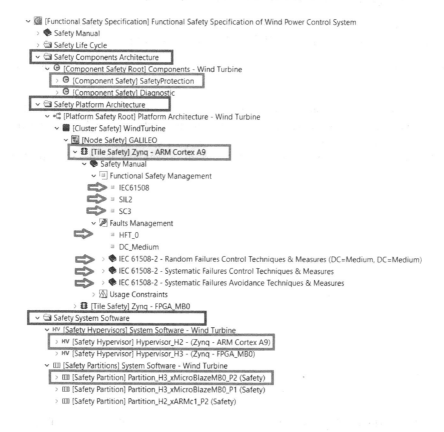

Figure 4.13: SCI – Safety Compliant Items Involved in the Model and Safety Attributes

4.4.3   Verification of Safety Solution Design

As mentioned at the beginning of this chapter, the safety model's main goal is the early detection of errors during the *realization* phase. Verification described in the following subsections *tries to find* inconsistencies in the safety model defined for the system.

Please note that this does not ensure that the safety model is correct, neither complete. Verification described below is just help for the designer to discover inconsistencies derived from the information defined in the model.

4.4.3.1   Verification Safety Rules

The goal of the DREAMS safety rules is to check if the SIL levels claimed by each Safety Function implemented by the system can be justified given the:

- Architectural style of the Safety Function

- The claimed SIL level of all SCI items involved in the Safety Function such as SW Components, hardware Nodes, hardware Tiles, SW Hypervisors and SW partitions.

4.4.3.2   Safety Functions Architectural Style Verification

This verification checks if the required SIL level declared for a Safety Function can be justified taking into account the following key features involved in the implementation of the function:

- `Architectural Style` as for example 1oo1, 1oo1D, 1oo2, 1oo2D, etc.

- `SW Components` involved in the implementation of the function.

- `SW Partition and SW Hypervisor` in which SW Components are running.

- `hardware Tiles and Cores` in which the SW Hypervisors and SW Partitions are running.

As an example, Figure 4.14 shows a) the architectural style of SafetyProtectionOverSpeed Safety Function, in this case, 1oo2D, b) the two channels implementing the function, and (c) the SW components of the first channel. The safety rule will check that:

- Channels are allocated in different tiles

- Each SW Component has the right SC level

- Each hardware Tile has the right SIL/SC level

The rule also recalls to the designer to check manually that tiles of each Channel must be arguably independent. The rule cannot verify this by itself.

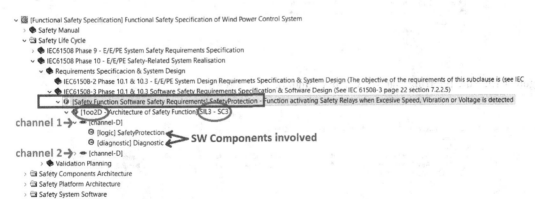

Figure 4.14: Architectural Style of SafetyProtectionOverSpeed Safety Function

4.4.3.3   Checking Whether SIL Claimed by Safety Compliant Items Is Achievable

Each SCI item involved in the allocation chain (SW-Component-Partition-Core-Hypervisor-Tile) must be able to justify the claimed SIL level. The SCRC checks if, according to the *IEC 61508-2 and IEC 61508-3 Techniques and Measures* implemented by the SCI, the claimed SIL level can be achieved. Figure 4.15 (and the text trace below in Figure 4.16) shows an example in which the set of *IEC 61508-2 Random Failure Control Techniques and Measures* (see tables A.2 to A.14 of IEC 61508-2 Annex A) implemented by the SCI produces a Diagnostic Coverage DC=[Medium, Medium] both for low and high demand mode respectively. This DC=Medium with a HTF=0 (Hardware Fault Tolerance) justifies at most a SIL 2 level according to IEC 61508-2 Table 3 (type B). However, the designer requires a SIL 3 level. This inconsistency is detected by the rule and shown in the trace below.

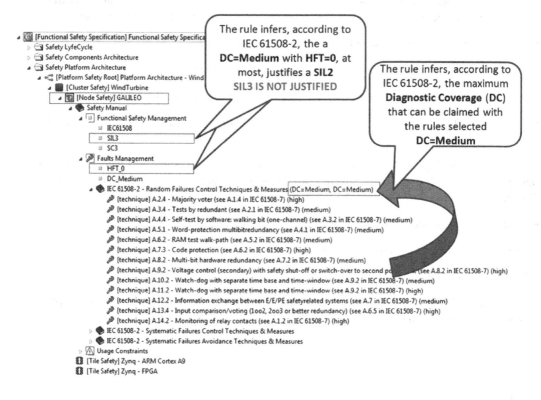

Figure 4.15: Rule Inferring that SIL Claim Cannot be Justified

```
Checking SfRule_IEC61508_2_Check_JustiableSILfromRandomFailure_Control
        Cluster: WindTurbine
            Node: GALILEO
                ---> Rule Warning: SIL3 cannot be claimed with DC = Medium and HFT_0 in Low Demand Mode if required.
                ---> Rule Warning: SIL3 cannot be claimed with DC = Medium and HFT_0 in High Demand Mode if required.
                Tile: Zynq - ARM Cortex A9
                Tile: Zynq - FPGA
---> Rule Fail.
```

Figure 4.16: Rule Showing Warning of a Rule that Is Failing

Figure 4.17: Example of Usage Constraint

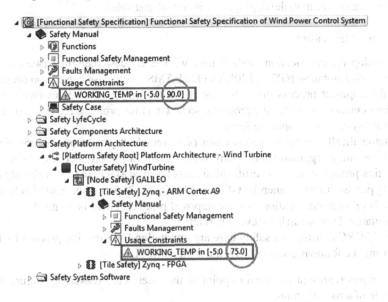

Figure 4.18: Usage Constraints of Subsystems not Met by the Compound Systems

4.4.3.4 System-Subsystem Composition and Usage Constraints

When a safety related subsystem is reused in a specific DREAMS system, it must be ensured that usage constraints specified by the subsystem in its safety manual are fulfilled by the system. In DREAMS the safety manual specifies a folder in which usage constraints can be specified. As a simple example, usage constraints about the possible values and ranges of a parameter called WORKING-TEMPERATURE are specified as shown in Figure 4.17. This means that the SCI that owns this manual requires that any SCI including this SCI must ensure that the WORKING-TEMPERATURE is maintained within this range.

The SCRC checks that any *parent* SCI in the hierarchy contains a list of usage constraints that takes into account all the usage constraints of its *child* SCIs. In Figure 4.18, the top-level system defines a usage constraint (WORKING-TEMPERATURE up to 90° C) that conflicts with the usage constraint of one of its sub-components (WORKING-TEMPERATURE up to 75° C).

4.4.3.5 Verification Safety Constraints for Deployment

Once the variability is resolved (i.e., the features are chosen), and a specific product variant is generated, candidate deployments are checked by the SCRC against a number of safety constraints (if any constraint is violated the deployment is discarded). These constraints check that:

- Safety software components are not allocated into non-safety partitions.

- Non-safety software components are not allocated into safety partitions.

- SW components of different SIL levels are not allocated together in the same safety partition.

- Safety software components are allocated isolated into one safety partition when specified by the designer.

- Safety software components are allocated into one given tile when specified by the designer.

- Safety software components are allocated into one given core when specified by the designer.

When all these constraints are met, the deployment is considered valid.

4.4.3.6 Safety Case Generation

DREAMS has developed a certification methodology with safety case patterns based on the well-known *Goal Structuring Notation (GSN)* [120]. The DREAMS safety case generation process takes advantage of a development process that relies on the composition of modules (with their own proven safety argumentation) to design the products, so that the final products reuse safety assurance artifacts either at subsystem or component levels.

The SCRC automatically composes safety cases of systems/subsystems in a recursive way and produces part of the safety argumentation of any valid variant of the product line. This is a cost-effective certification process that saves certification costs because a) it both reuses already certified modules (e.g., a hypervisor or a Commercial Off-The-Shelf (COTS) processor) and b) automatically constructs the basis of an argumentation of the composed product. There is no need to construct a complete argumentation from scratch for every variant.

The DREAMS SCRC constructs a safety case argumentation for a specific product by instantiating and composing the following sets of GSN diagram patterns:

- GSN root diagrams represent the starting point of the safety case reasoning. Figure 4.19 is a simple example of a root diagram.

- GSN diagrams represent safety constraints and safety-rule checks made by the SCRC.

- GSN diagrams serve for reused elements such as COTS processors, hypervisors, partitions and mixed-criticality networks.

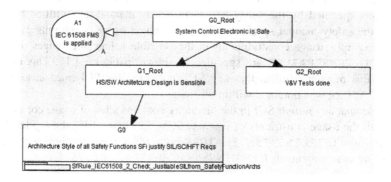

Figure 4.19: Simplified Example of Safety-Case GSN Root Diagram

The whole generation of the argumentation starts by traversing this diagram and calling subsequent linked diagrams via the *away goal node*. In Figure 4.19 the G0 node (the away goal in the bottom, left) calls another GSN diagram (see Figure 4.20) in which a safety rule checks the architectural style of all safety functions defined by the system (1oo1, 1oo1D, 1oo2, 1oo2D, etc.). Note the *option node* (a black diamond) which requires to choose one of several architectural options for each safety function.

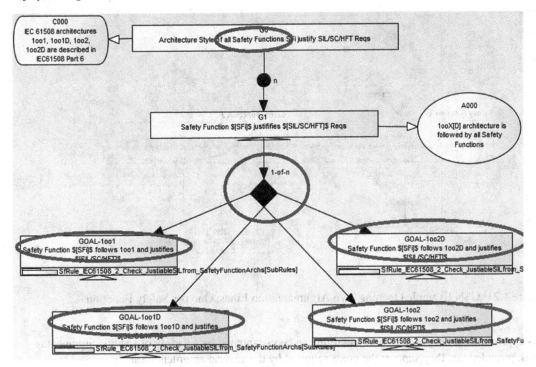

Figure 4.20: GSN Showing Architectural Style Options for Safety Functions

As a result of calling the diagram pattern in Figure 4.20, SCRC has instantiated diagrams in Figure 4.21 in which:

- There are two instances of G1, one per safety function, with the name of the safety function.

- For each safety function, the rule has instantiated the right architectural style chosen by the Design Space Exploration (DSE) tool.

Following the reasoning, the SCRC calls per safety function another GSN diagram (see Figure 4.22) to argue about 1oo2 and 1oo2D architectures (bottom nodes in Figure 4.22) instantiating the GSN to produce a GSN justification of architectures.

The whole recursive generation process finally reaches the GSN nodes linking the GSN argumentations for DREAMS Generic Hypervisor, Partitions and COTS processors. Specific argumentation (for example Xilinx ZynQ or XtratuM) are linked to generic argumentation via contracts modules.

Modular Safety Case Composition and Contracts

As mentioned before, the reuse of sub-systems is in the core of DREAMS safety certification methodology. WP5 has defined *generic* safety case argumentation for DREAMS COTS processors, hypervisors, partitions and mixed-criticality networks. For a specific product, these elements are instantiated with specific systems as for example ZynQ or XtratuM that, in turn, must provide their own safety case argumentation (in case they are certified). This argumentation must fulfill the generic safety case argumentation requirements defined by *generic* safety cases. The links between generic and specific argumentation are defined and *enforced* by contract modules (see Figure 4.23).

The contract mechanism of DREAMS has been developed following the approach in [121]. In

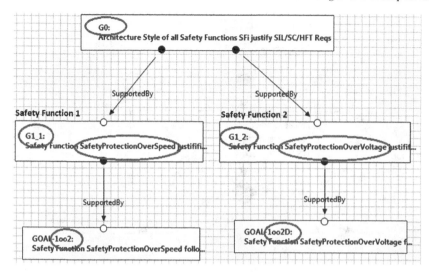

Figure 4.21: GSN Example Opening Two Argumentation Lines, One per Safety Function

this approach, a given generic argumentation can be instantiated by a variety of specific argumentations, provided that they support the goals required by the generic argumentation.

Safety Case Argumentation and Mixed-Criticality

One of the key aspects of the DREAMS project is mixed-criticality. This not only means that a given system can have a safety part and a non-safety part, but also that the safety functions of the same system can have different SIL levels. As mentioned above, SCRC copes with this situation and is capable of producing, for each safety function, a customized argumentation line. As shown in Figure 4.24, for systems that have several safety functions, the argumentation generation tool is capable of opening completely different argumentation lines.

4.4.3.7   Relationships to Certification and Mixed-Criticality Product Lines

A certification process according to a given standard (as for example IEC 61508) requires an argumentation that demonstrates that standard recommendations have been followed. In this sense, a solid and coherent argumentation of safety claims with the corresponding evidences must be provided to certification authorities. The SCRC, as described above, produces such pieces of evidence for each variant of a given product line. To integrate these checks as credible pieces of evidence in a certification process, there are two solutions: either a certification authority certifies the safety tool and the product line itself (which is a very costly process) or the tool produces (for each variant) a human readable document so that the certification authority can verify (in an argued way) the checks done by the tool. As seen in the previous subsections, the SCRC tool is capable of producing a readable document documenting all checks done. This document can be part of the argumentation to be submitted to the certification authority.

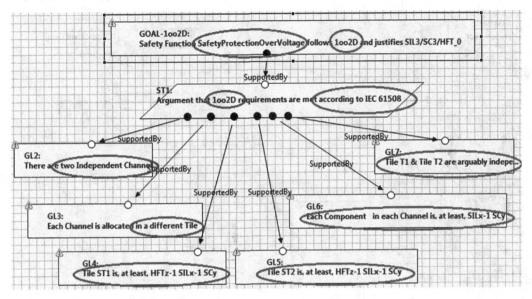

Figure 4.22: GSN Instantiated to Justify 1oo2D Architecture in SafetyProtectionOverVoltage Function

4.5 Deployment and Resource Allocation

This section elaborates how MDE supports to map mixed-critical applications onto the DREAMS platform. It is structured as follows: First, Sections 4.5.1 and 4.5.2 introduce metamodels required to specify how an application is mapped to the platform, and how this affects resource allocation, respectively. Then, we sketch the following two complementing processes that are required to transform the input models created by a system designer or integrator into deployable configuration artifacts for the DREAMS hardware/SW platform.

- The `basic scheduling workflow` introduced in Section 4.5.3 describes the steps that are required to derive configuration files from a single deployment (i.e., in case no adaptation strategy such as resource managment or reconfiguration is used).

- Section 4.5.4 presents the `resource management and adaptivity workflow` that is employed for offline preparation of system configurations for the resource management services.

4.5.1 Deployment Modeling Viewpoint

The deployment viewpoint provides metamodels to link the application model from the logical viewpoint (see Section 4.2.1.1) with the platform model defined in the technical viewpoint (see Section 4.2.4).

The metamodels defined in the deployment viewpoint provide the following information, which will be described in the subsequent sections.

- Mappings of logical components to execution units.

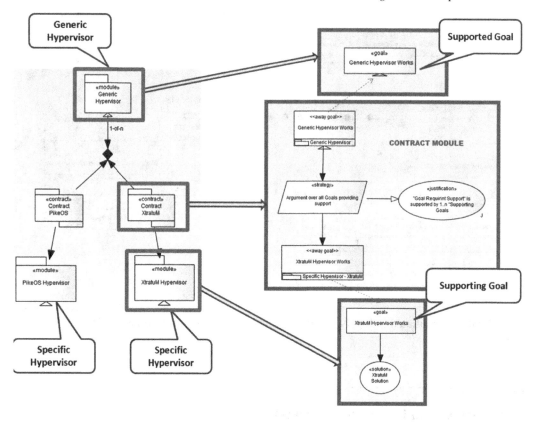

Figure 4.23: Contract Specification Between Generic Hypervisor and Specific Hypervisor

- Mappings of logical ports to the transceivers of execution units.

- Virtual links to represent multi-cast communication flows in the system.

4.5.1.1 Deployment Metamodel

The mapping model described in Table 4.21 maps a model element from the logical viewpoint to model elements of the technical viewpoint. Here, `ComponentAllocations` map logical `Components` to `Partitions`, and logical `Ports` to platform `Transceivers`.

Next, we discuss the mapping of logical `Ports` to the platform. A `Port` constitutes either an inbound or outbound interface of a `Component`, which can be either bidirectional or unidirectional. Hence, we distinguish the following cases:

- In a deployed application, communication between `Components` corresponds to the exchange of messages over `TransmissionUnits` connecting the corresponding `ExecutionUnits`. In case the `Transceiver` of a given `ExecutionUnit` is capable of performing bidirectional communication, the mapping of a `Port` (i.e., an `OutputPort` in case of a sending `Component`, and an `InputPort`, in case of a receiving `Component`) is described using a `TransceiverAllocation`.

- If the platform element to which a logical `Port` should be mapped to allows only unidirectional communication, the mapping is described using `Input-` and `OutputPortAllocations`. This is for instance the case if the `InputPort` (`OutputPort`) of a `Component` is mapped

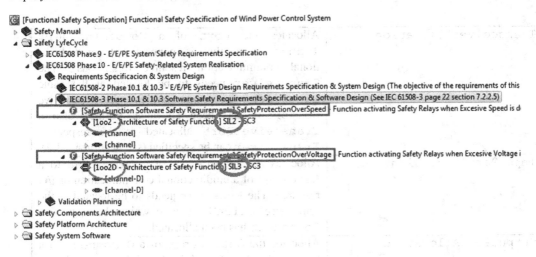

Figure 4.24: Safety Compliance Model Specifying Two Safety Functions with Different SIL Requirements

to the `Receiver` (`Transmitter`) provided by sensor (actuator) of the platform, or if the underlying `TransmissionUnits` provides (separated) unidirectional communication channels.

Mapping models can be instantiated either manually using a model editor, or automatically using a design-space exploration. In the latter case, the exploration evaluates the `Events` defined in temporal viewpoint that is associated to a given logical architecture (to decide, which `Components` to deploy, and to derive constraints), and the `DeploymentGranularity` annotation of `ExecutionUnits` to determine the deployment targets. We refer to Section 4.7.3 for a specification of the exploration process, and to Section 5.1.2.3 for a detailed treatment of the exploration's inputs (presented in the frame of the description of the underlying algorithm).

Table 4.21: Mapping Metamodel

Metaclass	Description
`Deployment`	Container for the logical architecture to platform architecture mapping.
`ComponentArchitecture-Reference`	References the `ComponentArchitecture` whose subelements are deployed onto a `PlatformArchitecture`.
`PlatformArchitecture-Reference`	References the `PlatformArchitecture` onto which a `ComponentArchitecture` is deployed.
`ComponentAllocation`	Maps a `Component` to the `Execution- Unit` that executes it.
`PortAllocation`	Marker interface for allocations of `Ports` and `Transceivers`.

`TransceiverAllocation`	Allocates the `Port` of a `Component` to a `Transceiver` that is connected to a bidirectional communication resource (e.g., bus). The `Transceiver` needs to be located at the same `ExecutionUnit` onto which the respective `Component` has been allocated. Further, also a `TransceiverPort` allocated to the respective `Transceiver` can be specified (see Section 4.5.1.2).
`InputPortAllocation`	Allocates the `InputPort` of a `Component` to a `Receiver` of a unidirectional communication or I/O resource. The `Receiver` needs to be located at the same `ExecutionUnit` onto which the respective `Component` has been allocated.
`OutputPortAllocation`	Allocates the `OutputPort` of a `Component` to a `Transmitter` of a unidirectional communication or I/O resource. The `Transmitter` needs to be located at the same `ExecutionUnit` onto which the respective `Component` has been allocated.

Another aspect of the deployment metamodel is the specification of deployment-specific parameters for logical components. A `Deployment` contains a map that allows to store a value for each element of the cross-product of the set of logical `Components` and `ExecutionUnits` in the system.

To ensure an extensible approach, the actual parameters are provided as annotations that are bound to *DeploymentParameterValue* objects which are contained in the `Deployment`'s `DeploymentKeyToDeploymentParameterValueMap`. The map identifies the parameters of a specific `Component-ExecutionUnit` pair that are referenced in a `DeploymentParameterKey` object (see Table 4.22).

Table 4.22: Deployment-Specific Parameter Map Metamodel

Metaclass	Description
`DeploymentKeyTo-` `DeploymentParameter-` `ValueMap`	Map that relates a (`Component`, `ExecutionUnit`) pair to the set of parameters describing its properties.
`DeploymentParameterKey`	Used to identify a deployment-specific parameter of a (`Component`, `ExecutionUnit`) pair.
`DeploymentParameter- Value`	Hold the value of a deployment-specific parameter of a (`Component`, `ExecutionUnit`) pair.

Table 4.23 lists the available parameters:

Table 4.23: Deployment-Specific Parameters

Metaclass	Description
`EnergyConsumption`	Contains the average energy consumption (in Joule) when executing the `Component` of the annotated `Component-ExecutionUnit` pair on the corresponding `ExecutionUnit`.

`Wcet`	Allows to define the WCET (in seconds) when executing the `Component` of the annotated `Component-ExecutionUnit` pair on the corresponding `ExecutionUnit`.

4.5.1.2 Virtual Link Metamodel

The virtual link metamodel introduced in this section allows to describe the characteristics of communication in the DREAMS platform and the corresponding routes. As defined in Section 2.1.3.2, virtual links are an abstraction over physical networks at different levels of the DREAMS platform architecture. In the MCS metamodel, they are defined as end-to-end multicast channels between the `OutputPort` of one logical sender `Component` and the `InputPorts` of multiple logical receiver `Components`. Each virtual link has a unique identifier (ID), and defines the message's route, traffic type (time-triggered or rate-constraint) and timing (period or minimum inter-arrival time).

Transceiver Ports

`TransceiverPorts` enable to describe platform-specific parameters of endpoints and waypoints of virtual links. For each virtual link, `TransceiverPorts` can be allocated to the `Transceivers` of the `TransmissionUnits` involved in the route that define parameters such as names, IDs, etc.

The following `TransceiverPort` specializations are used for the generation of platform configuration (see Section 4.6):

- `PartitionPorts` are allocated to the `InterPartitionComPort` of `Partitions` (see Section 4.2.4.6) and represent the communication ports of hypervisor partitions.

- `OnChipNetworkInterfacePorts` are allocated to the `BusOnChipNetworkExport` of `NetworkInterfaces` (see Section 4.2.4.5) and represent the communication ports of a Tile's network-interface, i.e., the on-chip LRS.

Mapping to DREAMS Namespace

The namespace defined in Section 2.1.3.1 defines the mapping between the logical architecture and the platform architecture, where the logical name of a message is defined as `<Criticality>.<Subsystem>.<Component>.<Message>`.

In order to map this to the MCS model, the following design rules must hold:

- The logical architecture is expressed as `ComponentArchitecture` that contains a logical `Component` for each *Subsystem*.

- Sub-system `Component` contains another logical `Component` to represent the *Component*.

- The message-based interface of `Components` is modeled using `OutputPorts`.

The logical name is defined as a 4-tuple of integers, where IDs are based on the following annotations introduced below:

- *Criticality*: annotated to logical `Component` representing the subsystem using the `SafetyIntegrityLevel` annotation (see Section 4.2.1.2).

- *Subsystem*: `ComponentId` (annotated to Subsystem logical `Component`).

- *Component*: ComponentId (annotated to Component logical Component).

- *Message*: MessageId (annotated to logical OutputPort).

For the physical name <Cluster>.<Node>.<Tile>.<Port> (see Section 2.1.3.1), annotations provide integer IDs to the respective model elements of the technical architecture level (Section 4.2.2).

The one-to-one mapping of the logical name to the physical name introduced in Section 2.1.3 is implemented using the TransceiverPort attribute of TransceiverAllocations (see Section 4.5.1.1). Since every logical Component is mapped to a Partition of a Hypervisor, the TransceiverAllocation maps the Component's InputPorts and OutputPorts to the respective PartitionPort allocated to the Partition's InterPartitionComPort.

The virtual link metamodel is integrated as follows into the deployment metamodel (introduced in the previous section):

- Since the communication characteristics depend on the Component-to-ExecutionUnit mapping, the Deployment contains a RoutingAllocation that defines the set of VirtualLinks required to represent the communication demands of the logical architecture.

- TransceiverPorts   are   contained   in   the   Deployment's TransceiverPortAllocation and support the definition of routes and physical names.

Virtual Links

In the following, we discuss the structure and attributes of virtual links. First, a VirtualLink is characterized by its endpoints:

- Given a VirtualLink, its sender / receiver PartitionPorts can be determined using its VlSender / VlReceivers annotation.

- Using the TransceiverAllocation, the resulting PartitionPorts can be resolved to the corresponding logical OutputPort / InputPorts for which the VirtualLink has been created.

Second, a number of parameters determine a VirtualLink's Quality of Service (QoS)-guarantees and its temporal properties:

- VlTrafficType defines the VirtualLink's type (time-triggered or rate-constraint).

- VlTempRepetition defines the temporal requirements onto the virtual link (period for time-triggered virtual links, minimum inter-arrival time for rate-constraint virtual links).

- VlPayloadSize defines the maximum payload size for the given virtual link.

Lastly, VirtualLinks represent multi-cast communication in a distributed system. Hence, its route is represented as a tree of Segments, where a segment represents one hop of a route, i.e., from one Transceiver to another. The tree representation assumes that there is only one possible route between each sender and receiver, which reflects the capabilities of the current implementation of the underlying platform services. However, the VirtualLink class could easily be changed to use a directed graph to represent routes (using Transceivers / TransceiverPorts as vertices, and Segments as edges). For example, during configuration generation, all routes could be explored by traversing this graph from the route vertex (Transceiver / TransceiverPort of sender) using a depth-first search.

- A `TransceiverPortsSegment` represents a `Segment` that ends in a `Transceiver` for which a dedicated `TransceiverPort` concretization has been defined in the metamodel (the corresponding `TransceiverPort` instance is allocated in the `TransceiverAllocation`).

- A `TransceiverSegment` represents a `Segment` that ends in a `Transceiver` for which no `TransceiverPorts` have been defined.

Virtual Links Example Instance

Virtual links provide a bridge between the data exchanged among logical `Components` and their realization as messages on the target platform. The corresponding elements in the logical application architecture are `OutputPorts`, which define logical senders, and connected `InputPorts` (via logical `Channels`, see Section 4.2.1.1), which receive the sent data. In the target platform, the equivalent to the mentioned Ports are the `Transceivers` that are defined in the port mapping of the `Deployment`. These `Transceivers` are the communication interfaces of the `ExecutionUnits` to which the `Components` containing the sender and receiver ports are deployed.

From the perspective of the target platform, a virtual link is a route from the sender resource (i.e., a `Partition`) through the involved communication resources to the target resources, e.g., the `NetworkInterface` and the `NocRouter` (see Section 4.2.2) denoted by ① in Figure 4.25.

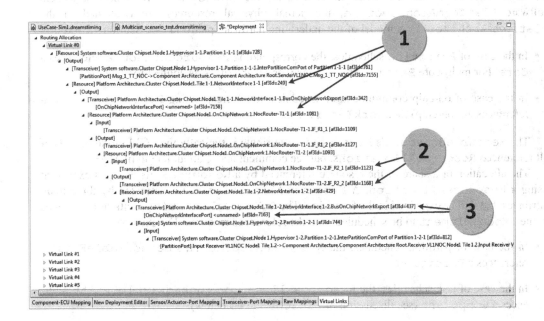

Figure 4.25: Virtual Links Example Instance

The actual route of virtual links is a tree of traversed `Transceivers` and `TransceiverPorts` (which reference the traversed `Transceivers` of the traversed communication resources) in the target platform, since the involved resources do not uniquely define the route of a `VirtualLink`. Hence, the in- and outputs of each traversed communication resource is explicitly defined. `TransceiverSegments` and `TransceiverPortSegments` encode `Transceivers` and `TransceiverPorts`, respectively, in the `VirtualLink`'s route. For instance, in Figure 4.25, the blue circle ② points to the `Transceivers` (resp. `TransceiverSegments`) of the communication resource *NocRouter-T1-2* that receive the virtual links and that emit it to the successor resources. Similarly, the blue circle ③ points to the `TransceiverPort` of the `TransceiverPortSegment` associated with the resource *NetworkInterface 1-2* and to the `Transceiver` referenced by the `TransceiverPort`.

4.5.2   Resource Allocation Modeling Viewpoint

4.5.2.1   Schedule Metamodel

The schedule metamodel is a device- and tool-independent metamodel to represent hierarchical schedules. It is a format that can capture the output of the different device-/platform-service specific scheduling tools and allows to exchange schedules between them.

A `SystemSchedule` collects the schedules for all involved resources in its `Deployment` (see Section 4.5.1). The Major Active Frame (MAF) is the period of time to be considered when computing the task allocation and scheduling. The MAF of the `SystemSchedule` is normally equal to the Least Common Multiple (LCM) of the periods of all `ResourceSchedules`, but can be less than this value under certain conditions such as harmonicity or geometricity of periods (see Section 5.3.1.2). The schedule of each individual resource is described using a `ResourceSchedule`. Its `resource` reference specifies the resource in the physical platform architecture contained by the `Deployment`'s `PlatformArchitecture` for which a schedule is defined Since in DREAMS, the `Deployment` references a system software `PlatformArchitecture`, the actual physical resources are contained in the `PlatformArchitecture` model of the underlying hardware platform (see Section 4.2.4.6).

- In the case of `Partition` schedules, the corresponding `ResourceSchedule` references a `Core`. For multi-core `Partitions`, a `ResourceSchedule` is defined for each `Core`.

- In the case of on-chip communication schedules, the corresponding `ResourceSchedule` references an (on-chip) `NetworkInterface`.

The `ResourceSchedule`'s `hyperperiod` attribute is typically the LCM of the periods of all referenced `ResourceAllocations`, but see comment above about length of MAF.

The allocation of a share of the resource referenced by a `ResourceSchedule` is expressed using a `ResourceAllocation`. A `ResourceAllocation` is characterized by three main attributes that will be explained in the following: The `SchedulableEntity` contains a reference to the `IModelElement` to be scheduled onto the referenced resource: For example:

- In the case of partition schedules, the `ResourceAllocation`'s `SchedulableEntity` references a `Partition`.

- In the case of task schedules, the `ResourceAllocation`'s `SchedulableEntity` references a `Component` (see discussion of `SubSchedules` below).

- In the case of communication schedules, the `ResourceAllocation`'s `SchedulableEntity` references a `VirtualLink`.

The `ResourceAllocation`'s `duration` attribute specifies the time for which the resource is reserved for the referenced `SchedulableEntity`. Lastly, the `Trigger` allows to depict the temporal activation pattern of the `ResourceAllocation` (see below for a discussion of triggers).

The schedule metamodel is designed to express hierarchical schedules: The share of a resource described by a `ResourceAllocation` allocated in a `ResourceSchedule` can be further sub-divided by declaring a `SubSchedule` for it. `SubSchedules` apply the aforementioned concepts recursively, i.e., they also contain `ResourceAllocations` that reference a `SchedulableEntity` and a `Trigger` object. A `SubSchedule` does not reference a platform resource, since it refines a `ResourceAllocation`. In DREAMS, this concept is used to define schedules for the tasks hosted a particular `Partition`.

`Triggers` are an extensible way to specify the temporal activation pattern of `ResourceAllocations`. In the scope of DREAMS, the following `Triggers` are relevant:

- `PeriodicTimeTriggers` allow to specify the period and phase of a `ResourceAllocation` and are intended for strictly time-triggered activities.

- `APeriodicTimeTriggers` provide the start time of aperiodic activities.

- `RateConstraintTriggers` define the maximum jitter and the minimum inter-arrival time of sporadic activities.

Schedule Model Example Instance

Figure 4.26 illustrates an example schedule model (the toolchain presented in Chapter 5 also provides a Gantt chart viewer to visualize and navigate schedule plans). It shows a `SystemSchedule` that is composed of `ResourceSchedules` for each of the `IPlatformResources` that are present in the target hardware platform. In the following, we discuss *Resource Schedule - Core 1-1-1* that specifies the schedules of the hypervisor `Partitions` that are allocated to `Core 1-1-1`. This `ResourceSchedule` contains `ResourceAllocations` of the `Partitions` (e.g., *Resource Allocation – Partition -1-1-1*) which are assigned to the corresponding `Cores` via a `ResourceLink` annotation in the *System Software Platform* model (c.f. Section 4.2.4.6). Each `ResourceAllocation` for a `Partition` contains a `SubSchedule` (e.g., *Tasks Partition 1-1-1*) that represents the schedule of the tasks executed within the `Partition`. These `SubSchedules` are again a list of `ResourceAllocations` that point to a `Component` in the logical architecture (e.g., *Resource Allocation – Component Sender VL1*, *Resource Allocation – Component Sender VL1NOC*, etc.).

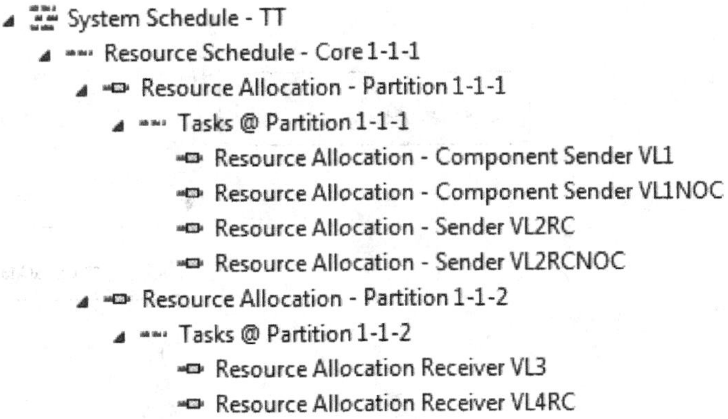

```
⊿ ⣿ System Schedule - TT
    ⊿ ▬ Resource Schedule - Core 1-1-1
        ⊿ ▭ Resource Allocation - Partition 1-1-1
            ⊿ ▬ Tasks @ Partition 1-1-1
                ▭ Resource Allocation - Component Sender VL1
                ▭ Resource Allocation - Component Sender VL1NOC
                ▭ Resource Allocation - Sender VL2RC
                ▭ Resource Allocation - Sender VL2RCNOC
        ⊿ ▭ Resource Allocation - Partition 1-1-2
            ⊿ ▬ Tasks @ Partition 1-1-2
                ▭ Resource Allocation Receiver VL3
                ▭ Resource Allocation Receiver VL4RC
```

Figure 4.26: Schedule Model Example Instance

4.5.2.2 Reconfiguration Metamodel

DREAMS applies global and local resource management to provide QoS guarantees and global as well as local recovery strategies to compensate for those classes of hardware-faults and run-time errors that are covered by the fault-assumption (see Chapter 9). Recovery strategies rely on global and local reconfiguration graphs that are computed at design-time (see Section 5.3.1). In this section, we introduce the reconfiguration metamodel, whose root element `ReconfigurationGraphs` clusters all reconfiguration graphs at the global (for the GRM, see Section 9.5) and local level (for the LRMs, see Section 9.4). A reconfiguration graph is described using a tree of `ConfigurationContainers` whose children may be of one of the following types:

- `ConfigurationContainer`: Further refinement of a `ReconfigurationGraph`'s structure. In DREAMS, one `ConfigurationContainer` is used to define the system's global reconfiguration graph, and there is a dedicated `ConfigurationContainer` for each of the system's `Nodes` that contains sub-containers for the local reconfiguration graphs of the respective `Tiles`.

- `LocalConfiguration`: Node in a local reconfiguration graph (see below).

- `CompositeConfiguration`: Node in a global reconfiguration graph (see below).

`LocalConfigurations` define the system's recovery strategy at the `Tile` level. For each `Tile`, the recovery strategy is encoded as a graph (typically a tree), whose vertices are `LocalConfigurations` that reference a `SystemSchedule` (see Section 4.5.2.1). The edges of the graph are represented by `Transitions` that specify the condition to switch from one local configuration to another (i.e., the list of failed resources).

Also at the system level, the recovery strategy is expressed as a graph (c.f. Sections 5.3.1 and 9.5) whose nodes are `GlobalConfigurations` and whose edges are `Transitions`. Vertices in the global reconfiguration graph reference `LocalConfigurations` to be activated by the corresponding LRMs in case the corresponding global mode is activated.

Reconfiguration Graph Model Example Instance

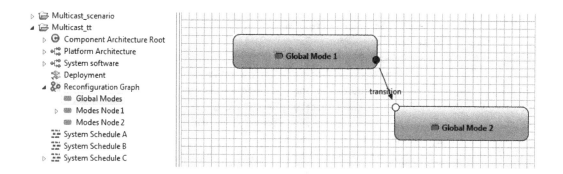

Figure 4.27: Global Reconfiguration Graph Example

Figure 4.27 shows the typical structure of a `ReconfigurationGraph` that is composed of `ConfigurationContainers`. As shown on the right-hand side of the diagram, one `ConfigurationContainer` contains the global reconfiguration graph that consists of the two global modes *Global Mode 1* and *Global Mode 2* between which a transition is defined. Further,

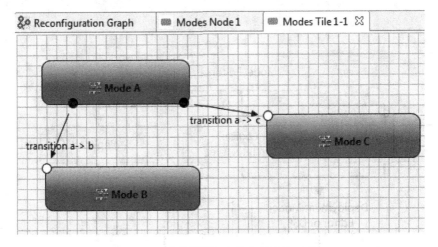

Figure 4.28: Local Reconfiguration Graph Example

there is one `ConfigurationContainer` for each `Node`, which can be seen in the tree-structure viewer on the left of the figure.

The `ConfigurationContainer` for *Node 1* contains two sub-`ConfigurationContainers` for its `Tiles` *Tile 1-1* and *Tile 1-2* (not shown in the figure). Figure 4.28 exemplary shows the `LocalConfigurations` (i.e., local modes) of *Tile 1-1*. There exist three local modes: *Mode A*, *Mode B*, and *Mode C*. *Transition*s exist from *Mode A* to *Mode B* as well as from *Mode A* to *Mode C* to which the condition when to switch the local mode is annotated.

4.5.3   Basic Deployment and Scheduling Workflow

The workflow illustrated in Figure 4.29 refers to one of the basic tasks in the development of any MCS, namely to determine feasible schedules for all resources in the system, and to derive the corresponding configuration files that ensure the segregation of applications of mixed criticalities and their communication. However, it is important to note that this workflow does not provide any support to provide QoS guarantees for low-critical tasks, or to mitigate hardware faults by means of recovery strategies.

Creation of Input Models

This first step bundles the modeling activities that are required to express the input artifacts for the `basic deployment and scheduling workflow`. The results are the following model from different MCS viewpoints (c.f. Section 4.1.1), which are typically created by different stakeholders.

- The architecture of the applications that are to be deployed to the DREAMS platform is modeled. The result is a logical architecture (see Section 4.2.1.1) that defines a `Component` hierarchy that is refined to the granularity of schedulable tasks. It also includes the criticalities of `Components`, their `Port` interfaces, and the logical channels that interconnect them.

- A timing model augments the logical architecture with temporal requirement for the execution of components and the transmission of messages in terms of events (e.g., periodicities). Further, the timing model is used to specify system-level timing requirements such as reaction constraints between a source and a target component (e.g., maximum latency from reception of sensor until emission of actuator control signal).

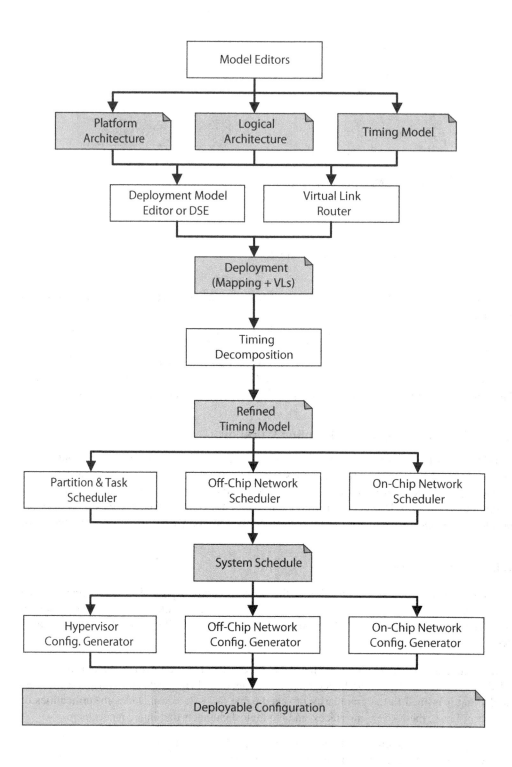

Figure 4.29: Basic Deployment and Scheduling Workflow

- The technical architecture model of the target platform complements the application model that results from the two previous steps. As pointed out in Section 4.2.2, the model is separated into two PlatformArchitectures, where the first one describes the topology and the resources of the physical platform. The second PlatformArchitecture depicts the configuration of the system software. It is linked to the physical platform architecture to describe which hardware resources are used or managed by it, i.e., in particular Hypervisors are mapped to Tiles, whereas time-space Partitions may optionally be linked to processor cores.

Deployment

The result of these steps are artifacts introduced in Section 4.5.1.1 that capture how the logical ComponentArchitecture is mapped to the system software PlatformArchitecture (which as pointed out above by itself references the physical platform).

As illustrated in Figure 4.29, the following sub models are created either by a human designer using a model editor, or using an automated design-space exploration procedure that determines optimized mappings based on the given set of optimization goals and constraints (see Section 4.7.3).

- The ComponentAllocation describes the mapping of logical Components to hypervisor Partitions.

- Input- and OutputPortAllocations specify the mapping of logical Ports to PartitionPorts (see Section 4.5.1.2) that according to the architectural style of DREAMS (see Chapter 2) serve as universal entry point at the platform level.

Lastly, the logical Channels of the input component architecture are transformed into VirtualLinks depending on the selected Component-to-Partition, considering the temporal requirements provided in the timing model.

Resource Allocation and Scheduling

This step of the process is concerned with the temporal allocation of system resources for a given Deployment, i.e., the computation of offline schedules. In order to split this highly complex problem into manageable pieces, in the first step, timing decomposition is used to break end-to-end timing constraints into sub-constraints for the different affected resource classes (see Section 5.2.1). The result is an updated timing model, in which the end-to-end budget defined in system-level event-chains is distributed into sub-chains for the different scheduling domains, i.e., application tasks running on processor cores, on-chip and off-chip communication.

In the next step, offline scheduling is performed using the tools introduced in Section 5.2, resulting into a SystemSchedule that contains a ResourceSchedule for each of the resources referenced by the given Deployment. For the task-domain, the outcome is hierarchical schedule whose first level specifies the partition schedule. Each of the ResourceAllocations at the partition level is refined into a SubSchedule depicting the schedule of the tasks hosted by the respective partition. The ResourceSchedules for the on-chip and the off-chip domain are flat and contain the ResourceAllocations for the VirtualLinks contained by the given Deployment.

Configuration Generation

In the last step, the configuration files for the DREAMS platform are generated by translating the relevant information that has been gathered in the preceding process into the format required by the respective hardware/SW services. Examples:

- Time-triggered schedules for hypervisors, on-chip and off-chip networks.

- Configuration of communication ports (e.g., names, addresses, etc.).

- Memory areas and communication routes.

In addition to the conversion, also default values for implementation-specific parameters that are not covered by the generic deployment and scheduling metamodels might be added and possibly adapted by the system integrator.

4.5.4   Adaptivity and Resource Management Workflow

The *adaptivity and resource management workflow* extends the process described in the previous section with the measures required to configure the resource management services (see Chapter 9) with reconfiguration graphs that enable the system to tolerate hardware faults and deadline overrun situations (as far as covered by the underlying fault hypothesis). We focus our subsequent discussions on the differences between the two workflows.

Creation of Input Models

Initially, the logical architecture, the timing model and the technical architecture model are created as described in Section 4.5.3. Since the resource management services (see Chapter 9) are implemented in software, they need to be considered in the resource allocation and configuration generation process, too. Therefore, for all LRMs and MONs managing the processing resources of the platform (i.e., the processor Core), a logical Component is instantiated. The same also applies for the GRM that is also implemented in software. The communication between the each of the LRMs and its associated MON as well as the GRM is modeled using a logical Channel. Further, it is important to note that the implementation of the resource management services requires the definition of timing constraints that ensure that the corresponding tasks are correctly aligned with the MAC. Lastly, a dedicated hypervisor Partition is defined for each of the resource management Components. Since the resource management software's architecture is aligned with the hardware platform, it can be instantiated automatically.

Deployment

In the deployment step, both the application and the resource management components are mapped to the target platform as pointed out in Section 4.5.3. The result is a Deployment for the system's *nominal model* (c.f. Section 5.3.2.2).

Generation of Nominal Partition Schedule

This step corresponds to the *resource allocation and scheduling* defined in Section 4.5.3. However, it is important to note that after the timing decomposition step, only the partition and task schedule of the nominal model is computed. As pointed out below, the resulting SystemSchedule contains only the ResourceSchedules for the computational resources and will be extended with appropriate communication schedules after the reconfiguration synthesis step (see below).

Synthesis of Local and Global Reconfiguration Graphs

In this step, the SystemSchedule for the nominal mode is used as an input to synthesize a ReconfigurationGraph defining global and local reconfiguration strategy that allow the resource management services to compensate hardware faults that are covered by the given fault-hypothesis (see Section 5.3.1). As explained in Section 4.5.2.2, each of the contained LocalConfigurations references a SystemSchedule that is based on a dedicated Deployment reflecting the re-scheduling of Partitions that is performed when the corresponding local mode is activated.

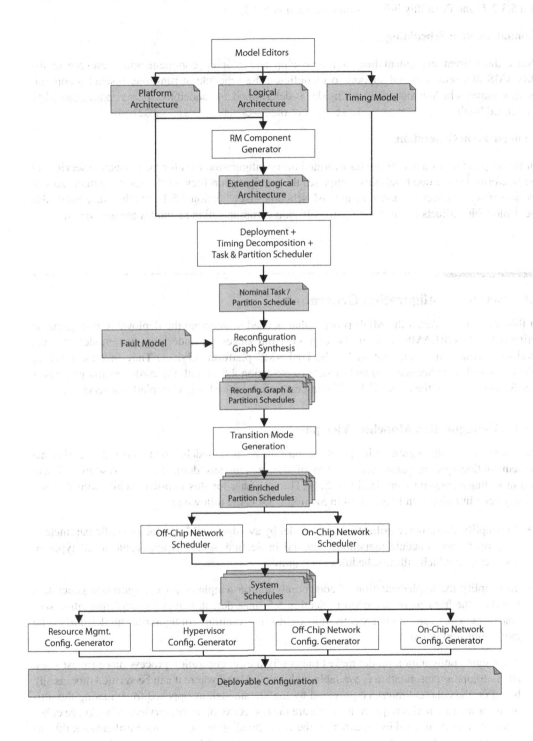

Figure 4.30: Adaptivity and Resource Management Workflow

Finally, the `ReconfigurationGraph` is enriched with transition modes (see Section 5.3.2.4) and flexibility information (see Section 5.3.3).

Communication Scheduling

Since the current implementation of the on-chip and off-chip communication resource of the DREAMS platform does not support to reconfigure the schedule at run-time, instead a communication super-schedule that covers all modes is determined and added to all `SystemSchedules` referenced by the `LocalConfigurations` of the `ReconfigurationGraph`.

Configuration Generation

In this step, all derived results are transformed into configuration files for the respective services of the platform. On the one hand, this comprises the configuration files for the resource management services that also cover the partition and task schedules (see Section 9.5.1). On the other hand, also the deployable artifacts for the on-chip and off-chip communication resources are generated.

4.6 Service Configuration Generation

In this section, we sketch the MDE process that is used to generate the deployable configuration artifacts for the DREAMS platform. Section 4.6.1 introduces the underlying metamodels that are used to represent the configuration for the DREAMS platform services. They link the resource allocation tools (see Section 5.6) and models (see Section 4.5.1) with the configuration generators (see Section 5.5) and the actual HW/SW implementation of the DREAMS platform services.

4.6.1 Configuration Modeling Viewpoint

The service configuration viewpoint contributes metamodels that provide additional implementation-specific parameters on top of the metamodels defined in the resource allocation modeling viewpoint (see Section 4.5.2). The rationale for this two-tier architecture that has already been introduced in Figure 2.34 in Section 2.3.4 is the following:

- To simplify the resource utilization metamodel by avoiding building block specific parameters. As a result, the schedule metamodel defined in Section 4.5.2.1 is applicable to all types of resources for which offline schedules are required.

- To simplify the implementation of code/configuration templates for configuration generators based on the framework described in Section 5.5 since default values for implementation specific parameters can be consistently injected to the resulting configuration model during the model-transformation.

- A separate configuration model makes the configuration generation process more robust since all configuration information is available at the model level where it can be verified more easily based on the coherent interface provided by the metamodel, as opposed to validating the ultimate configuration files required to configure the respective platform services. Checks are either automated, or performed by an expert in the corresponding model editors. Furthermore, this intermediate level helps mitigate changes in the configuration file format and enables the system integrator to perform manual adjustments using the respective configuration model editor.

4.6.1.1 Configuration Infrastructure Metamodel

The metamodel described in this section is the basis for a number of different configuration meta-models of a subset of the DREAMS services. Section 4.6.1.2 presents the configuration metamodel for the physical on-chip network interface (see Section 7.4) as a representative example. Other configuration metamodels, e.g., for the XtratuM hypervisor (see Section 6.3), or the TTEthernet off-chip network (see Section 8.1.1) are extensions of existing vendor-defined formats and hence are only loosely integrated into this infrastructure.

The configuration metamodel shown in Table 4.24 supports the generation of configuration files for component of the DREAMS platform. Hence, the `IConfiguration` interface provides a link to a reference element in the system model and defines the (relative) folder where the configuration should be generated. Since for a platform resource to be configured, potentially more than one configuration file is required, the names of the generated files are determined by the corresponding configuration generators code template (Section 5.5). The `Configuration` class is the base class for the concrete configurations defined for respective DREAMS services. `ConfigurationCollections` are used to cluster sets of `Configurations`, e.g., the configuration of the DREAMS virtual and physical platform (`VirtualPlatformConfiguration`, `PhysicalPlatformConfiguration`), or the root model element `ConfigurationProject` itself. The `ExternalConfiguration-Reference` can be used to link an external configuration artifact (e.g., an XML file in case of XtratuM or TTethernet). That way, the resulting `ConfigurationProject` clusters all configuration artifacts for a particular `Deployment`.

Table 4.24: Configuration Metamodel

Metaclass	Description
`Path`	Representation of file paths.
`IConfiguration`	Common interface of all configuration metamodels (including `Configuration- Collections`).
`ConfigurationCollection`	A collection of `IConfigurations`.
`VirtualPlatform-Configuration`	Configuration of DREAMS virtual platform.
`PhysicalPlatform-Configuration`	Configuration of DREAMS physical platform.
`ExternalConfiguration-Reference`	Reference to external configuration model.
`ConfigurationProject`	Root node of configuration model. It serves as entry point for configuration generation.
`Configuration`	Abstract base class for concrete configuration meta-models (that provide the platform-specific parameters required for the generation of configuration files).

4.6.1.2 Example: Physical On-Chip Network Interface Configuration Metamodel

As depicted in Table 4.25, the `OnChipNetworkConfiguration` collects information required to generate the configuration of the mixed-criticality on-chip network described in Section 7.4. It comprises the following sub-elements:

- Two `OnChipSchedParams` objects to define the scheduling parameters for time-triggered and rate-constraint on-chip network traffic.

- One `OnChipNiLrsConfiguration` for each of the on-chip `NetworkInterfaces` connected to the `OnChipNetwork` to be configured.

An `OnChipNiLrsConfiguration` references a `ResourceSchedule` (see Section 4.5.2.1) for the respective on-chip `NetworkInterface` and consists of the following sub-configurations:

- A `PortConfiguration` for all on-chip network interface ports to be defined for the given `NetworkInterface`. It is a collection of `PortConfigurationItems` that define low-level parameters such as buffer size, queue length, etc. for the referenced `OnChipNetworkInterfacePort` (see Section 4.5.1.1).

- A `TimeTriggeredCommunicationSchedule` that defines the configuration of the schedule for the time-triggered traffic handled by the on-chip `NetworkInterface` referenced by the `OnChipNiLrsConfiguration`. In order to match the configuration interface of the physical on-chip network interface LRS, time-triggered schedules are represented as a set of linked lists of `TimeTriggeredCommunicationScheduleEntrys` that provide operations to obtain the ID of the port to be scheduled, its index in the list, and its phase.

- `EventTriggeredCommunicationSchedules` define static schedules for rate-constraint traffic as linked lists (see above). However, instead of defining schedules for network interface ports, the `EventTriggeredCommunicationScheduleEntrys` control the on-chip interleaver of the on-chip network interface using operations that control the admission of event-triggered traffic using guarding windows (GW_OPEN, GW_CLOSE) and by-pass windows for the TTE and DDR controllers (BP_OPEN, BP_CLOSE).

Table 4.25: Physical On-Chip Network Interface Configuration Metamodel

Metaclass	Description
`OnChipNetwork-Configuration`	Hardware configuration of all LRS of a single physical `OnChipNetwork`.
`OnChipSchedParams`	Global scheduling parameters.
`OnChipNiLrs- Configuration`	Configuration of the on-chip `NetworkInterface`.
`PortConfiguration`	Port configuration parameters of on-chip LRS.
`PortConfigurationItem`	Parameters of a single port of the on-chip LRS (entry of `PortConfiguration`).
`TimeTriggered-CommunicationSchedule`	Cyclic lists of time-triggered message emission times.
`EventTriggered-CommunicationSchedule`	Cyclic lists of gate operations.
`ScheduleEntry`	Base class for a single entry into a `TimeTriggeredCommunication-Schedule` or `EventTriggered-CommunicationSchedule`.
`TimeTriggeredCommuni-cationScheduleEntry`	Schedule configuration for time-triggered traffic.
`EventTriggeredCommuni-cationScheduleOperation`	IDs of operations used by the ET-interleaver to eliminate the chance of temporal interferences for the time-triggered messages, by restricting the injection of event-triggered messages (e.g., rate-constraint or best-effort) and accesses to the TTE and DDR controller.

EventTriggeredCommuni- cationScheduleEntry	Schedule configuration for event-triggered traffic.

4.6.2   Model-transformations and Configuration Synthesis

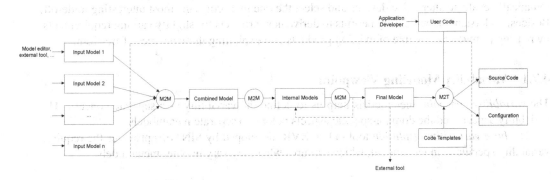

Figure 4.31: Model-Driven Configuration Generation

Figure 4.31 shows a general architecture of an MDE toolchain that implements the process steps discussed in the following. First, the input models are created using the model editors, c.f. step *Creation of Input Models* in the workflows described in Sections 4.5.3 and 4.5.4. Then, the input models are transformed in several steps into the final output model that is used to generate the configuration artifacts. As discussed in Sections 2.3.4 and 4.5.2.1, in the DREAMS MDE process, a layered approach is used. Initially, required design-time transformations are performed on a tool- and device-independent level (c.f. steps *Deployment* and *Resource Allocation and Scheduling* in Section 4.5.3 as well as steps *Deployment, Generation of Nominal Partition Schedule* and *Synthesis of Global and Local Reconfiguration Graphs* in Section 4.5.4). Afterward, the resulting models (system model, deployment, resource allocation, etc.) are transformed into configuration models (see Section 4.6.1) that provide a device-specific view of these artifacts that are typically also enriched with specific parameters such as IDs, hardware addresses, etc.

In order to obtain textual artifacts from the final model, template-based code generators rely on code templates that contain both hard-coded static contents of the output file, and dynamic parts that emit outputs based on references to objects of the input model. For the sake of clarity, the model-to-text transformation usually does not involve extensive calculations, but implements a more or less direct transformation of the final model to a corresponding textual representation by serialization.

In case the format of the final artifact is XML for which an XML schema or a DTD exist, template-based code generation is typically not applied. Instead, a model transformation is used, and the configuration is serialized using the service calls provided by the modeling framework (here: EMF).

4.7   Variability and Design Space Exploration

Another aspect of MCSs that DREAMS covers is "variability". Variability relates to the degrees of freedom that architects and developers have when they choose for instance, specific features, such as encryption to improve security, or specific implementations such as the Advanced Encryption Standard (AES) to implement encryption. Whereas analyzing requirements help uncover the constraints that govern what the system is meant to do, analyzing variability focus on the leeway we have when building MCS.

Understanding variability helps optimize the design and implementation, and also helps build families of similar products so called "product-lines". By understanding the degrees of freedom we have when building a system, and, conversely, what constraints the system must meet, we can automatically evaluate alternative designs and select the one that brings the most interesting trade-off. Besides, understanding variability permits to derive new products for slightly varying requirements, by reusing as much as possible of existing products and exploiting these degrees of freedom.

4.7.1   Variability Modeling Viewpoint

The *variability viewpoint* metamodels help us to define the variability as a separate concern: The variability aspects and the domain-specific aspects belong to separate metamodels.

The *base variability resolution* tool (a.k.a. BVR, developed by SINTEF) provides the required variability specification metamodel, which captures what may vary in other metamodels.

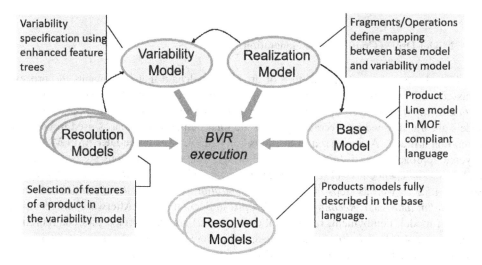

Figure 4.32: BVR Approach to Define Variability in a Separate Metamodel

Figure 4.32 sketches the approach to variability in BVR. The orange ovals as well as violet polygon represent BVR elements whereas the blue ovals depict models in any third-party language. The picture clearly shows that BVR does not amalgamate third-party languages (base models) with variability concepts rather defines variability in a separate model and links the base model by means of references. The BVR execution engine uses specified references to operate on base models to yield resolved models, i.e., products.

4.7.1.1   Modeling Variability

Intuitively, modeling variability is like creating a template document, where some well-identified parts are marked for later replacement with some ad hoc content. Templates are often initially derived from an existing document, where the parts to be specialized are stripped out or identified as samples sections. The process associated with the use of BVR follows a very similar scheme [122]:

1. *Preparing the product line* can be done by looking at existing models (i.e., products) and looking

for similarities and differences. The parts that vary often map to the variation points, as opposed to the parts that remain unchanged, which form the backbone of the future product line.

2. *Choosing a base model* consists in promoting one single existing product to be the matrix of subsequent products. BVR will use the model of this product, as input to a model-to-model transformation, which replaces each variation point with the associated model fragment and yields a valid new product, by substitution.

3. *Identifying a library of reusable fragments* significantly simplifies the use of the product lines, but enabling derivation of new products by feature selection. Variation points often have several possible solutions, which exist as model fragments, and which should be available for future injection even if they are not included into the selected base model.

4. *Creating a BVR model* formalizes the variation points, the base models, and the library of reusable fragments. It helps capitalize on the domain specific knowledge captured in the product line and to proceed with further product derivation.

5. *Generating products* is the final step, where one can generate new product by the sole prescription of the base model and the set of features to activate.

4.7.1.2 Exploiting Variability

Automated product derivation is the most emphasized feature of software product lines. By giving the user the possibility to select the features that one needs, it becomes possible to check the consistency of the whole product line, check the consistency of a given feature prescription, and to assemble the prescribed products. Checking the consistency of a product line as a whole consists in ensuring that there exists at least one single product that meets all the constraints embedded in the associated feature model. Interestingly feature models can be reduced to propositional logic formulae [123], and their validation thus boils down to the Satisfiability (problem) (SAT). Although SAT is well known to be a NP-Complete problem, recent advances in Software Product Lines (SPLs) [124] showed that industrial size SPLs form a very specific subset of SAT instance, which existing SAT solver can address.

Checking the validity of a specific feature prescription ensures that the prescribed features meet the constraints carried by the feature model. The prescription is valid if and only if the underlying variable assignment satisfies the associated logical formulae. SPLs thus permit to detect automatically invalid configurations that will not work in practice. Finally, assuming a given feature prescription is consistent with its enclosing SPL, it is possible to automate —possibly only partially— the construction and the validation of the associated products. This construction step is tightly coupled with the reuse capabilities of the underlying execution platform.

The BVR Tool Bundle currently supports product sampling based on generating so-called covering arrays [125]. The support is provided via the SPLCATool integrated into the BVR tool.

4.7.2 Variability Exploration Process

4.7.2.1 The Complexity of Product Lines

The construction of any complex system is greatly enhanced by designing loosely coupled parts, so that different products can emerge from assembling different set of parts, and even more so when it comes to software product-lines. As for verification, the systematic verification of every part however fails to guarantee the proper behavior of assemblies. Parts interact in unforeseen ways and, in turn, assemblies behave unexpectedly and eventually fail. In 1996, the European Ariane 5 space launcher exploded after about 40 s. of flight because of such an erroneous interaction [127], which

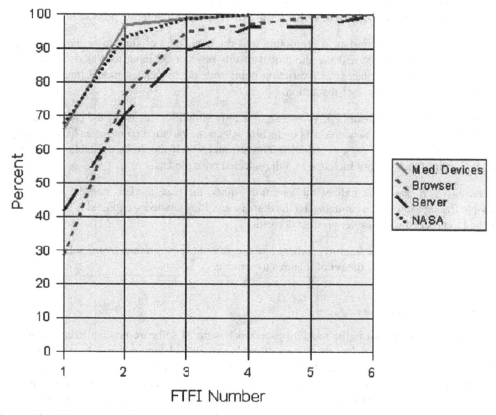

Figure 4.33: Percentage of Faults Detected Using Interaction Testing [126]

costs 500 000 000 USD. This feature interaction problem is characteristic of complex systems [128] and exacerbates the issue of their verification.

To be effective, verification thus has to cover every possible interaction. Empirical studies [126, 129, 130] show that the sole test of interactions between pairs of parts—so called 2-wise interactions test—already improves defect detection from about 50 % to 70 %. As shown on Figure 4.33, further including 3-wise interactions would then detect about 95 % of defects. All defects would eventually be found by investigating up to 6-wise interactions. Unfortunately, the number of possible interactions grows exponentially with the number of features. This is a key obstacle to the verification of industry-sized systems.

Feature interaction also impedes the verification of software product-lines. The number of systems that one can derive from a product-line grows exponentially with respect to its number of features. Deriving and verifying every single product is thus not feasible in practice.

4.7.2.2 Variability in Mixed-Criticality Systems

Many elements and properties can affect critical aspects of safety, timing and extra-functional properties such as reliability, energy consumption. It would be almost impossible and useless to list all the elements that can vary in DREAMS metamodels altering safety and timing. In the following, a list of a priori interesting variations is given (classified by metamodel), all of them heavily influencing safety and timing:

In Component model, the components of a system may vary in function of the selected features. For instance, variability emerges depending on:

- Presence/absence of components
- Management of logical ports connections of components
- Redundancy of components
- Different types of components (under replication or not)
- Versions of different operating systems
- Variations in the safety requirements (see example in Section 3.5.1.2)
- Variable requirements properties
  - Force to a tile/core
  - Isolation in one partition
  - Access rights to hardware elements

At the Platform model, the hardware elements can vary. Most interesting variability is:

- Presence/absence of hardware elements (nodes, tiles, cores, buses, watchdogs, clocks, RAM blocks, ROM blocks, etc.)
- Management of bus connections when eliminated
- Management of layout connections
- Variations in the safety requirements
- Variations in the safety manual

In the System software model, the hypervisors/partitions can vary. Interesting variability emerges from:

- Presence/absence of partitions
- Reallocation of hypervisors to other tiles
- Reallocation of partitions to other cores
- Versions and different operating systems supported by hypervisors
- Versions and different operating systems of the partition
- Variations in the safety manual

In the Safety model, all safety compliant items (SCI) corresponding to components, hardware elements, hypervisors and partitions can vary, as well as variations in the safety manual.

In the Timing model, the timing model itself may vary according to the selected software and software components.

The energy consumption model may vary according to the selected software and hardware components.

4.7.2.3 Exploring Business-level Variations

The DREAMS approach distinguishes between business and technical decisions, which resolve variability in what the system does and how the system does it, respectively.

Business decisions govern the features of interest, such as "all communication shall be encrypted". These decisions are first captured in a variability model, which documents how these decisions are realized into the system specification. We build product lines engineering tools to model variability using feature models, and to generate consistent sets of decisions using product sampling techniques.

Technical decisions govern the implementation of the features, for instance "the AES RC6 algorithm shall be used for encryption". These decisions are injected into the resulting specifications by evolutionary optimization techniques. Throughout this second refinement, extra-functional requirements are enforced using specific verification techniques such as the safety constraint checker.

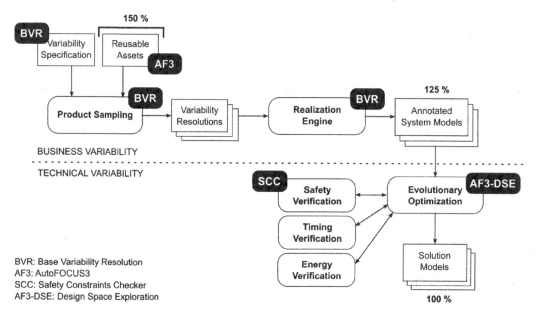

Figure 4.34: Overview of the Variability Management Process

The DREAMS variability management process exploits the inherent variability of MCS design models. As shown in Figure 4.34, it contrasts business variability capturing what the system does (the upper part), to technical variability covering how it does it (the lower part).

Variability is described by a variability specification (a BVR feature tree) and a set of reusable assets (mainly AF3 specifications), gathered into a so-called "150 %" model. We first address the business variability, by using product sampling techniques to extract a small set of products that maximize features interaction, and to derive the associated annotated AF3 system models, so called "125 %" models. The evolutionary optimization engine manages the remaining technical variability, especially replication and deployment schemes. It further explores variants of these "125 %" models, selecting those that best fit extra-functional constraints such as safety standard compliance, reliability or energy consumption. This process eventually yields a selection of solution models that fit both the extra-functional and the product lines constraints.

In Figure 4.34, we select a set of products, which are then fed as starting points for the evolutionary optimization. This selection includes sampling and realization. During sampling, we select specific products. Yet, our selection strategy eventually influences how successful is the design-space exploration. Interaction coverage for instance is one strategy, which ensures that verification

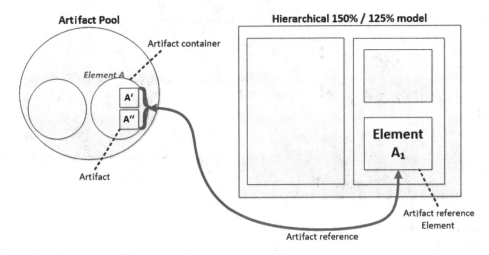

Figure 4.35: Element Library: Interlinking of the Artifact Pool with Model Elements [115]

techniques are effective, but does not help with the relevance of the final products. Uniform sampling is an alternative where each product has the same odds to be selected. Uniform sampling does not help relevance either, but it is a baseline against which other strategies may be compared. Diversity sampling is known to fit well evolutionary optimization, because it yields products that are different from one another, and therefore avoids premature convergence [131]. Then, during realization, we transformed these selected products into separate DREAMS system models. Sampling yields resolutions, which only describes the features included in each product. The realization of each product (see Section 5.6.3) implements the selected features, but if several implementations are possible, the final decision is delegated to the evolutionary optimization. In the DREAMS platform for instance, the deployment scheme does not alter the feature offered by the system, but affects the trade-off between extra-functional concerns. The optimization engine is therefore responsible for finding a relevant one.

4.7.3 Design-Space Exploration Process

The DSE process in DREAMS is based on multi-criteria optimization using genetic algorithms (GAs) to explore optimized mappings of a logical application architecture to a target platform with a focus on component-to-execution-unit mappings. Moreover, the DSE is capable of exploring suitable architectures of safety functions specified in a safety model (see Section 4.4.1) and considers diverse implementations of software components in the exploration. Finally, it also allows to explore redundant logical architectures including voter components that aggregate the outputs from replicated components.

Depending on the application and the product strategy, the DSE is used as a design aid in the early development process, or it is embedded in a workflow where it interacts with other development tools. In the context of the product-line exploration workflow, the DSE is executed for each "125 %" model where the business variability of the "150 %" model is resolved, but the technical variability remains. The unresolved variability points are additional optimization variables whose resolutions are explored by the DSE using feedback mechanisms.

In order to characterize technical variability, annotation-based and compositional variability is used [115]. The annotation-based variability typically are properties attached to model elements and whose values are not fixed, but resolved during the resolution of business or technical variability. Bounds may be given for these kinds of variability points that define the range from which the

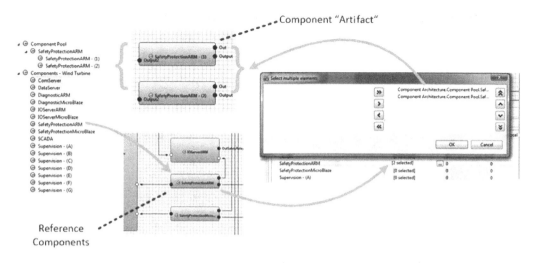

Figure 4.36: Example of Component Design Diversity [115]

resolution steps may instantiate values. Moreover, bounded variability points enable a multi-staged variability resolution where each step reduces the bounds until a final value for a product is instantiated. For instance, the replication bounds to increase the reliability of an application implement such variability bounds that can be reduced by BVR and instantiated to concrete replicas of components by the DSE. Here, a two-staged process is realized where the business variability resolution reduces the allowed minimal and maximal replication bounds while the technical variability resolution instantiates concrete values in the form of a deployment. Composition-based variability is defined such that a model element, e.g., a logical component, can be composed from linked model elements of a library. In DREAMS, the library is implemented by an *artifact pool* (see Figure 4.35) that is composed of *artifact containers*. They consist of model element artifacts that may substitute an *artifact reference element* from the actual model and are linked by an *artifact reference*. The artifact reference is a reference list that can be reduced by the business variability resolution, while the technical variability resolution instantiates the artifact reference element with a concrete artifact. This concept is used in DREAMS to express design diversity that is resolved by the DSE.

4.7.3.1   Architectures of Safety Critical Functions

One of the main features of the combined business and technical variability exploration workflow is the inclusion of safety modeling and automated safety checks (see [115]). In the following, a brief description of the deployment-centric requirements w.r.t. a system's safety functions that stem from the generic safety standard IEC 61508. This standard can be used directly for the certification of electrical systems or as a template for domain-specific standards, e.g., the ISO 26262 in the automotive domain. Thus, the developed generic exploration methods apply to wide range of different domains, as the basic safety considerations are similar.

Safety considerations in the development of a MCS include a risk analysis, the development of safety cases, and the realization of the safety requirements by safety functions. In a model-driven development process, safety functions are modeled as a part of the platform-independent model that is typically refined into a logical architecture model to be deployed onto the hardware/software execution platform. The realization of safety functions must satisfy a SIL that is determined by a risk analysis. The IEC 61508 standard, Part 2 [132] defines a table that associates a) the achievable SIL with the reliability of the hardware platform elements executing the safety function and b) the allocation of parallel channels of a (replicated) safety function to sufficiently independent hard-

Table 4.26: Achievable Safety Integrity Level of Safety Functions Related to Platform Element Failure Probability [132]

Safe Failure Fraction	HFTs		
	0	1	2
$X < 60\%$	-	SIL1	SIL2
$60\% \leq X < 90\%$	SIL1	SIL2	SIL3
$90\% \leq X < 99\%$	SIL2	SIL3	SIL4
$X \geq 99\%$	SIL3	SIL4	SIL4

ware elements (HFT; see Table 4.26). In addition, the IEC 61508 standard, Part 6 [133] describes architectures that realize safety functions and influence the achievable SIL of the safety function.

The architecture variants of safety functions differ in the number of replicas of their components and corresponding interconnects, the presence of additional diagnostic units, and the voting strategies. These properties of safety functions have a large impact on the non-functional properties that are explored by the DSE, aside from their impact on the safety of a system, e.g., its energy consumption and the reliability of sub-systems. Hence, the exploration of safety architectures for safety functions results in larger design spaces that can be explored to obtain "better" systems w.r.t. the defined overall design goals. Concretely, the selected architecture of safety functions, the design diversity of its constituting components, and their deployment impact the achievable SIL of a particular safety function and are inter-dependent with the aforementioned other non-functional properties of the system.

4.7.3.2 Input Specification

As mentioned before, the DSE operates on "125 %" models that are passed from the business variability resolution step for which it resolves the remaining variability. The DSE produces a set of optimized "100 %" models from which a promising product model can be extracted. In the following, we will consider the case where all DSE features described in Section 5.1.2.2 are enabled, although reduced invocations are of course possible. Namely, these features are enabled: the exploration of safety architectures, the exploration of software design diversity, the exploration of redundancy, and the exploration of component-to-execution-unit mappings.

The required input models are comprised of a logical component architecture (see Section 4.2.1.1), a timing model (see Section 4.3), a platform model (see Section 4.2.4), a safety model (see Section 4.4), and a deployment model (see Section 4.5) that carries the information about the deployment-specific parameters, e.g., the characteristic WCETs of components allocated to a particular execution unit. Furthermore, a specification of the exploration targets, i.e., design goals, in the form of objectives and constraints are required. These are specified using the Exploration metamodel [134] and typically defined in the "150 %" model. They are adjusted by the business variability resolution step, if required, such that the exploration targets are available in the "125 %" models.

4.7.3.3 Synthesized Artifacts

The set of output models of the DSE includes a deployment (see Section 4.5), a logical component architecture (see Section 4.2.1.1), a safety model (see Section 4.4), and a timing model (see Section 4.3). These models are "100 %" models, i.e., product models that do not contain any variability points and can be processed further by the DREAMS toolchain (see Section 5.6). Depending on the inputs to the DSE, different artifacts of these models are synthesized or existing models are reduced to match the technical choices of the exploration for a particular solution. Separated by the features of the exploration, the following list describes the output artifacts of the DSE.

- **Component-to-Execution-Unit Mappings**

  A mapping of components to execution units is the main result of the DSE and embedded into a deployment model as a list of `ComponentAllocations`. Moreover, the DSE calculates the required messages and their routes to derive the virtual links along with the respective `PortAllocations`. These two artifacts constitute the synthesized deployment model.

- **Safety Architectures**

  The exploration of safety architectures aims to explore the set of possible architectures for a safety function such that an optimal realization of the function in the synthesized logical application architecture is present that satisfies the function's safety requirements. This requires a matching component-to-execution-unit mapping that respects the required spatial isolation properties of the chosen architecture for each valid solution. Choosing a suitable safety function architecture implies many trade-offs, e.g., using more safety channels (see below) typically raises robustness but increases the energy consumption. A safety architecture is specified by its number of independent *safety channels*, its diagnostic, and voting units: For instance, a 1oo2D architecture describes a safety function that has two independent safety channels. One of these channels must send a trigger signal such that the voter signals that the safety function was triggered. Furthermore, this architecture specifies that the safety function is connected to some diagnostic unit that influences the voting results.

  The DSE uses the safety model to identify those components of the logical application architecture that are associated with the safety functions. In order to reflect the choices w.r.t. to the safety architectures, an additional logical component architecture is synthesized that contains the additional components, e.g., for the additional safety channel. The independence of the channels is guaranteed by the DSE by invoking the safety constraint checker (see Section 4.4.3) which checks each proposed component-to-execution-unit mapping and rates it according to its safety properties. For instance, a suitable mapping is characterized by allocating the components of different safety channels to sufficiently independent hardware.

  Furthermore, a safety model is synthesized that is a reduction of the input safety model. This model only defines the exact architecture choices for safety functions that were selected by the solution obtained from the DSE. Moreover, unneeded element descriptions, e.g., optional Partitions, are removed from the safety model if they are not used in the component-to-execution-unit mapping.

- **Software Design Diversity**

  In the introduction of this section, composition-based variability has been introduced, which is used in DREAMS to realize software design diversity by providing a set of components that must be used to instantiate a reference component, i.e., "interface" component, from the input logical application architecture. This concept is complemented by the DSE which is able to select suitable designs for reference components based on their characteristic properties from a library. In order to reflect the choices of the DSE, it synthesizes a logical component architecture that instantiates the reference components with a specific design from the referenced library.

- **Redundant Logical Architectures and Voting**

  If any component has been replicated, an additional logical component architecture is derived from the input architecture. In addition to the existing components, copies of the replicated components are added such that the synthesized model matches the information encoded in the internal component-to-execution unit mapping. Furthermore, voting components are instantiated that transform the outputs of all instances of a replicated component into a single signal that is passed to subsequent components.

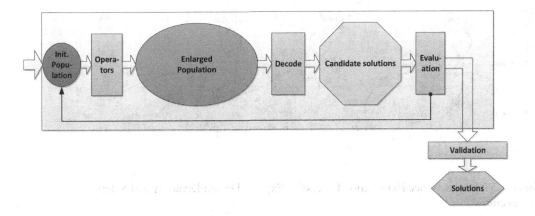

Figure 4.37: Overview of the Used Evolutionary Algorithm

In addition to the above mentioned artifacts, a timing model is synthesized along with a logical component architecture. This is triggered by any features that include the replication of model elements, or their instantiation: The exploration of safety architectures, the exploration of software design diversity, and the redundancy and voting exploration.

4.7.3.4 Technical Variability Exploration

The Technical Variability Exploration is based on the composite exploration framework discussed in Section 5.1.2 that enables a modular realization of DSE features (e.g., the exploration of safety architectures). In the following, we will explain the complete exploration process where all features are enabled (i.e., safety architecture exploration, component-to-execution unit mapping exploration including replication, and component instantiation). Detailed information about the DSE framework is given in Section 5.1.2.6 whereas the focus of this section is the use of the DSE framework to implement the technical variability exploration.

Overview. First, a brief overview of the base GA optimization framework is given which has been extended to the composite framework discussed in Section 5.1.2.6 (see Figure 4.37). In the starting phase of the algorithm, creators construct an initial population of *individuals* that consists of a *genotype*, a *phenotype*, and its evaluation results w.r.t. the defined objectives and constraints. In the starting phase, the individuals only have a genotype, while the phenotypes and evaluation results are added by subsequent steps of the algorithm. *Decoders* process the genotypes into phenotypes that can be rated by *Evaluators*. After applying the evaluators, the population is pruned to preserve only the fittest individuals w.r.t. the evaluation results and based on a given target size of the population, i.e., the number of individuals. If the target size of the population is not reached, no pruning is performed to maintain diversity in the population. Then, a *mating* mechanism generates the *offspring* individuals, which then consists only of a genotype, and applies operators to modify the genotypes, which are completed by decoders and evaluators. When the algorithm reaches the stopping criterion, the population is validated to contain only feasible solutions w.r.t. the constraint set, a Pareto-Front (see Section 5.1.2) is created and presented to the system designer.

In the following paragraphs, the DSE considers those components that shall be deployed to the target platform, as indicated by the given timing model, as tasks that shall be deployed, scheduled etc. When the DSE exports a solution of the exploration problem, the task representation is again transformed back to a component model with a matching deployment.

Technical Variability Exploration. Figure 4.38 shows the dependencies of the individual steps of the extended exploration process, as well as the operations that are performed to mod-

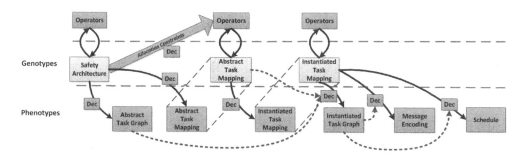

Figure 4.38: Design Space Exploration Process to Explore Design Diversity and Safety Architectures

ify genotypes ("Operators") and the operations to obtain phenotype encodings (decoders, denoted by "Dec"). In the following, when the exploration process starts, the initial population (set of individuals) must be created (see Section 5.1.2.6). Due to the dependencies within the composite genotype, which composes the overall safety architecture exploration problem, the creators for the sub-genotypes must be executed in a particular order: First, the safety architecture encoding is created, then the abstract task mapping, and finally the instantiated task mapping. These genotypes correspond to the exploration features "exploration of safety architectures", "exploration of design diversity", and "component-to-execution-unit mapping", which includes the "redundancy exploration".

In order to construct the safety architecture encoding, a logical input architecture and a corresponding timing model is required. Then, a decoder is used to derive an abstract task graph encoding from the safety architecture encoding and the input system models. The abstract task graph is an input encoding to the creator that constructs the abstract task mapping. It also uses an abstraction of the target platform to generate the task-to-execution unit mapping. Finally, the abstract task mapping is the input to a further creator that generates an instantiated task mapping such that all defined genotypes are created.

In the complete process, the safety architecture encoding is a pure genotype while the abstract and the instantiated task mappings are hybrid phenotypes (see Section 5.1.2.8), which are updated by the decoders to account for the changes that may have been applied to the equivalent genotypes. For instance, if an operator modifies the safety architecture encoding, these changes must be reflected in the abstract task graph that serves as an input to the decoder, which constructs the abstract task mapping. Here, the decoder only performs modifications to the existing hybrid phenotype such that previous optimizations (w.r.t. previous iterations) are preserved. The decoders executed last are responsible for generating an instantiated task graph encoding, a message encoding, and a strictly time-triggered schedule (in that order).

The evaluation of the obtained individual is performed by evaluators corresponding to the objectives and constraints defined by the input *exploration specification*. The available user-definable objectives and constraints are specified in Section 5.1.2. Based on the evaluation results, a set of individuals is selected, which serve as a basis to generate the population for the next iterations. The selection is based on the Pareto-optimality of the individuals comprising the population (see Section 5.1.2).

Operators modify the genotypes to derive alternate solutions. Subsequently, operators are applied to genotypes that are chosen by a *Selector* to generate new individuals that may be incorporated in the population in the next iteration based on their evaluation results. The operators implemented for the technical variability exploration are based on mutation and crossover operations which are the basic modification operation of genetic algorithms.

Once a fixed number of iterations is reached, the resulting Pareto-Front of individuals is calculated and presented to the user after a final constraint check that ensures only feasible solutions can be chosen. For selected solutions, the synthesized artifacts listed in Section 4.7.3.3 are exported as models compliant to the DREAMS Metamodels. These models are "100 %" models to which the process defined in Section 4.5.3 can be applied to obtain deployable configuration artifacts for each of the products.

5

Algorithms and Tools

J. Migge
RealTime-at-Work

P. Balbastre
Universitat Politécnica de Valéncia

S. Barner
fortiss GmbH

F. Chauvel
SINTEF

S. S. Craciunas
TTTech Computertechnik AG

A. Diewald
fortiss GmbH

G. Durrieu
ONERA

G. Fohler
Technische Universität Kaiserslautern

Ø. Haugen
SINTEF

A. A. Jaffari Syed
Technische Universität Kaiserslautern

C. Pagetti
ONERA

R. Serna Oliver
TTTech Computertechnik AG

A. Vasilevskiy
SINTEF

5.1		Variability and Design Space Exploration	166
	5.1.1	Variability Analysis and Testing Techniques	166
		5.1.1.1 Software Product Line Implementation Techniques	166
		5.1.1.2 Strategies for Software Product Line Analysis	166
		5.1.1.3 Software Analysis Techniques	167
		5.1.1.4 Testing of Software Product Lines	167
		5.1.1.5 Base Variability Resolution	169
	5.1.2	Multi-Objective Design Space Exploration	170
		5.1.2.1 Related Work	171
		5.1.2.2 DSE Features	172
		5.1.2.3 Input Specification	172
		5.1.2.4 Synthesized Artifacts	176
		5.1.2.5 Internal Representation	177
		5.1.2.6 Overview of the Composite DSE Framework	178
		5.1.2.7 Composite Encoding	179
		5.1.2.8 Composite Exploration Process	180
		5.1.2.9 Composite Safety Architecture and Mapping Exploration	182
5.2		Scheduling	184
	5.2.1	Timing Decomposition	184
		5.2.1.1 Scheduling Domains	185

		5.2.1.2	Clocks and Coordination of Schedules	186
		5.2.1.3	Timing Decomposition Heuristic	187
		5.2.1.4	Task Scheduling Domain	190
		5.2.1.5	On-Chip Communication Scheduling Domain	191
		5.2.1.6	Off-Chip Communication Domain	191
		5.2.1.7	Tool: RTaW-Pegase/Timing	192
	5.2.2	Partition Scheduling		192
		5.2.2.1	Model	192
		5.2.2.2	Allocation of Partition to Cores	194
		5.2.2.3	Core Schedule	194
	5.2.3	On-Chip Network Scheduling		194
		5.2.3.1	Model of STNoC Used in the DHP	194
		5.2.3.2	Scheduling Mechanisms and Timing Guarantees	196
		5.2.3.3	Approach for Safety-Critical On-Chip Communication	197
		5.2.3.4	Approach for Non Safety-Critical On-Chip Communication	198
		5.2.3.5	Tool: RTaW-Pegase/Timing	198
	5.2.4	Off-Chip Network Scheduling		198
		5.2.4.1	Network Model	200
		5.2.4.2	Formalization of Scheduling Constraints	202
		5.2.4.3	User Constraints	204
		5.2.4.4	SMT-Based Scheduling Algorithm	218
		5.2.4.5	Evaluation	219
5.3	Adaptation Strategies			221
	5.3.1	Recovery Strategies		222
		5.3.1.1	GRec Global Algorithm	223
		5.3.1.2	Constraint-Based Reconfiguration Graph Problem Formulation	225
	5.3.2	Comprehensive Offline Schedules		230
		5.3.2.1	Related Work	231
		5.3.2.2	Background	232
		5.3.2.3	System Model	234
		5.3.2.4	Methodology	234
		5.3.2.5	Tool: MCOSF	236
	5.3.3	Flexibility		236
		5.3.3.1	Related Work	237
		5.3.3.2	Terminology and Notations	238
		5.3.3.3	Job-Shifting Algorithm	239
		5.3.3.4	Efficiency Evaluation	245
		5.3.3.5	Overheads Evaluation	247
		5.3.3.6	Tool: MCOSF	250
5.4	Timing Analysis			250
	5.4.1	Problem Definition		250
	5.4.2	On-Chip Network		251
	5.4.3	Off-Chip Network		251
	5.4.4	Task Timing Analysis		251
	5.4.5	Composition		252
	5.4.6	Tool: RTaW-Pegase/Timing		252
5.5	Generation of Configuration Files			252
5.6	Toolchain			254
	5.6.1	Use Case 1: Basic Scheduling Configuration		255
	5.6.2	Use Case 2: Scheduling Configuration with Resource Management		257
	5.6.3	Use Case 3: Variability and Design-Space Exploration		259

In Chapter 4 the DREAMS meta-model and the model driven development process have been introduced. Figure 5.1 gives an overview of the DREAMS model driven development process. The first activity consists in the creation of "input models" that cover applications, the platform and a variability model that identifies feature trees and features selection rules. The resulting application and platform model "with variability" specify a product line. The "variability binding" activity consists in the selection of features to obtain concrete application and platform models. The following activity consists of a series of design decisions regarding deployment, scheduling, adaptation and produces the Platform Specific Model. Before its translation into configuration files in the last step, the Platform Specific Model needs to be verified, i.e., it needs to be checked that all requirements (e.g., timing, safety) are met.

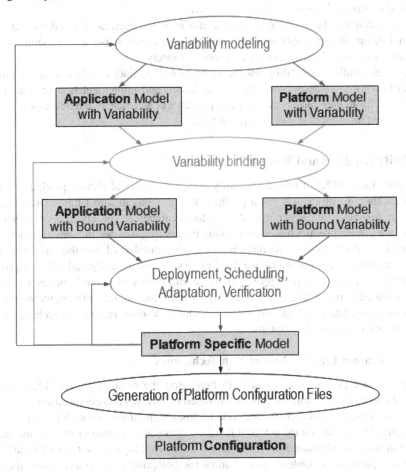

Figure 5.1: Workflow of the DREAMS Model Driven Development Process

This chapter introduces the algorithms and tools to support the design and verification activities of the model driven development process. In addition, the scheduling and configuration algorithms are described to support different scheduling domains of the DREAMS architecture. This chapter begins with Section 5.1.2, which describes variability and design space exploration in the design of mixed-criticality systems. In Section 5.2, scheduling algorithms at different levels, e.g., partition level, task level, on-chip and off-chip communication are elaborated. Adaptation strategies is another topic which is covered in Section 5.3. Recovery strategies, transition modes for faster switching and algorithms for online admission of tasks in offline scheduling tables are described in this section. Timing analysis is described in Section 5.4 at different levels and Section 5.5 describes the specification of the configuration files and how they are generated by the tools. At the end, in Section 5.6 three toolchain use cases are presented that help to apply the toolchain.

5.1 Variability and Design Space Exploration

Variability and requirements form the two faces of the same coin. Whereas requirements analysis uncovers the constraints and objectives that govern what the system must do, variability focuses on the degree of freedom that architects and developers have.

Understanding variability helps to better exploit the inherent degrees of freedom and search for a near-optimal design that not only meets functional requirements, but also maximizes extra-functional concerns such as reliability, energy efficiency or safety.

In addition, understanding variability also helps building "product-lines", which are families of similar products whose use-cases vary, but whose design and implementation remains highly similar. Variability helps to maximize reuse between products at all levels of the development cycle, and in turn reduce development cost and time to market.

5.1.1 Variability Analysis and Testing Techniques

A Software Product Line (SPL) of mixed-criticality systems is a set of similar products that still differ from one another. To produce a given product, the user has to select the desired features. Resolution and the associated tools generate the product realization. As with any other software engineering tool, a software product line must provide means to guarantee that any product is verified and validated. As the number of features increases, the number of possible products grows exponentially. Therefore, on the one hand, we need analysis tools to verify and validate products and, on the other hand a way to cope with the exponential number of possible products. As proposed in [135], software product lines encompass (i) product line implementation techniques, (ii) strategies for product line analysis and tool strategies, and (iii) software analysis techniques. The following sections describe these dimensions.

5.1.1.1 Software Product Line Implementation Techniques

Implementing a software product line consists in managing the product variety. There are two main implementation techniques, namely annotation-based techniques and composition-based techniques. In annotation-based techniques, source code fragments are annotated with features or combinations of features. Depending on the selected features, some code fragments may be included in the final product or may be eliminated. In composition-based techniques, the product results from the composition of separate executable units (e.g., modules, components, services), whose inclusion is governed by the selected features. In DREAMS, such units are software components, hardware platform elements, system software elements (e.g., hypervisors, partitions), safety compliant items and model elements like bus connections or their properties. A set of meta-models represents the composition of the product line and an external explicit variability model is in charge of eliminating elements not present in the final product.

5.1.1.2 Strategies for Software Product Line Analysis

Software product line analysis is the process of scrutinizing product lines and the set of products they define. The related analysis techniques focus either on a subset of representative products covering as many errors as possible, or take into account the variability and are able to analyze the family of products as a whole (these techniques are called family-based). To realize these techniques, tools employ mainly three approaches, namely product-based strategies, variability-aware strategies and the variability-encoding strategy.

In *product-based analysis*, the analysis techniques are applied to the generated products. In *variability-aware* strategies, the tooling is able to handle the product line as a whole, without deriving any final products. For example if the variants are represented with annotations (e.g., "#ifdef" directive in C/C++), these techniques evaluate the consequences of different sets of features. Finally, using *variability-encoding*, variability is encoded using conditional branching, so that the whole product line is a single complex executable product.

5.1.1.3 Software Analysis Techniques

These techniques are classified into three categories: (i) testing, (ii) verification and (iii) further analysis. In addition, there is the orthogonal category of sampling for the techniques that are applied to only a subset of products generated according to a given coverage of feature criteria.

Testing techniques imply executing the product to assess some properties. Testing techniques include test case generation, product sampling, and family-based testing. The use of test cases is common practice in Software Engineering. However, in the case of product lines, it may be necessary to generate customized test cases for each product, because features affect what can be tested. Test cases are in turn divided into unit-tests (to test individual functions/components), integration tests (to test compositions of units), and performance tests (to test performance properties such as response-time, accuracy, etc.). Techniques of product-reduction are able to reduce the number of products needed for a given test case. Finally, family-based testing techniques can execute all products in parallel, e.g., computing multi-value data for the combination of features.

In contrast with tests, verification techniques analyze the product without executing it. Verification techniques encompass type-checking, static analysis, model checking, theorem proving and consistency checks. Type checking makes sure that the source code of every possible generated product compiles with no errors. Static analysis operates at compile-time and predicts run-time behaviors without actual execution. Software-model checking translates the product line, when possible, into a state machine that is easier to analyze. The graph represents the whole family of products and different combinations can easily be checked. Theorem proving first translates the program and its specifications into logical formulas and uses deductive techniques to check its correctness. Finally, in consistency checking, the consistency of artifacts corresponding to different features is tested. For example, it is checked whether all involved artifacts are present and whether dead or superfluous source code is produced under some combinations of features.

Beyond the verification and testing of the functions provided by a product, extra-functional properties can be analyzed, as well as metrics reflecting the quality of its source code. One of the goals of the DREAMS platform is to optimize certain extra-functional properties (e.g., energy consumption, response time) while verifying others (e.g., safety claims, reliability). The analysis techniques are specific to each extra-functional meta-model of DREAMS and are a main research challenge.

5.1.1.4 Testing of Software Product Lines

The idea of automated testing is to invoke a piece of software with specific inputs for which the expected outputs are known a priori in order to detect discrepancies with actual outputs. We review below the main techniques developed to test product lines technologies. Interested readers may find a more comprehensive treatment in [136, 137]. We highlight two main techniques, namely the 150 % model and the coverage array techniques, jointly developed by SINTEF and University of Oslo.

Testing mainly varies depending on the scope of the System Under Test (SUT). At the finer scale, single routines are tested independently by so-called unit-tests. At a medium scale, the interaction between two or more components is performed by integration tests. Finally, the complete systems or products have to be tested by the end-to-end tests, checking specific usage scenarios.

Execution time, complexity and cost of maintenance all grow as scale does. Unit, integration and system tests represent different activities in the development process: end-to-end tests relate to general requirements, integration tests to the general design and unit tests to the detailed design. Since testing is a resource intensive activity, a "brute force" approach to SPL testing requiring to explicitly and separately test all possible products is not feasible. Testing SPL requires each activity to be reconsidered under variability: The "W" development model in Figure 5.2 is an attempt to adapt the well-accepted V-model to variability.

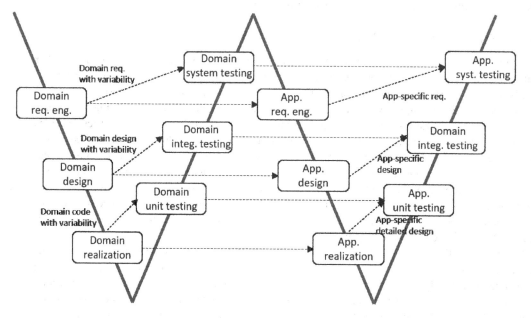

Figure 5.2: The "W Model" [136]: An Attempt to Formalize the Underlying Testing in the Context of Software Product Line

The W-model distinguishes between domain testing (i.e., product line) and application testing (i.e., product). Domain-centric activities result in test artifacts (e.g., test cases, test plan) which shall be reused across products from the family, whereas application-centric activities result in artifacts tailored for a single specific product. It is worth noting the importance of having system tests for each possible product. Having a set of well-tested components, each passing a separate large suite of tests, cannot detect issues occurring when two or more components interact. This problem, known as the feature interaction problem is one of the major challenges in SPL testing. We discuss below three main techniques addressing SPL testing: model-based testing, incremental testing and combinatorial testing. Interested readers may refer to [138] for additional details.

In Model-Based Testing (MBT), models are used to capture the desired behavior of the SUT, the testing strategies of interest, or both. For instance, one can describe the system as a finite state machine, capturing the set of legal inputs and the set of associated outputs. It is thus possible to generate a test suite (i.e., a set of test cases) to reach a specific coverage criterion. In the context of SPL, Cichos et al. [139] proposed to build for instance a 150 % model as a single state machine aggregating the behavior of all possible products. This 150 % model can thus be scoped down to generate tests covering any product resulting from the product line.

An alternative to minimize the cost of testing SPL is the use of incremental testing strategies. New test cases are generated based on the difference between the SUT and other products that have already been tested. Incremental testing exploits the relationship that binds SPL testing to regression testing: as regression testing aims at retesting a software piece that has changed, it can be used to test a new product, which was conceived as an extension of an existing one.

Another promising approach to SPL testing is the use of combinatorial interaction testing. The idea is to select a small subset of products, whose executions are likely to trigger feature interaction problems. As shown by Kuhn et al. [140], SPL follows some sort of 80/20 rule: most bugs are related to a few parameter configurations. SINTEF developed a technique to automatically select a minimal subset of products that maximizes the number of interactions exercised during testing [124], and in turn, the likelihood of detecting a feature interaction issue. The associated tool, called Covering

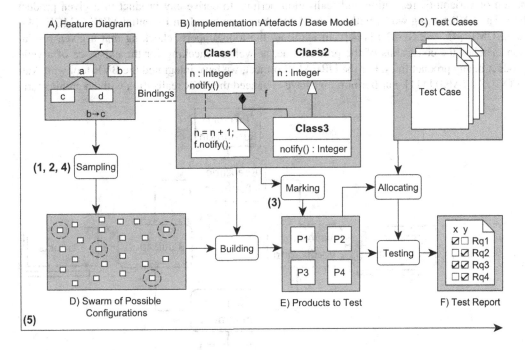

Figure 5.3: ICPL Approach to Testing of Software Product Lines

array generation algorithm for Product Lines (ICPL) [138] adopts the combinatorial testing of software product lines. The main contributions of this tool are summarized in Figure 5.3. Testing SPL requires three main inputs:

- A software system and its implementation artifacts.

- The feature model captures the inherent variability, and in turn the set of possible variants, which can be derived from the given software artifacts.

- A set of test cases used to validate products derived from the systems.

The first step consists in sampling the space of possible products, in order to cover the possible t-wise interactions between features (i.e., 1-wise ensures that all features are selected at least once, whereas 2-wise coverage ensures that each pair of features is selected at least once). The resulting products can thus be automatically built by assembling existing software artifacts, and tested using the provided test cases.

5.1.1.5 Base Variability Resolution

In DREAMS, variability modeling is supported by the Base Variability Resolution (BVR) tool [141] that is a reference implementation of the BVR language [142]. The BVR language promotes the orthogonal approach to variability [143] and therefore, it does not clutter a language of the product line with variability concepts. Figure 4.32 in Section 4.7.1 depicts the BVR methodology to specify variability in SPL. To define a software product line, BVR distinguishes between variability, resolution and realization models. A variability model outlines a set of all possible features with their constraints in a product line and therefore, it defines a family of all possible products. A resolution of the variability model specifies a set of features to include in a particular product. A realization of the features describes how to inject selected features into the product. In a nutshell, a combination of variability, resolution and realization permits to derive any product in a given product line. The BVR tooling works with any modeling language based on Essential MOF (EMOF) [44] (e.g., UML [45, 46], SysML [144]). In addition, the tool supports checking that a given resolution matches the constraints of the product line, as well as checking for the existence of a valid product in any product lines. For the DREAMS project, we have integrated the BVR tool with AutoFOCUS 3 (AF3) [111]. Furthermore, we have enhanced the BVR realization with an architectural

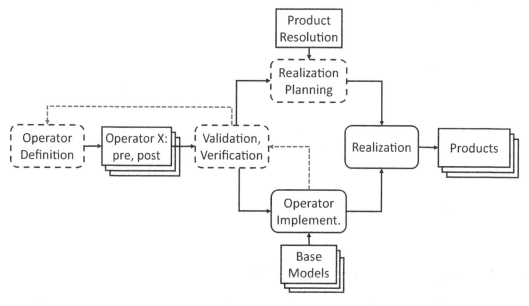

Figure 5.4: BVR Realization Process

pattern to cope with changes in software product lines throughout a product life-cycle. We break down our realization layer into a set of independently engineered operators. Each operator encapsulates a set of instructions to yield a new product. An operator describes assumptions on an initial state of a partially resolved product and guarantees a certain state of the product after an execution of the operator. These assumptions and guarantees are expressed in the form of pre-conditions and post-conditions and therefore, we can chain them into a sequence of the operators to derive a new product. Furthermore, we can use planning techniques to calculate this chain automatically. In addition, we propose a process to develop the realization layer. Figure 5.4 outlines this process where the steps can be parallelized and delegated to engineers with different competences to increase efficiency and scalability of the approach. An interested reader should refer to [145] for a comprehensive treatment.

5.1.2 Multi-Objective Design Space Exploration

The multi-objective Design Space Exploration (DSE) in DREAMS explores optimal deployments of selected components of a logical application architecture (equals tasks) to execution units (here: partitions) of a target platform architecture. The multitude of possible design goals induces to determine optimal deployments by means of a Pareto-Front[1] based on their evaluation w.r.t. the defined objectives, e.g., a system's energy consumption, and the properties of the elements composing the logical and platform architectures. Further, the DSE allows to define functional and non-functional constraint sets that must be satisfied by candidate solutions to obtain valid deployments for a system. The basic set of input models to the DSE is comprised of the set of objectives and constraints that are specified using the Exploration Meta-Model [134] and the input system models (see Section 5.1.2.3).

In order to explore the design space spanned by the logical and the platform architecture we employ a Multi-Objective Evolutionary Algorithm (MOEA), a genetic algorithm in particular. Its population-based approach, which maintains a set of candidate solutions, renders this exploration method an appropriate choice to explore the solution space spanned by multiple objectives. Although the latest generation of Satisfiability Modulo Theory (SMT) solvers also allows to optimize multi-objective optimization problems (see for instance [146]), the flexibility of evolutionary algorithms allows a straight forward realization of the framework presented in the following sections. Moreover, several optimizations of the algorithm enable to overcome the limitations of generic evolutionary algorithms w.r.t. combinatorial problems, e.g., the task allocation problem that is solved by the presented approach.

Depending on the application and the product strategy, the DSE is typically either used as a design aid in the early development process to reduce the engineering effort, or embedded in a workflow with a specific mission. It supports the system engineer by avoiding infeasible application designs for the mapping of the logical component architecture to the target platform, and to derive suitable application architectures (here with a focus on safety). Another scenario is the application in the variability exploration process discussed in Section 4.7.2 where the DSE is executed for a number of products to resolve technical variability.

In the following section, the related work is briefly summarized. Then, the in- and output specifications of the DSE are given that connect the DSE with the DREAMS common metamodel described in Chapter 4. Subsequently, the internal representation and the DSE algorithm are discussed.

5.1.2.1 Related Work

DSE has been applied to a large variety of design problems, with the goal to automate trade-off and dimensioning decisions in system design. DSE typically employs optimization or constraint solving techniques to derive designs that satisfy the system's requirements. Hence, a large number of generic and problem-specific approaches to explore the design space have been developed, e.g., in the area of the software architecture design and task allocation problem [147–149] or hardware/software co-design [150, 151].

Most exploration algorithms are based on generic optimization or constraint solving algorithms, often including custom algorithmic extensions and rarely use distinct search algorithms [147]. In [148], a SMT solver is used to obtain an initial solution for an Evolutionary Algorithm (EA) to optimize a deployment of software components to a hardware platform. For the chosen case study, that has a large and restrictive constraint set that is modeled using a design specific language, the initial solution was not further optimized by the EA. The meta-optimization framework Opt4J [152] includes solvers based on a genetic algorithm, SAT formulations, and others. It supports multi-objective optimization by means of a Pareto-Front and has been primarily applied to task allocation problems, e.g., [153]. Beyond the scope of the task allocation problem, [149] also explores the set of optimal hardware platform architectures by a step-wise increase of the platform's granularity and instantiating hardware elements by elements from a library using a framework incorporating different solver methods, e.g., mixed-integer linear programming.

Considering safety-critical systems, [154] uses a dedicated algorithm to explore the distribution

---

[1]Pareto-optimal solutions are characterized such that none of their evaluation results can be improved without worsening any other of their evaluation results.

of safety-relevant functionality to software components to reduce the costs associated with higher integrity levels. In order to increase the reliability of the resulting system, [155] proposes an EA-based method that allows the exploration of distributed embedded systems and their architecture using EAs and Binary Decision Diagrams for the reliability analysis. A similar approach to explore system designs w.r.t. their reliability is presented in [112] that combines a DSE with a model of the platform's capabilities. In the area embedded system design, [156] developed methods that consider DSE as a series of model-transformations. Similarly, [157] defines patterns of steps performed by a DSE, such as analyses or the creation of candidate solutions, and that enable tracing the artifacts derived in different steps of the DSE process.

5.1.2.2   DSE Features

The Composite DSE used in DREAMS targets mixed-criticality systems such that the synthesized deployments have to fulfill the requirements imposed by the critical subsystems. Nevertheless, the DSE focuses mainly on high-level aspects affecting the correct design of critical subsystems since it is employed in the early design phase. Hence, the exploration mechanism has been extended to combine the selection of suitable architectures for safety functions, e.g., 1oo2D (see IEC 61508, Part 2 [132] and Section 4.7.3.1). Furthermore, the exploration of redundant logical application architectures increases the reliability of selected components.

The aforementioned features are briefly summarized in the following:

- **Exploration of architectures of safety function:** This exploration feature requires a safety model (see Section 4.4.1) from which the DSE can extract a list of safety functions that are present in the application. Moreover, the model links the safety functions to its implementing components, the required safety level (assuming no redundancy), and a list of the achievable safety level of the execution units present in the target platform.

- **Software Design Diversity:** The exploration of alternate designs of software components requires a component library that contains the different component designs. It is passed as a secondary logical component architecture to the DSE. These alternate designs can be required by safety standards, or can be beneficial for the exploration by their differences in their properties.

- **Redundant Logical Architectures and Voting:** The exploration of redundancy depends on the provisioning of upper and lower bounds for the number of component instances in the annotations of the input logical component architecture model.

The listed exploration features are enabled based on the defined exploration targets and the available information in the input models. For instance, the safety architecture exploration is only executed if a safety compliance constraint is present in the *ExplorationSpecification* passed to the DSE. Based on the *ExplorationSpecification* and the input models, the DSE calculates Pareto-optimal solutions. These solutions can be exported from an internal format to models that are compliant with the DREAMS metamodel (see Chapter 4).

5.1.2.3   Input Specification

In the following sections, the DSE considers those components that shall be deployed to the target platform as tasks that shall be deployed, scheduled etc. When the DSE exports a solution of the exploration problem, the task representation is again transformed back to a component model with a matching deployment.

In order to perform the exploration of potential deployments of the logical component architecture to the target platform, a specification of the objectives and constraints of the exploration and models of the application are required.

Table 5.1: Used Symbols and Operators to Describe Exploration Targets

Symbol	Description	Symbol	Description
$\#(\ldots)$	Extracts the number of elements of a set	$C_{\text{inst}}(T_i)$	Identifies the non-abstract component[2] represented by the task T_i
$\rightarrow$	Operator that assigns a requester to a resource	e_i	Execution Unit with index i
T_i	Task with index i	$\mathcal{E}$	Set of execution units to which tasks can be deployed
$\mathcal{T}$	Set of tasks	$E_{\text{table}}(T_{c_i}, e_j)$	Extracts the stored energy value from a table (see Section 4.5.1)
$\mathcal{T}_{\text{RI}}$	Set of replicated and instantiated tasks	S	System Schedule
$\mathcal{T}(c_i)$	Set of tasks representing the component c_i	D	Set of deadline constraints
$TG(c_i)$	Task Graph containing the component c_i		

Design Goals - Objectives and Constraints

Objectives and constraints are defined using the Exploration Meta-Model (EMM) [134] in an *exploration specification* and are orthogonal to the exploration features discussed in the previous section. The DREAMS DSE is extensible such that new objectives and constraints can be realized by metamodel extensions and providing the corresponding evaluators. By default, the DSE offers the following objectives:

- **Energy Minimization:** Using a simplistic energy model, the DSE aims to minimize the overall energy consumption by exploiting the heterogeneity of the target platform. The energy model assumes no energy consumption of idle execution units, while the execution of components consumes a defined amount of energy. These values are specific for the target execution unit and may be an estimate or based on measurements. The use of this simplistic model is justified due to the early nature of the design phase in which the DSE is used and the unavailability of information that is required for a more precise calculation of the consumed energy.

$$\min \sum_{e_j \in \mathcal{E}} \sum_{T_i \in \mathcal{T}_{\text{RI}}} E_{\text{table}}(C_{\text{inst}}(T_i), e_j) \cdot (T_i \rightarrow j) + \sum_{m \in \mathcal{M}} E_{\text{comm}}(m) \qquad (5.1)$$

- **Failure Minimization:** Using a DSE-internal time-triggered schedule, and replication and voting mechanisms, the DSE tries to reduce the failure probability of given tasks by replicating selected tasks and their predecessors. Therefore, spacial and temporal replication is used such that a temporal or permanent fault in any of the aforementioned tasks does not negatively affect the proper functionality of the system. The reliability analysis that is used for this task is described in detail in [112].

$$\min \text{Pr}_{\text{fail}}(S, TG(c_i), c_i), \quad \text{(complex analysis, see [112]).} \qquad (5.2)$$

A constraint set defines the feasibility of proposed solutions for a system design. We categorize the constraints of an *exploration specification* into user-definable and derived constraints. The following types of user-definable constraints are provided by the standard configuration of the DSE:

- **Fixed Deployment Constraint**: Tasks representing the component c_i may only be allocated to execution units contained in the set

$$\mathcal{T}(c_i) \rightarrow \mathcal{E}_{\text{selected}}(c_i) \subseteq \mathcal{E}. \qquad (5.3)$$

- **Exclude Deployment Constraint**: Tasks representing the component c_i may *not* be allocated to execution units contained in the set $\mathcal{E}_{\text{selected}}(c_i)$.

$$\mathcal{T}(c_i) \nrightarrow \mathcal{E}_{\text{selected}}(c_i) \subseteq \mathcal{E}. \tag{5.4}$$

- **Safety Integrity Level Constraint**: The achievable Safety Integrity Level (SIL) of an execution unit $e_{\text{SIL},j}$ may not be lower than required SIL of the component c_i if its representing task is allocated to this execution unit.

$$c_{\text{SIL},i} \leq e_{\text{SIL},j}, \quad \mathcal{T}(c_i) \rightarrow e_j \tag{5.5}$$

This rather simplistic safety check, which does not support the replication of safety channels can be used if the architecture of safety functions is fixed and no safety model (see Section 4.4.1) is available.

- **Safety Compliance Constraint:** References a safety model (see Section 4.4) that contains the non-functional safety requirements and properties of the application. If it is added to the constraint set the DSE passes each proposed solution to a Safety Constraint Checker that classifies them by the number of violated safety checks. This constraint type highlights the DSE's extensibility as the constraint, the referenced model, and the corresponding evaluator are externally contributed.

Section 4.7.3 contains a more detailed description of this constraint since its main application is the technical variability resolution.

- **End-to-End Deadline Constraint:** Allows to define end-to-end deadlines that must be satisfied by the internally calculated schedule. The difference between the earliest start time of tasks representing the component c_{start} and the latest finish time of tasks representing the component c_{end} must be smaller than the defined deadline.

$$t_{\text{latency}}(S, c_{\text{start}}, c_{\text{end}}) = t_{\text{end}}(S, c_{\text{end}}) - t_{\text{start}}(S, c_{\text{start}}) \leq t_d, \quad \forall d \in \mathcal{D} \tag{5.6}$$

The corresponding evaluator compares the latency between the beginning of the execution of the *start* task with the end of the execution of the *end* task against the required deadline. If the deadline is violated, a penalty value is calculated from the delta of these values.

The aforementioned derived constraints are those constraints that are generated from the input models or during the exploration process. These constraints are used internally by the DSE in an extended exploration specification and are typically available to all building blocks of the DSE, e.g., operators and decoders. The following list summarizes the derived constraints available in the DSE.

- **Period Constraint**: These constraints are derived from the timing model (see Section 4.3) and are used to validate whether a proposed solution can schedule each component's task within their specified period before the next iteration of the schedule would be executed. The evaluation is based on a *strictly time-triggered schedule*.

$$t_{\text{end}}(S, T_i) \leq t_p(S, T_j), \quad \forall T_i \in \mathcal{T}_{\text{RI}}. \tag{5.7}$$

- **Disjoint Deployment Constraint**: Tasks representing the set of components $\mathcal{C}_i$ and $\mathcal{C}_j$ may not be allocated to the same execution units. The evaluation is based on an *abstract task mapping*.

$$\mathcal{T}(\mathcal{C}_i) \rightarrow \mathcal{E}(\mathcal{C}_i), \mathcal{T}(\mathcal{C}_j) \rightarrow \mathcal{E}(\mathcal{C}_j) : \mathcal{E}(\mathcal{C}_i) \cap \mathcal{E}(\mathcal{C}_j) = \emptyset, \tag{5.8}$$
$$\mathcal{E}(\mathcal{C}_i) \subseteq \mathcal{E}, \mathcal{E}(\mathcal{C}_j) \subseteq \mathcal{E}.$$

- **Joint Deployment Constraint**: Tasks representing the set of components C_i must be allocated to the given set of execution units $\mathcal{E}_{\text{selected}}(C_i)$. The evaluation is based on an *abstract task mapping*.

$$\mathcal{T}(C_i) \rightarrow \mathcal{E}_{\text{selected}}(C_i) \subseteq \mathcal{E}. \tag{5.9}$$

In addition to the exploration specification, the application models also define inherent constraints such as the execution order of tasks, or bounds for the number of replica of software components. These types of constraints are considered in the constructive and heuristic modules of the DSE, e.g., the internal schedule calculation. These implicit constraints are used only internally and are thus not part of an exploration specification as they provide no added value for system designers using the DSE, but are mentioned here to complete the presentation of the feasible set.

Application Models

A logical component architecture model (see Section 4.2.1.1), a platform model (see Section 4.2.4), and a timing model (see Section 4.3) are the main input sets to the DSE. These models carry the required functional and non-functional information to determine optimized mappings of the logical components to the target platform.

The logical component architecture contains information about the execution order of components, and the signals that are exchanged between them. An additional timing model describes the temporal properties and requirements of the DREAMS application and is used in the DSE to select the set of components that shall be deployed on the target platform and to calculate internal heuristic time-triggered schedules. Therefore, the *PeriodConstraints* are evaluated that are attached to components, while the *RateConstraints* and *AperiodicConstraints* of Ports determine the temporal behavior of the generated Virtual Links.

The given platform architecture provides the sets of available resources and their topology, e.g., which transmission units connect the processing resources. This model describes the system software and the target hardware platform and provides platform specific parameters such as the frequency at which processor operates (assuming a locked frequency). Some of these parameters are used within the DSE to determine the quality of the mapping decisions. Additionally, the target platform models contain the *DeploymentTarget* annotation (see Section 4.2.4.7) to define the set of valid target allocations for the tasks. Here, the set of target execution units consists of partitions only since DREAMS is about highly integrated mixed-criticality applications (see Section 4.2.4.6). Furthermore, a deployment model must be provided to extract deployment-specific parameters (see Section 4.5.1.1).

The DSE uses the following annotations depending on the exploration features, the defined exploration targets, and requested solution models (see Section 5.1.2.8):

- **Logical application architecture**

  - *MessageSize* Defines the size of (raw) data that is emitted by an output port in bytes. Used in the message calculation and Virtual Link generation.

  - *EventTriggerAnnotation* References a trigger event that has been defined for a component in the corresponding timing model of the logical architecture. Used to determine the components that shall be considered as tasks by the DSE.

  - *In-/OutputTriggerAnnotation* References an I/O trigger for (component) ports to describe their temporal behavior and reaction chains. They are used to determine the temporal characteristics of Virtual Links.

- **Technical architecture**

  - *DeploymentGranularity* Provides a predicate whether the annotated execution unit shall serve as a target for mapping tasks to it. Counterpart of the EventTriggerAnnotation.

- *FailureRate* The failure rate of the annotated platform element given as its failure probability λ.

- *TransmissionUnitBandwidth* The bandwidth of communication resources, i.e., Transmission- and GatewayUnits.

- *TransmissionUnitPower* Defines the power in Watts that is consumed on average by the annotated transmission unit when transmitting data.

- **Deployment**

  - *EnergyConsumption* Worst-case execution time of a given component on a given execution unit (in seconds). It is used by the DSE to determine the time to reserve at resources when executing a task on an execution unit during offline scheduling and to allow the evaluation of the timing requirements.

  - *WCET* Energy consumption (in Joule) when a task associated with the given component is executed on the given execution unit.

5.1.2.4   Synthesized Artifacts

The set of possible output models of the DSE includes a deployment (see Section 4.5), a logical component architecture (see Section 4.2.1.1), and a safety model (see Section 4.4). Depending on the inputs to the DSE, different artifacts of these models are synthesized or existing models are reduced to match the choices of the DSE for a particular solution.

- **Deployment (see Section 4.5)**

  A deployment is a mapping of a logical component architecture to a target platform model and consists of mappings of multiple different elements. The DSE synthesizes a tasks-to-execution-unit mapping and the Virtual Links (see Section 4.5.1.2) which map a signal between tasks to a multicast message in the target platform.

- **Logical Component Architecture (see Section 4.2.1.1)**

  The DSE creates a logical component architecture model that is generated from the input model to correctly reflect choices of the DSE affecting the logical application. This model is only synthesized if tasks were replicated, a safety function was instantiated with an architecture, or design diversity was resolved.

- **Safety Model (see Section 4.4)**

  In order to reflect the choices from the safety architecture exploration, a safety model is synthesized from the input model that contains only a single selected architecture for each safety function instead of a list of possible variants. Moreover, descriptions of platform resources that are not utilized are also eliminated such that the output safety model matches the resulting deployment.

- **Timing Model (see Section 4.3)**

  The timing model is closely coupled with the logical component architecture and synthesized whenever this architecture model needs to be generated.

5.1.2.5 Internal Representation

In order to efficiently explore the given optimization problem, the DSE transforms the previously described input models to a compact internal representation that contains only the required information for the algorithm. The synthesized artifacts listed in the previous section are generated from the internal representations and the input models. In the following the most relevant internal representations will be discussed that are also used, e.g., by externally contributed evaluators or exploration modules implementing additional features.

Task Graphs In order to calculate schedules or to derive routes for Virtual links, the causal relations of the tasks that shall be deployed must be known. Task graphs are extracted from the logical component architecture to obtain a flattened representation of the logical application that consists only of the tasks to be deployed and their exchanged signals, thus containing the causality information of the application. Each group of non-communicating tasks is represented as a separate task graph. Within a task graph, vertices represent tasks that are linked to components from the input logical architecture. A directed edge between two vertices implies a data dependency between connected tasks. Edges are extracted from the logical architecture by collecting the output ports of each component to be deployed. Each channel is followed until another deployable component is identified, traversing the different levels of the hierarchy.

Platform Communication Graphs The platform communication graph is a flat representation of a hierarchical DREAMS platform model (see Section 4.2.4). It contains all execution units that have been selected as deployment targets and all possible communication resources connecting those execution units. Consequently, it contains all possible communication paths between execution units. Nodes represent execution units and communication resources, whereas edges denote that two resources are physically connected. Typically, the leaves of this graph are execution units.

Mapping Representation A redundant mapping extends the search space towards more reliable

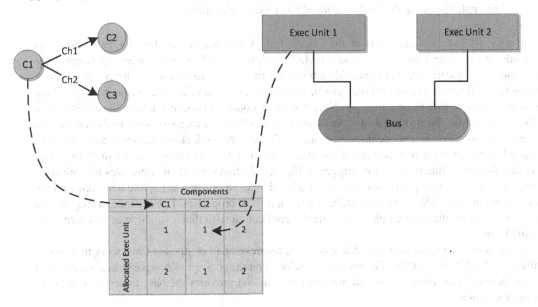

Figure 5.5: Allocation Table of Tasks to Execution Units

solutions of hardware/software systems [112, 158]. It allows allocating a task to more than one execution unit. The redundant mapping encodes both spatial and temporal redundant allocations. Spatial redundancy is achieved by allocating a replica of a task to a different execution unit than the original task. Temporal redundancy is allocating replicas of a task multiple times to the same execution unit. Hence, the relevant information for creating fault-tolerant mappings can be encoded in the mapping.

Messages Messages encode the route of a signal that is sent from a task A, hosted on execution unit 1, to a task B, hosted on execution unit 2 based on the *Platform Communication Graph*. If two communicating tasks are hosted on the same execution units, no messages are generated nor needed. The set of all messages encodes the complete communication visible to the execution platform. Additionally, messages reference their emitting time slot and its receiving time slots, which is needed for the scheduling of communication in the system-wide time-triggered schedule. Messages are only enriched with the time slot information if a schedule is actually calculated.

Safety Architecture Encoding Safety-critical functions demand to be designed such that they op-

	Safety Function		
	SF$_1$	SF$_2$	...
Num. Channels	2	1	...
Voting Channels	1	1	...
Diagnosis present	Yes, connected	-	...

Figure 5.6: Table Encoding the Architecture of Each Safety Function

erate correctly even if some parts of the system fail. A suitable architecture for these functions depends on the system type, i.e. fail-safe or fail operational, and the risk analysis performed for the function. Typically, standards provide advices for the architecture of safety functions although alternate architectures are valid if they can be proven to fulfill the safety requirements, e.g., if they have proven to be safe in practical use. Nevertheless, a common pattern found in the architectures of safety functions (be it in hardware, software, or a combination thereof) is the replication of the safety function across independent safety channels, the presence or lack of diagnostic units, and corresponding voting units that evaluate the signals from the channels and diagnostic units to decide whether the safety function shall be triggered. Figure 5.6 shows a table that encodes this information. The entries of each safety function can be altered by the algorithm to explore new architectures w.r.t. to given list of allowed safety architectures and their properties. For instance, it may not be allowed to neglect diagnostic units in a particular application such that a variation of this parameter would be useless.

In the following sections, we will first describe the developed composite framework that modularizes the MOEA-based DSE. The framework is based on the concept of *composite genotypes* [152] that structure the encoding of an optimization problem and provides the base to perform architectural exploration.

5.1.2.6   Overview of the Composite Design Space Exploration Framework

Figure 5.7 provides an overview of the base framework that is used for the DSE. The evolutionary algorithm first creates an initial population consisting of *individuals*. An individual consists of a *genotype*, a *phenotype*, and its evaluation results w.r.t. the defined objectives and constraints. *Decoders* transform genotypes into phenotypes that are processed by *evaluators*. After the evaluation phase, the fittest individuals (according to the evaluation results) are *selected* to be preserved in the

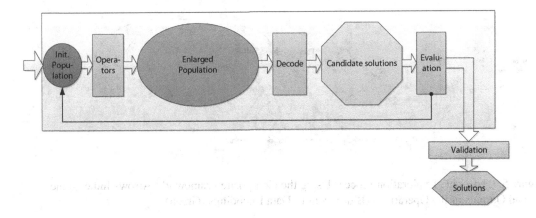

Figure 5.7: Overview of an Evolutionary Algorithm

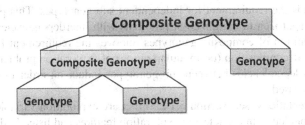

Figure 5.8: Composite Genotype Consisting of Sub-Genotypes

population for the next iteration. A *mating* mechanism uses these individuals to generate the *off-spring* (modified copies of existing individuals) by applying *operators* and adds these individuals to the population that is processed during further iterations.

Although the approach summarized above already provides some degree of modularity, especially by the separation of genotypes and phenotypes (operators and decoders, respectively), it focuses on the partial modularization of the exploration process. We found that our DSE would benefit from a more modular description of the DSE problems that would allow to describe a DSE problem as the union of its sub-problems. The sub-problems shall be loosely coupled as far as possible to enable a feature-wise modularization of the DSE. Consequently, dependencies must be respected by the framework in terms of required inputs and outputs of the sub-problems. Here, we use a constructive approach: our framework requires that the inter-dependencies between feature sets (i.e., a genotype and phenotypes generated from them) are optional such that the DSE operates correctly if only a sub-set of all available sub-problems is defined for a DSE instance. One example is provided by the exploration of optimal instantiations of *abstract* tasks with *concrete* tasks from a library that consists of the available implementation choices (see Section 4.7). If this "instantiation sub-problem module" is not added to an instance of the DSE, all tasks of the logical input architecture are considered to be concrete tasks and the DSE still generates the expected output (i.e., a deployment).

5.1.2.7 Composite Encoding

Genotypes are used in genetic algorithms to encode an optimization problem, i.e. they represent the optimization variables in the problem's context. The concept of composite genotypes is introduced in [152] and implemented in the author's meta-optimization framework Opt4J. In this approach,

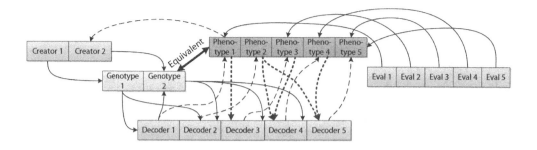

Figure 5.9: Example Exploration Process Using the Composite Framework: Arrows Indicate the In- and Outputs of the Operations (Blue) from/to Data Encodings (Green)

composite genotypes allow to separate a complex genotype into independent sub-genotypes. If applied recursively, this approach allows the use of hierarchies in genotypes to define DSE problems.

One major limitation is the requirement for independent sub-genotypes. This prevents the application of the existing approach to optimization problems with interdependencies between sub-problems. Thus, our definition of composite genotypes removes the requirement for independent sub-genotypes to allow a decomposition (or, modularization) of exploration problems with dependent variables. The inter-dependencies between sub-genotypes induce an order in which the sub-genotypes have to be processed.

Those internal representations (see Section 5.1.2.5) that are exploration variables are encoded as genotypes. They are coupled with the selected exploration features and listed in the following:

- Safety Architecture Encoding: Safety Architecture Exploration

- Abstract Task Mapping Encoding: Task/Component Instantiation

- Instantiated Task Mapping Encoding: Task/Component-to-Execution-Unit mapping and redundant logical architecture and voting exploration.

Further, we extend the phenotype definition to allow a) the composition of phenotypes equivalently to the composite genotype and b) to take the role of hybrid phenotypes that are also genotypes (indicated by the red arrow in Figure 5.9). The above re-definitions have a large impact on the exploration process that is described in the next section.

5.1.2.8 Composite Exploration Process

Allowing dependencies between the genotypes comprising a composite genotype enforces modifications to the create, decode, and evaluate operations that are also composable in our framework. Figure 5.9 illustrates the modified exploration process that uses composite genotypes, phenotypes, and operations. Each genotype (also the composite genotypes) must have a creator that is responsible for its default instantiation. They are executed automatically when the initial population is generated.

Each concrete genotype (i.e., a non-composite genotype) is associated with a decoder that transforms the genotype into a phenotype (i.e., a solution representation) which can also be composed to mirror the structure of the input composite genotype. The composite exploration process uses decoders to calculate phenotypes that may serve an input to subsequent decoding steps. Each decoder can access all independent and previously decoded phenotypes such that all possibly required information for the decoding of a genotype into phenotypes is accessible. Whenever a decoder produces a hybrid phenotype as its output, this phenotype is already present in the composite genotype, since

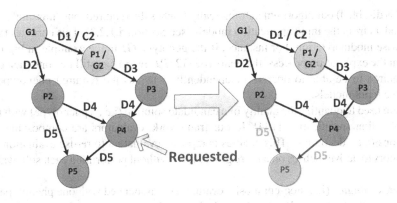

Figure 5.10: Dependency Graph that Determines the Execution Order of Decoders (Left: Complete Graph; Right: Relevant Sub-Graph if Only *P4* Is the Demanded Solution)

they are equivalents. This hybrid phenotype might be modified by operators before the decoding step has been launched to derive new variants of individuals for the exploration. In order not to lose these modifications, which are an essential part of every evolutionary algorithm, decoders producing hybrid phenotypes only perform *updates* of the already existing genotype, instead of overwriting them.

Our algorithm automatically determines the correct execution order of the decoders such that the dependencies introduced by the genotypes and phenotypes forming the required inputs for the decoders are respected. Therefore, our algorithm constructs a dependency graph before launching the exploration process that considers all encodings (genotypes and phenotypes) and decoders that the DSE is aware of (see the left graph in Figure 5.10). This graph is a directed acyclic graph (preventing circular dependencies) whose vertices are encodings (genotypes or phenotypes), while the edges represent decoders. All incoming edges of a vertex refer to the same decoder. The DSE processes the requested types of phenotypes (*P4* in Figure 5.10) that were either demanded explicitly (to obtain a certain solution model) or that are demanded by evaluators (see below) to derive the correct execution order for an instance of the exploration process. For instance, the right dependency graph shown in Figure 5.10 illustrates the case in which only *P4* is demanded as a solution from the DSE. Thus, decoders that are not required will also not be executed during the exploration process that increases the algorithm's performance (here, *D5* which generates *P5*).

The discussed modularity is utilized in the generation of Virtual Links for instance. They are generated by a model transformation that uses the DSE-internally calculated messages (see Section 5.1.2.5) which encode the routings within a DREAMS platform. Due to the fact that the routing is obtained by employing a shortest path algorithm, the scheduling information is not required for the Virtual Link Generation. Thus, only the *Message Encoding* (see Section 5.1.2.5) is requested from the DSE and the framework automatically detects that decoding into a *strictly time-triggered schedule* is not required and hence does not launch the scheduling decoder.

The concrete execution order of decoders (and implicitly also creators) is determined by iterating over the dependency graph from the requested encoding vertices (*P4* in Figure 5.10) to the root vertices. The traversal of the graph follows the *layers* encoded in the graph such that all outgoing edges of the required leave vertices are traversed before their predecessor vertices are traversed, and so on. During the iteration all decoders associated with the edges are collected in a list which defines the execution order defined for the process. It shall be noted that the encodings of disabled feature sets of a DSE instance are also not part of the dependency graph such that these unneeded decoders are also not executed.

Further, if some genotype is not enabled for an instance of the exploration process (i.e., some feature of the DSE is disabled) our algorithm automatically fetches the required, but "missing", input to subsequent decoders from the input application models (see Section 5.1.2.3). This enriches the DSE with a feature-wise modularization. For instance, if the genotype *G1* in the example in Figure 5.10 is excluded from the exploration process, the genotype *G2* (*P1*, resp.) can still be constructed, since creators are required to be able to operate independently, whereas the required phenotype *P2* is obtained from the input models.

Evaluators are used to quantify the quality of a candidate solution (i.e., a phenotype) with respect to an input exploration specification [134]. In our framework, evaluators are composable similar to other operations (e.g., decoding). This enables composite evaluators to re-use evaluation results from sub-evaluators to derive metrics on a more global level without performing each sub-evaluation again.

Each *concrete evaluator* (i.e., non-composite evaluator) is associated with one phenotype which can be also a composite phenotype. If an evaluator is defined for one concrete phenotype, it provides a quantification of the particular sub-solution, while an evaluator defined on a composite phenotype may also consider its (sub-)phenotypes. By these definitions, decisions that are made in the decoding phase to generate dependent genotypes (e.g., architectural decisions) can be also evaluated due to the fact that evaluators may be defined on any phenotype, not only *final phenotypes* (i.e., phenotypes that are not input to any decoder).

5.1.2.9   Composite Safety Architecture and Mapping Exploration

The MOEA-based DSE (see Section 5.1.2.6 and [112]) has been extended for DREAMS by the ability to optimize component instantiations and safety architectures (see Section 5.1.2.2), and to generate routes for DREAMS virtual links (see Section 5.1.2.4). The number and complexity of these features motivates the creation of the composite DSE framework, which is based on problem decomposition, in order to keep the DSE process flexible and extensible. Its modular architecture allows the DSE to execute only those operations that are needed to generate the output demanded by a user such that the DSE remains efficient when it is employed in different scenarios. For instance, the Virtual Link generation for existing deployments (see Section 4.5.1.2) does not require the internal scheduling, hence it is not executed in this scenario. Furthermore, the DSE's usability is increased by automatically triggering the required model transformations.

In this section, we exemplify the composite exploration framework introduced above by describing the realization of the Safety Architecture Exploration with a focus on the framework aspects. Each of the features listed in Section 5.1.2.2 is realized using the composite framework of the DSE. Consequently, a user of the DSE is enabled to specify whether, e.g., the exploration of safety architectures shall be performed by the DSE.

Encoding of the Safety Architecture Exploration

The safety architecture exploration problem uses three encodings that form the optimization variables: An encoding to describe the selected architecture of safety functions, e.g., 1oo2D, an encoding for possibly redundant *abstract* task-to-execution unit mappings, and an encoding that represents a possibly redundant mapping of *instantiated* tasks-to-execution units (see Section 5.1.2.7). In order to increase the modularity of the DSE, these encodings are realized as genotypes that comprise a composite genotype. These sub-genotypes have interdependencies as indicated in Figure 5.11, which must be respected by their creators and decoders (see below).

In the described use-case, the safety architecture encoding consists of the set of safety functions that are present in the input logical architecture. For each of the safety functions, the number of safety channels, the number of channels required to trigger the safety function, and the presence of diagnostic units connected to the voter (for the evaluation of the safety channels) are considered. Based on this encoding, an *abstract* task-to-execution unit mapping is derived. Its entries define the allocations of each task to an execution unit, whereby multiple entries for one task imply replication

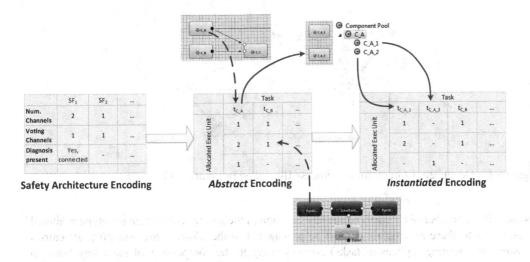

Safety Architecture Encoding   **Abstract** Encoding   **Instantiated** Encoding

Figure 5.11: Encodings Comprising the Composite Genotype for Safety Architecture Exploration. © IEEE 2016, Reprinted with Permission from [115]

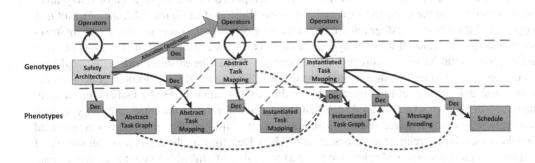

Figure 5.12: Modularized and Extended Exploration Process using the Composite Exploration Framework

of that particular task. For tasks related to safety functions, the entry number equals its channel number. This *abstract* task-to-execution mapping is an input to generate the *instantiated* task-to-execution unit mapping where all abstract tasks have been replaced by tasks referencing *concrete* components from a library. The allocation entries for abstract tasks are not altered and distributed over their *concrete* tasks.

Safety Architecture Exploration Process

The exploration process with all features enabled is illustrated in Figure 5.12. The safety architecture, abstract task mapping, and instantiated task mapping encodings are genotypes and hence modified by operators. They represent the optimization variables of the exploration process that is illustrated in Figure 5.12. The resulting dependency graph is shown in Figure 5.13.

Concrete phenotypes (i.e., non-hybrid phenotypes) can be calculated in a fixed manner by decoders and are not subject to optimization. The *abstract task graph*, the *message encoding*, the *instantiated task graph*, and the *schedule* are concrete phenotypes, whereas the *abstract task mapping* and the *instantiated task mapping* are hybrid phenotypes (see Section 5.1.2.8). Their decoders (i.e., decoders that have these phenotypes as their output) only perform updates on these phenotypes. For

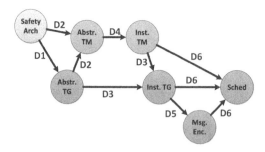

Figure 5.13: Dependency Graph for the Process Illustrated in Figure 5.12

instance, if the number of channels of a safety function in the *safety architecture* genotype is altered by an operator, these modifications must be propagated to the *abstract task mapping*. Its entries (possibly representing replicas of tasks) corresponding to the components of the safety function must be added or removed such that the number of entries equals the number of safety channels. Changes to the *abstract task mapping* are also propagated to the *instantiated task mapping* whose decoder only performs updates. Hence, optimizations from previous iterations are preserved.

The feature-wise modularization of the DSE is exemplified for the exploration of safety function architectures. In the complete exploration process illustrated in Figure 5.12, the *abstract task mapping* and the *abstract task graph* are the outputs of two decoders that use the *safety architecture encoding* as an input. If this genotype were disabled for an execution of the DSE, the *abstract task mapping* could be still constructed by its creator, whereas the *abstract task graph* could not be constructed due to the missing input to the decoder. Consequently, the decoder generating the *instantiated task graph* could not operate due to the missing input. Here, the fallback mechanism introduced in Section 5.1.2.6 is used such that the decoder for the *instantiated task graph* can alternatively retrieve the required *abstract task graph* via the input models (this graph is constructed from the logical input architecture, see Section 5.1.2.5). Deactivating the *safety function module* of the DSE has indeed a practical relevance since a) a number of embedded systems do not contain safety functions, and b) logical architectures may contain safety functions whose architecture is fixed, thus not requiring exploration.

5.2   Scheduling

In this section the scheduling and configuration algorithms are described, which support the different scheduling domains that have been considered for the DREAMS architecture. Section 5.2.1 presents a heuristic for decomposing end-to-end latency constraints into appropriate sub-constraints. Section 5.2.2 covers partition and task scheduling, whereas Sections 5.2.3 and 5.2.4 cover respectively on-chip and off-chip scheduling.

5.2.1   Timing Decomposition

End-to-end latency constraints impose bounds on the delay between the occurrence of stimulus and response events of timing chains. Examples are constraints on latencies between the acquisitions of sensor values and the application of the corresponding commands to an actuator. Since the sensor, the actuator and the functions that implement the control algorithm might all be located on different tiles and nodes, not only task execution delays but also communication delays may contribute to the end-to-end latencies. If furthermore different technologies and paradigms are used for the execution of tasks and the communication over on-chip and off-chip networks, then the overall scheduling problem becomes even more difficult. The goal of the *timing decomposition* approach introduced in this section is to divide the overall scheduling problem into sub-problems that can be solved independently and efficiently. For this purpose, the end-to-end latency constraints are subdivided into appropriate sub-latency constraints, so that each covers a different domain: task scheduling, on-chip or off-chip communication scheduling. If a solution is found for each sub-problem, i.e. for each scheduling domain, then their composition is an overall solution that satisfies the initial constraints.

The rest of the section is organized as follows. The notions of scheduling domain and coordination delay are introduced in Section 5.2.1.1 and 5.2.1.2. In Section 5.2.1.3 we explain the timing decomposition approach in general, followed by dedicated sections for its application to the different scheduling domains.

5.2.1.1 Scheduling Domains

For the precise definition of end-to-end latency constraints, we use the concept of timing event chains. As explained in Section 4.3.1, a timing event chain allows describing the path of the transformation of information through the component architecture, as for example from a sensor to an actuator. The timing chain mainly consists of an alternating sequence of task execution and communication segments. Figure 5.14 shows the example of a timing chain consisting of five tasks, name from "A" to "E" and four messages, name from "a" to "d". The delay between the execution start of the first task and the execution end of the last task should not exceed 53 ms.

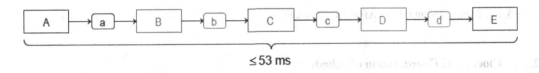

Figure 5.14: Timing Event Chain

After the deployment of the components to the execution units of the platform architecture (see Figure 5.15), the communication segments of the timing chain become either

- local intra-partition communication

- local inter-partition communication

- remote on-chip network communication

- remote on-chip and off-chip network communication

The local inter- and intra-partition communication takes place inside a tile. We suppose here that the corresponding communication delays are simply the result of the task scheduling, i.e. equal to the time interval between the execution end of the producing task instance and the execution start of the next instance of the consuming task. To be more precise one would have to add delays also for the (protected) access to the shared memory used for that kind of communication.

The subsets of tasks that are allocated to the same tile can be seen as sub-chains of tasks connected by their inter-task communication. The delays induced by these sub-chains depend on the configuration of the scheduling mechanism used at tile level. The communication delays over the on-chip and off-chip networks depend on the configuration of the scheduling mechanism used for each of these communication networks. Since the scheduling mechanisms and their configuration algorithms follow different logics, we see them as different *scheduling domains*.

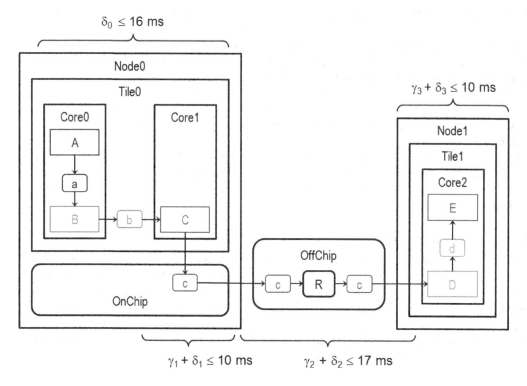

Figure 5.15: Timing Event Chain After Deployment

5.2.1.2   Clocks and Coordination of Schedules

In the previous section we have identified three scheduling domains, namely task scheduling, on-chip communication scheduling and off-chip communication scheduling, for which configurations are derived by algorithms. The schedules of these domains follow each their own logic but could be timely coordinated in order to achieve shorter end-to-end delays. Such time coordination could be achieved through fixed offsets between periodic activities of schedules in different domains. Such a coordination is however only possible if the following two conditions hold:

1. Clocks that drive periodic activities in the different schedules are synchronized.

2. Periods to be coordinated are harmonic, i.e. every smaller period is a divider of every larger period.

If clocks are not synchronized, then any initially established offset changes after a while due to clock drifts. The effect of non harmonic periods is similar to drifts, in the sense that the offset between the periodic activities is time dependent. Clock synchronization is foreseen by the DREAMS architectural style, but the clocks used by different platform building blocks have different granularities, which may make it difficult or even impossible to choose harmonic periods. It is not enough to have synchronized clocks. If two clocks are synchronized but their granularities are not harmonic, then only a limited set of common periods is possible, depending on the common dividers of the clock granularities. The clock of the on-chip communication driver of the DREAMS Harmonized Platform (DHP) (cf. Section 2.6) has a granularity equal to a negative power of 2. This implies that a period of 10ms, for example, is impossible because it is not a multiple of a negative power of two. Furthermore, the on-chip communication driver of the DREAMS Harmonized Platform (see Section 7.4) only allows periods exactly equal to a negative power of 2, i.e. not even multiples of

Table 5.2: Time Granularities

Building block	Time granularity
XtratuM	$1\mu s$
DHP On-chip Communication	2^{-n}s, for values smaller than 1s
TTEthernet	$20\mu s$

Table 5.3: Coordination Delays

Value of *Coordination Delays*	Use Case
Distance between the end of the last activity in the previous domain and the start of the first activity in the following domain.	Clocks are synchronized and periods are harmonic.
Period of first execution or transmission in the following domain.	Clocks are not synchronized.
	Clocks are synchronized but periods are not harmonic.

negative powers of 2. As a result, if a task has a period of 10ms, which cannot be changed then it is impossible to select a harmonic period for the on-chip communication and thus no coordination is possible between the task and the on-chip schedule for the concerned timing chain.

From the above discussion we can summarize that beyond the scheduling-domain internal delays, we also have to consider a *coordination delay*. The coordination delay accounts for the time it takes between the end of the last execution or transmission in the previous domain and the first execution or transmission of the next domain. Possible values of the coordination delay are summarized in Table 5.3.

Thus, if a timing event chain covers n scheduling domains, then the end-to-end delay is given by the following formula:

$$d = \delta_0 + \sum_{i=1}^{n}(\gamma_i + \delta_i), \qquad (5.10)$$

where

- δ_i is the scheduling induced delay of domain i. It spans from the start of the first execution or transmission in the domain, until the end of the last execution or transmission in the domain, with respect to the considered timing event chain.

- γ_i is the *coordination delay* between domain $i - 1$ and domain i, which accounts for the time it may take between the end of the last execution or transmission in the domain $i - 1$ and the first execution or transmission of the domain i.

Notice that the coordination delays γ_i only exist for $i > 0$. The end-to-end delay is illustrated in Figure 5.16, where 4 scheduling domains are present and thus $n = 3$.

5.2.1.3 Timing Decomposition Heuristic

The goal of a timing decomposition is to propose correct and efficient sub-latency constraints for the different scheduling domains such that

- the sum over all domains is smaller or equal to the end-to-end latency constraint

- and the configuration algorithms of the scheduling domains are able to find schedules that satisfy the corresponding sub-latency constraints.

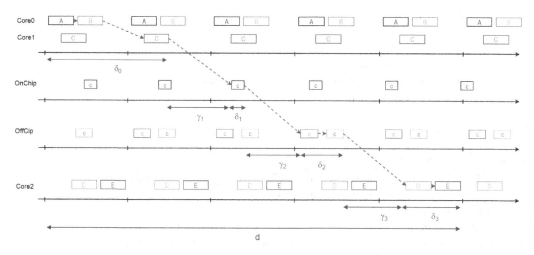

Figure 5.16: End-to-End Delay of an Event Chain Instance

Thus, the original end-to-end latency constraint is satisfied, if for each scheduling domain a solution is found that satisfies the sub-latency constraints. Such a decomposition is illustrated in Figure 5.17 by the red arrows.

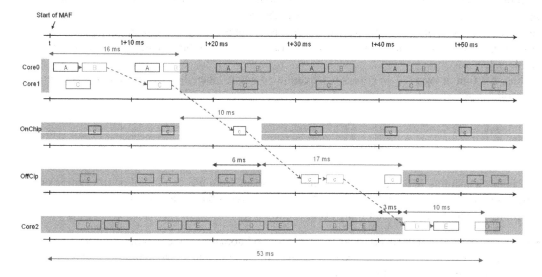

Figure 5.17: Timing Decomposition Example

The second property is related to the crucial question of how to best distribute the overall time budgets among the different scheduling domains. The goal is not to hinder the finding of an overall solution, when there is insufficient budget for one domain whereas more time than needed is allocated to other domains. In other words, we would like to achieve a distribution of the overall time budget that is optimal with respect to finding an overall-solution, if such a solution exists.

Some scheduling configuration algorithms try to regularly insert idle times to leave *unused zones*, where *best effort* tasks or packets may access the resource. This is possible, if latency constraints are loose enough to allow for slack time. In that case, the resulting schedules might not be work-conservative, i.e. not continuously treating pending work. If on the other hand latency constraints are tight, then in the extreme case, no unused zones can be inserted, because all entities must be scheduled as soon as possible to be able to meet the latency constraints. It means that for each scheduling domain there is some minimal latency which corresponds to a *minimal budget* below which no solution can be found for the scheduling problem.

If the sum of the minimal budgets is smaller than the end-to-end latency constraints, then we have slack time. The resulting question is how to distribute this slack time among the scheduling domains. A priori, each scheduling domain can profit similarly from the slack time and thus we decide to distribute it proportionally to the minimal time budgets. From the above discussion we derive the following goals:

1. Optimality of the distribution of time budgets with respect to finding an overall solution.

2. Distribution of slack time in proportion to minimal time budgets.

Priorities are often used with event-driven schedules to better respond to different levels of tightness of latency constraints. With time triggered schedules, a kind of "prioritization" can be achieved, by allocating the scheduling slots first to entities with tighter latency constraints and later to entities with looser latency constraints, so that the former can get the slots that are more suitable for achieving the tighter constraints, whereas the later get the remaining ones that may be less suitable and lead to longer latencies. However, the precise effect of these "prioritizations" is difficult to predict. For this reason we overestimate the minimal time budgets by considering the most unfavorable delay that would result from being served last, while all pending work is treated as soon as possible, i.e. without insertion of unused zones for best effort service. We call this delay *unfavorable minimal latency* and denote it $\hat{\delta}_i$. Furthermore we use $\hat{\gamma}_i$ for the coordination delays that cannot be influenced by the configuration of the domain scheduling. If no coordination is possible, it is set to the period $\hat{\gamma}_i = T_k$ of the first task or frame. If coordination is possible, we set $\hat{\gamma}_i = 0$, because the domain schedule has complete control over the coordination delay.

The sum is called *unfavorable minimal end-to-end delay* and denoted by $\hat{d}$:

$$\hat{d} = \hat{\gamma}_0 + \sum_{i=1}^{n} (\hat{\gamma}_i + \hat{\delta}_i) \tag{5.11}$$

If the end-to-end latency constraint D of the timing event chain is larger than the estimate, i.e. $D > \hat{d}$, then there is slack time that can be distributed. For the distribution of the slack time we have to remove the "incompressible" coordination delays captured by $\hat{\gamma}_i$.

We use delay estimates $\hat{\delta}_i$ that are scaled as follows:

$$\beta_i = \hat{\delta}_i \frac{D - \sum_{j=1}^{n} \hat{\gamma}_j}{\sum_{j=0}^{n} \hat{\delta}_j}$$

If these delay estimates are used as time budgets for the scheduling domains, then their sum, together with the coordination delay $\hat{\gamma}_i$, is equal to the end-to-end latency constraint of the timing event chain.

$$\beta_0 + \sum_{i=1}^{n} (\hat{\delta}_i + \beta_i) = D.$$

If $\hat{d} > D$, then $\beta_i < \hat{\delta}_i$, meaning that the proposed budgets are smaller than the delay estimates. Given that our estimates do not take into account prioritizations, the scheduling configuration algorithm of the different domains may still be able to find solutions where the delay is smaller or equal to β_i.

5.2.1.4 Task Scheduling Domain

To be able to propose appropriate sub-latency constraints for the task scheduling domain, we first have to understand the used scheduling algorithm and the delay it may induce on task chains.

The scheduling algorithms considered in DREAMS for the scheduling of tasks are based on static cyclic scheduling, i.e. without priorities and without preemption (see Section 5.2.2). Tasks are executed inside of scheduling slots of partitions. The task and partition schedules are designed so that at the end of a partition slot, all tasks scheduled inside the slot have finished their execution. With this kind of scheduling, the delay of the timing event sub-chain in the task scheduling domain depends strongly on the order and relative offsets of the executions of the tasks in the sub-chain. In the most unfavorable case, the distance between the execution end of a producing task and the start of a consuming task is almost equal to the period of consuming tasks. On the other hand, if all tasks of the sub-chain are executed one after the other in the correct order, then the total delay is equal to the sum of the WCETs of the task. Which kind of delay can be achieved depends on the logic of the scheduling configuration algorithms. The two main cases are discussed in the following sections.

In both cases, if the period of the last frame in the previous scheduling domain is not harmonic with the period of the tasks, then a coordination delay, equal to the period of the first tasks needs to be added $\hat{\gamma}_i = T_k$.

Without Latency Constraints

If a task/partition scheduling algorithm does not consider or does not know about timing event chains and their associated latency constraints, then the only timing constraints it can guarantee are the periods of the task executions. The distance between the execution end of a previous task and the start of the following task in a timing chain will have some value between 0 and the period of the following task. For the worst case analysis we have to consider the distance to be equal to the period of the following tasks, although in the average case it will be shorter. A scheduling algorithm will group the execution of tasks of an application and execute one after the other in a partition scheduling slot. However, if timing chains and their latency constraints are ignored or not known, then the order in which the tasks of a chain are executed might be exactly the opposite of the one suggested by the timing chain and thus the worst case distance between the consecutive executions of the tasks of a chain execution may occur. In this worst case the delay can be bounded by

$$\hat{\delta}_i = C_0 + \sum_{j=1}^{m} T_j + C_m$$

where T_k and C_k are respectively the period and the WCET of task k. Using this worst case delay as a minimal time budget for the task scheduling sub-chain is probably too pessimistic. One could consider a user-selected factor such as 75% of the worst case delay.

With Latency Constraints

Assuming that the task/partition scheduling algorithm considers timing chains with their associated latency constraints, then the algorithm is able to schedule tasks such that the delay between consecutive tasks is systematically smaller than the period of the following task and even equal to 0 in the extreme case, which would give the best case delay:

$$\hat{\delta}_i = \sum_{j=0}^{m} C_j$$

Notice that if other tasks than those of the considered timing chain are also executed on the same tile, then this best-case delay may be too optimistic. We have to distinguish two cases:

1. if no coordination with preceding or following schedules from other domains is possible or desired, then the execution start offset of the first tasks of the chains can be chosen freely and in a way that the workload is spread out and the tasks of each chain can be executed one after the other and the best-case delay can be achieved for each of them, unless two chains share tasks. Notice however that in this case the coordination delay is equal to the period of the first task, since this is the worst delay that can occur between the end of the last activity in the previous schedule and the execution of the first task.

2. if coordination with preceding or following schedules from other domains is possible and desired, then in the most unfavorable situation the input of all sub-chains might be ready at the same time and one of the sub-chains must wait until all others have finished their execution. If the previous schedules belong to the same domain then they might not all be able to end at the same time.

5.2.1.5 On-Chip Communication Scheduling Domain

A time-triggered scheduling algorithm is used for safety-critical on-chip communication (see Section 5.2.3) with NoC traversal times that are always equal to the basic network latency R_k^* defined by Equation 5.12. Thus, $\hat{\delta}_i$ is equal to R_k^*. If the period of the last task or frame in the previous scheduling domain is not harmonic with the period of the on-chip packet, then a coordination delay, equal to the period of the on-chip packet needs to be added $\hat{\gamma}_i = T_k$.

Recall that Rate-Constrained (RC) traffic is event driven. Unfortunately no worst case timing analysis is known that could help us to find quick estimates of network traversal times. However, if we see them as periodic with a period equal to the Minimum Inter-arrival Time (MINT) we can take them into account in the time-triggered scheduling configuration algorithm (see Section 5.2.3). Their entry time into the NoC can of course not be controlled, but if the algorithm finds a solution with all virtual links, Time-Triggered (TT) and RC, then it means that there is enough bandwidth available to transmit also the RC virtual links, and this within a time span that is smaller than the period. Thus, if the algorithm finds a solution, then the MINT is an upper bound on the traversal time that can be used as an estimation in the timing decomposition: $\hat{\delta}_i = MINT_k$, with $\hat{\gamma}_i = 0$.

5.2.1.6 Off-Chip Communication Domain

Recall that for the unfavorable minimal time budget, we consider the case where waiting frames are sent as soon as possible, without the insertion of unused zones for best-effort traffic. In case of time-triggered scheduling, even if entities are scheduled as soon as possible, the schedule contains idle times, due to the size of the frame windows, that must be larger than the transmitted frame to leave time for the transmission of clock synchronization frames with a higher priority. Time must also be left for the interference from lower priority frames in case the shuffling integration policy is used, as explained in the "Media Reservation" paragraph in Section 5.2.4. We can take this characteristic into account by not considering the actual length of the TT frames but the corresponding frame window length.

Since we assume that frames are forwarded as soon as possible, the scheduling behaves like a work-conservative scheduling policy. If we apply network calculus [159] to estimate network traversal times, then we obtain reasonable "unfavorable minimal time budgets" for the TT virtual links. They correspond to the unfavorable case where the considered TT virtual link is the last to be transmitted among those that share a link. The size of the frame windows should be chosen according to the integration policy, which in case of "blocking" leads to shorter windows and thus shorter budgets for the TT virtual links.

If on top of TT frame windows we add RC frames and apply again network calculus, then we obtain reasonable "unfavorable minimal time budgets" for the RC virtual links. They correspond to the unfavorable case where TT virtual links are all scheduled first just before the RC virtual links, and the considered RC virtual link is always the one served last among those that share the same link.

5.2.1.7   Tool: RTaW-Pegase/Timing

The above described heuristic is implemented in the RTaW-Pegase/Timing tool and integrated into the toolchain (see Section 5.6) as a context menu entry of "Deployments" in "Model Navigator" view of AF3. Figure 5.18 shows on the right the decomposition of the Timing Chain "SCADA" into sub-chains, for task and on-chip communication scheduling domains, which computes sub-latency constraints, consisting of a coordination delay (if applicable) and a budget for the resource scheduling.

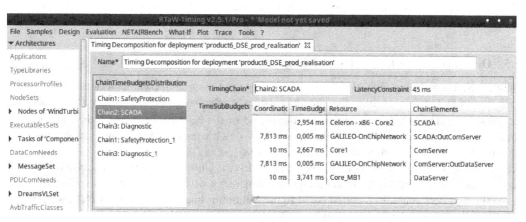

Figure 5.18: RTaW-Pegase/Timing Showing Timing Decomposition of Timing Chain of Wind Power Demonstrator (cf. Wind Power Demonstrator in Chapter 11)

5.2.2   Partition Scheduling

This section describes the partition scheduling model and the techniques used in DREAMS to generate static schedules.

5.2.2.1   Model

Partitioned systems for embedded real-time systems can be achieved by using virtualization techniques where a hypervisor plays the role of the virtualization layer 6. The hypervisor builds an environment offering virtual CPUs to execute several independent execution environments on the native hardware. Execution environments or partitions are a piece of software that includes the Operating System (OS) and the application. The virtualization layer is responsible for ensuring temporal and spatial isolation of partitions and for guaranteeing correctness in the resource access. Virtualization also serves to abstract from the real resources, which decouples the partition from the real resources, facilitating greater architectural flexibility and portability in system design.

A multi-core partitioned system can be seen as a system with three layers. The first layer is the hardware platform, the second layer is the hypervisor and the third layer is a set of partitions. Each partition consists in a runtime support or operating system and the application. From the scheduling point of view, only the second layer (hypervisor) and the third layer (partitions) require a scheduling policy.

The hypervisor manages several real CPUs and offers virtual CPUs to the partitions. From the point of view of scheduling, the following considerations should be taken into account:

- A partition is executed on top of the hypervisor in a similar way as it would be executed on the native hardware. It has an address entry point, when invoked by the hypervisor, boots and starts the partition operating system (guestOS). The guestOS can be a bare application (single-thread) or a minimal runtime with a cyclic scheduler, a real-time OS or a full environment (e.g., Linux).

- The hypervisor only knows the entry point of the partitions. Threads inside of the partition implemented by the guestOS are internal to the partition and are not seen by the hypervisor. Threads are handled by the guestOS and scheduled according the guestOS policy.

- A partition can be multicore (several VCPUs) or monocore (only one VCPU). We assume in the following that partitions are monocore.

- The allocation of a partition (VCPU) to a real CPU is configured at compilation time of the global system and specified in a configuration file that is an input to the hypervisor. Under this approach, one or several partitions (VCPU) can be allocated to one of the real CPUs and others to the rest of the real CPUs.

- The hypervisor scheduler requires to switch from one partition to another one by means of the partitions' context switch that can have a computation cost higher than the process context switch at operating system level.

- In order to save computation time and reduce problems related to the use of caches and memory tables per core, the core migration of a partition is not allowed. It means that the allocation of one partition (VCPU) to a real core is static and cannot be modified during the execution.

- The hypervisor implements a cyclic scheduling policy [160, 161]. Temporal windows in a Major Active Frame (MAF) are statically allocated to one partition.

- The hypervisor also can implement a priority based Periodic Server Policy (PPS) [162] where one or several partitions are allocated to a core and executed according the partition priority that defines a period and a budget.

- The hypervisor allows to define which scheduling policy is used in each core. For instance, in a quad-core, three cores can be scheduled under cyclic scheduling and the last one under a periodic server.

- Core scheduling policies, temporal allocation of partitions in cyclic scheduling or priorities, periods and budgets are statically defined in the configuration file. During the execution, the hypervisor maintains the static allocation of the cyclic scheduling core. It allows to change the period of the partitions allocated to the PPS.

The scheduling model is described in terms of a set of partitions (P) that are composed by the dupla $P_i = (\tau, L)$, where L is the level of criticality and τ is the set of tasks of the partition. A partition is defined by $\tau_j = (C_j, P_j, D_j)$ where C_j is the worst case execution time, P_j is the period and D_j is the deadline. Each task defines a processor utilization given by C_j/P_j. A partition has a utilization (U_i) calculated as the sum of the internal task utilizations.

To generate a schedule plan for the system, the following steps are defined:

Step 1 Allocation of partition to cores

Step 2 Schedule generation for each core

5.2.2.2   Allocation of Partition to Cores

The partition distribution is an allocation problem that consists in deciding on which processor a partition and its tasks should be executed.

The allocation problem is solved using typical memory allocators such as First Fit (FF), Worst Fit (WF), and Best Fit (BF). These heuristics can use a partition parameter, or a combination of several parameters, as the key for allocation. Partition parameters that can be used are partition load calculated as the sum of periodic task utilizations and partition criticality. For instance, First Fit Decreasing Utilization (FFDU) uses the partition utilization as a key for choosing the next partition to allocate.

In DREAMS a combination of the previous allocators, specifically the Worst Fit Decreasing Utilization (WFDU) is used, since the WF presents better results in balancing the load. WFDU maximizes the spare capacity in each processor [163].

After this step, each core allocates a partition subset.

5.2.2.3   Core Schedule

In order to generate a static schedule for each core, all partitions allocated to a core are seen as a flat task model. This approach consists of removing the frontiers defined by this partition subset and to consider all the tasks at once. Then a single global scheduler is in charge of managing all the tasks, and to conduct the corresponding schedulability analysis. At the end of the process, the solution is mapped back to the partitioned system by grouping the tasks of each partition in order to reduce the number of partition context switches.

To build the schedule, Extended Earlier Deadline First (EEDF) is used as a baseline for the proposed heuristic. EEDF consists in a modification of the EDF scheduling algorithm adapted for the temporal model defined. It also implements the Stack Resource Protocol (SPR) [164] for mutual exclusive access to shared resources, accounts for partition context switch overheads and copes with atomic tasks.

For the EEDF heuristic, the next job to execute is selected according to EDF and SRP, but the non-preemptability and context-switch overheads are properly taken into account in the generated schedule. This algorithm is executed off-line for each core and generates a static schedule for them if no missed deadlines are found while the plan is generated.

5.2.3   On-Chip Network Scheduling

In this section we describe the scheduling configuration algorithm developed for the safety-critical on-chip communication over the STNoC used in the DHP.

5.2.3.1   Model of STNoC used in the DREAMS Harmonized Platform

The following characteristics of the STNoC used in the DREAMS Harmonized Platform are relevant for the communication scheduling and its timing analysis:

- only unicast routes

- wormhole routing, applied to NoC packets consisting of a header flit followed by a certain number of payload flits

- two virtual networks with different priorities, sharing the same links but having each their own sets of flit buffers, managed so that flits in the lower priority network can never delay flits in the higher priority network

- payload flits contain up to 8 bytes of payload

- router input ports can store 2 flits per virtual network

- output ports of routers cannot store any flits. Flits are forwarded in the same cycle to the input port of the next router.

- routers forward flits as soon as possible (work-conservative) according to the following rules:

  - a flit can only be forwarded to an output port, if the credit is positive
  - with each transmission, the credit of an output port is decreased by 1
  - the credit is increased by 1 when a cell is freed at the target buffer, but the credit is only available in the next cycle
  - all flits of a NoC packet are transmitted before considering which packet to forward next
  - flits of the virtual network with the lower priority are only forwarded if there is no higher priority flit waiting for the same output

- a cycle lasts $T_{NOC} = 10ns$ (100 MHz).

- AXI3 interface, allowing for 128 byte NoC packets and thus virtual-link frames that are longer than 128 bytes are split into several NoC packets

Figure 5.19 illustrates the model of the NoC. In the shown scenario, two NoC packets A and B are sent through the same output port of a router, but since the transmission of packet B is already ongoing, flits of packet A are blocked and the "back pressure" builds up, i.e. all buffer cells on the route of A are being filled up.

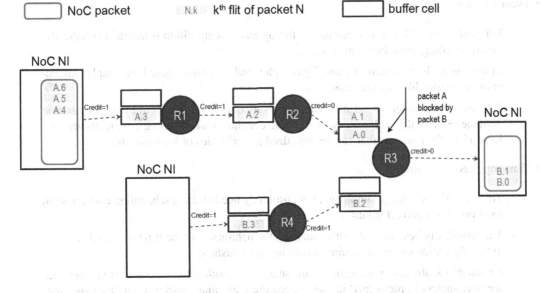

Figure 5.19: Model of STNoC Used in the DHP

Since the payload of a flit is at most 8 bytes, a chunk of 128 bytes of payload produces 17 flits, including the header flit. Thus the number of flits F_k produced by a virtual link payload of C_k bytes is

$$F_k = 17 \left\lfloor \frac{C_k}{128} \right\rfloor + \left\lceil \frac{C_k \bmod 128}{8} \right\rceil + 1$$

If the path through the NoC consists of H_k hops then the "basic network latency" is:

$$R_k^* = (F_k + H_k) T_{NOC}. \tag{5.12}$$

5.2.3.2 Scheduling Mechanisms and Timing Guarantees

The scheduling of safety critical communication must allow the provision of guarantees regarding timing constraints. In the off-chip network, time-triggered scheduling is used in routers and network interfaces in order to be able to provide these guarantees. On the other hand, the scheduling mechanisms of the STNoC (see Section 5.2.3.1) are event driven. Before describing the choice made for the on-chip communication in the DHP, let us first compare time-triggered and event triggered scheduling mechanisms and the different kinds of guarantees regarding delays. Time triggered and event driven scheduling mechanisms are based on different principles that have an impact on how guarantees on timing constraints can be provided. Let us recall them:

- Principles of time-triggered scheduling: Avoidance of run-time resource contention through separation in the time domain by the allocation of non overlapping time windows for serving resource requests.

- Principles of event driven scheduling: Run-time resource contentions are expected and solved through priorities or other arbitration mechanisms. Resource requests are served as soon as possible, in compliance with the arbitration mechanisms.

These principles lead to different characteristics of the configuration algorithms and the verification of timing constraints:

- Event driven scheduling

  - For verification of latency constraints, a timing analysis algorithm is needed to predict the delays resulting from the arbitration.

  - Timing analysis algorithms are usually complex and sometimes based on complex mathematical theories like network calculus.

  - Scheduling parameter optimization algorithms are often used to improve the configuration until the result of the timing analysis allows to conclude that all timing constraints are met. Examples of parameters that can be optimized are priorities or local offsets.

- Time triggered scheduling

  - Timing analysis is fairly simple and not absolutely needed since scheduling configuration tools produce a correct solution.

  - The complexity lies in the algorithm that sets the parameters of the time-triggered scheduling configuration so that all timing constraints are satisfied.

  - Although not absolutely needed, timing analysis algorithm for time triggered schedules are nevertheless implemented in tools, since these are much easier to qualify than time-triggered scheduling configuration tools. The qualification of tools is required for the use in a safety engineering process.

5.2.3.3 Approach for Safety-Critical On-Chip Communication

Since the scheduling mechanisms of the NoC are event-driven, we need a corresponding timing analysis. But no timing analysis for wormhole routing is known to us. [165] describes the timing analysis for a NoC with wormhole routing, but the assumption is made that every packet flow is transmitted in its own virtual network. This has the consequence of removing the cascade blocking phenomena that makes a timing analysis difficult. As a result, a variant of the classical analysis of non-preemptive priority based scheduling is possible. Unfortunately this cannot be applied in our context, where we only have two virtual networks with assigned traffic classes.

With event driven scheduling mechanisms of the NoC, it is not possible to control when flits are exactly sent over a link, but we can control when packets are sent into the NoC. Furthermore, we can compute the time required for a packet to traverse an idle NoC. In [165] it is called "basic network latency". In our context the basic network latency is given by Equation 5.12.

Thus a strategy could be the following:

- Strategy 1: Transmissions are scheduled such that the transmission of a packet starts only when the NoC is idle (again).

As a result, the "basic network latency" is the guaranteed delay, because no contention occurs inside the network. Notice that with this approach, the NoC is actually used like a bus, since only one transmission may take place at the same time. This is an unfortunate under utilization, given that the goal of having a network is precisely that of allowing concurrent transmissions. To get a feeling of the phenomena, we have generated a set of Virtual Links (VLs), as large as possible, without inducing overloaded links in the NoC topology used in the wind-power demonstrator, see Figure 5.20 (only tiles with VL communication are shown):

- For each tile between 55 and 60 VLs are created to send packets to other tiles, with a total of 305 VLs.

- The destination tile is chosen randomly, but must be different from the source tile.

- Periods are chosen randomly in the harmonic set {5ms, 10ms, 50ms, 100ms}.

- VL lengths are chosen randomly in the range [100 bytes, 1500 bytes].

Table 5.4 shows in the columns "All VLs" the link loads induced by the generated communication. As can be seen, some links are bottlenecks like $R2 \to R6 \to R5$, which have a load close to 100%, while others have quite low loads. These are the result of the used traffic generation strategy given the existing NoC topology, but such load distributions can also be observed in real use cases. The average load of all links is 55.33%. If strategy 1 applied, i.e. if the network is used as a bus, then many VLs cannot be scheduled and the average load of the link drops to 20%.

However, many of the pairs of VLs do not share any link. If two VLs do not share any link, then their transmissions could take place in parallel, while keeping the *basic network delay*. Based on this observation, an improved strategy would be the following:

- Strategy 2: Transmissions windows of VL packets that share a link are scheduled so as to have no temporal overlap.

When the second strategy is applied, almost all VLs can be scheduled as can be deduced from the corresponding link-load column in Table 5.4. In order to cross-check the scheduling configuration, we have used the timing accurate simulation or RTaW-Pegase/Timing (cf. Section 5.2.3.5), which is based on exactly the same NoC model as the configuration algorithm.

The second strategy has proven to be very efficient. Although the efficiency may vary a lot between use cases, we can nevertheless conclude that the second strategy is worth being used.

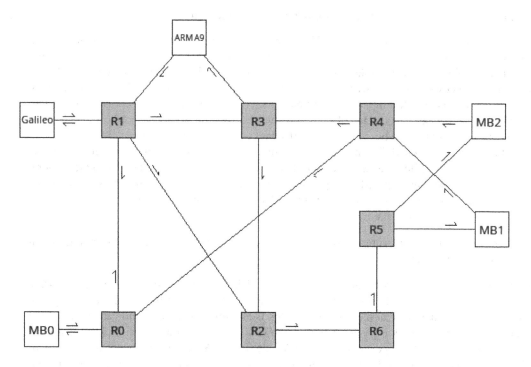

Figure 5.20: Wind-Power NoC Topopoly (cf. Wind Power Demonstrator in Chapter 11)

5.2.3.4   Approach for Non Safety-Critical On-Chip Communication

In the DREAMS architecture, non safety-critical communication is sent over the virtual network with lower priority than RC communication. Therefore, these packets can be delayed by safety-critical packets, but not the other way around. Messages on RC VLs are sent as soon as they arrive, unless they need to be delayed so as to satisfy their MINT. In any case, it is not foreseen to control the exact timing of their transmission. Thus we have an event driven scheduling scheme, where no parameters can be optimized.

5.2.3.5   Tool: RTaW-Pegase/Timing

The algorithm for computing transmission offsets for TT VLs that we have described in the previous sections has been implemented in the RTaW-Pegase/Timing tool. It is integrated into the toolchain (see Section 5.6) as a context menu entry of "System Schedules" and "Reconfiguration Graphs" in the "Model Navigator" view of AF3. An example outcome of the algorithm is shown in Figure 5.21.

5.2.4   Off-Chip Network Scheduling

In this section we present the scheduling problem resulting from both the DREAMS architecture and the TTEthernet off-chip network design used in the DHP.

In TTEthernet a global communication scheme, the *tt-network-schedule*, defines transmission and reception time windows for each time-triggered frame being transmitted between end-systems and switches in the network. The tt-network-schedule is typically built offline based on implicit technology constraints but also user constraints resulting from system requirements. In DREAMS, the upper scheduling layers will have specific schedules produced sequentially by dedicated schedulers. Each of these layers will produce constraints for the next underlying layers. Hence, the net-

Table 5.4: Link Loads

Source Node	Target Node	Link Loads		
		All VLs	Strategy 1	Strategy 2
Galileo	R1	52,22%	21,66%	51,31%
ARM A9	R1	44,13%	13,10%	42,59%
MB1	R4	56,44%	22,00%	56,04%
MB0	R0	52,75%	22,46%	51,76%
R1	Galileo	57,62%	24,06%	57,62%
R0	MB0	50,38%	19,00%	50,38%
R1	R0	20,08%	6,14%	20,08%
R0	R1	77,52%	31,56%	76,54%
R1	R3	28,72%	13,62%	28,72%
R4	R3	57,12%	17,94%	55,54%
R4	R0	55,07%	21,96%	55,07%
R3	ARM A9	55,29%	20,32%	55,29%
R3	R2	30,55%	11,24%	28,97%
R2	R6	98,00%	33,74%	92,99%
R6	R5	98,00%	33,74%	92,99%
R5	MB2	45,59%	19,84%	44,04%
R5	MB1	52,40%	13,90%	48,95%
MB2	R4	55,76%	17,90%	54,58%
R1	R2	67,45%	22,50%	64,02%
	Average	55,53%	20,35%	54,08%

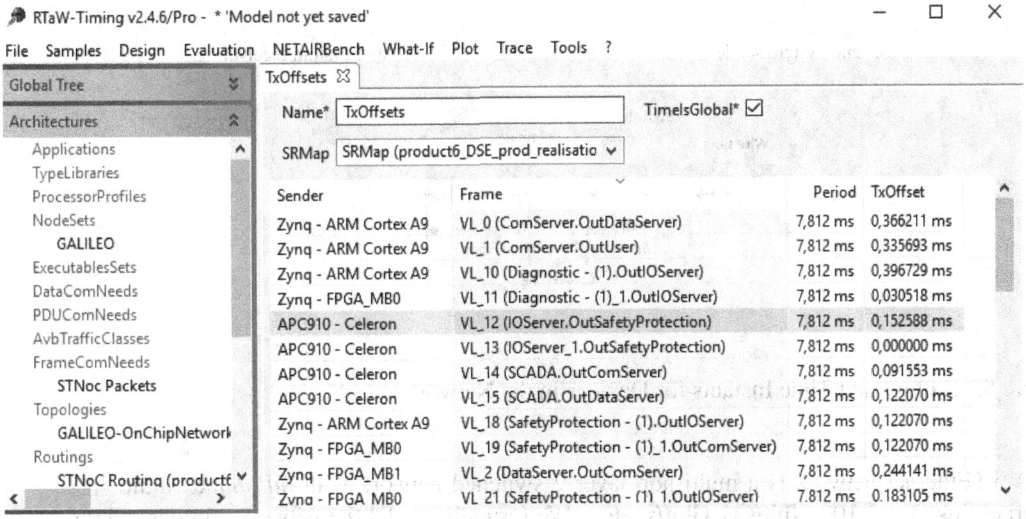

Figure 5.21: RTaW-Pegase/Timing Showing the Computed Transmission Offsets for TT On-Chip Communication

work layer, which is the last one to be scheduled in the process will be constrained by the schedules of all the upper layers. From the perspective of the separation of concerns, these derived constraints can be seen as system requirements for the network which are usually specified in the form of user constraints to the network scheduling tool.

Here, we formalize based on [166] and [167] the network model and describe both TTEthernet inherent constraints (adapted from [166] and [167]) as well as the basic user constraints which can be used by other scheduling layers to constrain the network schedule. Constraints specified by the user of TTE-Plan can be composed from these base user constraints which are designed to cover a large proportion of possible user constraints that result from the higher layers in the DREAMS architecture. In Section 5.2.4.1, we define the network model that we later use to formulate logical constraints inherent for TTEthernet (Section 5.2.4.2) and base user constraints (Section 5.2.4.3). We describe our incremental scheduling algorithm based on Satisfiability Modulo Theories solvers and discuss optimization opportunities resulting from the scheduling model. We also present a number of experiments designed to show the scalability of our approach using synthetic workloads.

Please note that the presented algorithm and constraints consider only the network level in isolation. For a more advanced method that schedules the network in conjunction with the task layer using a similar approach to the one presented here, we refer to reader to the research in [167] which we conducted in the context of the DREAMS project.

5.2.4.1 Network Model

The network model and formalization of frames described in this subsection follow the description from [167].

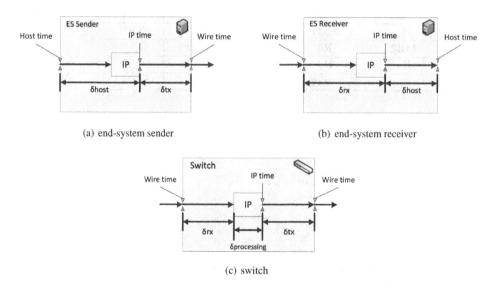

(a) end-system sender (b) end-system receiver

(c) switch

Figure 5.22: Important Time Instants for Devices in the Network

A TTEthernet network is a multi-hop layer 2 switched network with *full-duplex* multi-speed Ethernet links (e.g., 100 Mbit/s, 1 Gbit/s, etc.). We formally model the network, similar to [166], as a directed graph $G(\mathcal{V}, \mathcal{L})$, where the set of vertices ($\mathcal{V}$) comprises the communication nodes (switches and end-systems) and the edges ($\mathcal{L} \subseteq \mathcal{V} \times \mathcal{V}$) represent the directional communication links between nodes. Since we consider bi-directional physical links (i.e. full-duplex), we have that $\forall [v_a, v_b] \in \mathcal{L} \Rightarrow [v_b, v_a] \in \mathcal{L}$, where $[v_a, v_b]$ is an ordered tuple representing a directed logical link between vertices $v_a \in \mathcal{V}$ and $v_b \in \mathcal{V}$. In the implementation we also support half-duplex links which we do not formalize for simplicity.

We model time-triggered communication via the concept of TT virtual links (TT-VL)[3], where

---
[3]Please note that in this section we use virtual link to describe the network part of the DREAMS virtual links.

a TT virtual link is a logical data-flow path in the network from a sender node to one or multiple receiver nodes. The typical virtual link from the ARINC Specification 664 Part 7 defines virtual links as unidirectional paths from one sender node to multiple receiver nodes. The constraint formulation is agnostic to the number of sender nodes of one virtual link.

A typical TT virtual link $vl_i \in \mathcal{VL}$ from a sender end-system v_s to a receiving end-system v_r, routed through the nodes (i.e. switches) $v_1, v_2, \ldots, v_{n-1}, v_n$ is expressed, similar to [166], as

$$vl_i = [[v_s, v_1], [v_1, v_2], \ldots, [v_{n-1}, v_n], [v_n, v_r]].$$

We denote the set of links on which the sender nodes of a TT virtual link vl_i send a frame by $\mathcal{L}_i^s$. Conversely, the set of links on which receiver nodes of a virtual link vl_i receive frames is denoted by $\mathcal{L}_i^r$.

A TT virtual link vl_i will generate a frame instance on every link along the communication path (route). We denote a frame belonging to VL vl_i scheduled on a link $[v_a, v_b]$ with $f_i^{[v_a,v_b]}$.

We use a similar notation to [166] to model frames. A frame f defined by

$$\langle f.\phi, f.\pi, f.T, f.d, f.w, f.P_{tx}, f.P_{rx} \rangle,$$

where $f.\phi$ is the offset within the period on the tx side, $f.\pi$ is the initial period instance, $f.T$ is the period, $f.d$ is the frame duration on the link in nanoseconds, $f.w$ is the frame window duration, and $f.P_{tx}$ and $f.P_{rx}$ are the ports where the frame is sent and received, respectively.

Considering the minimum and maximum frame sizes in the Ethernet protocol of 84 and 1542 bytes (including the IEEE 802.1Q tag), respectively, the frame duration, for example, on a 1Gbit/sec link would be $0.672\mu sec$ and $12.336\mu sec$, respectively.

The initial period instance (denoted by $f.\pi$) is introduced to allow end-to-end communication exceeding the period boundary. We model the absolute moment in time when a frame is scheduled to be sent from the IP by the combination of the offset –bounded within the period interval– and the initial period instance, i.e., $f.\phi + f.\pi \times f.T$.

The frame window duration includes, beside the frame duration, the shuffling delay caused by lower-priority frames (usually the maximum Ethernet packet size of 1500 bytes leading to a delay of $12.336\mu sec$ on a 1Gbit/sec link) and the delay caused by higher-priority synchronization (PCF) frames. For now we do not go into details on how to compute the lower and higher priority shuffling duration as it is not necessary for the constraint definition but will return further below in the section with more insight when discussing media reservation.

A device v_x is defined by $\langle v_x.\delta_{host}, v_x.\delta_{processing} \rangle$ where $v_x.\delta{host}$ is the host delay of the device, and $v_x.\delta_{processing}$ represents the forwarding delay of the device. If the device is an end-system the forwarding delay will be 0.

A device can have multiple ports where each port P is defined by

$$\langle P.\delta_{rx}, P.\delta_{tx} \rangle,$$

where $P.\delta_{rx}$ is the receive delay of the port and $P.\delta_{tx}$ is the send delay of the port.

A link between nodes v_a and v_b, $[v_a, v_b]$, is defined by

$$\langle [v_a, v_b].mT, [v_a, v_b].\delta_t \rangle,$$

where $[v_a, v_b].mT$ is the macrotick in nanoseconds and $[v_a, v_b].\delta_t$ is the transmission delay on the wire for one bit. The macrotick is the time-line granularity of the physical link, resulting from e.g., hardware properties or design constraints. Typically, the TTEthernet time granularity is around $60ns$ [168] but larger values are commonly used. The transmission delay refers to the propagation of a bit on the medium, and depends on the link length and the link medium, i.e., copper or fiber.

Additionally, we denote the network precision with δ_{prec}. By network precision we mean the maximum difference between any two synchronized clocks in the network. At run-time, a network-wide fault-tolerant time synchronization protocol [169] guarantees that the clocks of the devices are synchronized with sub-microsecond precision [170, p. 186].

The offset for a frame represents the scheduled time for the frame in the IP (see Figures 5.22(a), 5.22(c), and 5.22(b)). Additionally there are two more important time points relative to the IP time, the wire time and the host time. The wire time refers to the time when a bit enters or leaves the port of the device in relation to the IP time. The host time represents the perspective of the host device when a frame has been either sent from the host layer to the IP or has been received at the host layer from the IP. While scheduled times always refer to IP time, user constraints can be given as either host or wire times.

For sender nodes (Figure 5.22(a)) the host delay (δ_{host}) is the delay from the host to the IP. The sending delay (δ_{tx}) represents the delay from the IP to the wire and is typically composed of the delay in the chip, the bus delay, and the PHY delay. For receiver nodes (Figure 5.22(b)) the host delay represents the time from IP to the host while the receive delay (δ_{rx}) is the delay from the wire to the IP, composed of the PHY delay, the bus delay, and the chip delay. For switches (Figure 5.22(c)) there is an additional delay, besides δ_{tx} and δ_{rx}, namely the processing delay ($\delta_{processing}$) within the IP, which is the delay to forward a bit from a receiving port to a sending port.

We now formulate the constraints in terms of inherent TTEthernet constraints (adapted from [167]) and in terms of the base user constraints in the following sections.

5.2.4.2 Formalization of Scheduling Constraints

Frame Constraints

For any frame scheduled on a link, the offset cannot take any negative values or any value that would result in the scheduling window to exceed the frame period. Therefore, we have

$$\forall vl_i \in \mathcal{VL}, \forall [v_a, v_b] \in vl_i, \forall f_i^{[v_a, v_b]} :$$
$$\left(f_i^{[v_a, v_b]}.\phi \geq 0 \right) \wedge \left(f_i^{[v_a, v_b]}.\phi \leq \left\lceil \frac{f_i^{[v_a, v_b]}.w - f_i^{[v_a, v_b]}.L}{[v_a, v_b].mT} \right\rceil \right).$$

Here, the offset refers to the IP time and can take values on the time-line scaled with the macrotick.

We also bound the initial frame period instance as follows

$$\forall vl_i \in \mathcal{VL}, \forall [v_a, v_b] \in vl_i, \forall f_i^{[v_a, v_b]} :$$
$$\left(f_i^{[v_a, v_b]}.\pi \geq 0 \right) \wedge \left(f_i^{[v_a, v_b]}.\pi \leq 2 \times HP \right),$$

where HP is the cluster cycle of the network, i.e., the hyper-period of all periods in the network. Please note that the upper bound on the initial period instance is not necessary for the correctness of the constraints or the schedule, however, the SMT solver will perform better if a bound is given. We can additionally put optimization objectives to minimize the period instance for the flows. Although this does not influence the correctness of the schedule, the resulting offsets for the sender frames of flows will be scheduled as closely as possible to the first period instance.

Link Constraints

The most essential constraint that needs to be fulfilled for time-triggered networks is that no two frames that are transmitted on the same link are in contention, i.e., they do not overlap in the time domain. Given two frames, $f_i^{[v_a, v_b]}$ and $f_j^{[v_a, v_b]}$, that are scheduled on the same link $[v_a, v_b]$ we need to specify constraints such that the frames cannot overlap in any period instance.

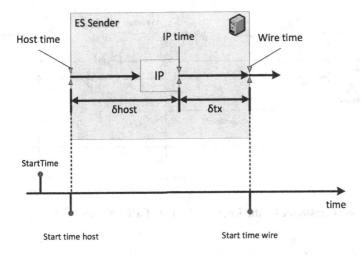

Figure 5.23: Start Time Constraint for the Sender Frame of a TT Virtual Link

$$\forall [v_a, v_b] \in \mathcal{L} : \forall f_i^{[v_a, v_b]}, \forall f_j^{[v_a, v_b]}, (i \neq j) \Rightarrow$$

$$\forall x \in \left[0, \frac{HP_i^j}{\frac{f_i^{[v_a, v_b]}.T}{[v_a, v_b].mT}} - 1 \right], \forall y \in \left[0, \frac{HP_i^j}{\frac{f_j^{[v_a, v_b]}.T}{[v_a, v_b].mT}} - 1 \right] :$$

$$\left(f_i^{[v_a, v_b]}.\phi + x \times \frac{f_i^{[v_a, v_b]}.T}{[v_a, v_b].mT} \geq f_j^{[v_a, v_b]}.\phi + y \times \frac{f_j^{[v_a, v_b]}.T + f_j^{[v_a, v_b]}.d}{[v_a, v_b].mT} \right) \vee$$

$$\left(f_j^{[v_a, v_b]}.\phi + y \times \frac{f_j^{[v_a, v_b]}.T}{[v_a, v_b].mT} \geq f_i^{[v_a, v_b]}.\phi + x \times \frac{f_i^{[v_a, v_b]}.T + f_i^{[v_a, v_b]}.d}{[v_a, v_b].mT} \right),$$

where $HP_i^j \stackrel{def}{=} lcm \left(\frac{f_i^{[v_a, v_b]}.T}{[v_a, v_b].mT}, \frac{f_j^{[v_a, v_b]}.T}{[v_a, v_b].mT} \right)$, is the hyper-period of the two frames being compared scaled to the macrotick. Although for a network link, the macrotick is typically set the granularity of the physical medium, we can use the macrotick property of the network link to reduce the search space and simulate what in [166] is called a scheduling "raster". Note that, increasing the macrotick for a network link will reduce the search space for that link, making the algorithm faster, but it will also reduce the solution space.

TT Virtual Link Constraints

We introduce TT virtual link constraints which describe the sequential nature of a communication from source to destination node. Here, we also consider, similar to [171], the precision of the network (δ_{prec}) which, as explained before, is a network constant and describes the maximum dif-

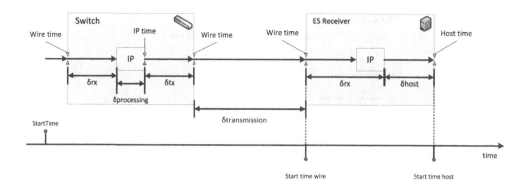

Figure 5.24: Start Time Constraint for the Receiver Frame of a TT Virtual Link

ference between the local clocks of any two nodes in the network.

$$\forall vl_i \in \mathcal{VL}, \forall [v_a, v_x], [v_x, v_b] \in vl_i :$$
$$[v_x, v_b].mT \times f_i^{[v_x,v_b]}.\phi + f_i^{[v_x,v_b]}.\pi \times f_i^{[v_x,v_b]}.T-$$
$$([v_a, v_x].mT \times f_i^{[v_a,v_x]}.\phi + f_i^{[v_a,v_x]}.\pi \times f_i^{[v_a,v_x]}.T) \geq$$
$$f_i^{[v_a,v_x]}.P_{tx}.\delta_{tx} + f_i^{[v_a,v_x]}.w + [v_a, v_x].\delta_t +$$
$$f_i^{[v_a,v_x]}.P_{rx}.\delta_{rx} + v_x.\delta_{processing} + \delta_{prec}.$$

The constraint expresses that, for a frame, the difference between the start of the transmission window on one link and the end of the transmission window on the precedent link has to be greater than the cumulated transmission delay for that frame plus the precision for the entire network and the delays of the devices.

Additionally, we introduce a constraint for the sender frames of each TT virtual link which ensures that the host time will be greater than or equal to 0.

$$\forall vl_i \in \mathcal{VL}, \forall [v_s, v_x] \in \mathcal{L}_i^s :$$
$$[v_s, v_x].mT \times f_i^{[v_s,v_x]}.\phi + f_i^{[v_s,v_x]}.\pi \times f_i^{[v_s,v_x]}.T \geq v_s.\delta_{host}.$$

5.2.4.3   User Constraints

Start Time Constraint

The start time constraint takes a set of TT virtual links to which it applies as its inputs. Here we simplify the formulation and look at one TT virtual link as input. However, the constraint can be readily extended and applied to each TT virtual link of the input set.

The start time constraint can be given either for the senders or receivers of a TT virtual link (TT-VL). In case of the senders (depicted in Figure 5.23) the start time constraint refers to the start of all sender frames of the given VL which have to be scheduled after the given start time r. Complementary, if the constraint is given for the receivers of a TT-VL (depicted in Figure 5.24) all receiver frames have to start being received after the given time. The constraint can be either given as host or wire time by the user.

Let vl_i be the given TT virtual link. If the start time r is given as host time, the send time

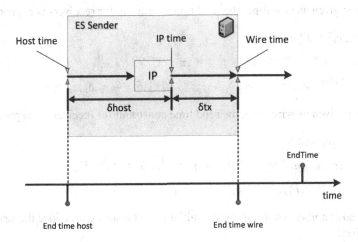

Figure 5.25: End Time Constraint for the Sender Frame of a TT Virtual Link

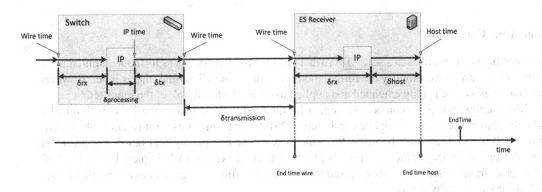

Figure 5.26: End Time Constraint for the Receiver Frame of a TT Virtual Link

constraint for senders is expressed as

$$\forall [v_s, v_x] \in \mathcal{L}_i^s :$$
$$[v_s, v_x].mT \times f_i^{[v_s,v_x]}.\phi + f_i^{[v_s,v_x]}.\pi \times f_i^{[v_s,v_x]}.T - v_s.\delta_{host} \geq r.$$

Conversely, if the start time r is given as wire time, the send time constraint for senders is expressed by

$$\forall [v_s, v_x] \in \mathcal{L}_i^s :$$
$$[v_s, v_x].mT \times f_i^{[v_s,v_x]}.\phi + f_i^{[v_s,v_x]}.\pi \times f_i^{[v_s,v_x]}.T + f_i^{[v_s,v_x]}.P_{tx}.\delta_{tx} \geq r.$$

For receivers, since we schedule the send times and not receive times for frames, we have to additionally include several delays to obtain either the start of the receiving at the wire or at the host. In both cases we have to add the transmission delay of the sending port of the frame. Moreover, we have to include the transmission duration of one bit on the wire. The delays within the receiving node have to be added in case the start time is specified as host time.

If the start time r is given as host time, the send time constraint for receivers is expressed as

$$\forall [v_x, v_r] \in \mathcal{L}_i^r :$$
$$[v_x, v_r].mT \times f_i^{[v_x,v_r]}.\phi + f_i^{[v_x,v_r]}.\pi \times f_i^{[v_x,v_r]}.T +$$
$$f_i^{[v_x,v_r]}.P_{tx}.\delta_{tx} + [v_x, v_r].\delta_t + f_i^{[v_x,v_r]}.P_{rx}.\delta_{rx} + v_r.\delta_{host} \geq r.$$

If the start time r is given as wire time, the send time constraint for receivers is expressed as

$$\forall [v_x, v_r] \in \mathcal{L}_i^r :$$
$$[v_x, v_r].mT \times f_i^{[v_x,v_r]}.\phi + f_i^{[v_x,v_r]}.\pi \times f_i^{[v_x,v_r]}.T +$$
$$f_i^{[v_x,v_r]}.P_{tx}.\delta_{tx} + [v_x, v_r].\delta_t \geq r.$$

Additionally, we add another constraint for the initial period instance to reduce the search space for the constraint solver:

$$\forall [v_s, v_x] \in \mathcal{L}_i^s : f_i^{[v_s,v_x]}.\pi \geq \left\lfloor \frac{r}{f_i^{[v_s,v_x]}.T} \right\rfloor .$$

End Time Constraint

The end time constraint takes a set of TT virtual links to which it applies as its inputs. Similar to the start time constraint we simplify the formulation and look at one TT virtual link as input. However, the constraint can be readily extended and applied to each TT virtual link of the input set.

The end time constraint can be given either for the senders or receivers of a TT virtual link (TT-VL). In case of the senders (depicted in Figure 5.25) the end time constraint refers to the receive time of all sender frames of the given VL which have to be completely sent before the given end time d. Complementary, if the constraint is given for the receivers of a VL, depicted in Figure 5.26, all receiver frames have to be fully received before the given time. The constraint can be either given as host or wire time by the user.

Let vl_i be the given TT virtual link. If the end time d is given as host time, the end time constraint for senders is expressed as

$$\forall [v_s, v_x] \in \mathcal{L}_i^s :$$
$$[v_s, v_x].mT \times f_i^{[v_s,v_x]}.\phi + f_i^{[v_s,v_x]}.\pi \times f_i^{[v_s,v_x]}.T +$$
$$f_i^{[v_s,v_x]}.w - v_s.\delta_{host} \leq d.$$

Conversely, if the end time d is given as wire time, the end time constraint for senders is expressed as

$$\forall [v_s, v_x] \in \mathcal{L}_i^s :$$
$$v_s, v_x].mT \times f_i^{[v_s,v_x]}.\phi + f_i^{[v_s,v_x]}.\pi \times f_i^{[v_s,v_x]}.T +$$
$$f_i^{[v_s,v_x]}.w + f_i^{[v_s,v_x]}.P_{tx}.\delta_{tx} \leq d.$$

If the end time d is given as host time, the end time constraint for receivers is expressed as

$$\forall [v_x, v_r] \in \mathcal{L}_i^r :$$
$$[v_x, v_r].mT \times f_i^{[v_x,v_r]}.\phi + f_i^{[v_x,v_r]}.\pi \times f_i^{[v_x,v_r]}.T + f_i^{[v_x,v_r]}.w +$$
$$f_i^{[v_x,v_r]}.P_{tx}.\delta_{tx} + [v_x, v_r].\delta_t + f_i^{[v_x,v_r]}.P_{rx}.\delta_{rx} + v_r.\delta_{host} \leq d.$$

If the end time d is given as wire time, the end time constraint for receivers is expressed as

$$\forall [v_x, v_r] \in \mathcal{L}_i^r :$$
$$[v_x, v_r].mT \times f_i^{[v_x, v_r]}.\phi + f_i^{[v_x, v_r]}.\pi \times f_i^{[v_x, v_r]}.T + f_i^{[v_x, v_r]}.w +$$
$$f_i^{[v_x, v_r]}.P_{tx}.\delta_{tx} + [v_x, v_r].\delta_t \leq d.$$

Additionally, we add another constraint for the initial period instance to reduce the search space for the solver.

$$\forall [v_s, v_x] \in \mathcal{L}_i^s : f_i^{[v_s, v_x]}.\pi \leq \left\lceil \frac{d}{f_i^{[v_s, v_x]}.T} \right\rceil .$$

Transmission Duration Constraint

The transmission duration constraint takes a set of TT virtual links to which it applies as its inputs. Again, we simplify the formulation and construct the constraints for one TT virtual link which can then be applied to each of the TT virtual links from the input set.

The transmission duration constraint, depicted in Figure 5.27, restricts the end to end latency of a VL to be between a given minimum and maximum. The user can specify that the end to end constraint applies either to host or wire time. If the user specifies that either the maximum or the minimum is -1 then the respective value will not be considered. In essence, the constraint states that the difference between the end of the receiver frames, either on the wire or at the host, and the start of the sender frames has to be either smaller than or equal to the maximum end-to-end latency allowed, greater than or equal to the minimum end-to-end latency or both.

Let vl_i denote the given TT virtual link to which the constraint applies and min_{E2E} and max_{E2E} the user-defined minimum and maximum end-to-end latency, respectively.

If the constraint applies to the wire time, we have the following constraint

$$\forall [v_s, v_a] \in \mathcal{L}_i^s, \forall [v_b, v_r] \in \mathcal{L}_i^r :$$
$$min_{E2E} + \delta_{prec} \geq$$
$$[v_b, v_r].mT \times f_i^{[v_b, v_r]}.\phi + f_i^{[v_b, v_r]}.\pi \times f_i^{[v_b, v_r]}.T +$$
$$f_i^{[v_b, v_r]}.w + f_i^{[v_b, v_r]}.P_{tx}.\delta_{tx} + [v_b, v_r].\delta_t -$$
$$([v_s, v_a].mT \times f_i^{[v_s, v_a]}.\phi + f_i^{[v_s, v_a]}.\pi \times f_i^{[v_s, v_a]}.T + f_i^{[v_s, v_a]}.P_{tx}.\delta_{tx})$$
$$\leq max_{E2E} - \delta_{prec}.$$

If the constraint applies to the host time, we have the following constraint

$$\forall [v_s, v_a] \in \mathcal{L}_i^s, \forall [v_b, v_r] \in \mathcal{L}_i^r :$$
$$min_{E2E} + \delta_{prec} \geq$$
$$[v_b, v_r].mT \times f_i^{[v_b, v_r]}.\phi + f_i^{[v_b, v_r]}.\pi \times f_i^{[v_b, v_r]}.T +$$
$$f_i^{[v_b, v_r]}.w + f_i^{[v_b, v_r]}.P_{tx}.\delta_{tx} + [v_b, v_r].\delta_t + f_i^{[v_b, v_r]}.P_{rx}.\delta_{rx} + v_r.\delta_{host} -$$
$$([v_s, v_a].mT \times f_i^{[v_s, v_a]}.\phi + f_i^{[v_s, v_a]}.\pi \times f_i^{[v_s, v_a]}.T - v_s.\delta_{host})$$
$$\leq max_{E2E} - \delta_{prec}.$$

Transmission Gap Constraint

The transmission gap constraint takes a set of TT virtual links to which it applies as its inputs.

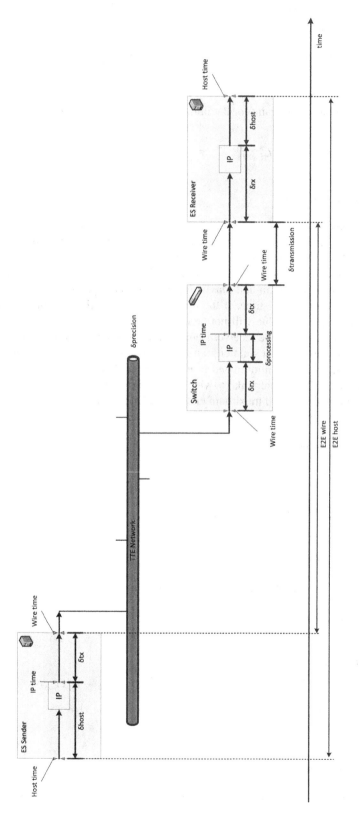

Figure 5.27: Transmission Duration Constraint for a TT Virtual Link

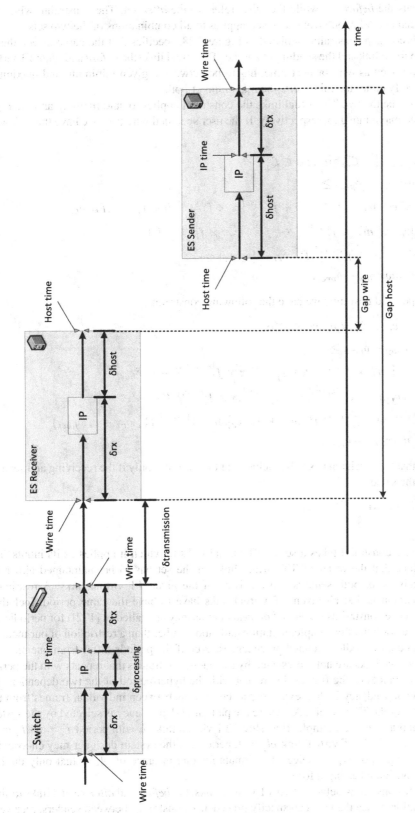

Figure 5.28: Transmission Gap Constraint Between Two TT Virtual Links

One set represents the *before* set while the other refers to the *after* set. The constraint, which we formulate between two VLs, one from each set, applies to all combinations of the two sets.

The transmission gap constraint, depicted in Figure 5.28, specifies that the gap between the receiving of a TT virtual link and the sending of another TT virtual link (the *before* and *after* TT virtual links), either expressed as wire or host time, has to be between a given minimum and maximum. The user can specify either the minimum, the maximum or both.

Let vl_i and vl_j be the two TT virtual links the constraint applies to, and min_{Gap} and max_{Gap}, the minimum and maximum gap, respectively. If the user specified wire time we have the following constraint:

$$\forall [v_a, v_r] \in \mathcal{L}_i^r, \forall [v_s, v_b] \in \mathcal{L}_j^s :$$
$$min_{Gap} + \delta_{prec} \geq$$
$$[v_s, v_b].mT \times f_j^{[v_s,v_b]}.\phi + f_j^{[v_s,v_b]}.\pi \times f_j^{[v_s,v_b]}.T + f_j^{[v_s,v_b]}.P_{tx}.\delta_{tx} -$$
$$([v_a, v_r].mT \times f_i^{[v_a,v_r]}.\phi + f_i^{[v_a,v_r]}.\pi \times f_i^{[v_a,v_r]}.T +$$
$$f_i^{[v_a,v_r]}.w + f_i^{[v_a,v_r]}.P_{tx}.\delta_{tx} + [v_a, v_r].\delta_t)$$
$$\leq max_{Gap} - \delta_{prec}.$$

If the user specified host time we have the following constraint:

$$\forall [v_a, v_r] \in \mathcal{L}_i^r, \forall [v_s, v_b] \in \mathcal{L}_j^s :$$
$$min_{Gap} + \delta_{prec} \geq$$
$$[v_s, v_b].mT \times f_j^{[v_s,v_b]}.\phi + f_j^{[v_s,v_b]}.\pi \times f_j^{[v_s,v_b]}.T - v_s.\delta_{host} -$$
$$([v_a, v_r].mT \times f_i^{[v_a,v_r]}.\phi + f_i^{[v_a,v_r]}.\pi \times f_i^{[v_a,v_r]}.T +$$
$$f_i^{[v_a,v_r]}.w + f_i^{[v_a,v_r]}.P_{tx}.\delta_{tx} + [v_a, v_r].\delta_t + f_i^{[v_a,v_r]}.P_{rx}.\delta_{rx} + v_r.\delta_{host})$$
$$\leq max_{Gap} - \delta_{prec}.$$

Please note that the precision has to be included in the formulas only if the receiving and sending devices are not the same.

Precedence Constraint

The precedence constraint takes a set of TT virtual links to which it applies as its inputs. The constraint specifies that the order of TT virtual links in the set has to be maintained either for the senders, receivers, or both senders and receivers of the given TT virtual links. A restriction of this user constraint is that all given TT virtual links have to have the same period. Including multi-rate precedence constraints (*extended precedences* as they are called in [172]) for periodic TT virtual links is a restriction of our implementation and model rather than a restriction of our method. Extending the model to handle extended precedences, even if the periods differ from one another, implies that TT virtual links are not represented by a sequence of frames that repeats with the period but by multiple instances of the frames that repeat with the hyper-period of the two dependent TT virtual links. The dependency is then expressed as constraints between individual frames from the multiple instances of the TT virtual links where the pattern of dependency is selected by the system designer. We elaborate on one example. If a "slow" TT virtual link vl_i with period $T_i = n \cdot T_j$ must be scheduled after a "fast" TT virtual link vl_j with period T_j, the system designer may choose, for example, that n outputs of vl_i are selected as inputs for one instance of vl_i or that only the last output of vl_j is considered as input for vl_i.

We specify the constraints between two TT virtual links, the *before* and *after* virtual links to simplify the formulation. Since the user can specify precedence constraints between senders, receivers,

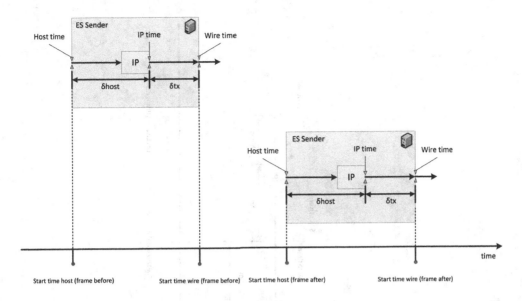

Figure 5.29: Precedence Constraint Between the Senders of Two TT Virtual Links

or both, we take each case in particular, noting that if both are specified, the constraints for senders and receivers are compositional, i.e., they can be decomposed into sender and receiver constraints separately. Again, the user can specify that the precedence constraint applies to either wire or host time.

For senders, we depict the constraint in Figure 5.29 and for receivers in Figure 5.30. Let vl_i and vl_j represent the *before* and *after* TT virtual links, respectively, i.e., TT virtual link vl_i has to be scheduled before vl_j.

For senders, the constraint must make sure that the *before* frame is fully sent, this includes the frame window, before the *after* frame starts being sent.

If the user specified precedence constraint for senders and wire times, we have the following constraint

$$
\forall [v_{s_i}, v_a] \in \mathcal{L}_i^s, \forall [v_{s_j}, v_b] \in \mathcal{L}_j^s, i < j :
$$
$$
[v_{s_j}, v_b].mT \times f_j^{[v_{s_j}, v_b]}.\phi + f_j^{[v_{s_j}, v_b]}.\pi \times f_j^{[v_{s_j}, v_b]}.T +
$$
$$
f_j^{[v_{s_j}, v_b]}.P_{tx}.\delta_{tx} - \delta_{prec} \ge
$$
$$
[v_{s_i}, v_a].mT \times f_i^{[v_{s_i}, v_a]}.\phi + f_i^{[v_{s_i}, v_a]}.\pi \times f_i^{[v_{s_i}, v_a]}.T +
$$
$$
f_i^{[v_{s_i}, v_a]}.P_{tx}.\delta_{tx} + f_i^{[v_{s_i}, v_a]}.w.
$$

If the user specified precedence constraint for senders and host times, we have the following constraint

$$
\forall [v_{s_i}, v_a] \in \mathcal{L}_i^s, \forall [v_{s_j}, v_b] \in \mathcal{L}_j^s, i < j :
$$
$$
[v_{s_j}, v_b].mT \times f_j^{[v_{s_j}, v_b]}.\phi + f_j^{[v_{s_j}, v_b]}.\pi \times f_j^{[v_{s_j}, v_b]}.T - v_{s_j}.\delta_{host} - \delta_{prec} \ge
$$
$$
[v_{s_i}, v_a].mT \times f_i^{[v_{s_i}, v_a]}.\phi + f_i^{[v_{s_i}, v_a]}.\pi \times f_i^{[v_{s_i}, v_a]}.T - v_{s_i}.\delta_{host} + f_i^{[v_{s_i}, v_a]}.w.
$$

Please note that, if the two sending devices are the same, the precision is not needed in the constraint formulation.

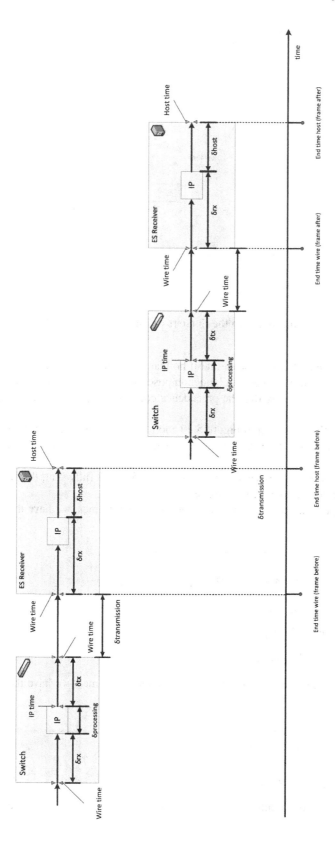

Figure 5.30: Precedence Constraint Between the Receivers of Two TT Virtual Links

Also note that, if the two sending ports are the same, and as a result the sending devices are the same (i.e., $v_{s_i} = v_{s_j}$), the constraint (for both host and wire time) simplifies to

$$\forall [v_{s_i}, v_a] \in \mathcal{L}_i^s, \forall [v_{s_j}, v_b] \in \mathcal{L}_j^s, i < j, v_{s_i} = v_{s_j}:$$

$$[v_{s_j}, v_b].mT \times f_j^{[v_{s_j}, v_b]}.\phi + f_j^{[v_{s_j}, v_b]}.\pi \times f_j^{[v_{s_j}, v_b]}.T \geq$$

$$[v_{s_i}, v_a].mT \times f_i^{[v_{s_i}, v_a]}.\phi + f_i^{[v_{s_i}, v_a]}.\pi \times f_i^{[v_{s_i}, v_a]}.T + f_i^{[v_{s_i}, v_a]}.w.$$

Additionally, if the two sending ports are the same, and the sending device supports FIFO queuing of frames, we do not need to consider the frame window ($f_i^{[v_{s_i}, v_a]}.w$).

For receivers, the constraint must make sure that the *before* frame is fully receives, this includes the frame window, before the *after* frame starts being received.

If the user specified precedence constraint for receivers and wire times, we have the following constraint

$$\forall [v_a, v_{r_i}] \in \mathcal{L}_i^r, \forall [v_b, v_{r_j}] \in \mathcal{L}_j^r, i < j:$$

$$[v_b, v_{r_j}].mT \times f_j^{[v_b, v_{r_j}]}.\phi + f_j^{[v_b, v_{r_j}]}.\pi \times f_j^{[v_b, v_{r_j}]}.T +$$

$$f_j^{[v_b, v_{r_j}]}.P_{tx}.\delta_{tx} + [v_b, v_{r_j}].\delta_t - \delta_{prec} \geq$$

$$[v_a, v_{r_i}].mT \times f_i^{[v_a, v_{r_i}]}.\phi + f_i^{[v_a, v_{r_i}]}.\pi \times f_i^{[v_a, v_{r_i}]}.T +$$

$$f_i^{[v_a, v_{r_i}]}.w + f_i^{[v_a, v_{r_i}]}.P_{tx}.\delta_{tx} + [v_a, v_{r_i}].\delta_t$$

If the user specified precedence constraint for receivers and host times, we have the following constraint

$$\forall [v_a, v_{r_i}] \in \mathcal{L}_i^r, \forall [v_b, v_{r_j}] \in \mathcal{L}_j^r, i < j:$$

$$[v_b, v_{r_j}].mT \times f_j^{[v_b, v_{r_j}]}.\phi + f_j^{[v_b, v_{r_j}]}.\pi \times f_j^{[v_b, v_{r_j}]}.T +$$

$$f_j^{[v_b, v_{r_j}]}.w + f_j^{[v_b, v_{r_j}]}.P_{tx}.\delta_{tx} + f_j^{[v_b, v_{r_j}]}.P_{rx}.\delta_{rx} +$$

$$v_{r_j}.\delta_{host} + [v_b, v_{r_j}].\delta_t - \delta_{prec} \geq$$

$$[v_a, v_{r_i}].mT \times f_i^{[v_a, v_{r_i}]}.\phi + f_i^{[v_a, v_{r_i}]}.\pi \times f_i^{[v_a, v_{r_i}]}.T +$$

$$f_i^{[v_a, v_{r_i}]}.P_{tx}.\delta_{tx} + f_i^{[v_a, v_{r_i}]}.P_{rx}.\delta_{rx} +$$

$$v_{r_i}.\delta_{host} + [v_a, v_{r_i}].\delta_t$$

Please note that, if the two receiving devices are the same, the precision (δ_{prec}) is not needed in the constraint formulation.

Also note that, if the two receiver frames are sent from the same port to the receiving node, and as a result the sending devices are the same (i.e., $v_a = v_b$), the constraint (for both host and wire time) simplifies to

$$\forall [v_a, v_{r_i}] \in \mathcal{L}_i^r, \forall [v_b, v_{r_j}] \in \mathcal{L}_j^r, i < j, v_a = v_b:$$

$$[v_b, v_{r_j}].mT \times f_j^{[v_b, v_{r_j}]}.\phi + f_j^{[v_b, v_{r_j}]}.\pi \times f_j^{[v_b, v_{r_j}]}.T \geq$$

$$[v_a, v_{r_i}].mT \times f_i^{[v_a, v_{r_i}]}.\phi + f_i^{[v_a, v_{r_i}]}.\pi \times f_i^{[v_a, v_{r_i}]}.T + f_i^{[v_a, v_{r_i}]}.w$$

Additionally, if the sending device supports FIFO queuing of frames, we do not need to consider the frame window ($f_i^{[v_a, v_{r_i}]}.w$).

Floating Window Constraint

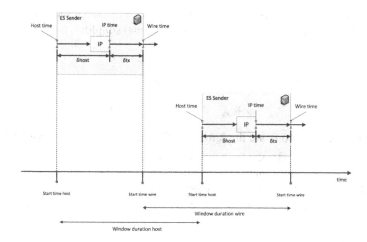

Figure 5.31: Floating Window Constraint Between the Receivers of Two TT Virtual Links

The floating window constraint takes a set of TT virtual links to which it applies as its inputs and expresses that all given TT virtual links have to be scheduled either to start (Figure 5.31) or to be received (Figure 5.32) within a given distance of each other, i.e., all TT virtual links have to be scheduled within a window which can be positioned anywhere. The user can specify either that the constraint applies to the senders or to the receivers of the TT virtual links. The window duration can again be specified either in terms of host or wire time. Let vl_i and vl_j be two TT virtual links out of the input set which have to be scheduled within the given window of size w_s.

If the constraint applies to the senders and wire time is specified, we have the following formulation

$$\forall [v_{s_i}, v_a] \in \mathcal{L}_i^s, \forall [v_{s_j}, v_b] \in \mathcal{L}_j^s :$$
$$([v_{s_i}, v_a].mT \times f_i^{[v_{s_i}, v_a]}.\phi + f_i^{[v_{s_i}, v_a]}.\pi \times f_i^{[v_{s_i}, v_a]}.T +$$
$$f_i^{[v_{s_i}, v_a]}.P_{tx}.\delta_{tx} + f_i^{[v_{s_i}, v_a]}.w -$$
$$([v_{s_j}, v_b].mT \times f_j^{[v_{s_j}, v_b]}.\phi + f_j^{[v_{s_j}, v_b]}.\pi \times f_j^{[v_{s_j}, v_b]}.T +$$
$$f_i^{[v_{s_j}, v_b]}.P_{tx}.\delta_{tx})$$
$$\leq w_s - \delta_{prec}) \wedge (w_s - \delta_{prec} \geq$$
$$[v_{s_j}, v_b].mT \times f_j^{[v_{s_j}, v_b]}.\phi + f_j^{[v_{s_j}, v_b]}.\pi \times f_j^{[v_{s_j}, v_b]}.T +$$
$$f_j^{[v_{s_j}, v_b]}.P_{tx}.\delta_{tx} + f_j^{[v_{s_j}, v_b]}.w -$$
$$([v_{s_i}, v_a].mT \times f_i^{[v_{s_i}, v_a]}.\phi + f_i^{[v_{s_i}, v_a]}.\pi \times f_i^{[v_{s_i}, v_a]}.T +$$
$$f_i^{[v_{s_i}, v_a]}.P_{tx}.\delta_{tx})).$$

If the constraint applies to the senders and host time is specified, we have the following formu-

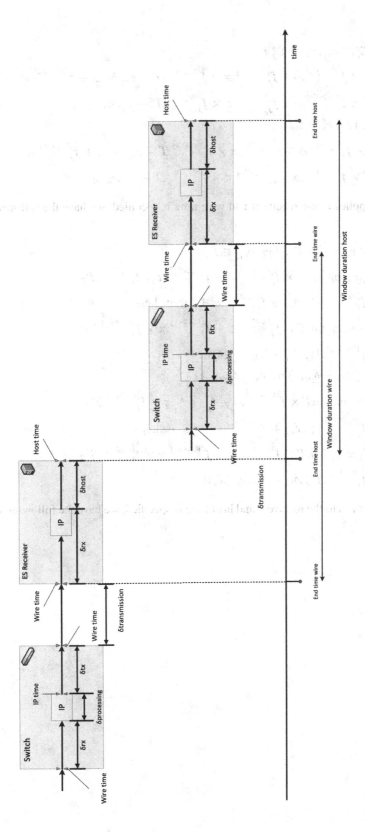

Figure 5.32: Floating Window Constraint Between the Receivers of Two TT Virtual Links

lation

$$\forall [v_{s_i}, v_a] \in \mathcal{L}_i^s, \forall [v_{s_j}, v_b] \in \mathcal{L}_j^s :$$

$$([v_{s_i}, v_a].mT \times f_i^{[v_{s_i}, v_a]}.\phi + f_i^{[v_{s_i}, v_a]}.\pi \times f_i^{[v_{s_i}, v_a]}.T - v_{s_i}.\delta_{host} + f_i^{[v_{s_i}, v_a]}.w -$$

$$([v_{s_j}, v_b].mT \times f_j^{[v_{s_j}, v_b]}.\phi + f_j^{[v_{s_j}, v_b]}.\pi \times f_j^{[v_{s_j}, v_b]}.T - v_{s_j}.\delta_{host})$$

$$\leq w_s - \delta_{prec}) \wedge (w_s - \delta_{prec} \geq$$

$$[v_{s_j}, v_b].mT \times f_j^{[v_{s_j}, v_b]}.\phi + f_j^{[v_{s_j}, v_b]}.\pi \times f_j^{[v_{s_j}, v_b]}.T - v_{s_j}.\delta_{host} + f_j^{[v_{s_j}, v_b]}.w -$$

$$([v_{s_i}, v_a].mT \times f_i^{[v_{s_i}, v_a]}.\phi + f_i^{[v_{s_i}, v_a]}.\pi \times f_i^{[v_{s_i}, v_a]}.T - v_{s_i}.\delta_{host}))$$

If the constraint applies to the receivers and wire time is specified, we have the following formulation

$$\forall [v_a, v_{r_i}] \in \mathcal{L}_i^r, \forall [v_b, v_{r_j}] \in \mathcal{L}_j^r :$$

$$([v_a, v_{r_i}].mT \times f_i^{[v_a, v_{r_i}]}.\phi + f_i^{[v_a, v_{r_i}]}.\pi \times f_i^{[v_a, v_{r_i}]}.T +$$

$$f_i^{[v_a, v_{r_i}]}.P_{tx}.\delta_{tx} + f_i^{[v_a, v_{r_i}]}.w + [v_a, v_{r_i}].\delta_t -$$

$$([v_b, v_{r_j}].mT \times f_j^{[v_b, v_{r_j}]}.\phi + f_j^{[v_b, v_{r_j}]}.\pi \times f_j^{[v_b, v_{r_j}]}.T +$$

$$f_j^{[v_b, v_{r_j}]}.P_{tx}.\delta_{tx} + [v_b, v_{r_j}].\delta_t)$$

$$\leq w_s - \delta_{prec}) \wedge (w_s - \delta_{prec} \geq$$

$$[v_b, v_{r_j}].mT \times f_j^{[v_b, v_{r_j}]}.\phi + f_j^{[v_b, v_{r_j}]}.\pi \times f_j^{[v_b, v_{r_j}]}.T +$$

$$f_j^{[v_b, v_{r_j}]}.P_{tx}.\delta_{tx} + f_j^{[v_b, v_{r_j}]}.w + [v_b, v_{r_j}].\delta_t -$$

$$([v_a, v_{r_i}].mT \times f_i^{[v_a, v_{r_i}]}.\phi + f_i^{[v_a, v_{r_i}]}.\pi \times f_i^{[v_a, v_{r_i}]}.T +$$

$$f_i^{[v_a, v_{r_i}]}.P_{tx}.\delta_{tx} + [v_a, v_{r_i}].\delta_t))$$

If the constraint applies to the receivers and host time is specified, we have the following formulation

$$\forall [v_a, v_{r_i}] \in \mathcal{L}_i^r, \forall [v_b, v_{r_j}] \in \mathcal{L}_j^r :$$

$$([v_a, v_{r_i}].mT \times f_i^{[v_a, v_{r_i}]}.\phi + f_i^{[v_a, v_{r_i}]}.\pi \times f_i^{[v_a, v_{r_i}]}.T +$$

$$f_i^{[v_a, v_{r_i}]}.P_{tx}.\delta_{tx} + f_i^{[v_a, v_{r_i}]}.w + [v_a, v_{r_i}].\delta_t +$$

$$f_i^{[v_a, v_{r_i}]}.P_{rx}.\delta_{rx} + v_{r_i}.\delta_{host} -$$

$$([v_b, v_{r_j}].mT \times f_j^{[v_b, v_{r_j}]}.\phi + f_j^{[v_b, v_{r_j}]}.\pi \times f_j^{[v_b, v_{r_j}]}.T +$$

$$f_j^{[v_b, v_{r_j}]}.P_{tx}.\delta_{tx} + [v_b, v_{r_j}].\delta_t +$$

$$f_j^{[v_b, v_{r_j}]}.P_{rx}.\delta_{rx} + v_{r_j}.\delta_{host})$$

$$\leq w_s - \delta_{prec}) \wedge (w_s - \delta_{prec} \geq$$

$$[v_b, v_{r_j}].mT \times f_j^{[v_b, v_{r_j}]}.\phi + f_j^{[v_b, v_{r_j}]}.\pi \times f_j^{[v_b, v_{r_j}]}.T +$$

$$f_j^{[v_b, v_{r_j}]}.P_{tx}.\delta_{tx} + f_j^{[v_b, v_{r_j}]}.w + [v_b, v_{r_j}].\delta_t +$$

$$f_j^{[v_b, v_{r_j}]}.P_{rx}.\delta_{rx} + v_{r_j}.\delta_{host} -$$

$$([v_a, v_{r_i}].mT \times f_i^{[v_a, v_{r_i}]}.\phi + f_i^{[v_a, v_{r_i}]}.\pi \times f_i^{[v_a, v_{r_i}]}.T +$$

$$f_i^{[v_a, v_{r_i}]}.P_{tx}.\delta_{tx} + [v_a, v_{r_i}].\delta_t +$$

$$f_i^{[v_a, v_{r_i}]}.P_{rx}.\delta_{rx} + v_{r_i}.\delta_{host}))$$

Please note that the precision has to be included in the formulas only if the sender frames of the two TT virtual links are not on the same sending device. Complementary, for receiver frames, if the sending device of both frames to each receiving device is the same, the precision can be excluded from the formula.

Media Reservation

In practical TTEthernet implementations, the scheduler does not schedule actual frames but rather frame windows that also include specific delays that a TT-frame can experience. This delay arises from the blocking time of lower or higher priority frames that were sent before the scheduled point of a TT-frame and have not finished transmission by the time specified in the schedule for the TT-frame. TTEthernet does not currently implement preemption, hence this delay must be considered when scheduling a TTEthernet network. Some mitigation methods like timely block or shuffling have been implemented in practice [173, p. 42-5] that can either eliminate or account for this blocking time. When shuffling is used, the duration of the scheduling window of TT-frames includes the worst case duration of any lower priority frame that can delay a TT-frame. The timely block method (also called media reservation) will prevent any low-priority frame to be sent if it would delay a scheduled TT-frame [173, p. 42-5], [174]. Additionally, synchronization frames always have higher priority with respect to TT-frames, hence, the higher-priority shuffling duration always needs to be considered by the scheduler.

An additional boolean variable has been introduced in the frame model representing whether to use media reservation or not. This variable is used in all constraints which include the frame window size. If media reservation is used, the window size will be calculated without the low priority shuffling size. If media reservation is not used, this low priority shuffling delay will be included in the window size.

We diverge for a moment to present the calculation of the window duration with and without media reservation. The frame duration d of a frame of size b bytes on a link of link speed s is calculated as

$$d = \frac{b \cdot 10^9 \cdot 8}{s}.$$

When media reservation is supported, the window duration w for a frame of size b bytes on a link of link speed s is calculated as

$$w = \frac{b + b_{PCF} \cdot N_{PCF}^{max}}{s},$$

where b_{PCF} is the size of a PCF frame and N_{PCF}^{max} is the maximum number of PCF frames that can interfere with the given frame. Please note that the higher priority shuffling duration is given by the formula $b_{PCF} \cdot N_{PCF}^{max}$. With shuffling, the maximum delay due to a BE or RC frame, i.e., the lower priority shuffling duration, has to be also considered. Hence the window duration for a frame is calculated as

$$w = \frac{b + b_{PCF} \cdot N_{PCF}^{max} + b_{max-frame} + \text{IFG}}{s},$$

where IFG is the inter-frame gap (12 bytes) and $b_{max-frame}$ is the maximum frame size on a TTEthernet network, including MTU (1500 bytes) and the Ethernet protocol overhead of 26 bytes.

For a frame instance of a flow vl_i on device $[v_a, v_b]$ the media reservation variable is defined by $f_i^{[v_a, v_b]}.mr$. Let $\delta_i^{[v_a, v_b]}$ and $\Delta_i^{[v_a, v_b]}$ be the low priority shuffling duration and high priority shuffling duration that the frame experiences on device $[v_a, v_b]$, respectively.

The frame window size will be calculated as follows

$$f_i^{[v_a, v_b]}.w = f_i^{[v_a, v_b]}.d + \Delta_i^{[v_a, v_b]} + f_i^{[v_a, v_b]}.mr \times \delta_i^{[v_a, v_b]}.$$

Using media reservation per frame can be enabled in different ways. Naturally, an instance of a frame on a device will only support media reservation if the device supports media reservation, otherwise it will be disabled by default.

5.2.4.4   SMT-Based Scheduling Algorithm

SMT is a method for checking the satisfiability of logic formulas in first-order formulation against a specific background theory, e.g., linear integer arithmetic ($\mathcal{LA}(\mathbb{Z})$) or bit-vectors ($\mathcal{BV}$) [175], [176]. A first-order formula consists of variables, quantifiers, logical operators, and functional and predicate symbols [177]. Scheduling problems can be naturally expressed in linear arithmetic and are thus suitable for solving with the use of SMT solvers (see for e.g., [178]).

As described in [167], the SMT-based scheduling algorithm generates assertions for the context of an SMT solver representing on the constraints defined in Section 5.2.4.2 where the offsets of frames and initial period instance are the variables of the formula. If the SMT context is satisfiable, the solver returns a solution for the set of variables (a so-called *model*) that conforms to the defined constraints.

SMT-based Incremental Schedule Synthesis

Our algorithm, which is based on the work in [166], uses SMT solvers to find out the schedule for a network based on the inherent TTEthernet and the user constraints which are given as an input by the toolchain. Our algorithm features an incremental approach which may improve scheduling time in the average case. Steiner [166] proposes an incremental backtracking algorithm which adds only a subset of the frames at a time to the SMT context using the push pop functionality. If a partial solution is found for the subset, additional frames and constraints are added until either the complete schedule is found or a solution to a partial problem cannot be found. In the case of in-feasibility, the problem is backtracked and the size of the added subset is increased. In the worst case the algorithm backtracks to the root, scheduling the complete set of frames in one step. Our algorithm adds each VL from the input to the SMT context in sequence following the algorithm described in [166].

Use of Optimization Objectives

We now briefly discuss how to introduce optimization criteria into the scheduling of the TTEthernet network. For a more in-depth discussion on MIP scheduling including experiments that compare the solution to SMT we refer the reader to [167].

SMT solvers return one solution out of potentially many valid solutions. Some systems have additionally requirements on properties of the retrieved solution which may not be hard constraints but are nevertheless important. Mostly these requirements come in the form of optimization objectives. In order to allow this additional constraint we can use either dedicated MIP optimization solvers, like Gurobi [179], or Optimization Modulo Theories (OMT), which add optimization capabilities to SMT solvers [146, 180–182].

In the case of MIP solvers, most constraints can be readily transformed to MIP constraints since they do not contain any logical clauses. Logical *either-or* constraints, like the ones used in the link constraints, can be transformed into a single inequality in MIP formulation by using the alternative constraints method (cf. [183, p. 278] or [184, p. 79]), similar to [171]. Consider, as before, two frames, $f_i^{[v_a,v_b]}$ and $f_j^{[v_a,v_b]}$, that are scheduled on the same link $[v_a,v_b]$ and cannot overlap in any period instance. For every contention-free assertion we introduce a binary variable $z \in \{0,1\}$, and formulate the link constraint as follows

$$\forall [v_a,v_b] \in \mathcal{L} : \forall f_i^{[v_a,v_b]}, \forall f_j^{[v_a,v_b]}, (i \neq j) \Rightarrow$$

$$\forall x \in \left[0, \frac{HP_i^j}{\frac{f_i^{[v_a,v_b]}.T}{[v_a,v_b].mT}} - 1\right], \forall y \in \left[0, \frac{HP_i^j}{\frac{f_j^{[v_a,v_b]}.T}{[v_a,v_b].mT}} - 1\right] :$$

$$\left(f_i^{[v_a,v_b]}.\phi + x \times \frac{f_i^{[v_a,v_b]}.T}{[v_a,v_b].mT} - z \times \Omega \geq \right.$$

$$f_j^{[v_a,v_b]}.\phi + y \times \frac{f_j^{[v_a,v_b]}.T + f_j^{[v_a,v_b]}.d}{[v_a,v_b].mT} \Bigg) \wedge$$

$$\left(f_j^{[v_a,v_b]}.\phi + y \times \frac{f_j^{[v_a,v_b]}.T}{[v_a,v_b].mT} + z \times \Omega \geq \right.$$

$$f_i^{[v_a,v_b]}.\phi + x \times \frac{f_i^{[v_a,v_b]}.T + f_i^{[v_a,v_b]}.d}{[v_a,v_b].mT} + \Omega \Bigg),$$

where $HP_i^j \overset{def}{=} lcm(f_i^{[v_a,v_b]}.T, f_j^{[v_a,v_b]}.T)$, is, as before, the hyper-period of the two frames and Ω is a constant that is large enough (in our case we choose the hyper-period HP_i^j) such that the first condition is always true for $z = 1$ and the second condition is always true for $z = 0$.

If we use an Optimization Modulo Theory (OMT), like Z3 v 4.4.1 [181] the optimization objectives can be readily integrated into the context without modification of the existing constraints.

5.2.4.5 Evaluation

For the experimental evaluation with SMT solvers, we have selected two SMT solvers. Based on our previous experiments [167] we have identified Yices v2.4.1 (64bit) [185] as the most high-performing SMT solver for NP-complete scheduling problems like the one discussed in the DREAMS project. Hence, we use Yices with linear integer arithmetic ($\mathcal{LA}(\mathbb{Z})$) without quantifiers as the background theory as the baseline solver for our algorithms.

As discussed before, typical SMT solvers like Yices return one arbitrary solution out of a set

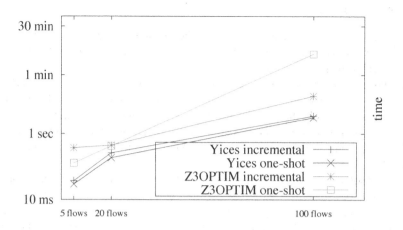

Figure 5.33: Runtime when Scheduling Flows with Periods $P_1 = \{10ms, 20ms\}$

of multiple valid solutions. Here we use Optimization Modulo Theories (OMT), namely Z3 v4.4.1 (64bit) [186], to show the scalability of scheduling networks with optimization objectives.

Our aim is to show that the incremental method improves the scalability of the tool compared to the non-incremental (one-shot) method, in which all flows are scheduled at once. We show the results of running both the incremental and non-incremental (one-shot) algorithm using both solver libraries running on a 64bit 8-core 3.40GHz Intel *Core-i7* PC with 16GB memory.

The scalability of network scheduling problems depends on several key factors, analyzed in detail in [167], the most important of them being the periods of the flows and the number of flows that need to be scheduled. For the experiments we use 3 different sets of flow periods and schedule sets of 5, 20, and 100 flows. The important aspect of the different period sets is the resulting hyper-period which greatly influences the runtime of the scheduler.

Figure 5.33 presents the experiment results when scheduling respectively 5, 20, and 100 flows where the flows are randomly assigned periods from the set $P_1 = \{10ms, 20ms\}$. The x-axis shows the number of flows scheduled and the logarithmic y-axis represents the runtime of the tool with $30min$ being the time-out bound. The hyper-period of the flows is $20ms$, hence the scheduling problem is relatively easy to solve. As we see, with Yices, the improvement between the incremental and one-shot method is relatively small. What can be seen is that when using optimization criteria with Z3, the runtime of the tool is improved significantly when scheduling 100 flows. On the other hand, for small number of flows, we can see that the incremental method is slower than the one-shot.

Figure 5.34 presents the experiment results when scheduling respectively 5, 20, and 100 flows where the flows are randomly assigned periods from the set $P_2 = \{10ms, 25ms, 100ms\}$. As before, the x-axis shows the number of flows scheduled and the logarithmic y-axis represents the runtime of the tool with $30min$ being the time-out bound. With a larger hyper-period resulting from the flow periods, we can see that the performance decreases in comparison to the previous experiment. Additionally, when scheduling 100 flows, the one-shot method with optimization (Z3OPTIM) reaches the time-out of 30 minutes and is interrupted. With the incremental method the same input set is solvable within 1 minute, representing a significant scalability improvement.

Figure 5.35 presents the experiment results when scheduling respectively 5, 20, and 100 flows where the flows are randomly assigned periods from the set $P_3 = \{20ms, 100ms, 1000ms\}$. Here, the hyper-period is again even greater than in the previous two experiments resulting in significantly harder problems to be solved by SMT solver, especially with increasing number of flows. As in the previous experiment we see that when scheduling 100 flows, the one-shot method with optimization

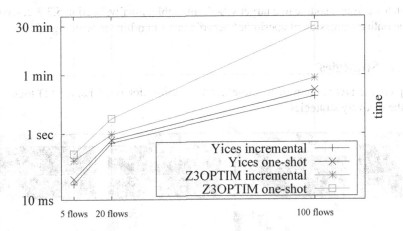

Figure 5.34: Runtime when Scheduling Flows with Periods $P_2 = \{10ms, 25ms, 100ms\}$

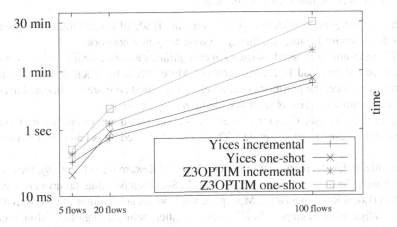

Figure 5.35: Runtime when Scheduling Flows with Periods $P_3 = \{20ms, 100ms, 1000ms\}$

(Z3OPTIM) reaches the time-out of 30 minutes and is interrupted. The incremental method with optimization finds a solution in under 5 minutes.

We have shown that in the average case, the incremental method significantly reduces the run-time of the tool, especially when scheduling a large number of flows even when optimization objectives are included in the problem.

5.3 Adaptation Strategies

In the previous Section 5.2 scheduling configuration algorithms for a single mode are described, which is typically the nominal mode. However, when cores fail on which safety critical functions are executed, it is interesting to foresee several scheduling configurations that guarantee the execution of the critical functions even if some cores are no longer available. Such recovery strategies and their configuration are described in Section 5.3.1. Section 5.3.2 considers Transition Modes that allow faster switching from a source to a target scheduling table. Finally, Section 5.3.3 describes an algorithm for the online admission of aperiodic/sporadic tasks in offline scheduling tables.

5.3.1   Recovery Strategies

Figure 5.36 depicts the architecture of a safe-critical demonstrator (cf. Chapter 11) used for the explanation of the recovery strategies.

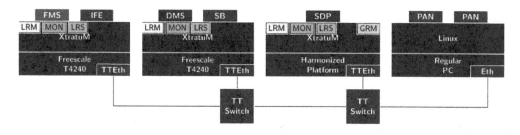

Figure 5.36: Safety-Critical Demonstrator Architecture

As seen in the above figure, considered architectures are made of several multicore processors (named *nodes* in the following) connected through a time-triggered network.

Two kinds of failures are considered: *permanent core failures* and *temporal overload situations*. Only permanent core failures lead to reconfiguration. More precisely, when a core is halted on permanent failure occurrence, the partitions executing on the failed core are re-allocated according to pre-computed configurations. This is what is called a *reconfiguration*.

A monitoring mechanism (see Chapter 9) must detect when such a permanent core failure occurs, and a recovery procedure must be applied. This recovery procedure is based on pre-computed mode changes.

Because of the hierarchical structure of the DREAMS middleware (see chapter 9), the reconfiguration strategies have to be defined hierarchically as well. So, when possible, failures are recovered locally by the LRM (Local Reconfiguration Manager), while when not, failures are recovered by the GRM (Global Reconfiguration Manager). In all cases, the latter maintains a global vision, regarding the current configuration of the whole architecture.

So, job scheduling reconfigurations mentioned above can be either *local* when a single node is involved in the reconfiguration, or *global* when several nodes are involved in the reconfiguration, and different decision levels are involved: either a local (LRM) or a global (GRM) reconfiguration manager.

A local reconfiguration can occur if:

* there are compatible spare time slots (**weak local reconfiguration**) on the still operational cores of the failed node.

* it is possible to re-allocate all applications, following a different slot mapping (**strong local configuration**) on the operational cores of the failed node

Otherwise, global reconfigurations are considered, where applications may be reallocated to others nodes under responsibility of the GRM.

The reconfiguration strategy is implemented as an offline process, which considers all possible failure configurations and tries for each of them to determine a possible (either local or global) reconfiguration. As a result, the process produces configuration files for the LRMs and for the GRM.

GRec

GRec (for **G**raph of **Rec**onfigurations tool) is a **constraint-based** tool, aimed at computing off-line job scheduling alternatives on a multicore architecture, to be used in case of core failure occurrence. It is based on the general principles stated above.

Symmetry

As *symmetry* is desired for computed reconfiguration solutions, the reconfiguration tree is in fact rather a graph than a pure tree, as seen in Figure 5.37.

no failure
P_i/c_k fails
P_i/c_j fails
P_i/c_k fails
P_i/c_j fails

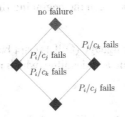

Figure 5.37: Symmetrical Reconfiguration in Case of Multiple Failures

Symmetry means that multiple failures involving the same cores in the same node, possibly with different failure sequences, must lead to the same reconfiguration solution. Since GRec is not able to automatically detect such symmetries, it is up to the scenario generator to produce only asymmetric scenarii (see below).

Scenarios

GRec attempts to process all possible asymmetric failure combinations, or scenarii, each scenario being represented as a character string formatted as seen at Figure 5.38. These scenarii are specific of a given system architecture, are computed by an auxilliary tool of GRec and are stored in an input file for GRec.

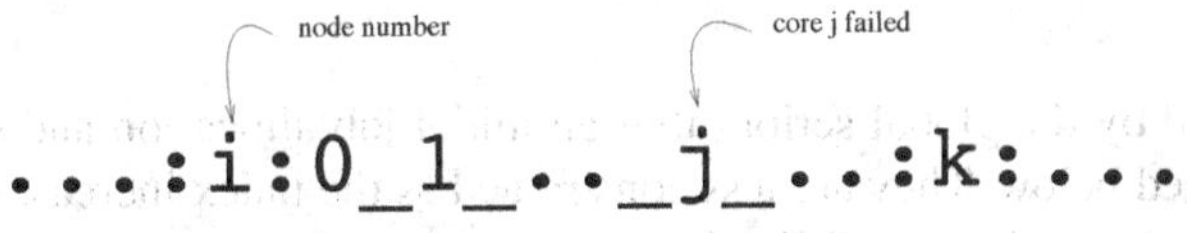
node number
core j failed
...:i:0_1_.._j_..:k:...

Figure 5.38: Example of Failure Configuration

Implementation

GRec is based on *IBM ILOG CPLEX Optimization Studio*, a proprietary (licensed) software package, and more specifically on the IBM ILOG CPLEX CP Optimizer for constraint programming.

5.3.1.1 GRec Global Algorithm

The applications to be executed on the hardware platform are modeled as sets of tasks. The parameters defining a task mainly are:

- its period.
- its WCET.

GRec computes the allocation and scheduling of a set of jobs, or periodic task invocations, on a time period Minor Frame (MIF) given as input parameter. During the configuration code generation, the calculated allocation and scheduling can be extended up to the MAF duration.

Job List

First, GRec builds the list of jobs to consider,

- from the parameters defining each task,

- from the required time window,

- the nth job being constrained to run between the beginning (its start time) and the end (its deadline) of the nth period.

A first Optimization Programming Language (OPL) script builds the list of jobs to be allocated and scheduled as provides name, start time and deadline, which is passed to the next step.

Search

The search for solutions is performed by two OPL scripts:

- a global script which

    - goes through the list of defined failure scenarios,
    - sets the search context according to the current scenario,
    - and asks for a solution to the search script.

    If no solution is found, the script successively tries less and less constraining strategies, first in the LRM area, second in the GRM area, until either a solution is found or the less constrained strategy fails.

- a script performing the search of a solution satisfying the set of constraints for a given strategy.

Strategies

The search strategies tried by the global script (after an initial job allocation and scheduling on a full safe hardware) are listed below. They are less constrained as the index increases.
Given a new failure configuration, the following actions are performed:

1. search first for a local reconfiguration.

2. if none is found, search for a global reconfiguration of non-critical applications on T4240 nodes.

3. if none is found, search for a global reconfiguration of non-critical applications including on the DREAMS Harmonized Platform node.

4. if none is found, search for a global reconfiguration including critical applications on T4240 nodes.

5. if none is found, search for a global reconfiguration including critical applications including on the DREAMS Harmonized Platform node.

6. if none is found, search for a global reconfiguration including critical applications including on the DREAMS Harmonized Platform node, dismissing one or more non-critical applications.

5.3.1.2 Constraint-Based Reconfiguration Graph Problem Formulation

Configuration Parameters

The master script mainly manages two configuration parameters:

- the current failure pattern is taken into account,
- the current strategy to apply in order to find a valid reconfiguration, given the hardware state.

The current failure pattern is processed by the main script, which consists in invalidating all slots hosted by the current failed cores.

The resolution strategy will be denoted Λ in the following. $\Lambda \in [0..5]$ is an index identifying one of the eight strategies defined above. Starting from 0, the master script increments Λ until either a reconfiguration solution is found for the current hardware state or no further strategy is available. As seen later, the value of Λ partially determines the set of constraints to be satisfied by the searched reconfiguration.

Another important parameter is the MIF, which defines the period of time to be considered when computing the task allocation and scheduling. The MIF is normally equal to the MAF, Least Common Multiple (LCM) of all task periods, but can be less than this value under certain conditions (e.g., harmonicity, geometricity of periods). The value of the MIF is given as an input.

System Model

The data modeling the system are detailed here:

- applications and tasks.
- nodes, cores and slots.

Nodes and Cores

The set of nodes is defined as an integer range:

$$\mathcal{N} = [0..n-1]$$

The set of cores in a node is defined as an integer range:

$$\mathcal{C} = [0..c-1]$$

Slots

The set of slots, or time intervals allocatable to applications, is defined as an integer range. As the decision variable array representing the allocation of jobs to slots uses these values, the low value of the range is 1, 0 having in this case a special meaning from the constraint point of view.

$$\mathcal{S} = [1..s]$$

Slot properties are specified as a record which gathers the following items:

$$\forall s \in \mathcal{S}, s \overset{p}{=} \ < b_s, l_s, n, c, 1_s >,$$
$$n \in \mathcal{N}, c \in \mathcal{C}, b_s \in \mathbb{N}, l_s \in \mathbb{N}, 1_s \in [\top, \bot, \mathrm{MON}, \mathrm{LRM}, \mathrm{GRM}]$$

- b_s is the start time of the slot within the MAF.

- l_s is the duration of the slot.

- n is the node to which the slot belongs.

- c is the core on which the slot is allocated.

- 1_s defines the availability of the slot (a slot is available for an application only if $1_s = \top$).

Applications

Applications considered are defined as a set of names:

$$\mathcal{A} = \{a_1, \cdots, a_{|\mathcal{A}|}\}$$

Application properties are specified as a record:

$$\forall a \in \mathcal{A}, a \stackrel{p}{=} <n, \sigma>, n \in \mathcal{N}, \sigma \in \{\text{critical}, \text{best\_effort}\}$$

- n is the node on which, under nominal conditions, all the tasks making up the application must be allocated. This requirement is relaxed when it is necessary to compute a global reconfiguration.

- σ is the status of the application.

Tasks

All tasks constituting the considered applications are defined as a set of names:

$$\Psi = \{\tau_1, \cdots, \tau_{|\Psi|}\}$$

Task properties are specified as a record which gathers the following items:

$$\forall \tau \in \Psi, \tau \stackrel{p}{=} <\text{T}, \text{WCET}, a>, a \in \mathcal{A}, \text{T} \in \mathbb{N}, \text{WCT} \in \mathbb{N}$$

- T is the period or the task.

- WCET is the worst-case execution time of the task.

- a is the application to which the task belongs.

Jobs

Tasks are periodically executed several times during the specified MAF. Each invocation is a job which has to be allocated and scheduled. As said above, the list of jobs to consider is elaborated during a preliminary phase, from the specified task set and from the value defined for the MAF. The result is a set of names:

$$\mathcal{J} = \{\tau_{1.0}, \cdots, \tau_{i.j}, \cdots, \tau_{|\Psi|.n}\}$$

- $\tau_{i.j}$ is the jth job (corresponding to the jth invocation) of the task τ_i.

- $n = \lfloor \text{MAF}/\text{T}_{|\text{T}|} \rfloor - 1$.

Job properties are specified as a record which gathers the following items:

$$\forall \tau_{i.j} \in \mathcal{J}, \tau_{i.j} \stackrel{p}{=} <\tau_i, \tau_{i.j-1}, r_{i.j}, d_{i.j}>, \tau_i \in \Psi, \tau_{i.j-1} \in \mathcal{J}, r_{i.j} \in \mathbb{N}, d_{i.j} \in \mathbb{N}$$

- $\tau_{i.j-1}$ is the predecessor of $\tau_{i.j}$ (used for expressing the constraint minimizing the number of slots used by a task: $\tau_{i.j}$ will be allocated, if possible, in the same slot as $\tau_{i.j-1}$)

- the release time and deadline of the job are computed as follows:

$$
\begin{aligned}
r_{i.j} &= j \times T_i \\
d_{i.j} &= (j+1) \times T_i
\end{aligned}
$$

Decision Variables

Decision variables are those variables the value determination of which is the goal of the constraint solving. In our case there are two decision variables:

- job allocation to available slots:

$$p[\tau_{i.j} \in \mathcal{J}] \in \mathcal{S}$$

- job time scheduling within the MAF:

$$s[\tau_{i.j} \in \mathcal{J}] \in [0..\mathrm{MAF}]$$

Constraints

This section is devoted to the constraints to be fulfilled by the decision variables. The constraints cover the following issues:

- basic rules, e.g., only one job active at a given time on a given slot.

- job related rules, e.g., the job release time precedes the job start time, which precedes the job end time, which precedes the job deadline.

- slot allocation related rules, e.g., only one application can be allocated to a slot.

- strategy related rules, e.g., for a given strategy, release some of the above constraints.

Some of these constraints are detailed below, in a mathematical fashion more concise and synthetic than the OPL one.

One job active in a slot at a given time

$$\forall \tau_{i.j}, \tau_{k.l} \in \mathcal{J}, \tau_{i.j} \neq \tau_{k.l} \wedge p_{i.j} = p_{k.l} \Rightarrow s_{i.j} \geq s_{k.l} + \mathrm{WCET}_k \vee s_{k.l} \geq s_{i.j} + \mathrm{WCET}_i$$

- τ_i is the task i, and $\tau_{i.j}$ is the jth job of the task i.

- $p_{i.j}$ is the slot allocated by GRec to the job $\tau_{i.j}$.

Only one application on a given slot

$$\forall \tau_{i.j}, \tau_{k.l} \in \mathcal{J}, \tau_{i.j} \neq \tau_{k.l} \wedge p_{i.j} = p_{k.l} \Rightarrow a_{\tau_i} = a_{\tau_k}$$

Slot start $\leq$ job start $\leq$ job end $\leq$ slot end & slot available

$$\forall \tau_{i.j} \in \mathcal{J}, \forall p_n \in \mathcal{S}, p_{i.j} = p_n \Rightarrow \mathrm{b}_n \leq s_{i.j} \wedge s_{i.j} + \mathrm{WCET}_i \leq \mathrm{l}_n - o_n \wedge \mathrm{1}_n = \top$$

Slot $\in$ node assigned to application

This is an example of constraint the validity of which depends on the applied strategy: for a strategy number less than 2 (see above), a critical application must be allocated to the node which it is initially assigned to.

$$\forall \tau_{i.j} \in \mathcal{J}, \forall p_n \in \mathcal{S}, \sigma_{a_{\tau_i}} = \mathrm{critical} \wedge \Lambda \leq 2 \Rightarrow n_{p_n} = n_{a_{\tau_i}}$$

Minimize job dispersion

This constraint specifies that once a job is assigned to a slot, the following jobs of the same task are assigned to the same slot until the end of the slot is reached.

$$\forall \tau_{i.j}, \tau_{k.l} \in \mathcal{J}, \forall p_n \in \mathcal{S}, p_{k.l} = p_n \wedge s_{i.j} + \mathrm{WCET}_i \leq \mathrm{l}_n - o_n \Rightarrow p_{i.j} = p_n$$

MON/critical application mutual exclusion

This is an implementation dependent constraint introduced for avoiding side effects on Worst Case Execution Times (WCETs) resulting from the parallel processing (on a given node) of a MON and of a job belonging to a critical application. This constraint thus prevents such a situation.

$$\forall \tau_{i.j} \in \mathcal{J}, \qquad \forall p_m \in \mathcal{S}, \mathrm{1}_m = \mathrm{MON} \wedge \sigma_{a_{\tau_i}} = \mathrm{critical} \wedge n_{p_m} = n_{p_{i.j}} \wedge$$
$$\neg \mathrm{dead}(p_m) \Rightarrow s_{i,j} \geq s_m + \mathrm{l}_m \vee s_m \geq s_{i,j} + \mathrm{WCET}_i$$

Overview of Constraint Expression in OPL Modeling Language

The above data and formulas translate quite directly in the OPL modeling language. Few examples are given here.

The basic types (task, job, slot, etc.) are defined either as sets or as ranges. For example, the involved applications and tasks are defined as a set:

```
{string} Applis = ...;
{string} Tasks = ...;
```

The available time slots are identified through a range:

```
range Slots = 1..nbSlots;
```

Properties of tasks, jobs, slots, etc. are specified through *tuples*. For example, the properties of tasks are defined as:

```
tuple Task {
int period;
int WCET;
string tset ;
}
```

Similarly, the properties of slots are defined as:

```
tuple Slot {
int start ;
int length ;
int node ;
int core ;
int availability ;
}
```

The set of applications is modeled as follows:

```
Applis = {
"ROSACE",
"MPEG",
"OrderGenerator",
};
```

The properties of applications are instantiated as follows:

```
// Definition of application properties: appli/node assignment,
// critical/best effort
AppliProps = #[
"ROSACE": <1,1>,
"MPEG": <1,0>,
"OrderGenerator": <0,1>,
]# ;
```

The set of tasks is modeled as follows:

```
Tasks = {
"elevator",
"q_filter",
"vz_filter",
"vz_control",
"engine",
"Mpeg2Server",
...
};
```

The properties of tasks are instantiated as follows:

```
// Definition of task properties: period, WCT, appli
TaskProps = #[
"elevator"     :< 50     ,1,   "ROSACE">,
"q_filter"     :< 100    ,1,   "ROSACE">,
"vz_filter"    :< 100    ,1,   "ROSACE">,
"vz_control"   :< 200    ,1,   "ROSACE">,
"engine"       :< 50     ,1," ROSACE">,
"Mpeg2Server"  :< 1000   ,100, "MPEG">,
...
]#;
```

The OPL expression of constraints is very close to the above equational specification. For example, the constraint ordering the release, start, end and deadline times is expressed as:

```
// First release, then start, then deadline
// for each task invocation
forall (j in Jobs) {
JobProps[j].release <= start[j] ;
start[j] + TaskProps[JobProps[j].task].WCET <=
JobProps[j].deadline ;
}
```

Similarly, the constraint "One job active in a slot at a given time" is expressed as:

```
forall (j in Jobs, k in Jobs: j != k) {
(p[j] == p[k]) =>
(start[j] >= start[k]+TaskProps[JobProps[k].task].WCET) ||
(start[k] >= start[j]+TaskProps[JobProps[j].task].WCET) ;
}
```

Data Produced by GRec

As said before, GRec first tries to compute a reconfiguration for each asymmetric failure pattern determined given the current system architecture.

From the obtained set of reconfigurations, GRec in a second step produces a YAML configuration file for each node of the system architecture, including:

- for each computed reconfiguration, a *hardware description* and a *partition description* of the plan assigned to each core of the node.

- a *local reconfiguration table* specifying, for each configuration, the new local configuration to use when a new core failure occurs.

- for the DREAMS Harmonized Platform node (hosting the GRM) a *global reconfiguration table* allowing, through messages specifying the current and expected next configuration of the different nodes, the GRM to perform the required global reconfigurations.

Figure 5.39 shows an example of reconfiguration graph, elaborated by GRec from a reduced version of the safety-critical demonstrator, in order to build the local and global reconfiguration tables for the different nodes.

5.3.2   Comprehensive Offline Schedules

Modern real-time systems require different set of tasks to be executed during different phases of operation. An avionics control system, for instance, needs to execute different sets of applications during take-off, normal flight and landing. Moreover, an aircraft may also define different flight plans for different routes or destinations, emergency landing plans for instance. When the system is designed without the knowledge of these (often) mutually exclusive operation phases, the higher priority tasks may suffer significant jitter and large latencies, which might (potentially) deem the system unfeasible. Moreover with such an all-in-one design, any minor change in the system (addition/modification of tasks/constraints) require the whole system to be rescheduled. On the contrary, when the information about the operation phases of the system is utilized, the system can perform better in terms of response-time, jitter and latencies. Moreover, more functionality in a single mode can be supported as the system utilization for a given mode would be significantly smaller compared to all-in-one design approach.

In literature (e.g., [187, 188]), these operation phases are termed as *modes*. In [187], Fohler used modes to handle mutually exclusive phases of operations which account for large system changes (as opposed to small system changes which are handled by the online scheduler). In [188], Kopetz *et al.* used modes to account for changes in scheduling tables (identifying different phases of operations) of TTP/TDMA based networks. In [189], Real *et al.* used modes to integrate time-triggered and priority-driven tasks/schedules to achieve lower response-time and jitter for control tasks and to handle system overload/emergency situations. In short, there exists no global notion of modes and therefore Graydon and Bate [190] classified modes based on their use. In the following, we provide a short summary of the classification of modes presented by Graydon and Bate [190]:

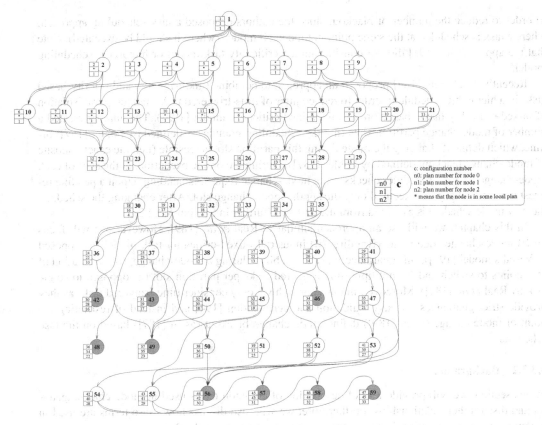

Figure 5.39: Graphical View of a Reconfiguration Graph Generated by GRec

Operation Modes: [190] An operation mode defines a system phase in which the system is being used by its operators (e.g., take-off, flight/auto-pilot, landing) including emergency situations.
Design Modes: [190] The modes in which the system is being designed or maintained are termed as design modes (e.g., normal, maintenance, firmware debugging, software debugging mode).
Service Modes: [190] A service mode defines a system phase in which the system is reconfigured for either survivability, after a loss of resource (e.g., core failure, deadline missed), or improving efficiency (e.g., processor sleep modes, frequency scaling modes).

In this section, we will focus on the service modes responsible for system survivability (i.e. fault-tolerance). In the following, we present a mode scheduling approach for mixed-critical hierarchic systems in order to provide fault-tolerance against core-failures.

5.3.2.1 Related Work

As the classification of modes is only dependent on the way they are used, all modes can be scheduled in similar ways irrespective of their classification. However, some scheduling approaches are more suitable for specific mode types. For instance, in [191] the authors used the modes as *operation modes* in MARS project and defined scheduling approach to maintain strict temporal and causal ordering of events by using sparse time-base [192]. In order to guarantee that the system remains schedulable during a change of mode (henceforth termed as mode-change), the authors in [191] defined blackout (further discussed in Section 5.3.2.2) slots (where a slot is the scheduling granule as defined by the sparse time-base [192]). This blackout annotation for a slot is calculated offline, where a slot is marked as blackout when a mode-change during this slot leads to a deadline miss.

In order to reduce the number of blackout slots, the authors proposed a task scheduling approach, where a task is scheduled at the same point in time in all modes (where it should be executed). Note that the approach in [191] did not consider mixed-criticality task model or hierarchic scheduling model.

Recently, Real *et al.* [189] proposed an approach to combine time-triggered and priority-based tasks in a hierarchic schedule in order to reduce jitter of time-triggered tasks and temporal isolation of mixed-criticality tasks (based on Vestal's mixed-criticality model [193]). The authors created a number of mode-change partition slots (not the scheduling granule, as defined by Fohler [187]) of-fline, which define if changing the mode during this partition slot is feasible (i.e., the exact opposite of mode-change blackout defined by Fohler [187]). During the online execution, at the start of each mode-change slot, the scheduler checks if the mode-change request is pending. Upon a pending request the mode change is executed only during the mode-change slot. After changing the schedule, the new mode schedule is executed from the first slot (further in Section 5.3.2.2).

In this chapter, we will use an approach similar to Real *et al.* [189]. However, we will focus on (i) mode-changes due to survivability, (ii) industrial mixed-criticality task model (as opposed to Vestal's model [193]), (iii) guaranteeing schedulability during a mode-change, (iv) estimation of safe points to switch and (v) multiple time-triggered tasks per partition slot (as opposed to single task by Real *et al.* [189]). Moreover, in our approach we only consider time-triggered tasks as they provide strict guarantees for easy validation and certification [194] and instead of restricting the point of mode-change (as in [189]) define mode-change blackout (as in [187]) based on the task schedule.

5.3.2.2 Background

In this section, we will provide a brief description of common terms used in mode-change analysis and discuss their relationships. Furthermore, we will also discuss how these terms are used in literature and what functionality is required in order to support mode-changes.

Mode-Change

The adaptability through mode-changes allows systems to adapt to major changes in the environment [187] or system state, e.g., fault recovery. Kopetz *et al.* [188] identified two different types of mode-changes which are triggered by a host node by requesting a mode change. In deferred mode-changes, all system nodes change their mode after the request at the end of the hyper-period. In immediate mode-changes, only one system node changes operation mode soon after the request, i.e. before the end of the hyper-period. The motivation for the former is consistency while decreased reaction time for the latter. Kopetz *et al.* [188] pointed out that immediate mode change may lead to consistency problems for the applications and therefore a number of tasks or messages must be aborted but should not go in an undefined state. This notion of an (immediate) mode change requires careful application design and leads to strong dependency between the modes and the scheduler. However, Fohler asserted that consistency is required in both mode change types to keep schedulability during and after the mode-changes [187]. This notion of a mode change decouples the application design from the scheduler and hence is more suited for complex systems. Note that the notion of *immediate* mode change used by Kopetz *et al.* [188] is quite different from the notion used by Fohler [187]. The former immediate mode change can occur at any point in time and always starts the new mode from offset 0. However, the latter immediate mode change can only occur at predefined points in time (i.e., not during a mode change blackout) and always executes the new mode from the same offset as in the currently executing mode.

Transition Modes

As pointed out by [187], a system may be schedulable in individual modes but not in the transition between them. To tackle this issue, transition of modes (or mode-change) is also considered as a separate mode so that offline guarantees and worst-case switching time between the modes can be provided. In [189], Real *et al.* pointed out for platforms with scarce resources, that it might not be feasible to start a new task from the destination mode immediately at the point of mode-change (as memory needs to be reserved). In such a situation, a transition mode can be used to prepare the platform for destination mode and/or reduce the mode-change delay.

The purpose of transition modes in DREAMS is to shorten the mode switching delay. This can only be achieved when immediate mode switching, as defined by Fohler [187], is employed. This type of switching requires mode change blackout information for each slot of all the partitions for all mode classes. This parameter is defined by the scheduler during a MAF. The target of the scheduler is to minimize the mode change blackout.

From the perspective of scheduling, the online scheduler (i.e. LRS) needs to distinguish between two classes of modes. A nominal mode is a mode which is executed cyclically until there is a switching event (e.g., core failure in DREAMS). This type of mode is defined by the system designer. On the other hand, a transition mode is a mode which is executed at most once and switches to a nominal mode when there is a safe point to switch. This type of mode can be defined by the scheduler to decrease the mode switching delay and/or by the system designer to initiate a service before the new nominal mode is established. The tools in DREAMS only support these two types of modes.

Mode-Change Blackout

As mentioned earlier, mode-change blackout interval defines if changing a mode during this time interval leads to consistency problems (defined by Kopetz *et al.* [188]). As the modes are only known to the hypervisor in DREAMS, instead of the schedulers inside the partitions, blackout information can only be defined for the elements scheduled by the hypervisor, i.e. the partition slots. This kind of blackout information is course-grained, as individual tasks can not be controlled based on the mode. Moreover, the hypervisors must be aware of the blackout slots. This means that the hypervisor must support immediate mode switching with an offset, as defined by Fohler [187], and the mode change can be deferred to a point in time before the end of the hyperperiod. This deferrable point in time is marked by the end of a mode change blackout.

For an independent task-set, safe points for mode-change can be trivially defined at time points when no task results in partial execution upon switching. Either a job should be completely executed or not executed at all during a mode change. However, for the case of communicating tasks, which is the case in DREAMS, defining this point in time is not as trivial. Although, the non-preemptive nature of tasks limits the number of safe points to switch, but it is still a difficult problem because of its strong co-relation with the schedule and the communicating tasks. In the case of DREAMS, the end of MAF is always a safe point to switch and hence the ends and the starts of MAF are never mode change blackouts. Furthermore, the resource management components must not limit the instants at which a mode may be switched immediately. On the contrary, a number of immediate switching points can be defined (e.g., by invoking LRM more frequently per MAF). As the implementation of mode change blackout requires a number of functionalities from different domains, the modification in the current strategies (if required by the application) is left up to the hypervisor designers and resource management service providers.

Modes and Mixed-Criticality

In the context of Vestal's mixed-criticality model [193], the mode is handled in two different approaches. The first one deals with each mode and each criticality separately (hence requiring a single scheduling table per mode per criticality) [195], while the second approach assumes criticality as the mode of the system (hence requiring only one scheduling table per mode) [196]. The first approach provides the flexibility of all classification of modes (see the introduction of the chapter) at the cost of increased maintenance overheads, while the second approach restricts the scope of the mode to the system criticality while keeping the maintenance overheads to the minimum.

In the context of industrial mixed-criticality model [3, 4, 56, 197], the mode defines a set of constraints for a defined set of tasks (depending on the mode classification). This notion of mode enables easier handling of tasks based on single worst-case execution time.

Modes and Hypervisors

Hypervisors are used to provide strong temporal and spatial isolation between partitions. A set of partitions are scheduled by the hypervisor according to an inter-partition scheduling strategy (defined as top level scheduler in hierarchical scheduling). Each partition can host a set of tasks which are scheduled according the intra-partition scheduling strategy (defined as low/bottom level scheduler in hierarchical scheduling). As we focus on safety-critical systems, from this point on we will only consider time-triggered scheduling as inter-partition scheduler while cyclic executive scheduler [197] for intra-partition scheduling.

In [198], Crespo *et al.* defined a *plan* in XtratuM hypervisor as an implementation of a mode that allows reallocation of the timing resources (the processor) in a controlled way. A plan defines how the partitions are scheduled (or the top most level in hierarchical scheduling). It is defined for a fixed length of time (MAF) and is repeated periodically, where the period is defined as LCM of the tasks' periods. The XtratuM plans may switch immediately or at the end of MAF depending on an event (i.e., user request and a fault, respectively). In XtratuM [198], plan 0 and plan 1 are reserved for system services, while plan $P > 1$ are defined to be user plans. Plan 0 is termed as initialization plan, where partitions are brought to operation. This then leads to switching to another plan $P \neq 1$. In case of an error detected by the XtratuM health monitor the plan switches immediately to plan 1.

Contrary to the XtratuM hypervisor, the Xen[4] hypervisor is only capable of executing cyclic executive for one core[5] and only support single mode. However, the PikeOS[6] hypervisor provides multi-core support with multiple modes defined as *schemes*. However contrary to the XtratuM hypervisor, the PikeOS hypervisor does not define an explicit mode-change state machine flow (except the `SCHED_BOOT` scheme which is similar to plan 0 in XtratuM).

5.3.2.3 System Model

In this section, we assume that the system is subjected to (temporal and spatial) partitioning enforced by the hypervisor. We utilize the industrial mixed-critical task model [3, 4, 56, 197] and therefore assume time-triggered (TT) scheduling approach for inter-partition scheduling and cyclic-executive for intra-partition scheduling. We execute the tasks on a multi-core platform, however, we do not consider the interferences due to shared caches. We assume that the system schedule is generated before mode-change analysis and blackout generation and the mode-change is implemented as an immediate mode switch as defined by Fohler [187]. Furthermore, we assume that the modes are service modes with fault-tolerance only against core-failures.

5.3.2.4 Methodology

In the following, we will distinguish between three different types of modes; nominal, degraded and transition. Transition mode is defined in Section 5.3.2.2. A nominal mode is defined as a mode which has all cores available to schedule tasks, while a degraded mode is defined as a mode which has one or more failed cores.

---

[4]https://xenproject.org/developers/teams/hypervisor.html
[5]https://blog.xenproject.org/2013/12/16/what-is-the-arinc653-scheduler/
[6]https://www.sysgo.com/products/pikeos-hypervisor/

Blackout Generation for Nominal Modes

As the partition slots stay the same between the source and the destination modes, it is quite trivial to assign mode change blackout intervals. For nominal and degraded modes, a mode change blackout interval for a source mode is defined equal to the sum of old and new partition slots in the destination mode. An example of the mode change blackout interval assignment is provided in Figure 5.40, where the black rectangles in Figure 5.40(a) define the blackout intervals. In the figure, the prefix C refers to the core and the prefix S refers to a partition slot. It is assumed that there is no danger of executing tasks more frequently. For example, when a mode is changed after slot S2 and before the end of slot S4, the slot S2 is executed twice. Note that the mode change blackout intervals are only required for modes which can be changed or non-leaf modes in the reconfiguration graphs.

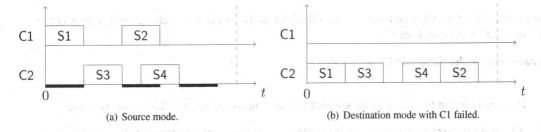

(a) Source mode.   (b) Destination mode with C1 failed.

Figure 5.40: Blackout for Nominal and Degraded Modes

Enrichment with Transition Modes

In the context of DREAMS, the Global Reconfiguration Graph (GRG) and Local Reconfiguration Graphs (LRGs) define the possible mode-changes. It is important to note here that defining transition modes for GRG can complicate and possibly increase the load on the resources. Moreover, the local changes (i.e., solid edges in the graphs in Figure 5.41) in the LRGs are not required to have transition modes because of the following reasons;

1. Any set of source and destination modes, defined by edges in the graph in Figure 5.41, has the same set of tasks/partitions but different numbers of resources.

2. There exists no intermediate task which needs to be executed after the source and before the destination modes.

3. Individual tasks can not be controlled based on the destination mode due to hierarchical scheduling.

The above mentioned issues lead to the decision that the GRGs and the local changes in the LRGs are not modified. Note that having no transition mode between local changes in LRGs does not eliminate the possibility of immediate transition and hence still gives a possibility to reduce the mode switching delay. The mode change blackout information is generated for all modes, which helps reduce the mode change delay in the average case.

In order to summarize the above discussion, a transition mode is added between two modes connected by a global change in the LRGs. An example of the modifications in LRGs is provided in Figure 5.41 where the prefix N refers to the nominal modes, prefix D refers to the degraded modes, prefix T refers to the transition modes, solid black edges refer to the local reconfigurations and red dashed edges refer to the reconfiguration enforced by the Global Resource Manager (GRM). Note that T1 in the figure is a transition mode, i.e. the entry in this mode is activated by an event (e.g.,

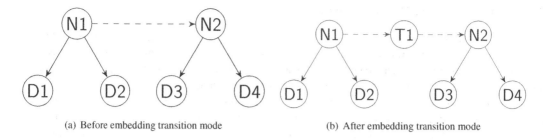

(a) Before embedding transition mode   (b) After embedding transition mode

Figure 5.41: Embedding Transition Modes in the Local Reconfiguration Graphs

core failure) but the exit is activated by the scheduler at the end of mode change blackout or at the end of MAF, whichever is earlier.

Transition Modes Schedule Generation

The generation of transition mode schedules is accomplished through the following steps:

1. All the common slots in source and destination modes are copied in the transition mode.

2. Long overlapping End-to End Flow (ETEF) chains in the source mode are detected.

3. ETEF chains are added to the transition mode such that no other chain starts before the chain under consideration finishes.

An example of the generated transition mode is shown in Figure 5.42. In the figure, the arrows between different slots reflect either a task preemption at the slot boundary or an incomplete ETEF. Assume that in the absence of a transition mode (see Figure 5.42(a)) a mode change request arrives just after starting the slot S1 in source mode. In this case, the mode change has to be postponed until the end of S7 otherwise the schedule will result in either an incomplete ETEF or an incomplete task. Upon such a switch, the slot S8 in the destination mode will execute during the next hyperperiod. On the other hand with transition modes (see Figure 5.42(b)), the slot S8 can be started earlier and hence the mode switching delay is decreased and new activities from the destination mode are started even before the destination mode is established.

5.3.2.5 Tool: MCOSF

Mixed-Criticality Offline Scheduler Framework (MCOSF) is a command-line tool integrated into the DREAMS toolchain. It calculates the blackout intervals for all destination modes, making sure that the scheduling during a mode-change is guaranteed. Moreover, the tool MCOSF estimates the schedules for transition modes, in order to decrease mode-change delay.

5.3.3 Flexibility

In this section, we describe the algorithm for online admission of aperiodic/sporadic tasks in offline scheduling tables. This algorithm provides a way to add flexibility to a static time-triggered schedule without compromising the schedulability of other tasks/jobs. Although trivial approaches to execute aperiodic/sporadic tasks are available and utilized in modern systems, these approaches are not efficient and pose several challenges when run-time guarantees need to be provided. For instance, interrupt mechanisms may lead to interference patterns which are difficult to analyze [199], especially for multi-core systems. Similarly, bandwidth reservation using server algorithms lead

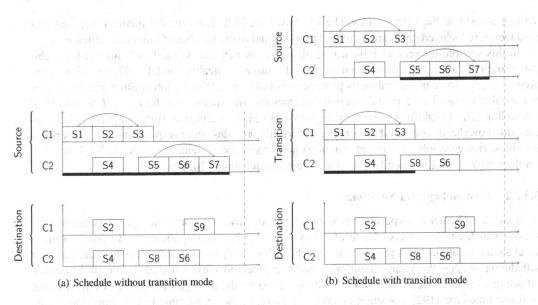

(a) Schedule without transition mode   (b) Schedule with transition mode

Figure 5.42: Enrichment with Transition Modes

to significant bandwidth loss when used with partitioned systems [200]. To resolve all these issues, we focus on the online admission paradigm and provide an algorithm which is efficient in admitting aperiodic and sporadic tasks and incurs low scheduling overheads; we call this algorithm 'Job-Shifting' [201].

5.3.3.1 Related Work

In 1992, Lehoczky and Thuel defined the slack-stealing algorithm to service soft [202] and hard [203] aperiodic tasks in fully-preemptive fixed-priority systems. In this algorithm, they pre-computed and stored the *slack function* of the task-set and used it online to service aperiodic tasks. Since, depending on the periods of the tasks, the size of the *slack function* table may be too large, the slack-stealing algorithm was later extended by Tia *et al.* [204] to efficiently compute the *slack function* online.

Similarly, Fohler [187] defined slot-shifting algorithm for aperiodic admission in fully-preemptive TT dynamic-priority systems (specifically EDF) with sparse time-base [192]. In this algorithm, *capacity (or slack) intervals* and their *spare capacities* are calculated offline and stored in a table to be used online to service firm and soft aperiodic and sporadic tasks [205]. The slot-shifting algorithm requires a significant amount of memory and utilizes an online mechanism to keep track of used resources activated at each *slot* boundary [192].

Recently, Schorr [206] extended the original slot-shifting algorithm to allow admission of non-preemptive aperiodic tasks in a preemptive schedule. The extension utilized the original precomputed offline scheduling table, however the aperiodic admission test (termed *acceptance test* [187]) for non-preemptive aperiodic tasks was modified to calculate enough consecutive slots to execute the released aperiodic job. Moreover, the *guarantee algorithm* [187] was modified to improve the response-time at the cost of flexibility [206]. Similar to the original slot-shifting algorithm, the extension by Schorr requires a significant amount of memory and uses an online slot based record keeping mechanism. Moreover, the slot-shifting extension proposed by Schorr has higher complexity and larger run-time overheads.

Similarly, Theis [195] extended the slot-shifting algorithm to support mixed-critical tasks (based on Vestal's mixed-criticality task model [193]) and mode-changes. Similar to the original slot-shifting algorithm, the extension by Theis is based on EDF, however the memory requirements (and therefore the overheads) are increased proportional to the number of modes or criticalities.

In this section, we present job-shifting algorithm, which can be used in industrial hierarchical mixed-critical systems (*not* based on Vestal's mixed-criticality model [193]) to admit non-preemptive aperiodic tasks. Unlike its predecessors [187, 195, 206], the job-shifting algorithm does *not* require (i) slot-based record keeping mechanism, (ii) sparse time-base [192] or (iii) EDF scheduling. The job-shifting algorithm respects separation of concerns (through partitioning), incurs small overheads compared to its predecessors and provides better response-times for aperiodic activities. However, job-shifting only supports non-preemptive scheduling, which is NP-Hard even for the simple case of independent tasks with implicit deadlines [207].

5.3.3.2   Terminology and Notations

A partitioning kernel or hypervisor is used to provide strict temporal and spatial isolation ([3, 4, 56]), which enables independence of safety functions between applications. In order to fulfill strict safety requirements, cyclic-executive scheduling [197] is assumed to be the inter-partition scheduling strategy. We assume a uni-processor non-preemptive execution environment. Moreover, all activities in the system are assumed to be triggered by the passage of time, i.e. TT with sparse or dense time-base [192]. For digital computing systems, the sparse time-base is implemented by the scheduler using a scheduling quantum significantly larger than the processor clock cycle length, e.g., a slot in slot-shifting [187]. However, when the scheduler does not enforce a sparse time-base, the system is subjected to dense time-base (aka 'as fast as possible') where the scheduling quantum is equal to the processor clock cycle length.

In order to apply job-shifting to a partition p_i in a partition set P, it is assumed that the intra-partition scheduling employs TT scheduling tables with a simple online dispatcher (relaxed in later sections), and contains both periodic and aperiodic tasks. The idle time inside a partition p_i is also assumed to be non-zero. The time when the partition p_i is not available to service applications is defined by the blocking set V_i. Each blocking $v \in V_i$ is defined by the tuple $\langle b, m \rangle$, where b and m represent the absolute beginning and termination time of blocking v, respectively.

A task-set Γ is defined as a collection of tasks. Each task $\tau \in \Gamma$ is defined by the tuple $\langle \phi, C, T, D, Y \rangle$, where ϕ represents the task phase, and C represents the worst-case execution time (WCET) of the task τ. When the task is periodic, the parameter T defines its period; otherwise, $T = \infty$. Moreover, D defines the relative constrained deadline (i.e., $D \leq T$), and Y represents the task criticality (i.e., the safety level, e.g., DAL-B). A task $\tau \in \Gamma$ consists of infinite jobs j, each of which is defined by the tuple $\langle r, d \rangle$, where r represents the absolute job release time, and d defines the absolute job deadline. The jobs j are assumed non-preemptive, however they can be paused at the partition boundary. It is necessary to mention here that in TT systems, only timer interrupts are enabled. The peripheral specific interrupt occurrence flags, in such systems, are used by the scheduler to check the occurrence of an event indicating the release of an aperiodic/sporadic task, e.g., a button is pressed, a message is received. However in TT systems, the interrupt flag does not lead to a preemption or branch to the interrupt service routine.

The scheduling table S_p for a partition p is constructed prior to the job-shifting offline phase for the length of the scheduling cycle (SC). The length of SC is defined by the following equation [208];

$$\text{SC} = \begin{cases} [0, \text{LCM}] & \forall \tau : \phi = 0 \\ [0, \phi_{max} + 2 \times \text{LCM}] & otherwise \end{cases} \tag{5.13}$$

where, ϕ_{max} represents the maximum phase ϕ of all tasks $\tau \in \Gamma$, while LCM represents the least common multiple of all the periods T of periodic tasks. For each job in SC, the scheduling table S_p defines the absolute job activation time a and the absolute job finish time f. It is assumed that the partition scheduling table S_p, available as input to job-shifting algorithm, is a valid feasible schedule. The tasks $\tau \in \Gamma$ may have precedence and/or mutual exclusion constraints. However,

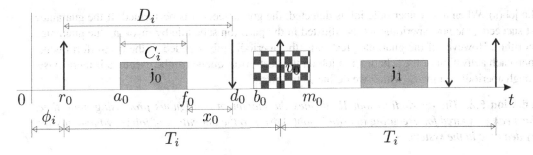

Figure 5.43: Model Parameters for a Task τ_i with Blocking v_0

Table 5.5: Summary of the Used Notations

Notation	Description	Notation	Description
j	Job	a	Job activation (absolute)
T	Task period	f	Job finish (absolute)
D	Relative task deadline	r	Job release (absolute)
C	Task WCET	d	Job deadline (absolute)
ϕ	Task phase	x	Job flexibility
R	Aperiodic room	v	Blocking interval

these constraints are assumed to be resolved either by modifying the job release time r and deadline d or by constructing the scheduling table S_p. An example scheduling table for a periodic task τ_i with blocking v_0 is shown in Figure 5.43, while a summary of the used notations is provided in Table 5.5.

5.3.3.3 Job-Shifting Algorithm

In this section, we provide details of the job-shifting algorithm to admit non-preemptive aperiodic tasks in flat and hierarchical scheduling models. Without loss of generality, in this section we assume that the sparse time-base [192] is used. Moreover, all the jobs are assumed to execute for the complete worst-case execution time C every time (relaxed later).

Methodology

In order to apply the job-shifting algorithm, a feasible offline scheduling table S_p for partition p is required. To construct S_p, the tasks $\tau \in \Gamma$ are allocated to the processing nodes and the partitions P. The periodic tasks are then unrolled to create jobs for the length of SC and the scheduling table S_p is constructed utilizing the desired scheduler such that the feasibility is ensured. For the offline scheduled partitions, this necessarily means that the values for a_i and f_i are defined for each job j_i inside the partition such that $\forall j_i \in S_p : r_i \leq a_i < d_i \wedge a_i < f_i \leq d_i \wedge f_i - a_i \geq C_i$.

Once the offline scheduling table S_p for a partition is ready, the job-shifting algorithm can be applied, which is divided in two phases: an offline phase and an online phase. In the offline phase, a parameter, we call the flexibility coefficient x_i, for each job $j_i \in S_p$ is calculated (for example, see x_0 in Figure 5.43). We define the flexibility coefficient x_i as:

Definition 5.1. *The flexibility coefficient x_i for job $j_i \in S_p$ defines the maximum delay, which can be added to the absolute activation time a_i of job j_i without changing the execution order of jobs in S_p and without missing any deadline.*

During the online phase of job-shifting algorithm, the scheduler checks the arrival of the aperiodic job(s). When a new aperiodic job is detected, the guarantee test is performed. If the guarantee test succeeds, the new aperiodic job is adjusted in the partition schedule by invoking the guarantee algorithm. However, if the guarantee test fails, the aperiodic job is added to the best-effort queue. Upon each activation of the scheduler, a job on the best-effort queue can be executed if there exists enough aperiodic room R_i, which we define as follows:

Definition 5.2. *The aperiodic room R_i defines the maximum contiguous processing node time, which can be spared for executing aperiodic job(s) prior to the activation of job j_i, without missing any deadline in the system.*

On account of the above definition, it can be observed that the best-effort queue is not a background queue. Instead, a job on the best-effort queue is executed as soon as enough aperiodic room R_i can be reserved.

Flat Scheduling Model

In this section, we assume that there exists only one partition which is available to service tasks at all times, i.e. $V_p = \emptyset$. Furthermore, the finish time f_i for a job j_i is not defined by the scheduling table S_p, instead f_i is directly calculated by the equation $f_i = a_i + C_i$.

Offline Phase:

As mentioned earlier, during the offline phase, the flexibility coefficient x_i for each job $j_i \in S_p$ is calculated starting from the job j_s with the maximum activation time a_s and ending at the job j_e with the minimum activation time a_e. Assuming $a_{s+1} = d_s$ and $x_{s+1} = 0$, for each job j_i the following steps are performed:

i) Calculate the parameter O_i, which defines the overlap between the flexibility windows $[a, d]$ of job j_i and job j_{i+1}, using the following equation;

$$O_i = max(0, d_i - a_{i+1}) \tag{5.14}$$

ii) Calculate the flexibility coefficient x_i using the following equation;

$$x_i = d_i - a_i - C_i - O_i + min(x_{i+1}, O_i) \tag{5.15}$$

Online Phase:

After finishing the offline phase, all the parameters for each job are passed to the online scheduler, which is activated at each job activation time a, finish time f and (when idle) at the release of a new aperiodic job.

The guarantee test is performed when the scheduler gets activated at time t and an aperiodic job release is detected. For the guarantee test, the job j_n is defined as the next job to be activated from the scheduling table S_p, while the job j_l is defined as the first job activated after the deadline of the released aperiodic job j_{ap}. During the guarantee test, the aperiodic room R_i for each job i from j_n to j_l is calculated using the following set of equations;

$$
\begin{aligned}
s_i &= \begin{cases} a_{i-1} + C_{i-1}, & n < i < l \\ t, & i = n \end{cases} \\
R_i &= a_i - s_i + min(d_{ap} - a_i, x_i)
\end{aligned}
\tag{5.16}
$$

where, s_i represents the start of the idle time before job j_i. The guarantee test passes when, for any job $j_i | n \le i < l$, the aperiodic room R_i is larger than or equal to C_{ap}. If the guarantee test passes, the aperiodic job j_{ap} can be guaranteed to finish before its deadline d_{ap} without missing any other deadline in the system. Once the guarantee test passes, the guarantee algorithm is activated. The guarantee algorithm requires the following steps to be performed in the defined order:

i) Insert the released aperiodic job j_{ap} in the schedule S_p before job j_i , i.e. $a_i = a_{ap} = s_{i+1}$.

ii) For each job j_k, such that $k > i \wedge a'_k > a_k$, perform the following steps starting from j_{i+1};

    (a) $a'_k = a_k + max(a_{k-1} + C_{k-1} - a_k, 0)$

    (b) $x_k = x_k - (a'_k - a_k)$

    (c) $a_k = a'_k$

iii) For each job j_p, such that $p \leq i$, calculate x_p similar to the offline phase starting from j_i and ending at j_n. The process of modifying x_p stops when the old x_p is equal to the new x_p.

At the scheduler activation time t, the scheduling decision can be made after handling the released aperiodic job(s). If the best-effort queue is empty, the next job with $a = t$ is scheduled from S_p. When there exists no such job, the processor is left idle. If there exists a job j_b in the best-effort queue and the next job aperiodic room $R_n \geq C_b$, the guarantee algorithm is performed for job j_b with $a_b = t$ and it is scheduled.

Hierarchic Scheduling Model

In this section, we assume that there exist multiple partitions in the system and the job-shifting algorithm is enabled in partition $p \in P$, i.e. $V_p \neq \emptyset$. For hierarchic scheduling model, we need to define two operators; the Blocking operator $B(p, q)$ and the Adjust operator $A(u)$. The blocking operator $B(p, q)$ returns the sum of the partition blocking duration between the interval $(p, q]$. The operator $A(u)$ modifies the time instant u such that it does not lie within the partition blocking time V. When the time instant u denotes the job finishing time f_i, the operator $A(u)$ also makes sure that the duration between the job activation time a_i and the job finish time f_i is enough to complete the job, i.e. $f_i = a_i + B(a_i, f_i) + C_i$. Unlike the flat scheduling model, the finish time f_i for a job j_i in the hierarchic scheduling model is also defined by the scheduling table S_p. Moreover, it is assumed that the parameters a_i and f_i are adjusted offline using the adjust operator $A(u)$.

Offline Phase:

Similar to the offline phase of flat scheduling model, the flexibility coefficient x_i for each job $j_i \in S_p$ is calculated during the offline phase of hierarchic scheduling model starting from the job j_s with the maximum activation time a_s and ending at the job j_e with the minimum activation time a_e. Assuming $a_{s+1} = d_s$ and $x_{s+1} = 0$, for each job j_i the following steps are performed;

i) Calculate the overlap parameter O_i using the following equation;

$$O_i = max(0, d_i - a_{i+1} - B(a_{i+1}, d_i)) \tag{5.17}$$

ii) Calculate the flexibility coefficient x_i using the following equation;

$$x_i = d_i - a_i - C_i - B(a_i, d_i) - O_i + min(x_{i+1}, O_i) \tag{5.18}$$

Online Phase:

After finishing the offline phase, all the parameters for each job and the partition blockings V_p are passed to the online scheduler. The online scheduler is activated at each job activation time a, finish time f, and when the partition is active *and* the processor is idle at the release of a new aperiodic job.

When the scheduler is activated at time t and an aperiodic job release is detected, the guarantee test is performed. Similar to the flat scheduling model, the job j_n is defined as the next job to be activated from the scheduling table S_p, while the job j_l is defined as the first job activated after the deadline of the released aperiodic job j_{ap}. During the guarantee test, the aperiodic room R_i for each job i from j_n to j_l is calculated using the following set of equations;

$$
\begin{aligned}
s_i &= \begin{cases} f_{i-1}, & n < i < l \\ t, & i = n \end{cases} \\
R_i &= a_i - s_i - B(s_i, a_i) + \\
&\quad min(d_{ap} - a_i - B(a_i, d_{ap}), x_i)
\end{aligned} \tag{5.19}
$$

The guarantee test passes if, for any job $j_i | n \le i < l$, the aperiodic room R_i is larger than or equal to C_{ap}. Once the guarantee test passes, the guarantee algorithm is activated. The guarantee algorithm requires the following steps to be performed in the defined order:

i) Insert the released aperiodic job j_{ap} in the schedule S_p before job j_i, i.e. $a_i = a_{ap} = s_{i+1}$ and $f_i = f_{ap} = A(a_{ap} + C_{ap})$.

ii) For each job j_k, such that $k > i \wedge a'_k > a_k$, perform the following steps starting from j_{i+1}:

   (a) $a'_k = A(a_k + max(f_{k-1} - a_k, 0))$
   (b) $f_k = A(f_k + (a'_k - a_k - B(a_k, a'_k)))$
   (c) $x_k = x_k - (a'_k - a_k - B(a_k, a'_k))$
   (d) $a_k = a'_k$

iii) For each job j_p, such that $p \le i$, calculate x_p similar to the offline phase starting from j_i and ending at j_n. The process of modifying x_p stops when the old x_p is equal to the new x_p.

During run-time, the jobs are scheduled similar to the flat scheduling model.

Job-Shifting Example

Offline Phase:

Consider two jobs j_1 and j_2 with release time r, activation time a, finish time f, deadline d, and worst-case execution time C as defined in Table 5.6. Assume that the partition blocking set V is defined to be $\{\langle 5, 8 \rangle\}$ as shown in Figure 5.44. During the offline phase of job-shifting algorithm, the flexibility coefficients x are calculated starting from job j_2 and ending at job j_1 as under:

For job j_2, $a_3 = d_2$ and $x_3 = 0$.

$$
\begin{aligned}
O_2 &= max(0, d_2 - a_3 - B(a_3, d_2)) \\
&= max(0, d_2 - d_2 - B(d_2, d_2)) = 0 \\
x_2 &= d_2 - a_2 - C_2 - B(a_2, d_2) - O_2 + min(x_3, O_2) \\
&= 10 - 2 - 2 - 3 - 0 + min(0, 0) = 3
\end{aligned}
$$

For job j_1, $a_2 = 2$ and $x_2 = 3$.

$$
\begin{aligned}
O_1 &= max(0, d_1 - a_2 - B(a_2, d_1)) \\
O_1 &= max(0, 8 - 2 - 3) = 3 \\
x_1 &= d_1 - a_1 - C_1 - B(a_1, d_1) - O_1 + min(x_2, O_1) \\
&= 8 - 0 - 2 - 3 - 6 + min(6, 6) = 3
\end{aligned}
$$

Online Phase:

Assume an aperiodic job j_{ap} with $C_{ap} = 2$ and $d_{ap} = 3$ is released at time $t = 0$ in the example shown in Figure 5.44. The online scheduler performs the guarantee algorithm with $n = 1$ and $l = 3$ (i.e., first job of the next SC). For job j_i with $i = 1$, $s_1 = t = 0$. The aperiodic room R_1 is obtained to be $R_1 = a_1 - s_1 + min(d_{ap} - a_1 - B(a_1, d_{ap}), x_1) - B(a_1, s_1) = 0 - 0 + min(3 - 0 - 0, 3) - 0 = 3$. Since $R_1 > C_{ap}$, the newly released job j_{ap} can be guaranteed to finish before its deadline d_{ap}. Therefore, the guarantee algorithm is triggered. The steps are shown below:

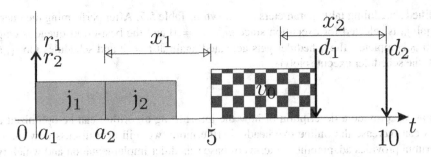

Figure 5.44: Example for Offline Phase of Job-Shifting

Table 5.6: Example Parameters After Offline Phase

j	r	a	f	d	C	x
1	0	0	2	8	2	3
2	0	2	4	10	2	3

Table 5.7: Example Parameters After Online Phase

j	r	a	f	d	C	x
0	0	0	2	3	2	1
1	0	2	4	8	2	1
2	0	4	9	10	2	1

i) Job j_{ap} is inserted at location 0, i.e. $a_0 = a_{ap} = s_1 = 0$, $f_0 = f_{ap} = A(a_{ap} + C_{ap}) = 2$, $C_0 = C_{ap} = 2$ and $d_0 = d_{ap} = 3$.

ii) For job j_1:

$$
\begin{aligned}
a_1' &= A(a_1 + max(f_0 - a_1, 0)) \\
&= A(0 + max(2 - 0, 0)) = 2 \\
f_1 &= A(f_1 + (a_1' - a_1 - B(a_1, a_1'))) \\
&= A(2 + (2 - 0 - B(0, 2))) = 4 \\
x_1 &= x_1 - (a_1' - a_1 - B(a_1, a_1')) \\
&= 3 - (2 - 0 - B(0, 2)) = 1 \\
a_1 &= a_1' = 2
\end{aligned}
$$

For job j_2:

$$
\begin{aligned}
a_2' &= A(a_2 + max(f_1 - a_2, 0)) \\
&= A(2 + max(4 - 2, 0)) = 4 \\
f_2 &= A(f_2 + (a_2' - a_2 - B(a_2, a_2'))) \\
&= A(4 + (4 - 2 - B(2, 4))) = 9 \\
x_2 &= x_2 - (a_2' - a_2 - B(a_2, a_2')) \\
&= 3 - (4 - 2 - B(2, 4)) = 1 \\
a_2 &= a_2' = 4
\end{aligned}
$$

iii) For job j_0:

$$
\begin{aligned}
O_0 &= max(0, d_0 - a_1 - B(a_1, d_0)) \\
&= max(0, 3 - 2 - B(2, 3)) = 1 \\
x_0 &= d_0 - a_0 - C_0 - B(a_0, d_0) - O_0 + min(x_1, O_0) \\
&= 3 - 0 - 2 - B(0, 3) - 1 + min(1, 1) = 1
\end{aligned}
$$

The modified scheduling table parameters are shown in Table 5.7. After performing the guarantee test, the job j_0 is selected for execution since $a_0 = t = 0$ and the best-effort queue is empty. When the job j_0 completes, the scheduler gets activated again at $t = 2$ and schedules job j_1, and then at $t = 4$, the scheduler executes job j_2.

Discussion

In this section, we provide a description of how the job-shifting algorithm can be optimized and which factors can increase the online overheads. Furthermore, we will also discuss how the job-shifting algorithm provides adaptability in terms of base scheduler implementation and which type of schedulers maximize the schedule adaptability.

Mixed-Critical Tasks:

It is important to note that neither flat nor hierarchical scheduling models take the task criticality Y into account. The reason for such a deliberate elimination is the use of the industrial mixed-criticality model defined by the standards IEC 61508 [56], DO-178C [3] and ISO 26262 [4]. In these standards, the task criticality Y is used to refer to the level of assurance applied to the development of software application and the different criticality tasks are segregated by allocating them to different partitions [194]. For such industrial standards, the tasks only define single WCET C, the higher criticality level of a task does not mean greater importance and, therefore, does not warrant rejection of lower criticality tasks. All these facts lead to a conclusion that the task criticality cannot be exploited by the scheduler (as long as critical and non-critical tasks stay in separate partitions). For more information on industrial mixed-criticality model, please refer to [194].

Offline Guarantee Analysis:

In order to design a robust and responsive real-time system, a usual practice is to guarantee the correct behavior of a sub-system or task offline. To guarantee some service to aperiodic activities, a certain amount of processor bandwidth is usually reserved for aperiodic tasks (or servers). Due to the limited system resources and large number of constraints, there is always a limit to the bandwidth which is considered a 'safe reservation' for handling aperiodic activities. In industry, such reservations are considered mandatory to guarantee system safety. However, reserving a specific bandwidth for such a case still limits the service for aperiodic execution. Moreover, the response-time of aperiodic tasks in such reservations also has a strong dependency on the employed reservation technique. In this work, we conjecture that, instead of reserving a limited amount of processor bandwidth, all free partition/processor bandwidth can be utilized to service aperiodic tasks. The following theorem provides the base for our conjecture:

Theorem 5.3.1. *For a given TT schedule S, if there exists a reservation such that a number of non-preemptive aperiodic jobs with a defined job release pattern can be executed feasibly, the job-shifting algorithm can also execute them without a reservation.*

Proof. Consider a TT scheduling table S with defined activation times a_i and finish times f_i for each periodic job j_i. A reservation window $\Lambda_{[t_1,t_2)}$ with window start time t_1 and end time t_2 is defined as contiguous *reserved* processing time for the execution of non-preemptive aperiodic job(s). According to Definition 5.2, the aperiodic room R_k is defined as the *maximum* room for executing non-preemptive aperiodic job, where $k \in \mathbb{N} | f_{k-1} \leq t_1 \wedge a_k \geq t_2$. When a non-preemptive aperiodic job j_{ap} with $C_{ap} = t_2 - t_1$ is released at $r_{ap} \leq t_1$, then by definition, the condition $\Lambda_{[t_1,t_2)} \leq R_k$ always holds. If the non-preemptive aperiodic job j_{ap} can be feasibly executed by $\Lambda_{[t_1,t_2)}$, R_k can also feasibly execute it without an offline reservation. The proof can be iteratively applied to multiple reservation windows. Hence, the theorem is proven.   □

In other words, the same practice of 'safe reservation' can be utilized to give guarantees for aperiodic execution in job-shifting without any *online* reservation. However with job-shifting, the bandwidth loss due to partitioning [200] can be significantly reduced and the aperiodic response-times can be improved.

Being a non-preemptive algorithm, a necessary condition for aperiodic guarantee can also be checked offline in job-shifting. Independent of the release time of the aperiodic job, there must exist enough aperiodic room R_i in the schedule S_p to accommodate complete aperiodic job. For an aperiodic task, if there exists no aperiodic room $R_i | R_i \geq C_{ap}$ for any $j_i \in S_p$, the aperiodic job will never be able to execute while keeping feasibility of other tasks in S_p. This necessary condition is checked between the offline and the online phases of the job-shifting algorithm.

Hypervisors and Clocks:

There exists a multitude of real-time hypervisors in the market today. Based on the type of interface they provide to the partition, the online complexity of the job-shifting algorithm can be modified. Some hypervisors, e.g., XtratuM [198], provide a partition local clock which ticks only when the partition is active. In such hypervisors, the equations for the online phase of the flat scheduling model can be used in a hierarchic design by defining a_i, r_i and d_i in terms of the partition local clock. For such an implementation, the adjust operator $A(u)$ can be completely eliminated and the blocking operator $B(p, q)$ is only required for the last step of guarantee algorithm (where the flexibility coefficient is calculated). However, there exist hypervisors which do not provide a partition local clock, e.g., PikeOS[7], and therefore need to implement both operators $A(u)$ and $B(p, q)$.

Optimizations:

At the start of this section, it was assumed that all the tasks execute for their complete worst-case execution time C. However, this assumption seldom holds during system operation. A task executing more than C time can lead to the violation of the temporal isolation between tasks of the same application. In the worst case, such violations may never lead to a processor yield for other tasks to execute. To protect from such a situation, a hardware timer can be programmed to generate an overrun interrupt which terminates the misbehaving job and returns the control to the scheduler. When the job executes less than C time, the overrun interrupt can be terminated (or rescheduled for the next job). In such a situation, the free processing time can be utilized by using a simple approach. At the observed job finish time t, if for the next job $a_n > r_n$, the flexibility coefficient x_n can be updated to $x_n + (a_n - max(t, r_n))$ and the job can be preponed to activate at $max(t, r_n)$. If for the next job $a_n = r_n$, the free resources can be accounted for by the aperiodic room R_n and therefore, require no further action.

When a large number of aperiodic jobs are released before the scheduler activation, performing guarantee algorithm for all of them may lead to large scheduler overheads. In the worst case, the reserved processing time to execute the next job might be reduced, leading to a job incompletion. To avoid such a scenario, Schorr [206] proposed a heuristic where at most one aperiodic job is guaranteed for each activation of the scheduler.

5.3.3.4 Efficiency Evaluation

In this section, we present the results of the experiments performed to evaluate efficiency of the job-shifting algorithm. First, we provide the description of the experimental setup. Later, we present and briefly discuss the obtained results.

Experimental Setup

In order to provide a reference point for the job execution guarantees provided by the job-shifting (JS) algorithm, we implemented a non-preemptive background scheduler (NP-BG). The NP-BG scheduler services the aperiodic jobs during the idle time available in the scheduling table, i.e. a non-preemptive aperiodic job is executed only when it can be guaranteed that the job will finish within the idle time. Note that any existing algorithm cannot be used to provide the reference point for comparison due to varying task models.

For the evaluation of job-shifting algorithm, 1000 task-sets were generated. The parameter

---

[7]https://www.sysgo.com/products/pikeos-hypervisor/

ranges for generating task-sets were selected as defined by Schorr [206] to capture the behavior of real applications. Each generated task-set consisted of at most one processor and one partition. The partition blockings V are generated with a strict period of 10 and a relative beginning time $b = 6$ of each blocking $v \in V$ until the end of the scheduling cycle SC.

Inside the partition, [1, 3] periodic tasks were generated with phases $\phi = 0$, WCET C in the interval [1, 15] and the period T in the interval [15, 30] (with implicit deadlines). The total utilization of periodic tasks was kept 25% and the task parameters were generated using UUniFast [209] algorithm to get uniform tasks distribution. The aperiodic tasks were generated with release time r in the interval [0, SC), the WCET C in the interval [5, 10] and the deadline $d = \text{DLX} \times C$, where DLX is the deadline extension factor. The factor DLX defines the tightness of the deadline compared to C. The generation process of the aperiodic tasks is stopped when the aperiodic utilization reaches the target utilization.

To distinguish the effects of varying urgencies, task-sets were generated for each different DLX factor in the list {4, 8, 12} and the aperiodic tasks utilization in the list {5%, 10%, 15%, 20%}. The guarantee ratios (i.e. the ratio of the number of guaranteed aperiodic jobs to the total number of aperiodic jobs) of the algorithms are also affected by the partition supply. Therefore, all the task-sets were generated for two partition utilizations in the set {70%, 50%}.

As mentioned in Section 5.3.3.3, the job-shifting (JS) algorithm requires a cyclic executive schedule for the partition. Therefore during the pre-processing stage, the partition schedule S_p is generated using the EDF scheduler. The task-sets which resulted in SC lengths outside of the interval [500, 5000) were rejected since the real-world tables are smaller [206]. Moreover, the task-sets were also filtered using the offline guarantee analysis method mentioned in Section 5.3.3.3.

Results and Discussion

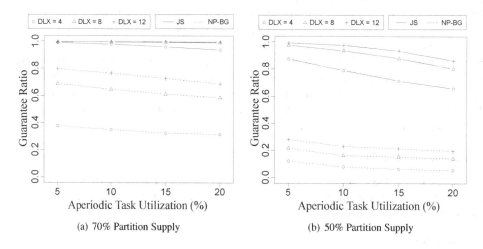

(a) 70% Partition Supply   (b) 50% Partition Supply

Figure 5.45: Job-Shifting Overheads Evaluation

Figure 5.45 shows the average guarantee ratio for the generated task-sets (i.e., each point in the figure represents the average guarantee ratio of 1000 task-sets) for varying aperiodic task utilizations. In the figure, the DLX factor is represented with different pointer types and the schedulers (JS and NP-BG) with different line types. For the non-preemptive background scheduler (NP-BG), the guarantee ratio defines the ratio of the number of aperiodic jobs which finished before their deadlines to the total number of aperiodic jobs.

Figure 5.45 shows that, for the background scheduling, the number of finished aperiodic jobs decreases to a great extent (e.g., from 0.7 to 0.2 for DLX=12 and aperiodic utilization $U_{ap} = 20\%$) due to just 20% decrement in the partition supply. However, for the job-shifting algorithm, the number of finished aperiodic jobs did not suffer as much (e.g., from 1.0 to 0.9 for DLX=12 and aperiodic utilization $U_{ap} = 20\%$). Moreover, notice in Figure 5.45 that the average guarantee ratio of the job-shifting algorithm is quite large compared to the background processing. Therefore, we conclude that job-shifting is more suitable for online aperiodic job admission (at least, in comparison to the background processing).

5.3.3.5 Overheads Evaluation

In this section, we present the results of the experiments performed to evaluate overheads incurred by the job-shifting algorithm. We start by providing the description of the experimental setup. Later, we present and briefly discuss the obtained results.

Experimental Setup

In order to evaluate the job-shifting overheads, we implemented the job-shifting algorithm on an ARM platform. In the following, we provide a description of the hardware and software platforms and the job-shifting integration strategy in DREAMS.

Application Architecture:

The system for overheads evaluation is composed of four different applications (APP1–APP4) communicating with each other as shown in Figure 5.46. Each application consists of a number of periodic and aperiodic tasks as defined in Table 5.8. The tasks sending a message to other task(s) are executed synchronously (i.e., periodic tasks), during which they decide if data needs to be communicated; while the tasks receiving messages from other task(s) are executed asynchronously[8] (i.e., aperiodic tasks) when data is received on a defined virtual network port. Apart from the communicating tasks, there exist a number of tasks which are activated by external events, e.g., push of a button. Figure 5.46 shows that a number of tasks in the application APP1 are triggered by external inputs (see Table 5.8) which are managed by aperiodic tasks.

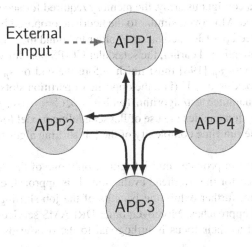

Figure 5.46: Application Communications (Solid Arrows) and External Inputs (Dashed Arrow)

Hardware & Software Platforms:

In order to evaluate the overheads incurred by the job-shifting algorithm, the Xilinx Zynq ZC-706 FPGA platform is used. The Zynq platform provides a dual-core ARM Cortex-A9 processor

---

[8] With the exception of the communication from APP3 to APP4, to evaluate overheads for partitions without any aperiodic task.

Table 5.8: Task-Set for the Overhead Evaluation of Job-Shifting

Application	Num. of Periodic Tasks	U_p	Num. of Aperiodic Tasks	
			Comm	External Input
APP1	6	0.021	1	17
APP2	10	0.545	1	0
APP3	5	0.055	3	0
APP4	1	0.015	0	0
Total	22	0.636	5	17

Table 5.9: Application Deployment Parameters

Application	Flexibility Coeff. (x)		Activation Probability	
	min (ms)	max (ms)	min	max
APP1	13	30	0.008	0.35
APP2	0	4	NA	NA
APP3	0	7	NA	NA
APP4	0	2	NA	NA

running at 400Mhz. From the software front, the XtratuM hypervisor [198] was used to provide temporal and spatial isolation. To implement the online scheduler of job-shifting algorithm, the local resource scheduler (LRS) for application partitions was modified to admit aperiodic tasks online with a cyclic-executive intra-partition scheduler (CEIPS [197]) with dense time-base [192]. As CEIPS does not make use of a ready-queue, a simple job dispatch table for each partition slot [198] is specified in the Partition Configuration File (PCF). To detect the aperiodic job activation, the sources of the aperiodic task activation are also specified in the PCF (i.e., virtual network ports for message activated tasks, and interrupt sources for tasks activated due to external inputs). As the job dispatch table is implemented using a contiguous array, the memory required for accommodating new aperiodic jobs is statically reserved. Moreover, similar to the heuristic proposed by Schorr [206], at most one aperiodic job is guaranteed per LRS activation in order to bound the scheduler overheads.

Besides the constraints mentioned earlier, the scheduler CEIPS puts forward another constraint: A job started in a partition slot s_p [198] must finish before the end of s_p. To accommodate this constraint, each partition blocking $v \in V$ (i.e., the opposite of partition slots in XtratuM [198]) can be considered a job for the guarantee test/algorithm (with $a = b, C = m-b, f = m$ and $x = 0$). The advantage of such an assumption enables the use of flat scheduling model for the implementation in the LRS, in turn reducing the run-time complexity of the job-shifting algorithm.

System Deployment

Although the ZynQ platform provides multiple cores, only one of the ARM cores running the XtratuM hypervisor was used for the overhead evaluation. This approach enables evaluation with large enough core utilization, further exhibiting benefits of the job-shifting algorithm over reservation based or background approaches. Moreover, other DREAMS services (e.g., global resource management) were disabled as their focus is orthogonal to the overheads evaluation of the job-shifting algorithm.

For the overheads evaluation of the job-shifting algorithm, the four applications mentioned in Table 5.8 were deployed each in its own user partition on a single ARM core. Moreover, the external inputs to activate aperiodic tasks were triggered with a defined probability to simulate event occurrence. This process enables more frequent aperiodic activations compared to generating the events manually or through measurements in the environment. The system was deployed as shown in Figure 5.47, including the resource management system partitions for resource management services RM1 and RM2. Due to large range of the task periods, the generated schedule resulted in a

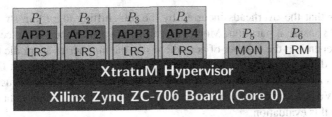

Figure 5.47: Demonstrator Deployment

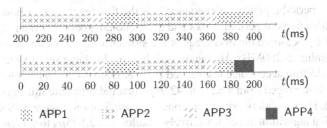

Figure 5.48: Hypervisor Schedule for Safety-Critical Demonstrator

scheduling cycle length SC = 40s. To give a rough idea, the initial part of the generated hypervisor schedule is shown in Figure 5.48, while the range of flexibility coefficients and the activation probabilities for tasks activated by external events are provided in Table 5.9. In the table, the activation probability of 1 means that a job of the aperiodic task is released every second.

Results & Discussion

Figure 5.49 shows the bar plot of the measured maximum overheads for each partition utilizing job-shifting scheduler within 1000 scheduling cycles SC. The error bars in Figure 5.49 represent the 98% confidence interval for the cumulative overheads. The legends in Figure 5.49 represent the overheads distribution, where 'Queue Shifting' overheads represent the time elapsed in inserting a newly released aperiodic job in the job dispatch table and the 'Rest Overheads' represents the rest of the scheduling overheads (e.g., next job selection). It is important to mention here that the graph in Figure 5.49 includes the logging overheads and excludes the overheads due to the detection of aperiodic task arrival, since these overheads are dependent on the aperiodic activation source, the hyper-calls implementation, and the platform architecture. For the aperiodic tasks activated by the reception of a message on a virtual network port, the overheads were measured to be approximately $13\mu s$ per aperiodic task. Whereas the detection of the aperiodic tasks triggered by the external events required approximately $2\mu s$ per aperiodic task.

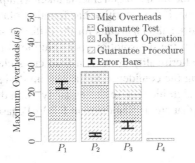

Figure 5.49: Job-Shifting Overheads on Zynq Board (Error Bars Represent 98% Confidence Intervals for Cumulative Overheads)

Figure 5.49 shows that the overheads incurred by the job-shifting algorithm strongly depend on the number of tasks inside a partition. Moreover, the overheads of the guarantee test and the guarantee algorithm depend on the number of periodic tasks, the number of aperiodic tasks and the flexibility of guaranteed tasks inside a partition. It is important to mention here that the overheads due to queue shifting can be significantly reduced by implementing the schedule using a double linked list. Since the V&V efforts for such a modification in a legacy partition data structure are higher, it is avoided in this evaluation.

It was also observed that, on average, 22.6 aperiodic tasks were released per second during the execution of 1000 scheduling cycles SC. Furthermore, none of the released aperiodic jobs missed its deadline and all the aperiodic jobs were guaranteed for the first invocation of Equation 5.16 due to relatively smaller utilizations of periodic tasks and good enough aperiodic room R.

In 2015, Schorr [206] measured the overheads of the slot-shifting algorithm on a cycle accurate MPARM simulator running at 200MHz. He mentioned that, for his large set of synthetic task-sets, the overheads for slot-shifting acceptance test ranged from $8.695\mu s$ to $23.7\mu s$, while the overheads for the guarantee algorithm ranged from $9.82\mu s$ to $42.99\mu s$ (excluding queue shifting overheads, see experiment 3 in [206]). For our safety critical demonstrator application, Figure 5.49 shows that the overheads of job-shifting algorithm are comparatively smaller (i.e. $3.825\mu s$ to $9.17\mu s$ for guarantee test and $7.71\mu s$ to $12.3775\mu s$ for guarantee algorithm). The smaller overheads for job-shifting on the Zynq ZC706 platform were partly expected as the operating frequency of the platform is twice compared to the MPARM. However, this comparison manifests that the overheads incurred by the job-shifting algorithm are quite comparable to preemptive algorithms (e.g., slot-shifting) despite the NP-hard nature of non-preemptive scheduling problem.

5.3.3.6 Tool: MCOSF

In order to execute the offline phase of job-shifting algorithm for partition p – which utilizes the job-shifting algorithm – the tool MCOSF (Mixed-Criticality Offline Scheduling Framework) is used. The MCOSF tool calculates the flexibility coefficients x for each job of partition p from the application, platform and system software model defined by the DREAMS metamodel. The generated information is forwarded to the Local Resource Scheduler (LRS [210]) of partition p. For more information on the tool MCOSF, please see Section 5.3.2.5.

5.4 Timing Analysis

Configuration algorithms for time-triggered schedulers usually produce configurations that are "correct by construction", meaning that the timing requirements taken as input by these algorithms are satisfied by the resulting configuration, if the algorithm has found a solution. Furthermore, the timing decomposition algorithms produce sub-constraints for the different scheduling domains so that "by construction", the combination of the sub-solutions is a solution of the initial overall problem. But since we are considering mixed-criticality systems a final cross-check of latency constraints is a useful contribution to increasing the confidence in the system design. Furthermore, the event driven schedules used for non-critical traffic require timing analysis. In this section we describe the relevant timing analysis algorithms that are implemented in RTaW-Pegase/Timing for this purpose.

5.4.1 Problem Definition

The DREAMS platform is mainly composed of distributed tasks that are executed in partitions, possibly located in different tiles and nodes and therefore with data exchange over on-chip and/or off-chip networks. End-to-end latency constraints are expressed for timing event chains, which define the path of transformation of information through the component architecture, typically from a sensor to an actuator. As discussed in Section 5.2.1, the resulting end-to-end delay is the sum of scheduling domain internal delays δ_i and coordination delays γ_i between domain schedules, see Equation 5.10. In the following sections we explain how δ_i is computed for each scheduling domain.

5.4.2 On-Chip Network

As described in Section 5.2.3, with the configuration algorithm used for TT virtual links, the delay is equal to the basic network latency: $\delta_i = R_k^*$. For RC virtual links that are sent over the virtual network with the lower priority and may therefore also be delayed by TT virtual links, no timing analysis is known, see Section 5.2.3.3. Thus we we propose to use timing accurate simulation (i.e., RTaW-Pegase/Timing) to estimate quantiles of communication delays.

5.4.3 Off-Chip Network

For TT VLs, the network traversal time can be determined by matching the start and end times of successive transmission window instances in the hops along the routing path through the network.

For RC VLs we use the network calculus based bounds from [159], which are able to account for the effects of the time-triggered higher priority traffic on the rate constrains traffic with lower priority.

5.4.4 Task Timing Analysis

The used task and partition schedules are based on static cyclic tables, consisting of slots where task instances are executed one after the other, without preemption.

If the period of a task is shorter than the MAF, then several instances appear in the slots of the cyclic table. Every instance of the first task of the sub-chain is the beginning of an instance of the sub-chain. If the consecutive instances of a task are placed in slots with a distance exactly equal to the period of the tasks and if in a slot the tasks are always executed in the same order, then it is enough to consider the first instance of the sub-chain. Otherwise, the delays may be different for each sub-chain instance and all instances starting within one MAF need to be analyzed.

In order to compute the delay of an instance of an event sub-chain, it is necessary to iterate over all tasks of the sub-chain and identify step by step the slot in which the next instance of the next task in the chain is executed, see Figure 5.50. Notice that the last slot found this way for the last task of the sub-chain, may be at a distance of several MAFs to the first slot. The delay of the sub-chain instance is equal to the distance between the start of the slot with the instance of the first task and the execution end of the instance of the last task in the slot derived as explained above.

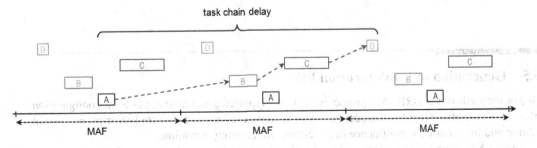

Figure 5.50: Task Chain Delay

5.4.5 Composition

When composing worst-case analysis results of different scheduling domains, two aspects need to be considered. The first is the coordination delay γ_i discussed in Section 5.2.1.2. The second one is inter-arrival jitter that appears with event driven schedules and might be propagated from one scheduling domain to other ones. But in the DREAMS architecture this jitter is shaped out by imposing the MINT at the entrance of the on-chip network as well as at the entrance of the off-chip network. Thus the timing analysis of the communication can be based on the MINT of the virtual links.

The input shaping may delay a frame f_i, but the increased delay cannot exceed the network traversal time of the previous frame f_{i-1}. If f_i enters the previous network exactly MINT time units after f_{i-1}, then its network delay is increased exactly to the one of f_{i-1}. If f_i enters the previous network more than MINT time units after f_{i-1} then it is delayed by less time and there is no impact on the worst-case traversal times.

5.4.6 Tool: RTaW-Pegase/Timing

In the previous sections we have described how upper bounds can be derived for end-to-end delays of timing chains that span the processor tiles and on-chip and off-chip networks of the DREAMS architecture. The corresponding algorithms have been implemented in the RTaW-Pegase/Timing tool.

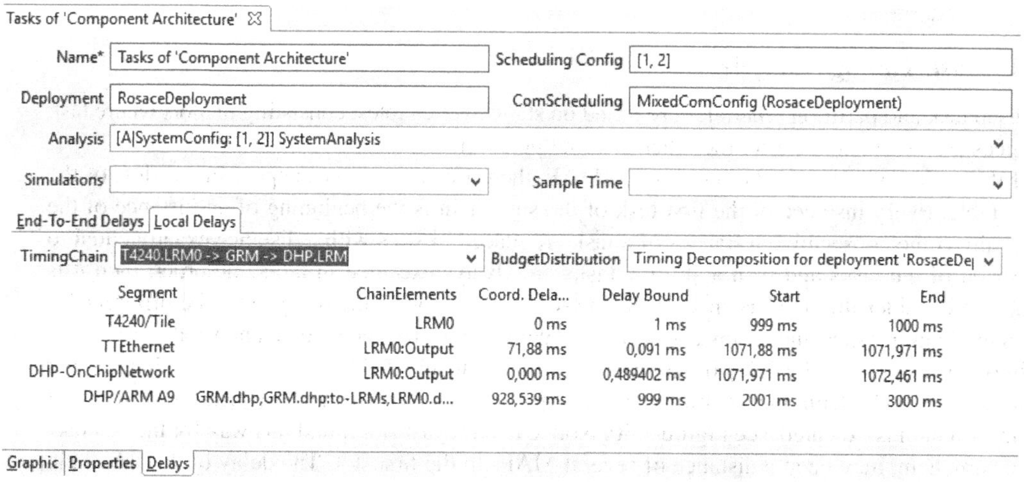

Tasks of 'Component Architecture' ▦

Name*	Tasks of 'Component Architecture'	Scheduling Config	[1, 2]
Deployment	RosaceDeployment	ComScheduling	MixedComConfig (RosaceDeployment)
Analysis	[A|SystemConfig: [1, 2]] SystemAnalysis		⌄
Simulations	⌄	Sample Time	⌄

End-To-End Delays | Local Delays

TimingChain [T4240.LRM0 -> GRM -> DHP.LRM] ⌄ BudgetDistribution [Timing Decomposition for deployment 'RosaceDe| ⌄

Segment	ChainElements	Coord. Dela...	Delay Bound	Start	End
T4240/Tile	LRM0	0 ms	1 ms	999 ms	1000 ms
TTEthernet	LRM0:Output	71,88 ms	0,091 ms	1071,88 ms	1071,971 ms
DHP-OnChipNetwork	LRM0:Output	0,000 ms	0,489402 ms	1071,971 ms	1072,461 ms
DHP/ARM A9	GRM.dhp,GRM.dhp:to-LRMs,LRM0.d...	928,539 ms	999 ms	2001 ms	3000 ms

Graphic | Properties | Delays

Figure 5.51: RTaW-Pegase/Timing Showing Computed Bounds on Delays of a Timing Chain

5.5 Generation of Configuration Files

One of the goals of the DREAMS project is the tool supported generation of platform configuration files out of a verified DREAMS model of a system configuration, so as to avoid error-prone manual editing and to increase the confidence in the correct functioning at runtime.

Figure 5.52 gives an overview of the related artifacts and processing steps. The starting point is the *DREAMS model* of a (complete) system configuration, which consists of rather high-level information, if compared to the deployable device configuration files. A configuration file *genera-*

tor takes the *DREAMS model* as input and generates human readable configuration files, based on *generation templates* that may either be explicitly available as files or integrated into the code of the generators. Some generators create first an intermediate configuration model, before the actual generation of the configuration files. Certain vendor specific or very low-level configuration parameters are not covered by the DREAMS metamodel and need to be provided separately, as fixed values in the templates, as the result of some algorithm or as user input to the generator. They may also be set through manual post-processing of the configuration files. The generated *configuration files* are the inputs to the vendor specific *platform service configurator*, which usually creates some binary files (called *deployable configuration*) that need to be stored/uploaded to the service instance (e.g., non-volatile memory of the corresponding device memory such that the configurations are available at boot and run time).

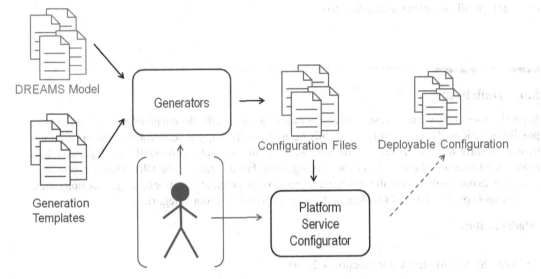

Figure 5.52: Overview of Model-Driven Generation of Platform Configuration Files

Model-to-Text (M2T) transformations are a generic approach that allows to generate, for instance, configuration files tailored to a particular target system from a given input model. The configurations to generate follow a determinate target format: comma separated values (CSV), XML, proprietary text formats, etc.

The generation of text artifacts is based on a framework for Model-to-Text (M2T) transformations that provides an automatic transformation of a model into text. The output of text ranges from source code to configurations, depending on the nature of the input model and the implementation of the M2T transformation. Depending on the requirements, the resulting output file may require further manual modifications, resulting into a semi-automatic overall transformation process.

In the presented framework, the most common generation approach is based on generation templates that combine static parts of the configuration files to be generated with dynamic parts that are determined during the generation. This approach has the advantage that generation templates can easily be created from representative handcrafted configuration file examples, by augmenting them with dynamic parts that are evaluated by the generation engine. Hence, a generation template considers the following input:

- Parameters extracted from DREAMS model constitute the main part of the dynamic parts considered during the generation. Examples include the platform architecture model (see Section 4.2.2), a deployment model (see Section 4.5.1) or a configuration model that is specific building block of the DREAMS platform (see Section 4.6.1).

- User-provided runtime parameters: If user input is required for the generation process, the developer of a configuration generator can provide contributions to the configuration user interface (UI) asking for these specific parameters.

- Static parameters (applicable to all generated files) can be directly "hardcoded" in the code templates, because they will not vary among the generated configuration files.

In the Eclipse IDE that is used to implement the configuration generation modules for the DREAMS platform, a number of M2T transformation engines are available. The current state-of-the-practice encourages the usage of the Acceleo [211], since it is based on the OMG approved specifications MOFM2T [212] and OCL [117]. In addition, a comfortable editor with syntax highlighting and checker as well as the debugger that is integrated into the Eclipse IDE ease the creation of model specific templates using Acceleo.

5.6   Toolchain

In previous sections, we have described algorithms to solve specific design problems and/or to verify possible solutions. These algorithms can only be applied efficiently to real-world problems if implemented in software tools. A set of software tools can only be applied efficiently if integrated into a toolchain that ensures the seamless flow of design data. For this reason, the DREAMS metamodel is used as backbone onto which all tools are hooked through automated data-exchange functionalities.

The tools of the DREAMS toolchain [38] are grouped into four categories:

Model Editors

- AF3 / System Model Editor (Section 4.2.1.3)
- Timing Model Editor (Section 4.3.1)
- Safety Model Editor (Section 4.4.1)
- BVR Variability Editor (Section 4.7.1)

Design Tools

- AF3 / DSE (Section 5.1.2)
- BVR Product Generator (Section 5.1.1.5)
- RTaW / Timing Decomposition (Section 5.2.1.7)
- RTaW-Pegase / On-chip TT Sched (Section 5.2.3.5)
- TTE-Plan (Section 5.2.4)
- Xoncrete (Section 5.2.2)
- GRec (Section 5.3.1)
- MCOSF (Section 5.3.2)

Verification Tools

- RTaW-Pegase / Evaluation (Section 5.4.6)

- Safety Constraint Checker (Section 4.4.3)

Configuration File Generators

- XtratuM Hypervisor

- TTEthernet off-chip network

- On-chip Network Interface layer

- Resource Management

In the following sections, we describe three use cases of the tools introduced in this chapter, which increased complexity of the goals to achieve.

Section 5.6.1 describes a minimal use case with the goal of creating an operational configuration for the nominal mode of one variant of the system.

The second use case, described in Section 5.6.2, aims at creating an operational configuration with resource reconfiguration (global and local resource management). These two use cases cover the right part (see Figure 5.1) of the proposed DREAMS model driven development process, where all variability is bound.

The last use, see Section 5.6.3, case covers all parts of the workflow. Its goal is the creation of operational configurations of products drawn from a product line, with the optional exploration of the product space in order to find products that are optimal with respect to certain combinations of criteria.

5.6.1 Use Case 1: Basic Scheduling Configuration

The use case described in this section covers the minimal sequence of steps to define and configure one variant of a system (product) with one resource configuration (nominal mode). The focus is on the tools used to perform the steps of the workflow described in Section 4.5.3. Figure 5.53 gives a schematic overview. Arrows represent input/output relations between the tools and the artifacts involved in the use case. The arrows have two kinds of labels:

- Sequence numbers of the process steps.

- Acronym of the involved format or model, in parenthesis, preceded by the type of transformation: model-to-model (M2M) and model-to-text (M2T). In case of a model-to-model transformation there is actually always also a model-to-text transformation, but the later is not explicitly depicted, in order to preserve the readability of the diagram.

If an arrow is not drawn continuously, then the sequence number is repeated on each part. This is the case of step 13. In the following list, the corresponding workflow steps are grouped into four macro-steps. Furthermore, the tools that are used in each micro-step are given in brackets.

I Applications and Resources

1) Modeling of the Logical Architecture [AF3]

2) Modeling the Timing Requirements [AF3 plug-in]

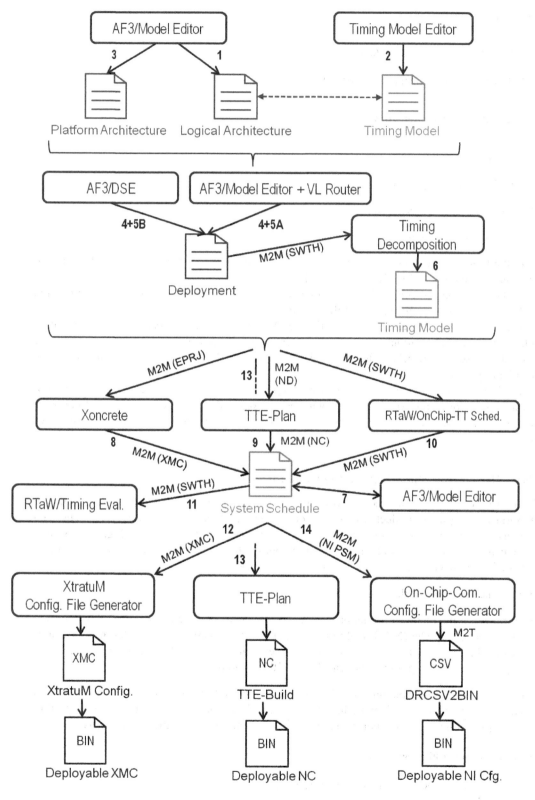

Figure 5.53: Overview of the Tool Workflow in "Use Case 1"

3) Modeling the Platform Architecture [AF3]

II Deployment

4) Modeling the System Software model [AF3]

5A) Defining the Deployment manually [AF3]

5B) Defining the Deployment through DSE [AF3]

III Configuration of Schedulers

6) Timing Decomposition [RTaW-Pegase/Timing]

7) Creating an empty System Schedule [AF3]

8) Adding partition/task schedules [Xoncrete]

9) Adding on-chip transmission phases for time-triggered virtual links [RTaW-Pegase/Timing]

10) Adding off-chip communication schedules [TTE-Plan]

11) Timing Analysis [RTaW-Pegase/Timing]

IV Platform building block configuration file generation

12) Generation of XtratuM configuration files [AF3 plug-in]

13) Generation of on-chip network communication configuration files [AF3 plug-in]

14) Generation of TTEthernet configuration files [TTE-Plan]

5.6.2 Use Case 2: Scheduling Configuration with Resource Management

The use case described in this section extends "Use Case 1" by the design of resource scheduling reconfigurations that defined how the resource managers (GRM/LRM) react to the failures of certain processor cores by switching scheduling plans (see also Section 4.5.4 for details of the workflow from the modeling point of view).

Figure 5.54 depicts the part of the tool workflow that differs from the one of "Use Case 1" (see Section 5.6.1 for meaning of the displayed information). The main differences can be found in the macro-step "Configuration of Schedulers":

1. Configuration of Schedulers

7A) Creating a System Schedule with partition scheduling slots. [Xoncrete]

7B) Manual creation of a System Schedule with partition scheduling slots. [AF3]

8) Adding Scheduling reconfigurations for failure modes [GRec]

9) Adding transition modes and flexibility parameters [MCOSF]

10) Adding off-chip communication schedules [TTE-Plan]

11) Adding on-chip transmission phases for time-triggered virtual links [RTaW-Pegase/Timing]

12) Timing Analysis [RTaW-Pegase/Timing]

Here, the tool GRec (see Section 5.3.1) is used to generate reconfigurations of task schedules for recovering from permanent core failures and the tool MCOSF (see Section 5.3.2) to add transition modes. Furthermore, the tools used to generate the communication schedules TTE-Plan and RTaW-Pegase/Timing take into account the possible changes of communication paths due to reconfiguration.

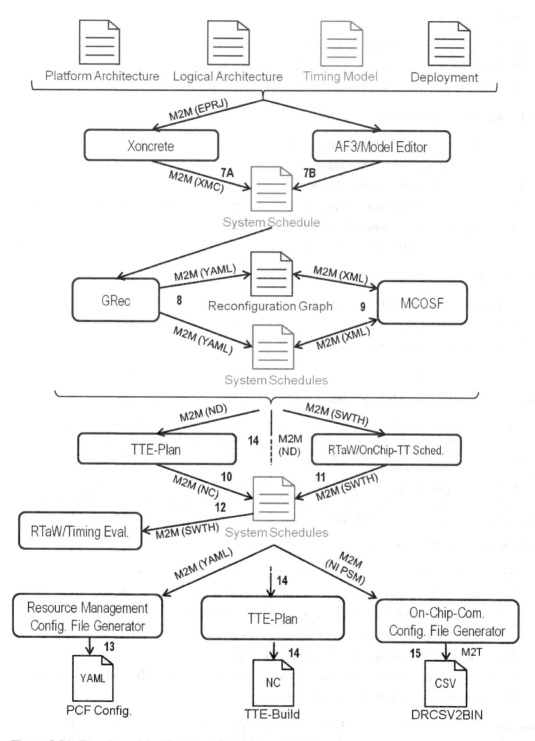

Figure 5.54: Overview of the Tool Workflow in Use Case 2

5.6.3 Use Case 3: Variability and Design-Space Exploration

This scenario describes how to build and exploit a product-line of mixed-criticality systems, that is, a family of related systems that differ in the features they offer, but share a common architecture. Product-lines help derive new systems that offer a previously untried combination of features (see Section 5.1).

The construction of a product-line differs from the design of a single product, because it assumes the existence of several products (here, DREAMS system models), from which the product-line will be factored out. It consists of the following steps:

1. Collect the system models of the various products of interest into a 150 % model that includes all required artifacts such as application, platform, timing and deployment models. Note that the combined 150 % model itself is not necessarily valid w.r.t. to the semantic of the product model since it may for instance aggregate model fragments that are mutually exclusive.

2. Identify the variation points (i.e., features) that will lead to new products. These variation points are summarized in a BVR variability specification (VSpec) (see Section 4.7.1).

3. For each variation point, define the transformation to apply to the 150 % model in order to enable (or disable) them. The BVR fragment substitution editor supports their definition as model fragment substitutions.

At this point, the product-line is operational: the BVR realization engine can derive a DREAMS system model from the sole prescription of the feature it must offer. Now the designer may either simply *derive a single product* and then proceed with the configuration of schedulers and the generation of platform configuration files using either of the previously introduced use cases of the toolchain. Alternatively, the designer may want to *explore the product-line*, and search for the most relevant products, with respect to design goals by applying the variability exploration process defined in Section 4.7.2:

1. Generate a set of products using the BVR product sampler (see Section 5.1.1.5). The current strategy yields the minimal set of products that covers n-ary feature interactions (all possible combinations of $2, 3, \ldots,$ or n features).

2. Construct the associated DREAMS system models for every resulting product using the BVR realization engine, as explained for the derivation of a single product.

3. Evolve the set of previously sampled products (so-called 125 % percent models, see Section 4.7.2) using the AF3 DSE that optimizes the application-to-platform deployments such that the design objectives are maximized (e.g., energy efficiency, reliability), and the defined design constraints are met (e.g., safety).

6

Execution Environment

A. Crespo

Fent Innovative Software Solutions, S.L.

P. Balbastre

Fent Innovative Software Solutions, S.L.

K. Chappuis

Virtual Open Systems

J. Coronel

Fent Innovative Software Solutions, S.L.

J. Fanguède

Virtual Open Systems

P. Lucas

Virtual Open Systems

J. Perez

IK4-Ikerlan

6.1   Virtualization Technologies 262
        6.1.1   Introduction 262
        6.1.2   Basic Implementation Types: Bare-Metal and Hosted 264
        6.1.3   Provided Virtual Environment - Full Virtualization and Para-Virtualization
                  Technology 265
                  6.1.3.1   Full Virtualization Technology 265
                  6.1.3.2   Para-Virtualization Technology 266
6.2   Execution Architecture 267
        6.2.1   Hardware Layer 267
        6.2.2   Virtualization Layer 268
                  6.2.2.1   Partitioning Kernel 268
                  6.2.2.2   Interrupt Virtualization Layer 269
        6.2.3   Runtime Layer 270
        6.2.4   Application Layer 271
6.3   Bare-Metal Hypervisor: XtratuM Case 273
        6.3.1   Overview 274
        6.3.2   System and Partitions Operation 275
                  6.3.2.1   System Operation 276
                  6.3.2.2   Partition Operation 276
        6.3.3   Partitions 277
                  6.3.3.1   Names and Identifiers 278
                  6.3.3.2   Partition Scheduling 278
                  6.3.3.3   Memory Management 279
                  6.3.3.4   Inter-Partition Communication 280
        6.3.4   Health Monitor 281
        6.3.5   Access to Devices 282
        6.3.6   Services 284
        6.3.7   Configuration 284
                  6.3.7.1   Hardware Description 286

		6.3.7.2	Scheduling Plans Description	287
		6.3.7.3	Partition Description	287
		6.3.7.4	Channels Description	287
	6.3.8	Deployment	292
		6.3.8.1	Virtualization Layer Configuration and Distribution	292
		6.3.8.2	System Resource Allocation	293
		6.3.8.3	Application Development	293
		6.3.8.4	System Integration	294
6.4	Operating System Hypervisor: Linux-KVM Case	294
	6.4.1	Overview	295
	6.4.2	Scheduling and Coordination for Linux-KVM	295
	6.4.3	'Memguard' to Boost KVM Guests on ARMv8	299
	6.4.4	Secure Monitor Firmware	304
		6.4.4.1	Overview	304
		6.4.4.2	Architecture	307
		6.4.4.3	Services	310
		6.4.4.4	Performance	311

This chapter describes the DREAMS mixed-criticality layered execution environment with a special focus on virtualization technologies such as hypervisors (e.g., XtratuM and Linux-KVM). Virtualization is the foundation for the deployment of partitioned systems that support the development of mixed-criticality systems. A physical computer is partitioned into several logical partitions, each of which looks like a real computer. Each of these partitions can have an operating system installed on it, and execute as if it was a completely separate computer machine. This foundation is nowadays also present in current state-of-the-art operating systems where processor and memory are multiplexed to processes.

Section 6.1 reviews the overall virtualization technology state-of-the-art with a focus on embedded systems and serves as an introduction for the following sections. Section 6.2 describes the layered execution architecture defined in DREAMS, based on an application layer, runtime layer, virtualization layer and hardware layer. Finally, two specific hypervisor solutions extended to provide DREAMS support are described, XtratuM (see Section 6.3) and Linux-KVM (see Section 6.4).

6.1 Virtualization Technologies

The current state of virtualization technology is the result of a convergence of technologies such as operating system design, compilers, interpreters, and hardware support among others. This heterogeneous origin, jointly with a growing evolution, has caused a confusion on the terminology. This section provides a review of virtualization technologies with a focus on embedded systems.

Two specific hypervisor cases, XtratuM and Linux-KVM, are described in Sections 6.3 and 6.4 respectively.

6.1.1 Introduction

A Virtual Machine (VM) is a software implementation of a machine (computer) that executes programs like a real machine. Hypervisor (also known as Virtual Machine Monitor (VMM) [213]) is a layer of software (or a combination of software and hardware) that allows to run several independent execution environments in a single computer platform.

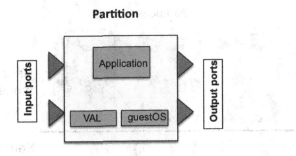

Figure 6.1: Application Partition

The separation kernel proposed in [214] established a combination of hardware and software to allow multiple functions to be performed on a common set of physical resources without interference. The Multiple Independent Levels of Security and Safety (MILS) initiative was a joint research effort between academia, industry, and government to develop and implement a high-assurance, real-time architecture for embedded systems. The technical foundation adopted for the so-called MILS architecture was a separation kernel. Also, the Aeronautical Radio INC. (ARINC) [215] standard uses these principles to define a baseline operating environment for application software used within Integrated Modular Avionics (IMA), based on a partitioned architecture.

The idea behind a partitioned system is the virtualization. Although virtualization has been used in mainframe systems since the 60s, the advances in the processing power of the desktop processors in the middle of the 90s opened the possibility to use it in the PC market. The embedded market is now ready to take advantage of this promising technology. Most of the recent advances on virtualization have been done for desktop systems, and transferring these results to embedded systems is not as direct as it may seem.

The key difference between hypervisor technology and other kinds of virtualization (e.g., Java VM, software emulation) is performance. Hypervisor solutions have to introduce a very low overhead; the throughput of the virtual machines has to be very close to that of the native hardware. The hypervisor has to be adapted and customized to the requirements of the target application, and can also be designed to meet real-time, power efficiency, dependability and security properties.

In general, virtualization provides a complete or partial re-creation of the hardware behavior of a native system to each of the partitions. The objective is that the partition should not be able to detect whether the system is real or virtual. On the other hand, specific partitions require complete knowledge of the environment in order to perform system monitoring and control the execution of other partitions.

When the native hardware is a multi-core system, the hypervisor can offer as many Virtual CPUs (vCPUs) as real Central Processing Units (CPUs) to the partitions in order to develop applications that can require multi-core solutions. So, in this context, partitions can be mono or multi-core according to the application needs.

As shown in Figure 6.1, a partition encapsulates an application with its runtime support that can include two components: a guest Operating System (OS) that provides the mechanisms to execute multiple threads in the application and a Virtualization Abstraction Layer (VAL) that provides the services related to the virtualization. For simple applications, the VAL provides the minimum services for executing single thread code avoiding the use of a guest OS. A partition can communicate with other partitions through defined Input / Output (I/O) communication ports.

The following sections describe major hypervisor types with respect to implementation and provided virtual environments:

- Implementation (see Section 6.1.2): Bare-metal (type 1) and hosted (type 2)

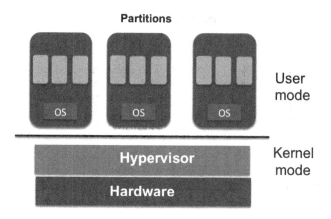

Figure 6.2: Bare-Metal Hypervisor (Type 1)

- Provided virtual environment (see Section 6.1.3): full virtualization and para-virtualization technology

6.1.2 Basic Implementation Types: Bare-Metal and Hosted

According to the resources used by the hypervisor there are two basic types of hypervisors as described below:

- Type 1 hypervisors (bare-metal) run directly on the native hardware. The hypervisor is in charge of the hardware initialization, system booting and partition execution with associated dependability, real-time and security properties. XtratuM [216,217], AIR [218] and Hyper-V [219] are bare-metal hypervisors examples and Figure 6.2 shows the generic software architecture for this approach.

- Type 2 hypervisors (hosted) are executed on top or inside of an operating system. The native operating system is called host operating system and the operating systems that are executed in the virtual environments as partitions are called guest operating systems. VMWare, VirtualBox, QEMU are hosted hypervisor examples and Figure 6.3 shows the generic software architecture for this approach.

On the other hand, there are additional virtualization technologies that do not meet previous classification, e.g.,

- Xen [220] and Kernel-based Virtual Machine (KVM) [221] hypervisors are installed as modules in the Linux operating system. Therefore, they can be considered hosted hypervisors as they are installed in an OS that performs the system initialization and then the module is loaded. But in practice, they could be considered bare-metal because after initialization they take control of the native hardware.

- Microkernel technology was developed as a paradigm to implement a single operating system. However, services provided by the microkernel can be used to build several different operating systems, resulting in a virtualized system. The evolution of some microkernels has followed a direction close to the hypervisors, e.g., PikeOS [222].

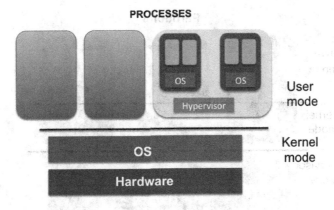

Figure 6.3: Hosted Hypervisor (Type 2)

6.1.3 Provided Virtual Environment - Full Virtualization and Para-Virtualization Technology

Considering the virtual environment provided by the hypervisor, there are two basic types of hypervisors: full virtualization and para-virtualization.

6.1.3.1 Full Virtualization Technology

Full virtualization provides a complete re-creation of the hardware behavior of a native system to each of the guests. The objective is that the guest system should not be able to detect whether the system is real or virtual. A full virtualizer is able to run unmodified versions of the software. Figure 6.4 shows the generic software architecture for this approach.

There are basically two techniques that can be used to implement a full virtualizer: code analysis with binary translation and hardware assistance.

- The code analysis technique and binary translation is used for the emulation of some instructions or a complete set of instructions. Thus, it allows executing unmodified guest OSs by emulating one instruction set by another through translation of code. It was developed considering that: 1) the amount of code that uses conflicting instructions is small; 2) it is highly unlikely to execute self modifying code. The hypervisor scans (disassembles) the code searching for conflicting instructions and performs a binary translation using virtualized instructions. A cache of already scanned code is maintained to speed up the process. The hypervisor is not very efficient at startup time, but once enough code has been scanned, the performance is increased. The hypervisor can detect when a privileged instruction is executed and perform the binary translation. The main drawback of this approach is that some conflicting instructions in some processors are not privileged and can not be captured by the hypervisor.

- Hardware support virtualization is not a new concept. It was first implemented when the Virtual Machine Facility/370 became available in 1972 for the IBM System/370. Recently, processor vendors such as AMD and Intel introduced virtualization extensions to their processors. Hardware support for virtualization implements a specific processor mode (hypervisor) with a higher privileged mode in the processor architecture.

The CPU extensions for virtualization support allows executing unmodified guest OSs in kernel processor mode and applications at user processor mode. Hardware assisted virtualization enhances CPUs to support virtualization without the need of binary translation. The advantage is

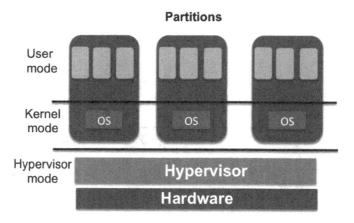

Figure 6.4: Hardware Assisted Full Virtualization Software Architecture

that this technique reduces the overhead caused by the trap-and-emulate model. Hypervisor processor mode offers a set of instructions to deal with CPU contexts, interrupts tables and memory maps that can be handled by the hypervisor. The hypervisor is executed in this processor mode and prepares appropriated context to each guest OS that runs at kernel processor mode.

6.1.3.2   Para-Virtualization Technology

Para-virtualization is a virtualization technique where the conflicting instructions are explicitly replaced by functions provided by the hypervisor. In this case, the guest system has to be aware of the limitations of the virtual environment and use the hypervisor services. Figure 6.5 shows the generic software architecture for this approach.

The para-virtualization technique greatly improves the performance and simplifies the hypervisor. With this technique it is possible to develop very compact and efficient hypervisors. The main drawback of this technique is that the guest operating system must be modified or adapted to be executed on top of the hypervisor. The hypervisor has to provide an Application Programming Interface (API) to be used by the adapted environment to replace the conflicting instructions. If the hypervisor API is close to the native processor operation, then the porting may require the modification of only a small number of instructions.

The hypervisor is in charge of managing the hardware resources and enforce the required spatial and temporal isolation of the guests. The para-virtualization technique fits basic requirements for embedded systems (e.g., low overhead, small footprint). And the customization (para-virtualization) of the guest operating system can be performed if the required OS source code or the hardware access layer is available.

Para-virtualization abstracts the CPU resources to build virtual machines (e.g., network, memory, disk) that can be virtualized or controlled by guest OSs. In para-virtualization, the second option is commonly used in order to avoid the inclusion of a huge number of drivers at hypervisor level and increase its complexity. In that case, the hypervisor controls the access to the device from a partition by enabling the set of addresses of the device or providing specific services for that purpose.

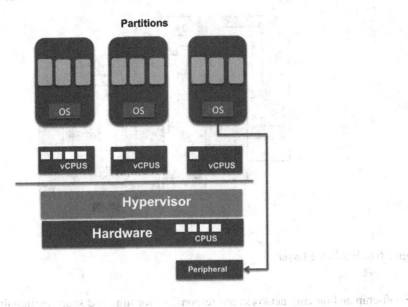

Figure 6.5: Para-Virtualization Software Architecture

6.2 Execution Architecture

This section describes the layered execution architecture defined in DREAMS:

- Application layer (see Section 6.2.4) that contains the developed application functionality.
- Runtime layer (see Section 6.2.3) that includes all required mechanisms to boot the application.
- Virtualization layer (see Section 6.2.2) that provides hardware virtualization to applications.
- Hardware layer (see Section 6.2.1) structured in clusters composed of interconnected nodes based on multi-core devices.

6.2.1 Hardware Layer

The DREAMS platform described in Chapter 2 is physically structured as follows:

- A set of clusters, where each cluster consists of nodes that are interconnected by a real-time communication network and associated topology.
- Each node is a multi-core chip containing tiles that are interconnected by a Network On Chip (NoC).
  - Each tile provides a Network Interface (NI) to the NoC.
  - The NI offers ports, each of which serves for the transmission or reception of the NoC's messages.
  - A tile can be a processor cluster with several processor cores, caches, local memories and I/O resources. Alternatively, a tile can also be a single processor core or an IP core (e.g., memory controller that is accessible using the NoC and shared by several other tiles).

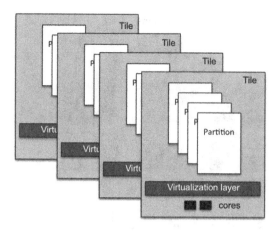

Figure 6.6: Hardware Layer

- Off-chip and on-chip networks are responsible for time and space partitioning between nodes or tiles respectively.

Based on the previous description, from the execution point of view, the hardware layer can be abstracted in a set of interconnected multi-core tiles as shown in Figure 6.6.

6.2.2   Virtualization Layer

The virtualization layer (see Section 6.1) is a software layer that provides hardware virtualization to the applications. Two complementary approaches are considered in DREAMS depending on the application constraints, partitioning kernel and interrupt virtualization layer, based on XtratuM and Linux-KVM respectively.

6.2.2.1   Partitioning Kernel

The partitioning kernel approach is based on the bare-metal XtratuM hypervisor. It provides virtualization of the hardware resources by defining a set of services that are used by the partitions to access the virtualized resources. The partitioning kernel provides spatial and temporal isolation to the partitions. The DREAMS virtualization layer is provided by the XtratuM hypervisor as shown in Figure 6.7.

Conceptually, XtratuM abstracts the underlying hardware and provides virtualization of the CPUs, allowing the execution of multiple isolated partitions as virtual machines. Partitions can use one or more vCPUs and allocate them to real CPUs. XtratuM can offer to the partitions as many vCPUs as real CPUs exist in the tile. Additionally, it offers additional services to partitions: e.g., get the system or partition status, control the execution of partitions, manage faults, inter-partition communication. The basic properties that the virtualization layer shall accomplish are:

- Spatial isolation: The hypervisor has to guarantee the spatial isolation of the partitions. A partition is completely allocated in a unique address space (code, data, stack) that is not accessible by other partitions. The system architect can relax this property by defining specific shared memory areas between partitions.

- Temporal isolation: The hypervisor has to execute partitions under a scheduling policy that guarantees the execution of partitions according to this schedule (e.g., period, offset, execution time slot).

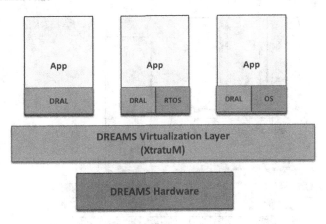

Figure 6.7: XtratuM Based DREAMS Virtualization Layer

- Fault isolation and management: Faults and errors, when occur, have to be detected and handled properly in order to isolate them, avoid propagation and perform required actions. A fault model to deal with the different types of errors is to be designed (see Section 2.5). The hypervisor implements a fault management model and allows partitions to manage the errors involved in the partition execution.

- Predictability: A partition with real-time constraints has to execute its code in a predictable way. It can be influenced by the underlying layers of software (guest OS and hypervisor) and by the hardware. From the hypervisor point of view, the predictability applies to the provided services, the operations involved in the partition execution and the interruption management of the partitions.

- Security: All the information in a system (partitioned system) has to be protected against access and modification from unauthorized partitions or unplanned actions. Security implies the definition of a set of elements and mechanisms that permit to establish the system security functions. This property is strongly related with the static resource allocation and a fault model to identify and confine the vulnerabilities of the system.

6.2.2.2 Interrupt Virtualization Layer

This layer virtualizes the host OS interrupts. The main objective is to preserve Real-Time Operating System (RTOS) timing guarantees by taking the control of hardware interrupts from the host OS and handling them in a thin layer. As shown in Figure 6.8, an interrupt virtualization layer is introduced below the host OS and real-time partition to prioritize the RTOS. This approach is implemented using the OS hypervisor Linux-KVM.

KVM converts host (Linux) processes into virtual machines, and re-uses most of the common features provided by host OS: e.g., process scheduling, memory management, interrupt handling. In order to support a hard real-time partition, it is either possible to introduce a thin interrupt virtualization layer below the host kernel or modify most of the host kernel sub-systems. The former approach is considered a better option, such as using Adaptive Domain Environment for Operating Systems (ADEOS) or equivalent, thanks to its smaller Trusted Computing Base (TCB). For example, ADEOS 'nanokernel' is composed of a few Kilo Lines of Code (KLOC) for ARM processors as opposed to a fully featured host OS such as Linux, which has a very large TCB. Thus, an interrupt virtualization layer along with the KVM hypervisor is necessary for realizing the RTOS - GPOS co-existence use-case.

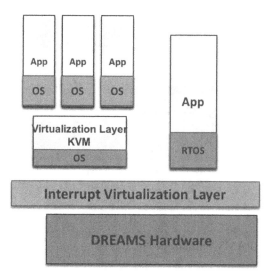

Figure 6.8: DREAMS Virtualization Layer: Linux-KVM

The interrupt virtualization layer schedules multiple operating system instances running above it and supports the co-existence of multiple prioritized domains (real-time and non real-time). This layer implements an interrupt management scheme, which allocates specialized interrupt handlers for the host OS and RTOS. The RTOS-specific interrupts are given higher priority to ensure real-time behavior. KVM will run on the host OS (within the non-real-time domain) and create multiple virtual machines.

6.2.3   Runtime Layer

A partition encapsulates the application code and the runtime layer that can include all the mechanisms to boot the applications. From the point of view of the virtualization layer, a partition is an auto-contained software with an entry point to initialize it. From this perspective, when the virtualization layer starts the execution of a partition, it jumps to this entry point and the partition starts as it was on top of a native hardware.

The partition entry point corresponds to the boot procedure to initialize the internal data structures and the virtualized hardware abstractions. In general, it is performed by the operating system that initializes the internal data structures and then calls the application code to execute it. In DREAMS, three different runtime layers are defined:

- DReams Abstraction Layer (DRAL): It is a minimum software component that includes the partition boot and several basic partition services.

- Guest operating system (guest OS): An operating system that is executed in a partition. Depending on the virtualization technology used, it might require to be adapted to be executed as partition.

- Linux-KVM: KVM specific boot and services.

In DREAMS, a partition can use a runtime component as support for the execution. Several schemes are possible:

- Bare Partitions: Partitions execute as if they were executing on top of the native hardware. The

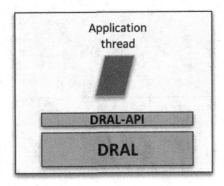

Figure 6.9: Bare Partition Using One Virtual CPU

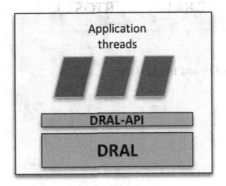

Figure 6.10: Bare Partition Using Three Virtual CPUs

application code can be a single thread executed in one core or several single-threads as shown in Figures 6.9 and 6.10. In both cases, the runtime uses a standalone version of DRAL that includes the partition booting procedures and several basic services for the partition. Multi-core partitions that use more than one vCPU have the same scheme but with as many application threads as defined vCPUs.

- Real-Time Partitions: These partitions shall contain a real-time operating system adapted to be executed on top of the DRAL as shown in Figure 6.11. Additionally, it can include the DRAL that complements the RTOS services with specific services for partitioning. The partition boot is managed by the RTOS.

- General purpose partitions: These partitions shall contain a full featured operating system (e.g., Linux) that offers the OS services to the partitions as shown in Figure 6.12. Additionally, it can include the DRAL layer that complements the OS services with specific services for the partition. The partition boot is managed by the OS.

6.2.4 Application Layer

The application layer is built with the application code developed by the developers. A set of tools included in the virtualization layer distribution allows to build partitions with the required components and application runtime.

A mixed-criticality system is composed by a set of partitions where the applications have been

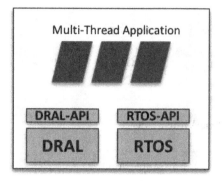

Figure 6.11: Partition with Guest OS and DRAL Layer

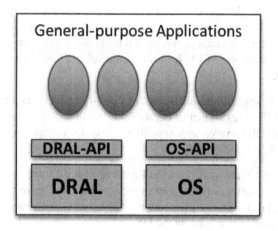

Figure 6.12: Partition with a General Purpose Guest OS

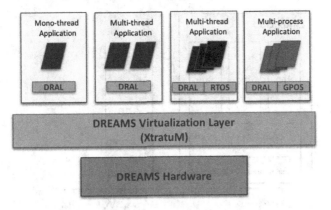

Figure 6.13: A System with Four Partitions in One Tile

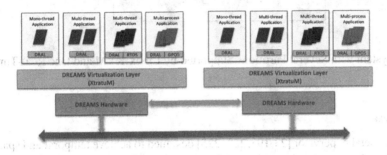

Figure 6.14: A System with Two Tiles

split in different partitions based on the criticality of the functionalities. Chapter 2 defines this as the logical model of the applications. This model can be used to represent systems developed with different design options, e.g.,

- Figure 6.13 shows a system composed by four applications that are executed in a multi-core tile. Each partition can have different runtimes: a stand-alone DRAL with one and two vCPUs, a RTOS with DRAL and a General Purpose Operating System (GPOS) with DRAL.

- Figure 6.14 extends the number of tiles and communicates them through the STMicroelectronics NoC (STNoC) and TTEthernet devices.

- Figure 6.15 represents a second approach developed in DREAMS, based on the KVM hypervisor at Linux level. Several partitions can be implemented in the Linux environment supported by KVM. The system is completed at low level by an Interrupt Virtualization Layer (IVL), based in ADEOS, which redirects the interrupts to multiple domains. One of these domains is Linux and the other is the real-time application.

6.3 Bare-Metal Hypervisor: XtratuM Case

This section describes the XtratuM hypervisor and developed DREAMS extensions to support the development of mixed-criticality systems.

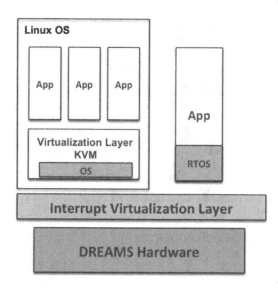

Figure 6.15: A System with Several Partitions Supported by Linux-KVM and One Real-Time Partition

6.3.1 Overview

XtratuM is a bare-metal hypervisor [23, 198, 223–225] designed to achieve temporal and spatial partitioning for safety-critical applications. Figure 6.16 shows the main components of the architecture:

- Hypervisor: XtratuM provides virtualization services to partitions. It is executed in supervisor mode of the processor and virtualizes the CPU, memory, interrupts, and some specific peripherals. The internal XtratuM architecture includes the following components:

  - Memory management: XtratuM provides a memory model for the partitions enforcing the spatial isolation. It uses the hardware mechanisms to guarantee the isolation.
  - Scheduling: Partitions are scheduled by using a cyclic scheduling policy.
  - Interrupt management: Each interrupt can be configured to be exclusively used by one partition. The hypervisor also provides to partitions an interrupt model that extends the concept of processor interrupts by adding 32 additional sources of interrupts (events).
  - Clock and timer management.
  - Inter-Partition Communication (IPC): This communication supports information exchange between two partitions or between a partition and the hypervisor. The hypervisor implements a message passing model which highly resembles the one defined in the ARINC. A communication channel is the logical path between one source and one or more destinations. Two basic transfer modes are provided: sampling and queuing. Partitions can access to channels through access points named ports. The hypervisor is responsible for encapsulating and transporting messages.
  - Health Monitor (HM): The HM detects and reacts to anomalous events or states as defined during design stage. Some of these events are captured by partitions that must communicate the event to the hypervisor. The purpose of the HM is the early detection of errors, confine the faulty subsystem and react with associated actions defined at design stage.
  - Tracing facilities: The hypervisor provides a mechanism to store and retrieve the traces generated by partitions and the hypervisor itself. Traces can be used for debugging, during

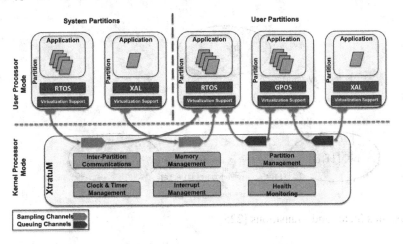

Figure 6.16: XtratuM Architecture

the development phase of the application, but also to log relevant events or states during the production phase.

- API: Defines the para-virtualized services provided by the hypervisor. Those services are provided by means of *hypercalls*.

- Partitions: A partition is an execution environment managed by the hypervisor that uses the provided virtualized services. Each partition consists of one or more concurrent processes (concurrency must be implemented by the operating system of each partition because it is not directly supported by the hypervisor), that share access to processor resources based upon the requirements of the application. The partition code can be: an application compiled to be executed on a bare-machine; a real-time operating system (or runtime support) and its applications; or a general purpose operating system and its applications.

Partitions need to be *virtualized* in order to be executed on top of a hypervisor. Depending on the type of execution environment, the virtualization implications in each case can be summarized as:

- Bare application: The application has to be virtualized using the services provided by the hypervisor. The application is designed to run directly on the hardware and therefore it has to be adapted to use the para-virtualized services.

- Mono-core operating system application: When the application runs on top of a (real-time) operating system, it uses the services provided by the operating system and does not need to be directly virtualized. However, the operating system has to deal with the virtualization and be virtualized (ported on top of the hypervisor).

- Multi-core operating system application: A guest OS can use multiple virtual CPUs to support multi-core applications.

6.3.2 System and Partitions Operation

This section describes the system and partitions operation by means of state-machines.

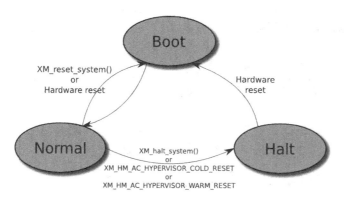

Figure 6.17: System's States and Transitions [225]

6.3.2.1 System Operation

System states and its transitions are shown in Figure 6.17 and described below [225]:

- Boot: At boot time, the resident software loads the image of XtratuM in main memory and transfers control to the entry point of the hypervisor. The period of time between starting from the entry point and the execution of the first partition is defined as 'boot' state. In this state, the scheduler is not enabled and the partitions are not executed.

- Normal: At the end of the boot sequence, the hypervisor is ready to start executing partition code. The system changes to 'normal' state and the scheduling plan is started. The transition from 'boot' to 'normal' state is performed automatically (the last action of the set up procedure).

- Halt: The system can switch to 'halt' state by the Health Monitor (HM) system in response to a detected error or by a *system partition* invoking the service *XM_halt_system*. In the 'halt' state the scheduler is disabled, the hardware interrupts are disabled, and the processor enters in an endless loop. The only way to exit from this state is via an external hardware reset.

It is possible to perform a warm or cold (hardware reset) system reset by using the hypercall (see *XM_reset_system*). On a warm reset, the system increments the reset counter, and a reset value is passed to the new rebooted system. On a cold reset, no information about the state of the system is passed to the new rebooted system.

6.3.2.2 Partition Operation

Once XtratuM is in 'normal' state, partitions are started. Partition states and transitions are shown in Figure 6.18 and described below [225]:

- Boot: On start-up each partition is in 'boot' state. The hypervisor prepares the virtual machine to be able to run the applications. For example, if the partition code is composed of an operating system and a set of applications: it sets up a standard execution environment, creates the communication ports, requests the hardware devices (I/O ports and interrupt lines), etc. Once the partition has been initialized, the state transits to 'normal' state.

- Normal: The partition receives information from XtratuM about the previous executions, if any. From the hypervisor point of view, there is no difference between the 'boot' state and the 'normal' state. In both states the partition is scheduled according to the fixed plan, and has the same capabilities. Although not mandatory, it is recommended that the partition emits a partition's state-change event when changing from 'boot' to 'normal' state. The 'normal' state is subdivided in three sub-states:

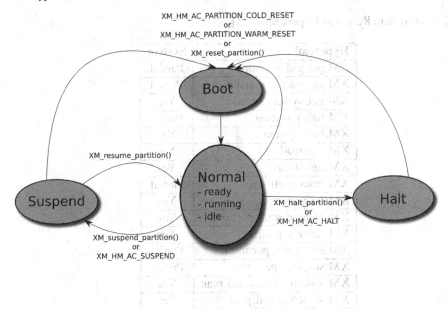

Figure 6.18: Partition States and Transitions [225]

- Ready: The partition is ready to execute code, but it is not scheduled because it is not in its time slot.
- Running: The partition is being executed by the processor.
- Idle: If the partition does not want to use the processor during its allocated time slot, it can relinquish the processor and wait for an interrupt or for the next time slot (see *XM_idle_self*).

- Halt: A partition can halt itself or be halted by a system partition. In the 'halt' state, the partition is not selected by the scheduler and the time slot allocated to it is left idle (it is not allocated to other partitions). All resources allocated to the partition are released. It is not possible to return to normal state.

- Suspend: In suspended state, a partition will not be scheduled and interrupts are not delivered. Interrupts raised while in suspended state are left pending. If the partition returns to normal state, then pending interrupts are delivered to the partition. The partition can return to ready state if requested by a system partition by calling *XM_resume_partition* or *XM_resume_imm_partition* hypercall.

6.3.3 Partitions

XtratuM defines two types of partitions: *normal* and *system*. System partitions are allowed to manage and monitor the state of the system and other partitions. Some hypercalls cannot be called by a normal partition or have restricted functionality.

Note that system partition rights are related to the capability to manage the system, and not to the capability to access directly to the native hardware or to break the isolation: a system partition is scheduled as a normal partition; and it can only use the resources allocated to it in the configuration file.

Table 6.1 shows the list of hypercalls reserved for system partitions. A hypercall labeled as

Table 6.1: List of System Reserved Hypercalls [225]

Hypercall	System
XM_get_gid_by_name	Partial
XM_get_partition_status	Partial
XM_get_system_status	Yes
XM_halt_partition	Partial
XM_halt_system	Yes
XM_hm_read	Yes
XM_hm_status	Yes
XM_reset_partition	Partial
XM_reset_system	Yes
XM_resume_imm_partition	Yes
XM_resume_partition	Yes
XM_shutdown_partition	Partial
XM_suspend_partition	Partial
XM_switch_imm_sched_plan	Yes
XM_switch_sched_plan	Yes

'partial' indicates that a normal partition can invoke it if a system reserved service is not requested. In order to avoid name collisions, all the hypercalls of XtratuM contain the prefix 'XM'. Therefore, the prefix 'XM', both in upper and lower case, is reserved.

6.3.3.1 Names and Identifiers

Each partition is globally identified by a unique identifier *ID*. Partition identifiers are assigned by the integrator in the 'XM_CF' file. XtratuM uses this identifier to refer to partitions. System partitions use partition identifiers to refer to the target partition. The 'C' macro 'XM_PARTITION_SELF' can be used by a partition to refer to itself.

These *IDs* are used internally as indexes to the corresponding data structures. The fist ID of each object group shall start in zero and the next ID's shall be consecutive. It is mandatory to follow this order in the 'XM_CF' file.

The attribute *name* of a partition is a human readable string. This string shall contain only the following set of characters: upper and lower case letters, numbers and the underscore symbol. It is advisable not to use the same name on different partitions. A system partition can get the name of another partition by consulting the status object of the target partition.

6.3.3.2 Partition Scheduling

XtratuM schedules partitions in a fixed and cyclic basis (ARINC scheduling policy). This policy ensures that one partition cannot use the processor for longer than scheduled to the detriment of other partitions. The set of *time slots* (start time and a duration) allocated to each partition is defined in the 'XM_CF' configuration file during the design phase. Within a time slot, XtratuM allocates the processor to the partition. Figure 6.19 shows how the modes have been considered in the XtratuM scheduling. However, in some cases, a single scheduling plan may be too restrictive. For example:

- Depending on the guest operating system, the initialization can require a certain amount of time and can vary significantly. If there is a single plan, the initialization of each partition can require a different number of slots due to the fact that the slot duration has been designed considering the operational mode. This implies that a partition can be executing operational work whereas others are still initializing its data.

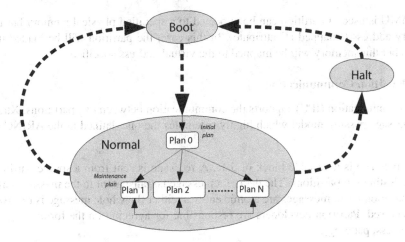

Figure 6.19: Scheduling Modes [225]

- The system can require to execute some maintenance operations. These operations can require allocating other resources different from the ones required during the operational mode.

In order to deal with these issues, XtratuM provides multiple scheduling plans that allows to reallocate the timing resources (the processor) in a controlled way. In the scheduling theory this process is known as mode changes. Plans are defined in the 'XM_CF' file and identified by an identifier. The scheduler (and so, the plans) is only active while the system is in 'normal' mode. The plan switch that occurs as a consequence of an immediate action is synchronous: the current slot is terminated and the next plan is started immediately. System partitions can request a plan switch using one of these hypercalls:

- 'XM_switch_sched_plan': The plan switch is not immediate; all the slots of the current plan will be completed, and the new plan will be started at the end of the current one.

- 'XM_switch_imm_sched_plan': The current slot is terminated, and the new plan is started immediately. This type of change will interrupt the execution of all the partitions on each core in the system and it will force the beginning of a new Major Active Frame (MAF) for the new scheduling plan.

On the other hand, if there are several concurrent activities in the partition, the partition shall implement its own scheduling algorithm. This two-level scheduling scheme is known as *hierarchical scheduling*. XtratuM is not aware of the scheduling policy used internally on each partition.

6.3.3.3 Memory Management

Partitions and XtratuM core can be allocated at defined memory areas specified in the XM_CF during the design phase. A partition can be assigned several memory areas. Each memory area can define some attributes or rights to permit other partitions to access their defined areas or allow the cache management. The following attributes area allowed:

- 'Unmapped': Allocated to the partition, but not mapped by XtratuM in the page table.

- 'Mapped at': It allows to allocate this area from a virtual address.

- 'Read-only': The area is write-protected to the partition.

- 'Uncacheable': Memory cache is disabled.

When the MMU is used, a partition can be allocated to a specified physical memory but using a virtual memory address ('mapped at' attribute). In this case, the partition will be loaded in the physical address but this memory will be mapped to the virtual address specified.

6.3.3.4   Inter-Partition Communication

Inter-Partition Communication (IPC) supports the communication between two partitions. XtratuM implements a message passing model which highly resembles the one defined in the ARINC standard.

- Message: A message is a variable block of data. A message is sent from a source partition to one or more destination partitions. The data of a message is transparent to the message passing system. At partition level, messages are atomic entities: either the whole message is received or nothing is received. Partition developers are responsible for agreeing on the format (e.g., data types, endianness, padding).

- Communication channel: A communication channel is the logical path between one source and one or more destinations. Partitions can access to channels through access points named ports. The hypervisor is responsible for encapsulating and transporting messages that have to arrive to the destination(s) unchanged. Channels, ports, maximum message sizes and maximum number of messages (queuing ports) are entirely defined in the configuration file.

XtratuM provides two basic transfer modes, *sampling* and *queuing*, and when a new message is available in the channel, XtratuM triggers an extended interrupt to the destination(s).

- Sampling port: It provides support for broadcast, multicast and unicast messages. No queuing is supported in this mode. A message remains in the source port until it is transmitted through the channel or it is overwritten by a new occurrence of the message, whatever occurs first. Each new instance of a message overwrites the current message when it reaches a destination port, and remains there until it is overwritten. This allows the destination partitions to access the latest message.

  A partition's write operation on a specified port is supported by *XM_write_sampling_message* hypercall. This hypercall copies the message into an internal XtratuM buffer. Partitions can read the message by using *XM_read_sampling_message* that returns the last message written in the buffer. XtratuM copies the message to the partition space.

  Any operation on a sampling port is non-blocking: a source partition can always write into the buffer and the destination partition(s) can read the last written message.

  The channel has an optional configuration attribute named 'refresh period'. This attribute defines the maximum time that the data written in the channel is considered 'valid'. Messages older than the valid period are marked as invalid. When a message is read, a bit is set accordingly to the valid state of the message.

- Queuing port: It provides support for buffered unicast communication between partitions. Each port has a queue associated where messages are buffered until they are delivered to the destination partition. Messages are delivered in FIFO order. Sending and receiving messages is performed by two hypercalls: *XM_send_queuing_message* and *XM_receive_queuing_message*, respectively. XtratuM implements a classical producer-consumer circular buffer without blocking. The sending operation writes the message from partition space into the circular buffer and the receiving one performs a copy from the XtratuM circular buffer into the destination memory.

6.3.4   Health Monitor

The Health Monitor (HM) is the XtratuM component that detects and reacts to anomalous events or states. The purpose of the HM is to discover errors at an early stage and try to solve or confine the faulting subsystem in order to avoid a failure or reduce the possible consequences.

It is important to clearly understand the difference between 1) an incorrect operation (e.g., instruction, function, application, peripheral) that is handled by the normal control flow of the software, and 2) an incorrect behavior which affects the normal flow of control in a way not considered by the developer or which can not be handled in the current scope.

An example of the first kind of errors is the 'malloc()' function that returns a null pointer because there is not memory enough to attend the request. This error is typically handled by the program by checking the return value. As for the second kind, an attempt to execute an undefined instruction (processor instruction) may not be properly handled by a program that attempted to execute it. The XtratuM HM component manages those faults that cannot, or should not, be managed at the *scope* where the fault occurs.

The XtratuM HM component defines four different execution scopes, depending on which part of the system has been initially affected: process (partition process or thread), partition (partition operating system or run-time support), hypervisor (XtratuM code) and board (resident software: BIOS, boot ROM or firmware).

Some events are detected directly by the hypervisor and some hardware events are received directly by the partitions that generate them. Partition events must be handled by the partitions and those events are not detected by the hypervisor. Therefore, the hypervisor provides to the partition mechanisms for the notification of events, which will be processed based on actions configured in the configuration file. These services of notification should be used by the partitions in order to maintain the scheme proposed by the hypervisor to detect and react to anomalous states.

Since HM events are, by definition, the result of an unexpected behavior of the system, it may be difficult to clearly determine which is the original cause of the fault, and so, what is the best way to handle the problem. XtratuM provides a set of 'coarse grain' actions that can be used at the first stage, right when the fault is detected. Although XtratuM implements a default action for each HM event, the integrator can map an HM action to each HM event using the XML configuration file. Once the defined HM action is carried out by XtratuM (e.g., stop the faulty partition and log the event), a HM notification message is stored in the HM log stream (if the HM event is marked to generate a log). A system partition can then read those log messages and perform a more advanced error handling. The XtratuM HM component is composed of four logical blocks:

- HM event detection: To detect abnormal states, using logical probes in the XtratuM code. When a HM event is detected, the relevant information (e.g., error scope, offending partition identifier, memory address, faulty device) is gathered and used to select the appropriate HM action. There are three sources of HM events:

    - Events caused by abnormal hardware behavior: Most of the processor exceptions are managed as health monitoring events.

    - Events detected and triggered by partition code: These events are usually related to checks or assertions on the code of the partitions.

    - Events triggered by XtratuM: Caused by a violation of a sanity check performed by XtratuM on its internal state or the state of a partition.

- HM actions: A set of predefined actions to recover from the fault or confine an error. Once an HM event is raised, XtratuM has to react quickly to the event. The set of configurable HM actions is listed in Table 6.2.

- HM configuration: To bind the occurrence of each HM event with the appropriate HM action.

Table 6.2: Configurable Health Monitor Actions

Action	Description
XM_HM_AC_IGNORE	No action is performed.
XM_HM_AC_SHUTDOWN	The shutdown extended interrupt is sent to the failing partition.
XM_HM_AC_COLD_RESET	The failing partition / processor is cold reset.
XM_HM_AC_WARM_RESET	The failing partition / processor is warm reset.
XM_HM_AC_PARTITION_COLD_RESET	The failing partition is cold reset.
XM_HM_AC_PARTITION_WARM_RESET	The failing partition is warm reset.
XM_HM_AC_HYPERVISOR_COLD_RESET	The hypervisor is cold reset.
XM_HM_AC_HYPERVISOR_WARM_RESET	The hypervisor is warm reset.
XM_HM_AC_SUSPEND	The failing partition is suspended.
XM_HM_AC_HALT	The failing partition / processor is halted.
XM_HM_AC_PROPAGATE	No action is performed by XtratuM. The event is redirected to the partition as a virtual trap.
XM_HM_AC_SWITCH_TO_MAINTENANCE	The current scheduling plan is switched to the maintenance one.

Note that the same HM event can be binded with different recovery actions in each partition HM table and in the XtratuM HM table. The HM system can be configured to send a HM message after the execution of the HM action. There are two tables to bind the HM events with the desired handling actions:

- XtratuM HM table: Defines the actions for those events that must be managed at board or hypervisor scope.

- Partition HM table: Defines the actions for those events that must be managed at hypervisor or partition scope.

- HM notification: To report the occurrence of the HM events. It is possible to select whether an HM event is logged or not.

6.3.5 Access to Devices

Access to devices is performed directly on the bare-metal without para-virtualization. However, which partitions access which devices (see Figure 6.20) must be configured in the XtratuM configuration file. The partition is in charge of handling properly the device and no driver is provided by XtratuM. The configuration file has to specify the interrupt lines that will be used by each partition and the device identifier. XtratuM is responsible for mapping the areas of memory where the device is placed. This mapping is performed so that the virtual addresses match the physical device addresses. This does not prevent that user additionally could map the device in virtual memory space as needed.

Two partitions can use neither the same device nor an interrupt line. When a device is used by several partitions, a user implemented I/O server partition (see Figure 6.21) may be in charge of the device management. An I/O server partition (see Section 10.4.1.4) is a specific partition that accesses and controls the devices attached to it, and exports a set of services via the inter-partition communication mechanisms provided by XtratuM sampling or queuing ports, enabling the rest of partitions to make use of the managed peripherals. The policy access (e.g., priority, FIFO) is implemented by the I/O server partition. Additionally other partitions can access devices that have been assigned exclusively.

Note that the I/O server partition is not part of XtratuM. It should, if any, be implemented by the user of XtratuM (see Section 10.4.1.4). XtratuM does not force any specific policy to implement I/O server partitions but a set of services to implement it.

There are two clocks per partition, both virtual and provided and managed by XtratuM. Only

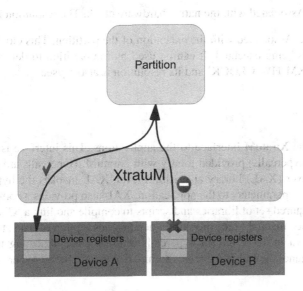

Figure 6.20: Memory Access and Exceptions

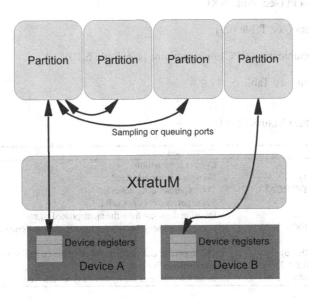

Figure 6.21: I/O Server Scheme

one timer per partition can be armed for each clock. The partition runtime should provide as many timers as needed to the application based on the timer provided by XtratuM.

- XM_HW_CLOCK: Associated with the native hardware clock. The resolution is 1 μsec.

- XM_EXEC_CLOCK: Associated with the execution of the partition. This clock only advances while the partition is being executed. It can be used by the partition to detect overruns. This clock relies on the 'XM_HW_CLOCK' and its resolution is also 1 μsec.

6.3.6   Services

This section describes the XtratuM interface to the applications. This interface is composed by a set of hypervisor calls (hypercalls) provided jointly with XtratuM. Hypercalls are provided by the XtratuM Abstraction Layer (XAL) library at partition level. XAL invokes the hypercall from the application and returns the parameters to the application. XAL also provides a compact developing environment with the required set of libraries and scripts to compile and link a 'C' application.

The XtratuM reference manual [226] provides a detailed description of data structures and hypercall profiles. As shown in Table 6.3 all hypercalls return a code indicating the result of the hypercall. On the other hand, the following tables summarize most relevant hypercalls classified by topic:

- Time management (see Table 6.4)

- System management (see Table 6.5)

- Health Monitor (HM) management (see Table 6.6)

- Partition management (see Table 6.7)

- Multi-core management (see Table 6.8)

- Schedule management (see Table 6.9)

- Inter-Partition Communication (IPC) management (see Table 6.10)

- Interrupt management (see Table 6.11)

Table 6.3: Hypercall Return Values [225]

Return Code	Description
XM_OK (0)	Correct termination
XM_NO_ACTION (-1)	No action was performed
XM_UNKNOWN_HYPERCALL (-2)	The hypercall does not exist
XM_INVALID_PARAM (-3)	Some parameter is not valid
XM_PERM_ERROR (-4)	Partition does not have the appropriated rights
XM_INVALID_CONFIG (-5)	Some parameter does not conform to the configuration parameters
XM_INVALID_MODE (-6)	The hypercall can not be invoked in the current state
XM_NOT_AVAILABLE (-7)	The resource is empty
XM_OP_NOT_ALLOWED (-8)	In the current state the hypercall is not allowed

Table 6.4: Time Management Hypercalls [225]

Hypercall	Description
XM_get_time	Retrieve the time of the clock specified in the parameter.
XM_set_timer	Arm a timer.

Table 6.5: System Management Hypercalls [225]

Hypercall	Description
XM_get_system_status	Returns the current status of the system.
XM_halt_system	The board is halted immediately. The whole system is stopped and interrupts are disabled.
XM_reset_system	The system is reset immediately. There are two ways to reset the system depending on the 'reset mode' parameter: warm reset and cold reset.

Table 6.6: Health Monitor Management Hypercalls [225]

Hypercall	Description
XM_hm_raise_event	The invoking partition generates a new HM event.
XM_hm_read	Read one or more HM log entries.
XM_hm_status	Get the status of the HM log stream.

Table 6.7: Partition Management Hypercalls [225]

Hypercall	Description
XM_get_partition_mmap	This function returns a pointer to the Memory Map Table (MMT).
XM_get_partition_status	Get the current status of a partition.
XM_halt_partition	Terminates a partition.
XM_idle_self	Idles the execution of the calling partition.
XM_params_get_PCT	Return the address of the PCT.
XM_reset_partition	Reset a partition.
XM_resume_partition	Resume the execution of a partition.
XM_shutdown_partition	Send a shutdown interrupt to a partition.
XM_suspend_partition	Suspend the execution of a partition.

Table 6.8: Multi-Core Management Hypercalls [225]

Hypercall	Description
XM_get_vcpuid	Returns the vCPU for the calling partition.
XM_halt_vcpuid	Halts the execution of the 'vcpu id' vCPU for the calling partition. The vCPU identified in the calling parameter is set to halt state.
XM_reset_vcpuid	Resets the vCPU 'vcpu id' of the calling partition. The vCPU that is reset will start its execution from the entry point identified in the calling parameters 'entry point'.

Table 6.9: Schedule Management Hypercalls [225]

Hypercall	Description
XM_get_plan_status	Return information about the scheduling plans.
XM_switch_imm_sched_plan	The new is scheduled to be started immediately, no waiting to the end of the current MAF.
XM_set_plan	Request a plan switch at the end of the current MAF.

6.3.7 Configuration

This section describes the mechanism for specifying the system configuration. The system is configured statically using a XML Configuration File (CF) ('XM_CF'), based on a XML schema

Table 6.10: Inter-Partition Communication Hypercalls [225]

Hypercall	Description
XM_create_queuing_port	Create a queuing port.
XM_create_sampling_port	Create a sampling port.
XM_get_queuing_port_info	Get the info of a queuing port.
XM_get_queuing_port_status	Get the status of a queuing port.
XM_get_sampling_port_info	Get the info of a sampling port.
XM_get_sampling_port_status	Get the status of a sampling port.
XM_read_sampling_message	Reads a message from the specified sampling port.
XM_receive_queuing_message	Receive a message from the specified queuing port.
XM_send_queuing_message	Send a message in the specified queuing port.
XM_write_sampling_message	Writes a message in the specified sampling port.

Table 6.11: Interrupt Management Hypercalls [225]

Hypercall	Description
XM_clear_irqmask	Unmask interrupts.
XM_clear_irqpend	Clear pending interrupts.
XM_disable_irqs	Disable interrupts.
XM_enable_irqs	Enables interrupts.
XM_set_irqmask	Mask interrupts.
XM_set_irqpend	Set some interrupts as pending.

type [227] that defines the data structures and types required to configure a DREAMS system (see Figure 6.22 and general overview of the 'XM_CF' in Listing 6.1). The XML CF is used by the system integrator to specify all the required allocation of system resources. The deployment of a system shall include the CF, and the virtualization layer will take it as a contract to guarantee defined resource allocation constraints. Prior to the deployment of the CF, this has to be validated to ensure the correct usage of the defined resource allocation. A CF shall describe among others the:

- Hardware: Describes the hardware platform.

- Devices: Details the devices in the platform.

- Schedule: Defines the scheduling plan(s).

- Partitions: Defines the partitions and their attributes.

- Channels: Defines the channels that will support the IPC.

The CF is compiled during the system compilation process to obtain a binary file as a result of the compilation of the XML file. This binary file is called configuration vector. The configuration vector generation is performed by a script included in the XtratuM distribution called 'xmcparser'. This script checks the correctness of the configuration analyzing the following information: e.g., memory area overlapping, memory region overlapping, memory area inside any region, duplicated partition's name and ID, allocated CPUs, replicated port's names and ID, cyclic scheduling plan, cyclic scheduling plan slot partition IDs, hardware IRQs allocated to partitions, I/O port alignment, I/O ports allocated to partitions, allowed HM actions.

6.3.7.1 Hardware Description

The hardware description details the information of the board used as target for the system deployment. The hardware section specifies the memory blocks available in the board, the scheduling plan of the hypervisor and the devices available in the board. Figure 6.23 shows a graphical view of the hardware description schema, and Listing 6.2 shows an example hardware description with three memory regions, a cyclic scheduling plan and three devices: serial, DRNoC and TTEthernet device with two ports.

6.3.7.2   Scheduling Plans Description

The scheduling plan is defined in the hardware description. However, as there are multiple configuration options, this section describes some of the different scheduling options. A scheduling plan can include: multiple cores schedules and multiple plans. Figure 6.24 shows the cyclic plan scheduling configuration schema and Listing 6.3 shows a specific scheduling configuration with two cores and three execution modes.

6.3.7.3   Partition Description

The partition configuration section in the XML file allows to specify the list of partitions in the system. A partition entry in the file specifies the partition identifier (numeric) that the partition will use in execution, the name, the device where the output will be sent and a list of flags to define if the partition is a system partition or if the partition has to be booted the first time it is scheduled. If 'boot' attribute is not set, the partition is not booted until another partition explicitly performs a reset action on it. Moreover, the partition specifies the set of memory areas that the partition will use. Memory areas are specified by the initial address and the size. Figure 6.25 shows a graphical representation of the partition configuration schema and Listing 6.4 shows an example partition configuration that defines 5 partitions ('Partition0' is a system partition).

6.3.7.4   Channels Description

Channels section in the XML file allows to specify the list of communication channels that support the Inter-Partition Communication (IPC). Figure 6.26 shows a graphical representation of the channels configuration schema and Listing 6.5 shows an example IPC channels configuration.

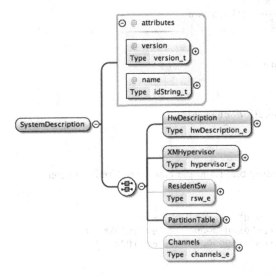

Figure 6.22: XM_CF Schema [227]

Listing 6.1: XML CF Schema Definition [227]

```
<SystemDescription xmlns="http://www.xtratum.org/xm-3.x" version="1.0.0"
name="Test">
<HwDescription>
<MemoryLayout>
— describes the memory blocks in the platform
</MemoryLayout>
```

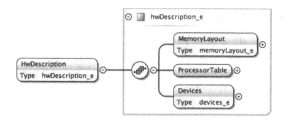

Figure 6.23: Hardware Description Schema [227]

Figure 6.24: Cyclic Plan Scheduling Configuration Schema [227]

```
<ProcessorTable>
— defines the scheduling plans
</ProcessorTable>
<Devices>
— Describes the devices in the platform
</Devices>
</HwDescription>

<XMHypervisor console="Uart">
— describes the hypervisor allocation
</XMHypervisor>

<PartitionTable>
— describes the partitions
</PartitionTable>

<Channels>
— describes the communication channels
</Channels>
</SystemDescription>
```

Listing 6.2: Example Hardware Configuration

```
<HwDescription>
<MemoryLayout>
<Region type="rom" start="0x0" size="1MB" />
<Region type="sdram" start="0x00100000" size="1023MB" />
<Region type="sdram" start="0x60000000" size="1024MB" />
</MemoryLayout>
<ProcessorTable>
<Processor id="0" frequency="400Mhz">
<CyclicPlanTable>
<Plan id="0" majorFrame="4s">
<Slot id="0" start="0s" duration="1950ms"
partitionId="0" />
<Slot id="1" start="2s" duration="1950ms"
partitionId="1" />
<Slot id="2" start="4s" duration="1950ms"
partitionId="2" />
</Plan>
</CyclicPlanTable>
</Processor>
</ProcessorTable>
```

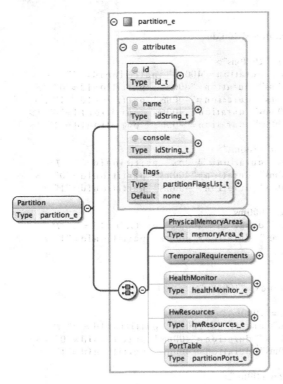

Figure 6.25: Partition Configuration Schema [227]

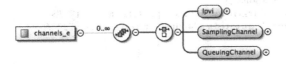

Figure 6.26: Channels Configuration Schema [227]

```
<Devices>
<Uart id="1" baudRate="115200" name="Uart" />
<DRNoC name="DRNoC_1" baseaddress="0x40000000">
<Port name="DSTP_CA9_0_MB0_6" type="TT"
direction="destination" offsetAddr="0x00000000" size="64B" />
<Port name="DSTP_CA9_1_MB1_6" type="TT"
direction="destination" offsetAddr="0x00010000" size="64B" />
<Port name="DSTP_CA9_2_MB0_8" type="RC"
direction="destination" offsetAddr="0x00020000" size="64B" />
</DRNoC>
<TTEthernet name="TTEthernet_1" baseaddress="0x50000000">
<Port name="DTTP_CA9_0" type="TT" direction="source"
ttEtherPortId="0" />
<Port name="DTTP_CA9_1" type="RC" direction="source"
ttEtherPortId="1" />
</TTEthernet>
</Devices>
</HwDescription>
```

Listing 6.3: Example Scheduling Configuration

```
<ProcessorTable>
<Processor id="0" frequency="400Mhz">
<CyclicPlanTable>
<Plan id="0" majorFrame="2000ms">
<Slot id="0" start="0ms" duration="400ms" partitionId="2" />
<Slot id="1" start="400ms" duration="400ms" partitionId="0" />
<Slot id="2" start="800ms" duration="400ms" partitionId="1" />
<Slot id="3" start="1200ms" duration="400ms" partitionId="0" />
<Slot id="4" start="1600ms" duration="400ms" partitionId="1" />
</Plan>
<Plan id="1" majorFrame="1200ms">
<Slot id="0" start="0ms" duration="400ms" partitionId="2" />
<Slot id="1" start="400ms" duration="400ms" partitionId="0" />
<Slot id="2" start="800ms" duration="400ms" partitionId="1" />
</Plan>
<Plan id="2" majorFrame="800ms">
<Slot id="0" start="0ms" duration="400ms" partitionId="2" />
<Slot id="1" start="400ms" duration="400ms" partitionId="1" />
</Plan>
</CyclicPlanTable>
</Processor>
<Processor id="1" frequency="400Mhz">
<CyclicPlanTable>
<Plan id="0" majorFrame="1000ms">
<Slot id="0" start="0ms" duration="200ms" partitionId="2" />
<Slot id="1" start="200ms" duration="400ms" partitionId="0" />
<Slot id="2" start="600ms" duration="400ms" partitionId="1" />
</Plan>
<Plan id="1" majorFrame="800ms">
<Slot id="0" start="0ms" duration="400ms" partitionId="0" />
<Slot id="1" start="400ms" duration="400ms" partitionId="1" />
</Plan>
<Plan id="2" majorFrame="800ms">
<Slot id="0" start="0ms" duration="400ms" partitionId="1" />
<Slot id="1" start="400ms" duration="400ms" partitionId="2" />
</Plan>
</CyclicPlanTable>
</Processor>
</ProcessorTable>
```

Listing 6.4: Partition configuration example

```
<PartitionTable>
<Partition id="0" name="Partition0" flags="system boot" console="Uart">
<PhysicalMemoryAreas>
<Area start="0x10000000" size="1MB" />
</PhysicalMemoryAreas>
<PortTable>
<Port type="sampling" direction="source" name="CA9_0_MB0_6" />
<Port type="queuing" direction="source" name="CA9_4_MB0_10" />
</PortTable>
</Partition>

<Partition id="1" name="Partition1" flags="boot" console="Uart">
<PhysicalMemoryAreas>
<Area start="0x11000000" size="1MB" />
</PhysicalMemoryAreas>
<PortTable>
<Port type="sampling" direction="destination" name="MB0_0_CA9_6" />
<Port type="queuing" direction="destination" name="MB0_4_CA9_10" />
</PortTable>
</Partition>

<Partition id="2" name="Partition2" flags="boot" console="Uart">
```

```
<PhysicalMemoryAreas>
<Area start="0x12000000" size="1MB" />
</PhysicalMemoryAreas>
<PortTable>
<Port type="queuing" direction="source" name="portQ"/>
<Port type="sampling" direction="source" name="portS"/>
</PortTable>
</Partition>

<Partition id="3" name="Partition3" flags="boot" console="Uart">
<PhysicalMemoryAreas>
<Area start="0x13000000" size="1MB" />
<Area start="0x70000000" size="256MB" />  <!-- shared -->
</PhysicalMemoryAreas>
<PortTable>
<Port type="sampling" direction="source" name="portS"/>
</PortTable>
</Partition>

<Partition id="4" name="Partition4" flags="boot" console="Uart">
<PhysicalMemoryAreas>
<Area start="0x14000000" size="1MB" />
<Area start="0x60000000" size="256MB" />
<Area start="0x70000000" size="256MB" />  <!-- shared -->
</PhysicalMemoryAreas>
</Partition>
</PartitionTable>
```

Listing 6.5: IPC channels configuration example

```
<Channels>
<SamplingChannel maxMessageLength="64B">
<Source partitionId="0" portName="CA9_0_MB0_6" />
<DRNoCDestination portName="DSTP_CA9_0_MB0_6" />
</SamplingChannel>
<SamplingChannel maxMessageLength="64B">
<Source partitionId="0" portName="CA9_2_MB0_8" />
<DRNoCDestination portName="DSTP_CA9_2_MB0_8" />
</SamplingChannel>
<QueuingChannel maxMessageLength="64B" maxNoMessages="16">
<Source partitionId="0" portName="CA9_4_MB0_10" />
<DRNoCDestination portName="DSTP_CA9_4_MB0_10" />
</QueuingChannel>
<SamplingChannel maxMessageLength="64B">
<DRNoCSource portName="DSTP_MB0_0_CA9_6" />
<Destination partitionId="1" portName="MB0_0_CA9_6" />
</SamplingChannel>
<SamplingChannel maxMessageLength="64B">
<DRNoCSource portName="DSTP_MB0_2_CA9_8" />
<Destination partitionId="1" portName="MB0_2_CA9_8" />
</SamplingChannel>
<QueuingChannel maxMessageLength="64B" maxNoMessages="16">
<DRNoCSource portName="DSTP_MB0_4_CA9_10" />
<Destination partitionId="1" portName="MB0_4_CA9_10" />
</QueuingChannel>
<SamplingChannel maxMessageLength="64B">
<TTEtherSource portName="DTTP_CA9_0" />
<Destination partitionId="2" portName="TT4" />
</SamplingChannel>
<SamplingChannel maxMessageLength="64B">
<Source partitionId="2" portName="TT1" />
<TTEtherDestination portName="DTTP_CA9_2" />
</SamplingChannel>
<QueuingChannel maxMessageLength="64B" maxNoMessages="16">
```

```
<TTEtherSource portName="DTTP_CA9_1" />
<Destination partitionId="3" portName="TT3" />
</QueuingChannel>
<QueuingChannel maxMessageLength="64B" maxNoMessages="16">
<Source partitionId="3" portName="TT2" />
<TTEtherDestination portName="DTTP_CA9_3" />
</QueuingChannel>
</Channels>
```

6.3.8   Deployment

Deployment phase describes the different steps to be performed to build a system on a specific target. This section details these steps describing which roles have to be defined in order to carry out the whole process. The development process of a partitioned system involves several roles in order to cover the different development phases. These roles are:

- System Architect: It is responsible for defining the overall system requirements and the system design, including the optimal decomposition into hosted partitions jointly with the detailed resource allocation per partition.

- System Integrator: It is responsible for verifying the feasibility of the system requirements defined by the 'system architect', as well as responsible for the configuration and integration of all components. This role can be assumed by the 'system architect'.

- Application Suppliers: They are responsible for the development of an application according to the overall requirements defined by the 'system architect' and the 'system integrator'. 'Application suppliers' shall verify that the allocated budget and safety parameters are respected. Assuming that each application is located in a partition and a partition can have only one application.

From the software development point of view, the following steps should be performed. These steps are detailed in the next sections and more detailed information can be found in the hypervisor user manual [225]: virtualization layer configuration and distribution for a specific platform, system resource allocation for partitions according to the application requirements, application development and validation, and system integration and global validation.

6.3.8.1   Virtualization Layer Configuration and Distribution

The final system will be executed in a computer platform with a set of devices according the system requirements. For this platform, the 'system architect' and the 'system integrator' shall configure the different configuration parameters of the virtualization layer. Example configuration parameters are: processor and board to be used, available board memory, number of cores of the platform and maximum number of virtual cores, memory management type, devices to be virtualized and communication mechanisms.

Once XtratuM has been configured, partitions can be compiled and the binary distribution generated as shown in Figure 6.27. The binaries generated are packed in an auto-executable file that can be distributed by the 'system architect' to the partition providers and can be installed in each development site. All partition providers share the same virtualization layer configuration. Several tools are available for this process:

- 'xmcparser': It translates the XML configuration file containing the system description into binary form that can be directly used by XtratuM. It performs a set of non-syntactical checks, e.g., memory areas defined, names and identifiers, CPU allocation to partitions, cyclic plan(s) specified.

- 'xmpack': It creates a system image container including the partitions, the configuration file and XtratuM.

- 'rswbuild': It generates a bootable system image adding a bootloader to the container.

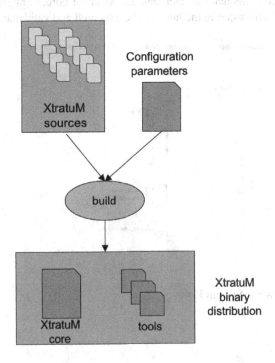

Figure 6.27: XtratuM Compilation and Binary Distribution Generation Phase

6.3.8.2 System Resource Allocation

The system resource allocation is performed by the 'system architect', who defines the resources to be allocated to each partition. In this phase, the system CF is defined allocating the available system resources to the partitions. Example resources are: vCPUs allocated to each partition, permission right attributes of each partition, memory areas where each partition will be executed or can use, detailed information of the scheduling plan(s), memory allocation for communication ports, communication ports between partitions, virtual devices and physical peripherals allocated to each partition and HM configuration of XtratuM and partitions.

The result of this process is the CF (detailed in Section 6.3.7). This file is distributed partially to each partition provider with the specific information related to the partition to be developed.

6.3.8.3 Application Development

The development of the application is carried out by the 'application supplier' that according to the requirements received from the 'system architect' develops the application. Figure 6.28 shows the inputs and outputs of this process. The 'system architect' or 'system integrator' sends the binary distribution generated for the specific platform and the CF to the 'application supplier'. The supplier installs the package, develops and validates the application. The result is the application partition that is sent to the 'system integrator'.

6.3.8.4   System Integration

As shown in Figure 6.29, the 'system integrator' receives the application partitions from the 'application provider(s)' and builds and validates the final system on the final target. The result of this process is the system file container that includes the XtratuM core, configuration vector and partitions. This container is then loaded in the target to be executed and validated.

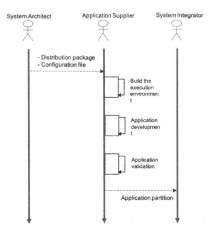

Figure 6.28: Interactions Between System Roles

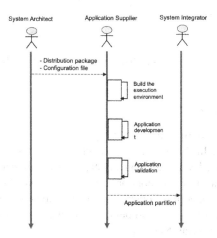

Figure 6.29: System Integration

6.4   Operating System Hypervisor: Linux-KVM Case

This section describes the extensions developed in the KVM hypervisor in order to enhance isolation and performance for the execution of soft real-time applications in KVM guests.

6.4.1 Overview

The Linux KVM is an established system virtualization solution, implemented as a driver running within Linux, which effectively turns the Linux kernel into a hypervisor. This approach takes advantage of the existing infrastructure within the Linux kernel, including the scheduler and memory management. This results in the KVM code base to be very small compared to other hypervisors; this has allowed KVM to evolve with an impressive pace and become one of the most well regarded and feature full virtualization solutions. KVM works by exposing a simple interface to the user-space, through which a regular process can request to be turned into a virtual machine. Usually Quick Emulator (QEMU) is used on the user-space side to emulate I/O devices, with KVM handling vCPUs and memory management. Through this interface, regular Linux processes will be turned into virtual machines, with threads acting as vCPUs. KVM will handle the context switching of the processor when the process of a virtual machine gets scheduled by Linux, using hardware virtualization support in order to virtualize the processor and the memory. To virtualize I/O devices, such as network interfaces and storage, an interface to user-space exists so those can be emulated by the application setting up the virtual machine (usually the QEMU emulator). Therefore, KVM exploits the Virtualization Extensions (e.g., ARM-VE) to execute guest's instructions directly on the host processor and to provide VMs with an execution environment almost identical to the real hardware. Each guest runs in a different instance of this execution environment, thus isolating the guest operating system. For this reason, this isolation has been used for security purposes [228] in many scientific works. In the ARM architecture, the KVM isolation involves CPU, memory, interrupts and timers [221].

For the purpose of the Healthcare demonstrator (see Section 11.3), different scheduling enhancements have been implemented in the KVM hypervisor. Firstly, storage I/O and task scheduling extensions have been developed and integrated in the Healthcare demonstrator, essentially targeting the Hospital server. In addition, the memory bandwidth scheduler ('Memguard') has been updated to support ARMv8 architecture and extended to enable memory bandwidth regulation between guests that run on top of Linux-KVM. The main idea is that the host scheduler is aware of guest prioritization and memory bandwidth needed. This can be achieved by enabling the guest to communicate with the host and inform dynamically when it needs to be prioritized as described in the following section.

6.4.2 Scheduling and Coordination for Linux-KVM

The solution chosen as hierarchical scheduling is a coordinated scheduling mechanism. It is a kind of para-virtualized scheduling that supports the communication between the guests and the host, so that a host scheduler is aware of guest actors in the system.

This solution has been implemented on a Linux kernel (host and guest OS), but could be ported to other guest operating systems. The goal of the coordinated scheduling (for simplicity also named 'co-scheduling') is to achieve, among others, low latency and reasonable time to start for interactive application and soft real-time tasks. The word 'scheduler' is used, here, to designate a generic scheduler that could be a storage-I/O scheduler or a task scheduler.

The main idea is that the host scheduler is aware of guest prioritization. This can be achieved by enabling the guest to communicate with the host and inform dynamically when it needs to be prioritized. Figure 6.30 describes the principle of 'co-scheduling'. Each time the guest OS, decides that its priority needs to be changed or when it executes a real-time program, it sends a request to the host that will modify the current scheduling policy.

Guest scheduling problem

In the case of storage I/O and task scheduling, any heuristics implemented in the guest are transparent to the host, failing to affect the overall scheduling of the guest by the host. This can be seen with

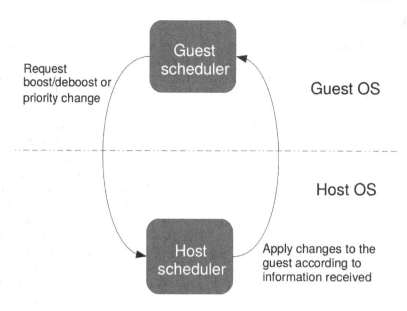

Figure 6.30: Coordination Scheduling Overview

a simple example. Consider a system running a guest operating system, say guest G, in a virtual machine. Application A, is being started (loaded) in guest G while other applications are already executing without interruption in the same guest. Such a system is represented in Figure 6.31. In such conditions, the scheduling patterns of guest G, as seen from the host side, may exhibit no special property that allows the scheduler in the host to realize that an application is being loaded in the guest. Hence, the scheduler in the host may have no reason for privileging the requests coming from guest G. In the end, if also other guests or applications of any other kind, are executed in the host then guest G may receive no help to be prioritized.

Implementation

For the implementation of the 'co-scheduling' mechanism on ARMv7 / ARMv8 architectures, a communication method is required between the host and the guest. For ARM, this can be achieved with the Hypervisor Call (HVC) instruction that is considered a hypercall. The HVC instruction can have an argument that can be used to pass different types of information. With this argument different requests can be defined that will be handled differently by the host.

The 'co-scheduling' mechanism has been implemented on the Linux kernel for the host (use of KVM on ARM) but also for the guest. These modifications imply to modify the kernel code for both guest and host schedulers. On the guest side, the scheduler has to be modified in order to extend any heuristics or scheduling policies with hypercalls in mind. This was done by utilizing 'paravirt-ops' ('pv-ops'), the hypervisor-agnostic Linux interface. Thus, new 'paravirt-ops' configurations regarding the I/O and task scheduler have been introduced in the Linux kernel, making it easy for a user to take advantage of the coordinated scheduling.

'Paravirt ops' Interface for KVM on ARM

Linux already provides a way to perform some para-virtual actions through an infrastructure named 'paravirt-ops' ('pv-ops' for short) [229]. This API is used to run para-virtualized virtual machines on multiple hypervisors with the same kernel binary. That is to say the same kernel binary can run

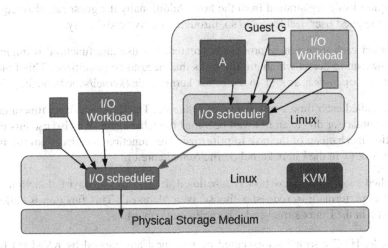

Figure 6.31: Example Highlighting the Missing Connection between Schedulers in Virtualized Environments

on bare hardware, or on hypervisors, such as VMWare, VNI or Xen, it can be para-virtualized or full virtualized. This infrastructure exists for multiple architectures and hypervisors, but not for KVM on ARM, one of the virtualization solutions for DREAMS. Therefore, a basic 'paravirt' operator was developed in order to implement the coordinated scheduling extensions with a ready to use infrastructure, providing better flexibility and maintainability for new features. The approach is based on a previous work that enables 'paravirt-ops' for Xen on ARM/ARM64 [230], thus we focused on the implementation for KVM on ARM.

Para-virtual functions require a hypercall implementation, in order to be able to send information to the host system. Therefore, hypercall functions specific to KVM have been implemented into the KVM code base of the Linux kernel. These functions use the HVC (hypervisor call) instruction of the ARM architecture, with the immediate argument of the HVC instruction being a constant integer used to recognize a 'paravirt' call (from a power state coordination interface call, for instance, which can also use an HVC instruction). The parameters of the hypercall are passed through the scratch registers, $r0$ contains the identification number of the hypercall and registers $r1$ to $r3$ represent the potential arguments for this hypercall.

Those hypercalls are called from the 'paravirt-ops' implementation of each para-virtualized sub-system. In our case it consists of pointers to functions, stored in a structure that represents the 'paravirt' subsystem. Those functions are called if the 'paravirt-ops' infrastructure is enabled for the hypervisor on which the virtual machine is running. For our needs a 'paravirt-ops' interface has been added for each scheduler, that is to say, the I/O scheduler and the task scheduler. For instance the 'paravirt' interface in the guest for the task scheduler coordination contains five functions:

- Register VM: During boot the virtual machine issues a register call to the hypervisor, if the hypervisor does not support the 'paravirt' interface, then all coordinated functionality is disabled. The registration procedure, if successful, also enables a 'sysfs' entry in the Linux host, where the user can selectively enable / disable or even fine tune the priority of a guest.

- De-register VM: At any point in time, the host might decide that coordination is no longer desirable during runtime. If a hypercall attempt for the guest is denied the next action of the guest is to request its de-registration from the host. Additionally the guest can also request de-registration if the guest user decides to do so, through a Linux 'sysfs' entry.

- New task: Called each time a new process is created. We use this function to implement a heuristic mechanism to detect which are the tasks that need to be prioritized. This function is called from `wake_up_new_task()` in the Linux kernel code (kernel/sched/core.c) [231].

- Activate task: Called each time a task becomes runnable. That is to say, each time a task that was waiting voluntary or due to an I/O wait becomes runnable again. We also use this function for the detection mechanism of the task to prioritize. This function is called from the function `activate_task()` in the Linux kernel (kernel/sched/core.c).

- Schedule: Called each time a new task is scheduled. It is in this 'paravirt' function that the hypercall call is performed; to request a 'boost' or a 'deboost'. This function is called from `__schedule()` in the Linux kernel code (kernel/sched/core.c).

On the host side, HVC instructions executed by the guest are trapped by KVM (in function `handle_hvc()` in arch/arm/kvm/handle_exit.c), and thus, can be handled correctly, the immediate argument of the HVC instruction is also checked to be sure that it is a hypercall and not something else (e.g., a Power State Coordination Interface (PSCI) call, or an invalid call). Then, KVM can perform the corresponding action to this hypercall according to the value retrieved from $r0$.

Host 'sysfs' user interface

In order for the user to have more control over the host system, a set of Linux 'sysfs' entries are created when a guest is using the coordinated scheduling extensions. By default, the Linux host / guest provide a 'sysfs' interface to the user in which coordination can be enabled or disabled for each type of scheduler. For example in the case of disk I/O scheduling the following entry is created when each system boots: `/sys/block/sda/queue/vbfq/enable`

The user can then issue a simple command to completely enable or disable the coordination enhancements of the scheduler:

```
# echo 1 > /sys/block/sda/queue/vbfq/enable
# echo 0 > /sys/block/sda/queue/vbfq/enable
```

When a guest is booted, the first step of the respective scheduler is to issue a registration hypercall to the host system. If registration is successful, then the guest is in a position to continue with normal coordination hypercalls. If registration is denied, or if any coordination attempts fail to be completed, then the guest scheduler is no longer issuing any hypercalls, avoiding unnecessary context switches for handling the hypercall between the host / guest.

From the host side, once a guest is successfully registered another set of 'sysfs' entries are populated, where the user can have a more fine-grained control for each registered guest. Each registered VM has its own entry in 'sysfs' with a listing of all VM processes that are involved, along with an entry to selectively enable or disable coordination for a particular VM. Additionally a priority entry is provided where the user of the host system can select if a guest should be further prioritized among other different guests that use coordination (by default they share the same priority). In the following 'sysfs' example 10301 is the Process Identification Number (PID) of the registered guest:

```
/sys/block/sda/queue/vbfq/10301/enable
/sys/block/sda/queue/vbfq/10301/priority
```

Performance

Those scheduling enhancements have shown significant improvement in latency and overall application responsiveness for virtual machines. They are integrated in the healthcare demonstrator.

Overall I/O guest scheduling is improved for applications that are interactive in nature, making

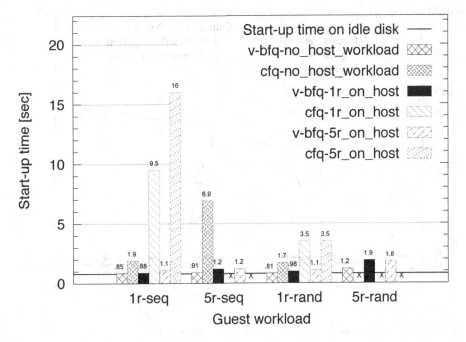

Figure 6.32: Start-Up Latency Results Compared between the Default Linux I/O Scheduler (CFQ) and the 'Co-Scheduler' V-BFQ [232]

them more responsive. Start-up application latency in synthetic benchmarks dropped to host idle levels, while aggressive I/O workloads were also present. In contrast the default scheduling solution for the guest needs extreme amounts of time to finish the same test (or even fail), as seen in Figure 6.32. Additionally in real scenarios, video playback was improved with no or less stuttering artifacts compared to defaults.

For task scheduling and interrupt latency, the improvement is also significant, where with coordination the guest latency is dropped nearly to host levels (virtualization overhead still present). Coordinated scheduling can dynamically change the priority of the guest when needed, minimizing unnecessary context switches during the execution for a critical task in the guest. This can be seen in Figure 6.33 where the interrupt latency with cyclic test is ranging from 2000 to 9000 μs in default situations, while with scheduling enhancements is kept at a steady 100 μs.

6.4.3 'Memguard' to Boost KVM Guests on ARMv8

'Memguard' [234,235] is a memory bandwidth aware scheduler, it distinguishes memory bandwidth in two parts, guaranteed and best-effort. It provides guaranteed bandwidth for temporal isolation and best-effort bandwidth to use as much as possible the spare bandwidth [236] (after all cores are satisfied). 'Memguard' is designed to be used on actual systems using DRAM as main memory.

The common DRAM architecture consists of banks with different rows / columns. Therefore, maximum memory bandwidth can be achieved in the case where data are located in different banks, in other cases the memory bandwidth can be limited and to address this bottleneck a solution named 'Memguard' is used, which takes care of scheduling the memory bandwidth to provide the desired Quality of Service (QoS).

'Memguard' is implemented as a Linux kernel module, which is based on the use of the Performance Monitoring Unit (PMU). It captures the memory usage of each core by reading the Perfor-

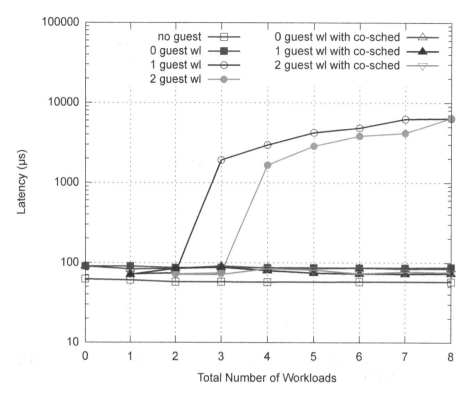

Figure 6.33: Interrupt Latency Results with CFS and 'Co-Scheduling' Mechanism [233]

mance Counter Monitor (PCM) (reading memory requests if used with PCM version up to 2.4 and memory reads and writes with more recent version of PCM).

The module architecture is based on two parts (see Figure 6.34), the first being the reclaim manager, which stores and provides bandwidth allocation to all per-core Bandwidth (B/W) regulators, while the other part is the per-core B/W regulator that monitors (thanks to the PCM) and regulates the memory bandwidth usage of each core. 'Memguard' is linked to physical cores, the regulation process works only at the core level. Due to this architecture, regulating a process running on several cores at once is not easily feasible. The 'Memguard' architecture can be described as follows:

- The global budget manager also known as reclaim manager: It handles the memory budget on each core of the CPU. Every scheduler tick ($1\ ms$) if the predicted budget of each core is under the assigned (fixed) budget of overall system, a memory budget tank is set to give more bandwidth during the future time slice if a task need to access to more B/W than required (and some B/W is available in the reclaim manager).

- The per core B/W regulator: It handles the memory management for each core, updating the actual used budget with the PCM value, configuring the PCM to generate an overflow when all memory budget is used and reclaiming more bandwidth from the reclaim manager if needed.

Beside the overall architecture, 'Memguard' has different features. Its major functionality is bandwidth management limiting, allowing a user to set a limit (in MB/s, 'weight' or in percent). Another feature is the per-task mode, it uses task priority as a core's memory weight. The last major feature is the 'reclaim bandwidth' functionality, distributed any leftover bandwidth that was not consumed. This last feature enables to use as much as possible memory bandwidth. When not in use, the available bandwidth is equal to the max-bandwidth setting set at start (or updated later).

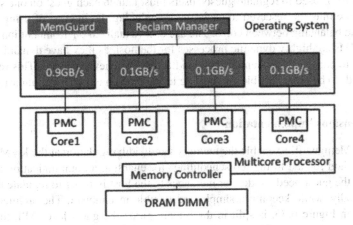

Figure 6.34: 'Memguard' Architecture Overview

'Memguard' can be used in different ways. The simplest use of 'Memguard' is to balance workloads, reducing the memory bandwidth of a task to preserve memory-bandwidth for others. 'Memguard' usage is linked to the physical cores of the CPU, consequently the application level use is complicated and must be done manually. 'Memguard' requires setting the bandwidth manually, as such users must be careful on which core, applications are running on, and adjust meaningfully each application's B/W needs.

Guest bandwidth problem

New kind of needs appear, requiring powerful embedded systems with a large number of connection interfaces. In the near future, most actual multi-chip embedded systems will be replaced by a central unit performing all computation and networking tasks. This embedded systems architecture direction raises the problem of mixed-criticality which is at the heart of the system. If a single platform is used to run different criticality software, some requirements are needed. Indeed, mixed-criticality means running some hard-real-time application with soft-real-time or standard application at the same time on the same processing unit. Also this kind of system, needs to provide security separation between tasks to ensure data / program isolation. Cooperation between hard and soft real-time processes pulls all the software interface to be more cooperative and resource need aware. Each program needs to exchange information in order to provide QoS.

Virtualization is the last component of future unified embedded systems architecture. VMs give the possibility to ensure the security and resource isolation between tasks. Each task, for example a video processing task (capture video from a sensor and process the image to find particular patterns) could be executed at the same time as video playback and / or a more critical task. Each task can then be executed in a separated VM with all the software needs and the correct amount of processing / memory bandwidth reserved.

In this context, the memory bandwidth management becomes the bottleneck of the system not only because all cores use the same memory but also because all different VMs are running simultaneously. Each VM handles a certain software environment, with a specific memory bandwidth need. Since with QEMU / KVM a virtual machine is seen as an additional task to schedule in the host, then the memory-bandwidth bottleneck becomes a limiting factor. Every guest is using the same memory bandwidth and no hierarchy is implemented (like in a CPU scheduler) between guests.

This memory bandwidth bottleneck can eventually affect the performance of guests in scenarios where memory is aggressively utilized.

When 'Memguard' is used to regulate guests, users must launch each guest on one specific core (or several but, a core must be reserved to each guest), reducing the interest of using Linux with KVMs, with the load balancing between cores. The use of a virtualized environment introduces also another use-case, VMs are highly dynamic processes for the host, as they have dynamic workloads and there is a need to change their memory bandwidth limit whenever needed. This results in the need for 'Memguard' to be more flexible and be able to regulate on a process granularity instead of cores.

'Memguard' Extensions Implementation

The extensions of 'Memguard' to enable the memory bandwidth regulation at the KVM guest level is similar to the method related to the 'co-scheduling' for enabling a communication from guests to the host. Indeed, the guest needs to deliver messages to the host in order to regulate the memory bandwidth dynamically, while keeping a simple and flexible mechanism. The architecture of the solution, described in Figure 6.35, is split in three main parts: the guest level API, the hypercall exchange mechanism and parts of 'Memguard' linked to Completely Fair Scheduling (CFS).

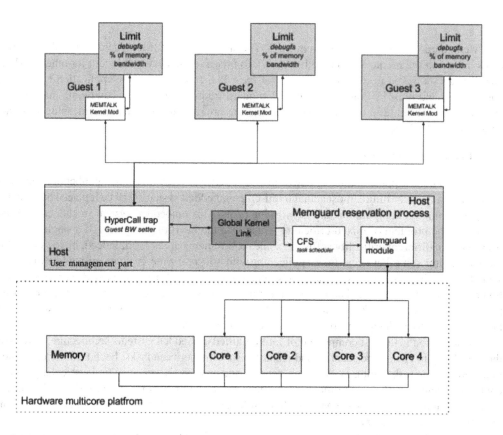

Figure 6.35: 'Memguard' with Virtualization Extension Overview [235]

The first part is composed of a simple 'debugfs' interface, allowing user to write / read from a simple file in order to set the needed memory-bandwidth value. Moreover, it allows setting the memory bandwidth from another program such as a local resources manager. Every call is made with:

- Request ID: Host is aware that this call is a guest request.

- Request type: Host is notified if guest wants to update the bandwidth or be removed from the guest reservation process.

- Value: A 64-bit variable to exchange information (e.g., bandwidth need 70%)

The second part is based on the same principle as I/O and task scheduling mechanism with the hypercall module [237], which is processed by KVM in the host. When the guest issues a hypercall, KVM traps it and the memory-bandwidth request is stored in the host kernel. The kernel structure for the information needed is composed of:

- 'Memguard' scheduling guests: Number of guests executed with 'Memguard' reservation enabled

- 'Memguard' scheduling PID: List of guests PIDs

- 'Memguard' scheduling B/W: Bandwidth request of guest

- 'Memguard' update bandwidth: A pointer to the 'Memguard' callback function

The third part is the mechanism, which regulates the bandwidth by applying the requested bandwidth. This part is composed of two components CFS (Linux scheduler) and 'Memguard' (kernel module) regulating memory-bandwidth at the core level. When CFS has scheduled the next task, a callback to 'Memguard' is executed, which then enforces the memory bandwidth regulation. The following pseudo code snippet is a method that calls 'Memguard' when the guest vCPU process is being scheduled:

```
function memguard_guest_update(cpu_number)
    if next_task = a_guest_in_the_list
        callback_to_memguard()
```

When CFS has scheduled the next task, a callback to 'Memguard' is executed, which then enforces the memory bandwidth regulation. It is also worth mentioning that 'Memguard' had to be also modified in order for it to handle the callback from CFS. This function in 'Memguard' updates the memory bandwidth of the core corresponding to the linked guest.

```
update_budget_sched(int cpu_n, long bw_n)
    convert_bandwidth_to_cache_event()
    set_the_core_budget()
    initialize_the_memguard_statistics()
```

Performance

This section provides some tests and benchmarks executed on the ARM Juno R1 platform, which show how the 'Memguard' extensions can be used to get better performance. Figure 6.36 shows the problem of the memory bottleneck. When two guests are running on the same core, both tasks are limited. As in the first set of tests, in the virtualized environment the bottleneck remains the same.

Figure 6.37 highlights the gains of the 'Memguard' extensions. Initially, both tasks are bandwidth limited, the first guest at 70% of the guaranteed bandwidth while the second guest at 20%. When the 'Memguard' module is disabled (between 13th and 20th second) the first guest can reach

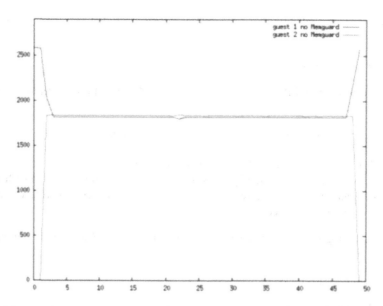

Figure 6.36: Memory Bandwidth Sharing between Two Guests (X Axis in Seconds, Y Axis in MB/s) [235]

the maximum bandwidth. After 20 seconds the first guest requests more bandwidth, which results for less bandwidth for the second guest. The interesting point is that both guests are executed with a different bandwidth percentage, which allows for a hierarchical differentiation between them.

Figure 6.38 demonstrates the memory separation between guests. The first guest initially is running without a limit, after 17 seconds, a limit is enforced, a second guest is launched after 33 seconds with a limited bandwidth. The memory-bandwidth fall is due to the CPU time shared between both guests (running on the same core). The extended version allows 'Memguard' to regulate memory bandwidth from a core level to the guest level.

Finally, the 'Mplayer' benchmark was done with a decoding process per guest. The results are following ones produced without the guest environment. The current implementation is giving at least the same results as standard 'Memguard'. The actual implementation has several benefits. The first one is the limited overhead due to a change in the memory bandwidth requested by the guest, as a hypercall is performed only when needed, reducing the total time processing the bandwidth modification. The second benefit relates to the use of the CFS scheduler. This significantly reduces the complexity of integrating the solution, and the overhead is kept to a minimum. The last benefit comes from the 'Memguard' callback, which provides memory bandwidth reservation and limitation functionalities.

6.4.4   Secure Monitor Firmware

This section describes the secure monitor firmware (i.e., VOSYSmonitor [238]) that has been developed in the context of DREAMS in order to consolidate mixed-criticality systems on a single multi-core heterogeneous platform.

6.4.4.1   Overview

With the emergence of multi-core embedded System-on-a-Chip (SoC), the integration of several applications with different levels of criticality on the same platform is becoming increasingly popular.

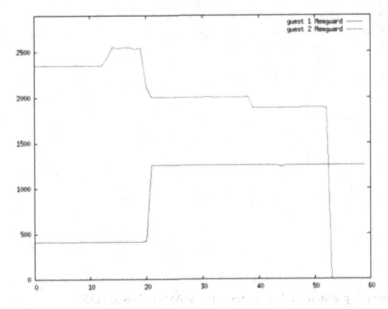

Figure 6.37: Memory Regulation between Guests (X Axis in Seconds, Y Axis in MB/s) [235]

Table 6.12: Different Rendering Times in Guests with Different Memory B/W Settings [235]

Memory bandwidth setting	Rendering Time
Plain Linux 2 cores executing the same video task (mplayer)	Guest 1: 62.112*s* Guest 2: 67.968*s*
'Memguard' with under estimated bandwidth: 20 MB/s on all cores	Guest 1: 386.893*s* Guest 2: 384.655*s*
'Memguard' with correct estimated bandwidth for core 0 (250 20 20 20)	Guest 1: 57.947*s* Guest 2: 312.014*s*
'Memguard' with correct estimated bandwidth for core 0 and best-effort policy	Guest 1: 60.911*s* Guest 2: 97.665*s*

These platforms, known as mixed-criticality systems, need to meet numerous requirements such as real-time constraints, OS scheduling, memory and OSs isolation.

To construct mixed-criticality systems, various solutions, based on virtualization extensions, have been presented where OSs are contained in a VM through the use of a hypervisor. However, such implementations usually lack hardware features to ensure a full isolation of other bus masters between OSs (e.g., Direct Memory Access (DMA) peripherals, Graphics Processing Unit (GPU)).

In this context, a 'secure monitor' firmware layer has been developed for the purpose of the Healthcare demonstrator. This software technology runs in the 'secure monitor' mode (EL3) of ARMv8-A processors, which is the highest secure operating mode available on ARM processors. As shown in Figure 6.39, it enables the native concurrent execution of two operating systems, such as for example a safety critical RTOS and a GPOS with the option to use virtualization extensions, such as Linux-KVM, in order to instantiate a variety of different non-critical VMs. Since VOSYS-monitor is a 64-bit monitor layer, it allows the execution of both 32-bit and 64-bit applications. By leveraging on ARM TrustZone [239], VOSYSmonitor guarantees peripherals and memory isolation between both OSs, while providing, at the same time, functions to enable a safe and secure communication between them. Therefore the RTOS, running in the 'secure world', is totally isolated from

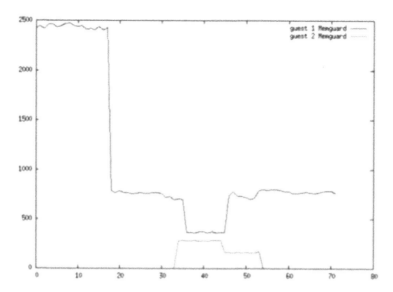

Figure 6.38: Memory Separation (X Axis in Seconds, Y Axis in MB/s) [235]

applications executing in the 'normal world' (see Section 6.4.4.1). Finally, VOSYSmonitor manages the context switching between the two worlds by triggering a Secure Monitor Call (SMC) instruction or by hardware exception mechanisms, such as interrupts. VOSYSmonitor oversees these exceptions in order to ensure a correct operation for each world.

TrustZone

ARM TrustZone is a hardware security extension, which provides a system-wide security approach by integrating protective measures into ARM processors, bus and system peripherals. The security of the system is achieved by partitioning the hardware and software resources in two compartments: the 'secure world' and the 'normal world'. The 'secure world' is usually used during the boot process in order to enforce a chain of trust. Indeed, starting with an implicitly trusted component, every other runtime binaries can be authenticated before their execution. In this context, some security specific configuration as well as sensitive data and peripherals can be only accessible from the Secure world. On the other hand, the 'normal world' is intended to host a rich operating system (e.g., Android or Linux). Security sensitive operations, such as the access to a private key or the interaction with a real time task, are provided to the non-secure application running in this compartment by the services run in the 'secure world'. Moreover, TrustZone enables a single core to safely and efficiently execute code from both worlds, allowing to save silicon area and power since a dedicated security processor is not needed.

While TrustZone is present since ARMv6 architecture [240], ARMv8-A provides a new architecture related to the TrustZone management. Indeed, a new highest Exception Level (EL) called EL3 (i.e., secure monitor mode) manages the interaction between these two compartments. This layer is always considered secure regardless of the current CPU state. Moreover, it manages the context switch between both worlds by triggering hardware exceptions, such as interrupts, synchronous and asynchronous events. The kernel level (EL1) and user level (EL0) are available in both worlds. This allows the execution of Trusted Execution Environment (TEE) in the 'secure world', while the 'normal world' is expected to run a rich OS with the option to use virtualization extension, such as Linux-KVM, since the hypervisor mode (EL2) is only available in the non-secure side.

The isolation provided by TrustZone is stronger than virtualization technology since this latter is

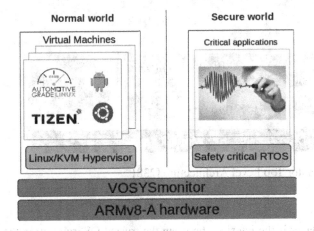

Figure 6.39: VOSYSmonitor Overview [238]

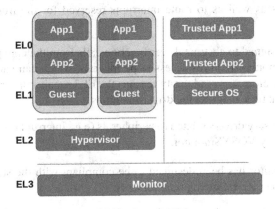

Figure 6.40: ARMv8-A Execution Level Overview [238]

restricted to the processor through the implementation of a hypervisor. Any other bus masters in the system, such as DMA peripherals or GPUs, can bypass the protections provided by the virtualization layer. In fact, ARM processors supporting security extensions along with TrustZone compliant Memory Management Unit (MMU) ensure the isolation of CPU execution mode, interrupts, hardware peripherals, memory and caches. However, the hypervisor can manage these peripherals for security reasons at the cost of introducing overhead.

6.4.4.2 Architecture

VOSYSmonitor is designed to ease the support of new hardware platforms, as well as the integration of critical operating systems in the 'secure world'. Depending on the hardware requirements, VOSYSmonitor can reuse software components, such as common peripheral drivers, in order to minimize the integration effort. The top level architecture of VOSYSmonitor is seen in Figure 6.41, where different sub-systems of the firmware are depicted, including:

- EL3 Monitor Layer, mostly coded in ARM assembly, is specifically implemented targeting the ARMv8 architecture. It handles the world context switch operations triggered by SMC instruction or hardware exception mechanisms.

VOSYSmonitor

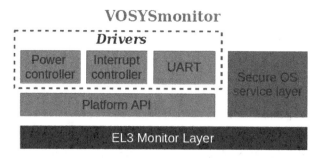

Figure 6.41: VOSYSmonitor Top Level Architecture [238]

- Secure OS service layer is the interface between the trusted OS, running in the 'secure world' (S-EL1), and VOSYSmonitor. The secure OS service layer is in charge of the secure context initialization before jumping on the trusted OS entry-point. Moreover, it is responsible for dispatching SMC services as well as to route interrupts reserved for the trusted OS during its runtime.

- The platform API is designed to abstract driver function calls from the core part of the monitor layer. Indeed, VOSYSmonitor requires access to peripherals, which can vary according to the hardware platform. Therefore, for modularity reasons the EL3 monitor layer uses these generic API functions, which, in turn, call specific driver functions.

- Drivers include all necessary drivers related to peripherals (e.g., interrupt controller, power controller), which are used by VOSYSmonitor.

In addition, VOSYSmonitor has been designed to be compliant with the stringent ISO 26262 certification as well as to meet the following requirements:

- Functional requirement: Safety critical RTOS and GPOS (e.g., Linux-KVM) co-execution on the same platform.

- Functional requirement: Isolation of safety critical OS resources (e.g., memory, peripherals) from GPOS illegal access.

- Functional requirement: Preserve the context of each OS during a switching operation.

- Performance goal: A strong requirement for the RTOS in automotive is related to the boot time, which must be limited even in a worst-case scenario. Since VOSYSmonitor adds a software layer before the execution of RTOS, it directly impacts the RTOS boot time. In this context, VOSYSmonitor setup must be achieved in less than 1% of the full RTOS boot time. For instance, a RTOS boot time of 60 ms implies a VOSYSmonitor setup which has to be performed in less than 600 μs regardless of the platform.

- Performance goal: The overhead added by the co-execution of software applications must be optimized to meet real-time constraints. In this context, the overhead due to the context switching in order to forward an interrupt to the RTOS, running in the 'secure world', has to be lower than 1 μs. This requirement is self-imposed and concerns a standard context switch where only the general-purpose registers and ARM systems registers are saved (see Section 6.4.4.2 for more details). This overhead can also be used for estimating the Worst Case Execution Time (WCET) of a task in the RTOS.

Context Switch

As ARMv8-A CPU execution is split in two parts ('secure world' and 'normal world') and some registers are not banked, VOSYSmonitor has to preserve the world context during the switching operation. This part is the most performance related function of the system, as execution and RTOS interrupt latency are directly affected by the time consumed for the context save / restore operations. For this reason, most of the code executed during the context switching is written in ARM assembly in order to reach the best performance. In the current implementation, VOSYSmonitor only saves the vital registers, such as general-purpose registers and some ARM system registers, needed for the correct RTOS / GPOS co-execution in each CPU operating mode. Indeed, for the test purpose, RTOS does not use advanced CPU features as the floating point context. However, a memory segment in the backup context structure is reserved to save additional registers during the context switching in order to extend VOSYSmonitor functionalities.

VOSYSmonitor periodically transfers the execution from one world to the other. As the RTOS/GPOS shares only a specific core, the context switch is currently tied to this processor and may take place in two main cases: an interrupt assigned to the 'secure world' is triggered during the execution of the 'normal world' application; one world requests a context switch (e.g., secure services, no RTOS workload) by calling VOSYSmonitor service through an SMC. In this context, it first saves the current state of the world suspended, then restores the state of the other world.

Scheduling Policy

On multi-core architectures, VOSYSmonitor is able to dynamically share a core between both worlds by operating under the assumption that the 'secure world' tasks should be prioritized over the 'normal world' execution. This means that once a core is assigned to the secure RTOS tasks, the 'normal world' applications can use it only if the RTOS, running in the 'secure world', has decided to release the core resource; something that happens when there is no real-time task to schedule. For this implementation, a minor update of the RTOS is needed in order to call VOSYSmonitor service, to schedule the 'normal world' execution, when the RTOS workload is null. Generally, this could be achieved through the creation of an 'idle task' with the lowest priority, which will be scheduled only if no other tasks need to be executed.

While giving the full priority to the 'secure world' may seem simple and an important impact on the 'normal world' execution, VOSYSmonitor has been implemented under the assumption that tasks performing in the 'secure world' take precedence over the 'normal world' execution. For instance, tasks whose failure to meet a real-time requirement could arise to a life-critical situation, should be executed in the 'secure world'.

Although multiple scheduling methods have been proposed for mixed-criticality systems [241], such a design has several benefits related to the execution of the safety critical RTOS. First of all, critical interrupts dedicated to the RTOS systematically preempt the 'normal world' execution in order to be handled by the RTOS with a minimum overhead (see Section 6.4.4.4), thus ensuring to meet real-time constraints. Moreover, there is no risk to preempt the RTOS execution during a critical operation since the RTOS releases the core only when it wants to. Finally, the core usage is optimized since VOSYSmonitor enables the scheduling of the 'normal world' application when the RTOS has no operations to execute. Indeed, real-time tasks are usually used to perform a brief action bounded in terms of time that could imply a low workload of the RTOS depending on the use case. However, this solution could lead the 'normal world' to the impossibility to run its applications if the RTOS execution monopolizes the core. This is mitigated through the assignment of other cores to the 'normal world' GPOS.

ARMv8-A architecture including the hardware security extension TrustZone has been designed to support two interrupt types: the Fast Interrupt Request (FIQ) for low latency interrupt handling, and the more general Interrupt Request (IRQ), which is commonly available also in other architectures. The former has higher priority (IRQs are automatically masked by the CPU core when an FIQ occurs) and can directly use some banked registers without the overhead of saving / restoring them.

VOSYSmonitor takes advantage of the ARM architecture [242], which offers the ability to trap IRQ and FIQ directly in its exception layer (EL3) without intervention of code in either world, thus allowing for the creation of a flexible interrupt management for safety critical RTOS tasks. Indeed, FIQs are considered as secure interrupts when TrustZone is supported, meaning that the configuration of FIQ cannot be altered by 'normal world' accesses. Therefore, VOSYSmonitor sets the 'secure world' (RTOS) to respond only to FIQs and the 'normal world' (GPOS) to handle IRQs. This design allows critical applications to benefit from fast and high priority interrupts, while isolating them from the 'normal world'. In addition, it minimizes the interrupts latency when the world is already scheduled since the interrupts are directly caught in the interrupt handler of the current world, avoiding any overhead due to VOSYSmonitor operations.

6.4.4.3   Services

VOSYSmonitor is a low latency software monitor layer developed for the 64-bit ARMv8-A architecture and it already supports the last generation of ARMv8 platforms such as:

- ARM Juno Development board (2 Cortex-A57 + 4 Cortex A-53) [243]

- Renesas R-Car H3 board (4 x A57 + 4 x A53): ISO 26262 ASIL B compliant [244]

- Renesas R-Car M3 board (2 x A57 + 4 x A53): ISO 26262 ASIL B compliant [244]

- NVIDIA Jetson TX1 board (4 x Cortex-A57 in big.LITTLE configuration) [245]

'Secure World' Failure Handling

As mentioned in Section 6.4.4.2, the 'secure world' execution is always prioritized over the 'normal world'. However in some scenarios, the secure RTOS might crash because of an internal failure and never giving the control back to the 'normal world'. To prevent such case, VOSYSmonitor proposes a 'secure world' failure handler, based on the ARM physical secure timer. The handler execution is similar to a watchdog: if after a specific timeout the 'secure world' has not given back the control, VOSYSmonitor preempts the core and restarts the 'secure world'. The handler timeout is reset at each context switch. However this feature is an option in VOSYSmonitor for two reasons:

- Huge overhead of the context switch: When performing a context switch from the 'secure world' to the 'normal world', the secure timer must be disabled. Likewise when returning to the 'secure world', the timer must be enabled and the timeout value calculated. It has been measured that adding the secure timer management doubles the context switch latency.

- All interrupts in the 'secure world' are FIQs: Only the interrupt corresponding to the secure timer is configured as an IRQ, so it can be trapped by VOSYSmonitor. However this means IRQs are no longer masked during the 'secure world' execution.

  When the system is mono-core, there are no issues as other IRQs can be disabled. Nevertheless, if the 'normal world' is running on a multi-core system, secondary cores might forward IRQs to the primary core which is in the 'secure world', causing undefined behavior and possibly a crash of the RTOS. This could be avoided by updating the GPOS source code however this is against the policy of VOSYSmonitor which should be able to run a GPOS with minimal modifications. Furthermore, this would create a vulnerability exploitable from the 'normal world'.

  In platforms supporting GICv3, interrupts are classified in three types, where one can be reserved to the VOSYSmonitor execution. In this configuration, it is possible to use the secure timer while retaining multi-core support for the 'normal world'. However all platforms currently supported are using GICv2. Thus for the time being the secure timer is only proposed as an optional feature.

ARM Convention Compliance

VOSYSmonitor is compliant with several ARM standards in order to ease the integration of this software component in a full system.

SMC Calling Convention (SMCCC) [246] specifies the calling procedure (e.g., registers used as parameters and return arguments) of the SMC instruction, used in the ARMv7 and ARMv8 architectures. This convention simplifies the integration between different software layers, such as operating systems, hypervisor and secure monitor. Moreover, it categorizes SMC service providers to allow the coexistence of services in the secure monitor firmware (e.g., ARM, OEM, trusted OS).

PSCI [247] defines a standard interface for power management that can be used by software working at different ARM privilege levels. During a power management operation, rich OS, hypervisor, VOSYSmonitor and trusted OS must not conflict each other. In this context, PSCI aims to standardize the communication between supervisory software to arbitrate power management requests. As a matter of fact, Linux kernel 'AArch64' relies on PSCI calls for powering up / down secondary cores avoiding the need of platform dependent drivers. VOSYSmonitor is compliant with the PSCI convention v1.0, which requests the support of mandatory functions related to the power management such as power up / down a core and suspend core execution.

6.4.4.4   Performance

In this section, the performance of VOSYSmonitor is benchmarked with the ARM Trusted Firmware (ATF). The evaluation uses the ARMv8 PMU [248]. The tests have been performed on the ARM Juno board R1, which plays the role of hospital server in the Healthcare demonstrator, and on the Renesas R-Car H3. Both boards have a Cortex A-57 and a Cortex A-53 cluster. As results may vary depending on the core where VOSYSmonitor is operating, all tests are performed on both A-53 and on A-57.

On top of VOSYSmonitor a bare-metal application is executed in the 'normal world', which constantly loops a NOP instruction in order to have a minimal impact on the test execution. On the other hand, the 'secure world' is hosting a FreeRTOS version 8.2.3 modified by Virtual Open Systems in order to use FIQ for interrupt processing. While it is not safety compliant, FreeRTOS is a widely used real-time kernel and is the base for an ISO 26262 certified RTOS called SAFERTOS [249].

Since VOSYSmonitor targets the automotive market, the compiler used to create the binary file should be qualified according to the ISO 26262 standard. For this reason, the ARM Compiler [250] with compilation optimization (-O2) is used since functional safety support package [251, 252] is provided with the version 6.6.1, thus avoiding further qualification activities when the recommendations and conditions are respected. Similarly, the upstream version of the ATF has been pulled from GitHub (v1.3) and also compiled with -O2 optimizations.

Context Switch Latency Using Performance Monitoring Unit (PMU)

VOSYSmonitor aims at having small impact on the co-execution of the 'normal world' and the 'secure world'. The context switch is the most critical aspect in the VOSYSmonitor execution, therefore it is of utmost importance to implement it in an efficient way. To assess the time induced by a context switch from the 'normal world' to the 'secure world', a timer in FreeRTOS is configured to generate an FIQ every 1 μs and then gives the control to the non-secure world. On Juno, the timer used is the SP804 dual-timer, on R-Car H3 the 'compare match timer'.

The timer interrupt is triggered while the processor execution is in the 'normal world' and thus it is trapped in VOSYSmonitor vector table. The PMU counter is started there and stopped before giving the control back to FreeRTOS, then the number of clocks cycles is displayed. The same test is performed with the ATF in order to compare the results. Adding the PMU breaks the code execution, therefore, the board has to be manually powered on / off after each measurement inducing a small sample size.

Although VOSYSmonitor includes the interrupt handler described in Section 6.4.4.2, the context switch latency has also been measured without this handler in order to gain time execution and assess the optimum value possible. This implementation makes that any FIQ trapped, when the processor execution is in the non-secure world, causes a request for a context switch. While lacking flexibility, this implementation may be of interest for scenarios where the RTOS response must be as fast as possible.

Figure 6.42 shows the context latency expressed in microseconds on the Juno board and Figure 6.43 for the Renesas R-Car H3 board. The results for the Juno board, shows that, in the worst case (VOSYSmonitor with interrupt handler on Cortex-A57), VOSYSmonitor is 118% faster than ATF and almost twice as fast if VOSYSmonitor without interrupt handler is executed on an A-53 core. On the R-Car H3, the comparison is in favor again of VOSYSmonitor, which can be more than thrice faster than ATF. However, it should be noted that the ARM trusted firmware used here is a software updated by Renesas, and thus may include features not present in the upstream version. VOSYSmonitor aims at having a context switch faster than 1 μs. Although on Cortex-A53 the requirement is met on VOSYSmonitor with or without an interrupt handler, a context switch on Cortex-A57 barely misses the prerequisite on Juno.

In this measurement, the interrupt handler called by VOSYSmonitor is a minimal implementation, which consists in fetching the pending interrupt ID, then after comparison, jumps to the context switch macro. Therefore, a decrease in performance can be expected if a more complex interrupt management is requested. However, the same can be said for ATF and the latency difference between both ATF and VOSYSmonitor should remain as it is.

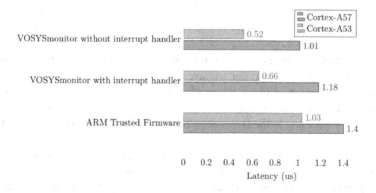

Figure 6.42: Juno Context Switch Latency [238]

Context Switch Including the Hardware Latency

Although the previous measurement allows to measure the exact number of clock cycles consumed during a context switch, a second test has been performed to overcome some limitations of the first one. Indeed, VOSYSmonitor is running with both instructions and data caches enabled, which drastically improves performance. However, caches misses can cause a decrease in performance in worst-case scenarios, which can not be estimated unless the sample size is big enough.

In this context, a second performance test has been performed in order to take into account the FIQ latency induced by the hardware, i.e., the time between an FIQ triggering and the beginning of

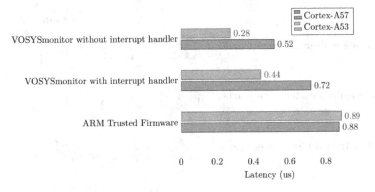

Figure 6.43: R-Car H3 Context Switch Latency [238]

its processing by the software, on an important sample size. To achieve this, the ARM EL1 physical timer is used to trigger an FIQ while the core is in the 'normal world' as in the previous test. The ARM timer is composed of three registers:

- CNTP_CTL_EL0: Used to enable / disable the timer.

- CNTP_TVAL_EL0: Contains the current value of the timer.

- CNTP_CVAL_EL0: Holds the compare value. When equal to CNTP_TVAL_EL0, the interrupt is triggered.

When the context switch occurs, the timer interrupt trapped in VOSYSmonitor is forwarded to the FreeRTOS vector table. There, the very first instruction consists in reading the value of CNTP_TVAL_EL0, the timer is disabled and we jump to FreeRTOS handling routine. Finally, VOSYSmonitor latency between an FIQ triggering and the interrupt handler can be deduced by subtracting from CNTP_CVAL_EL0. Since VOSYSmonitor code has been untouched, we are able to run this test for as long as necessary and without any instrumentation of the code execution.

As before VOSYSmonitor is tested with and without the interrupt handler. The results with a sample size of 2048 contexts switch are presented on Figures 6.44 and 6.45. By taking into account the FIQ latency, it confirms that VOSYSmonitor is still faster than the ATF version in all scenarios. Moreover, it is better, in terms of performance, to have the RTOS executing on an A-53 core since the context switch is at least twice as fast as an A-57 for VOSYSmonitor. In this test, an issue has been the inability to run FreeRTOS and a non-Secure OS in co-execution with ATF. However, the full context switch with hardware latency has been extrapolated by adding the previous results of ATF measured in Section 6.4.4.4 and the FIQ hardware latency shown in Table 6.13.

Table 6.13: FreeRTOS Fast Interrupt Request Latency [238]

Platform	Processor	Average (ns)	Min (ns)	Max (ns)
Juno	Cortex-A53	220	180	280
	Cortex-A57	220	200	260
R-CarH3	Cortex-A53	240	120	360
	Cortex-A57	240	120	360

VOSYSmonitor and ATF Booting Time

The performance analysis also compared the setup time of VOSYSmonitor and ATF. The setup time

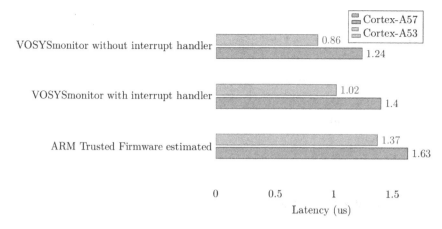

Figure 6.44: Juno Board Context Switch with Hardware Latency [238]

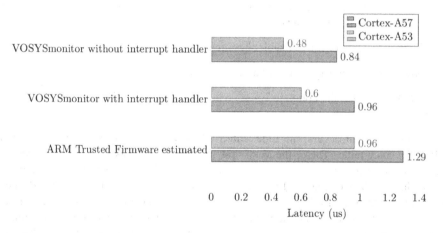

Figure 6.45: R-Car H3 Board Context Switch with Hardware Latency [238]

corresponds to the time consumed between VOSYSmonitor / ATF entry point and the first jump to the 'secure world'. In this test, PMU has been used since the Generic Interrupt Controller (GIC) and the timers are not yet configured. As for the previous tests, the measurements have been performed on both A-53 and A-57 cores. Table 6.14 shows that on Juno VOSYSmonitor is able to achieve its setup configuration within 18 μs, while ARM trusted firmware needs around 870 μs on Cortex-A53 and 920 μs on Cortex-A57. VOSYSmonitor is able to outperform ATF because the page tables, used by the memory management unit, are defined by the developer and included statically during the compilation, whereas ATF generates the pages during the setup time.

Although it requires more effort and reduced flexibility, it is worth the trade-off since VOSYS-monitor impact is significantly reduced during the setup and meets the 600 μs boot time requirement presented in Section 6.4.4.2.

OSs Co-Execution Workload and Multi-Core Support

RTOS / GPOS co-execution is ensured by sharing a core between these two OSs. The full priority is given to the RTOS, running in the 'secure world', in order to meet real-time requirements. With

Table 6.14: VOSYSmonitor and ARM Trusted Firmware Setup Time [235]

Platform	Processor	ARM Trusted Firmware (*us*)	VOSYSmonitor (*us*)
Juno	Cortex-A53	853.480	17.746
	Cortex-A57	1119.815	31.004
R-CarH3	Cortex-A53	1661.806	44.493
	Cortex-A57	1119.815	31.004

this implementation, the GPOS execution can be impacted depending on the RTOS workload. In order to measure this impact, the *hackbench* [253] benchmark has been used in three scenarios:

- Linux standalone.

- Linux + FreeRTOS low workload. FreeRTOS requests a context switch every 1 μs and gives back the control immediately after.

- Linux + FreeRTOS 60% workload. The CPU will spend around 60% of the runtime executing FreeRTOS code.

Table 6.15 shows that adding FreeRTOS with a low workload has no impact on the performance, whether using one or more cores. However, a deterioration can be noticed when FreeRTOS is executing 60% of the time on a core. By executing the test on a multi-core configuration (four Cortex-A53 and two Cortex-A57), we observe that the performance degradation is minimal between the Linux standalone and the 60% workload scenarios. Although a core is busy executing 'secure world' application, the other cores are nonetheless able to continue performing.

Table 6.15: VOSYSmonitor Hackbench Result [235]

Linux one core			
Hackbench(s)	Linux standalone	Linux + FreeRTOS low workload	Linux + FreeRTOS 60% workload
	3.411	3.431	14.251
Linux multi-core			
Hackbench(s)	Linux standalone	Linux + FreeRTOS low workload	Linux + FreeRTOS 60% workload
	0.500	0.498	0.596

7

Chip-Level Communication Services

M. Grammatikakis

TEI of Crete

H. Ahmadian

Universität Siegen

M. Coppola

STMicroelectronics

S. Kavvadias

TEI of Crete

A. Mouzakitis

Virtual Open Systems

K. Papadimitriou

TEI of Crete

A. Papagrigoriou

TEI of Crete

P. Petrakis

TEI of Crete

V. Piperaki

TEI of Crete

M. Soulié

STMicroelectronics

G. Tsamis

TEI of Crete

7.1	Bandwidth Regulation Strategies in Linux	318
	7.1.1 Genuine MemGuard Principles and Extensions	319
	7.1.2 Genuine vs. Extended MemGuard (MemGuardXt)	319
	7.1.3 NetGuard Extension (NetGuardXt)	322
7.2	Hardware MemGuard: Bandwidth Control at Target Devices	323
	7.2.1 Limitations of Hardware MemGuard	324
	7.2.2 Architecture of the Hardware MemGuard	325
	7.2.3 Synthetic Traffic Evaluation	325
	7.2.4 NoC-Based Evaluation	327
7.3	Hardware Support at Network-on-Chip Level	330
	7.3.1 STNoC Implementation of Address Interleaving	330
	7.3.2 Evaluation Framework: Performance and Power Consumption	332
7.4	Mixed-Criticality Network-on-Chip	333
	7.4.1 Support for Mixed-Criticality	333
	7.4.2 Support for Heterogeneous Communication Paradigms	333
	7.4.3 Overall Architecture	333
	7.4.4 Network Interface	334
	7.4.5 Core Interface Using Ports	335
	7.4.6 Mixed-Criticality Unit	337
	7.4.6.1 Operational Description	337
	7.4.6.2 Integration of Time-Triggered and Event-Triggered Traffics	338
	7.4.7 Back-End	339

This chapter elaborates on chip-level solutions in mixed-criticality systems, which are categorized in memory and network bandwidth regulations, address interleaving and support for mixed-criticality at the Network On Chip (NoC) level. In Section 7.1, MemGuardXt and NetGuardXt are introduced as memory and network bandwidth management policies to improve communication-intensive memory-bound applications. Section 7.2 elaborates on the architecture and evaluation of the hardware MemGuard and in Section 7.3 address interleaving at the NoC-level is discussed. In Section 7.4, requirements, architecture and operational specification of the time-triggered extension layer at on-chip Network Interface (NI) are described. This hardware building block serves as an extension to STMicroelectronics NoC (STNoC) to support heterogeneous types of communication in mixed-criticality systems.

7.1   Bandwidth Regulation Strategies in Linux

Memory and network bandwidth management policies can improve communication-intensive memory-bound applications running on distributed embedded multi-core systems-on-chip. In particular, memory bandwidth management schemes can allocate memory resources to applications in a fair manner, avoiding local saturation or monopoly phenomena, while network bandwidth regulation apply packet monitoring and control to avoid filling the network capacity and to utilize the available budget more efficiently. Moreover, by combining these policies with CPU bandwidth scheduling and the provided functionality by the Linux kernel, it is possible to develop a holistic approach for the system resource management.

While previous research on bandwidth regulation has relied on specialized hardware to successfully manage shared resources, e.g., at network interface and memory level [254–256], we concentrate on operating system support without additional hardware. In critical hard real-time operating systems deployed in space, transportation or medicine, it is obligatory for certification reasons to completely avoid interference. In less critical systems running Linux, it is often enough to ensure that such disruptions are not harmful. Thus, existing regulation policies in Linux manage interference so that critical application tasks in a mixed-criticality environment achieve a sufficient and predefined performance.

MemGuard [85,234] provides an algorithm to perform memory bandwidth management at CPU-level by using hardware performance counters to monitor the number of last-level cache misses, or accesses via the memory interconnect. We present an extension to MemGuard algorithm (known as MemGuardXt) which supports a Violation Free (VF) mode for placing hard guarantees on the traffic rate, and improves prediction of future bandwidth requirements using Exponentially Weighted Moving Averages (EWMA). In addition, by improving code modularity, we allow the core of our MemGuardXt algorithm, defined as a self-contained IP, to be used in one or more Linux Kernel Module (LKM) instances.

Notice that network bandwidth regulation can apply packet monitoring and control to avoid filling the network capacity and more efficiently utilize available budget. Hence, relying on the same methodology, we implement two different LKMs, a memory bandwidth regulation module running MemGuardXt, and a new network regulation module on top of netfilter (NetGuardXt) that uses a similar algorithm to MemGuardXt.

In Chapter 11, the LKMs in correlation to real-time are evaluated, using an actual mixed-criticality use-case with two types of tasks, a distributed embedded soft real-time Electrocardiogram (ECG) processing application and an incoming best-effort video traffic. This evaluation focuses on how simultaneous fine-grain control of network/memory bandwidth via NetGuardXt/MemGuardXt can improve the QoS of soft real-time ECG application when MemGuardXt operates in VF mode, rather than in the Non-Violation Free (Non-VF) mode defined originally in MemGuard.

7.1.1 Genuine MemGuard Principles and Extensions

Genuine MemGuard allows sharing guaranteed bandwidth over several cores using a dynamic reclaim mechanism. Using this mechanism, cores are allocated at the beginning of each period, a part (or all) of their assigned bandwidth (according to history-based prediction) and donate the rest of their initially assigned bandwidth to a global repository (called G). Then, during the period, a core may obtain additional budget from G based on the past traffic demand (history) and residual guaranteed bandwidth. This self-adaptive reclaim mechanism avoids over-provisioning, improves resource utilization and is similar to extended self-adaptive Dynamic Weighted Round-Robin (DWRR) [257].

Since guaranteed memory bandwidth within a period under worst-case (r_min) is much smaller than maximum attainable bandwidth (e.g., usually 20%), MemGuard allows Best-Effort (BE), i.e., traffic in excess of r_min. Thus, once all bandwidth has been exhausted within a period, MemGuard supports two methods to generate BE bandwidth, which refers to the bandwidth used after all cores (i in total) have utilized all their assigned budgets, before the next period begins. First, it allows all cores to freely compete for guaranteed bandwidth, posing regulation until the end of the period. Second, it applies sharing of BE bandwidth proportionally to reservations. There is no explicit provision for BE traffic sources. As long as r_min is not exhausted, genuine MemGuard allows sources with a zero reservation (or sources that have otherwise exceeded their reservation), to repeatedly extract guaranteed bandwidth from G, up to the configurable minimum allocation (Q_{min}).

7.1.2 Genuine vs. Extended MemGuard (MemGuardXt)

The genuine MemGuard algorithm [85, 234] targets the average bandwidth instead of peak bandwidth reservation, which limits its use in real-time applications. More specifically, a Rate-Constrained (RC) flow may steal guaranteed bandwidth from other RC flows and even exhaust the global repository, while other RC-flows have not yet demanded their full reservation potentially leading to guarantee violations. Although the genuine MemGuard supports a reservation-only mode that removes prediction and reclaiming and allocates to RC traffic sources their full reservation in each regulation period, this mode performs poorly, since it cannot retrieve budget from ($r\_min - \Sigma Qi$), if $\Sigma Qi < r\_min$.

The proposed MemGuardXt algorithm provides a hard guarantee option on the bandwidth rate (called VF), which is important for real-time applications. This extension is implemented in `overflow_interrupt_handler` and restricts reclaiming from G by one or more rate-constrained cores, in case it can lead to a guarantee violation for another RC-flow within the same period. More specifically, for the VF mode, as shown in Figure 7.1, we first calculate the donated bandwidth to G and then serve only from the remaining rate-constrained bandwidth that will not be needed by any core (and not the whole amount).

```
int donated = 0;
if( mg_input.VF == true )
{
    for(int count=0; count<mg_input.i; count++)
        if( (mg_state.qi[count] < mg_state.Qi[count]) && (mg_state.crti[count] == true) )
            donated += mg_state.Qi[count] - mg_state.qi[count];
}
if( mg_input.VF == false || ( donated + min( mg_input.Qmin, mg_stats.G ) <= mg_stats.G ) )
    mg_state.qi[core] += min( mg_input.Qmin, mg_stats.G );
```

Figure 7.1: MemGuardXt Violation Free Mode, for State Definitions also Refer to Figure 7.2

Finally, note that the genuine MemGuard supports limited adaptivity for predicting memory bandwidth requirements. MemGuardXt supports a general EWMA scheme, which computes a weighted average of all past periods based on parameter (λ), which determines the impact of history. EWMA prediction is pre-calculated for each core when a new period starts using the following equation:

$$z_t = \lambda * x_t + (1 - \lambda) * z_{t-1},$$

where $t > 1, 0 \le \lambda \le 1, z_1 = x_1$, Z_t is the predicted bandwidth for the next period $(t + 1)$ and x_t is the consumed bandwidth from the core at the end of the current period (t). This formula better adapts to traffic perturbations between small bandwidth fluctuations and abrupt changes (e.g., periodic data sent from sensors), while its computation costs are similar to those in MemGuard.

```
struct MG_Input {         struct MG_State {         struct MG_Stats {          struct MG_Metrics {
    int i;                    int *Qi;                  int *access_stats;        int Int_counter;
    int Qmin;                 int *qi;                  int *z;                   int Ui_counter;
    int r_min;                int *Qi_predict;          int *z_prev;              int BE_counter;
    int period;               int *ui;                  int *x;                   int GV_counter;
    bool VF;                  bool *crti;               int Lamda;                }mg_metrics;
}mg_input;                 }mg_state;                int period_current;
                                                     int previous_period;
                                                     int period_unit;
        (a)                       (b)               int G;       (c)                    (d)
                                                     }mg_stats;
```

Figure 7.2: MemGuardXt Input Parameters, System State, Statistics and Metrics

Unlike the genuine MemGuard, the implementation of the proposed MemGuardXt module is on x86_64 and ARM platforms (32 and 64-bit) and modular and supports multiple parallel instances. Its core functionality is implemented as a separate self-contained package (.c and .h files) written in ANSI C. As shown in Figure 7.2, MemGuardXt supports four data structures:

- MG_INPUT data structure with input parameters $i, Q_{min}, r\_min, period$, and VF; this info can be dynamically modified via `debugfs` and update action is taken when the next period starts.

- MG_STATE with initial, current, predicted and total used bandwidth and a criticality flag (set to true for RC traffic and false otherwise) $Qi[], qi[], Qi\_predict[], ui[]$, and $rc\_flag[]$. If left uninitialized, Qi for all cores automatically takes the value of r_min/i. MemGuard algorithm distinguishes between RC and BE cores using the $rc\_flag[]$ array, which denotes the criticality level of each core. Notice that both RC and BE cores can consume guaranteed and generate best-effort traffic.

- MG_STATS is related to EWMA prediction algorithm with $z_t, z_{t-1}, x_t, \lambda, previous\_period(t-1), current\_period(t), period\_unit$, and G. Notice that for optimization reasons, it is called from the kernel mode, EWMA bandwidth prediction is implemented using integer numbers only (instead of double).

- MG_METRICS with the number of interrupts, i.e., requests for reclaim, used bandwidth (from all cores), best-effort bandwidth, and number of guarantee violations. An example is where guaranteed bandwidth has been donated and already consumed by others. Notice that if VF is set, then $GV = 0$.

The above data structures and functions are used from an LKM, which provides the necessary monitors and actuators, e.g., timers, cache metrics and throttle mechanisms for MemGuardXt, or timers, bandwidth metrics and accept/drop functionality in NetGuardXt. Basic interactions between MemGuardXt LKM (left) and the core algorithm (right) are shown in Figure 7.3. During module insertion and removal (`insmod`/`rmmod`), kernel driver functions `init_module` and `cleanup_module` invoke core functions for initialization and memory cleanup. Periodically, `Prediction_EWMA` function is called to update the bandwidth consumed by each core and `periodic_timer_handler` resets statistics and reassigns the estimated bandwidth per core. This information is extracted from MemGuardXt core by calling `make_traffic` from `period_timer_callback_slave` when the period starts; it is increased on-the-fly by asynchronous calls to `make_traffic` from `MemGuard_process_overflow` function of LKM, which also informs `Prediction_EWMA` that previously assigned bandwidth is already consumed.

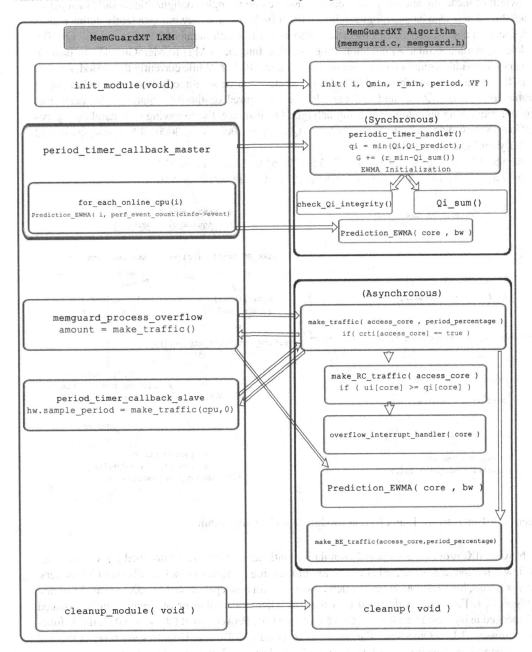

Figure 7.3: MemGuardXt: Linux Kernel Module and Core Algorithm

7.1.3   NetGuard Extension (NetGuardXt)

NetGuardXt is an incarnation of extended MemGuard as an LKM that uses custom netfilter hooks, the packet filtering framework built around Linux kernel `sk_buff` structure to drop, accept or buffer network packets. This module allows independent kernel-level monitoring and control of network bandwidth of incoming and outgoing network flows using a single configurable module (a separate kernel thread may also be used for each flow). Each such instance may independently define source or destination client Intellectual Properties (IPs) and bandwidth parameters, e.g., r_min, Qi. The module supports a similar Application Programming Interface (API) to MemGuardXt to provide network bandwidth regulation on Linux on x86_64 and ARMv7. While currently the period, number of traffic sources per interface and EWMA parameters can be set directly from the module, other parameters (r_min, Q_{min}, and Qi) can also be configured on-the-fly, separately for each flow direction (incoming and outgoing) using debugfs. For instance, the following command configures NetGuardXt outgoing traffic to r_min, Q_{min}, Q_0, $Q_1 = 7000, 500, 3000, 400$ bytes/period (and similarly for incoming traffic).

```
echo ''7000 500 3000 4000 15000 500 5000 10000''
> /sys/kernel/debug/netguardxt/netguardxt_config
```

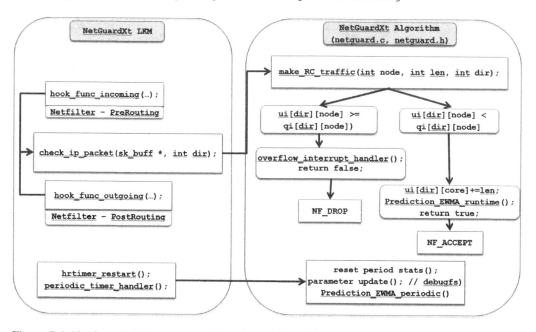

Figure 7.4: NetGuardXt: Linux Kernel Module and Core Algorithm

NetGuardXt provides instant and cumulative statistics of accepted or dropped packets or bytes per flow direction and connected client. Without focusing on Linux kernel details (network drivers, netfilter hooks, high resolution timers, debugfs etc), main concepts of NetGuardXt are as illustrated in Figure 7.4. Each packet destined to a network client (incoming or outgoing) can be counted and checked using `bool make_rc_traffic` function. Packet is sent (NF_ACCEPT), if this function returns a TRUE. Otherwise, the packet is dropped (NF_DROP). Counters are reset at the end of each period via function `period_timer_handler()`. A high resolution timer (`hrtimer`) implements the period. Similar to MemGuardXt, the functions `Prediction_EWMA_periodic` and `Prediction_EWMA_runtime` are used to adjust the predicted bandwidth per client. When requested bandwidth exceeds the given "budget" `overflow_interrupt_handler` is called to reclaim unused bandwidth from the global repository where donations may occur.

7.2 Hardware MemGuard: Bandwidth Control at Target Devices

Multi-Processor Systems-on-a-Chip (MPSoCs) offer tremendous potential in embedded systems, due to combining computational capacity and energy efficiency. However, current MPSoC architectures have significant limitations in supporting safety-critical applications in space, avionics and transportation industry [258]. Thus, modern shared memory-based MPSoCs must efficiently integrate mission-critical and non-critical subsystems on a same platform to support Quality of Service (QoS) features that provide latency and bandwidth guarantees in an automated way [259].

In this section, we focus on the design of a hardware bandwidth regulation module, which is implemented by adapting and extending the algorithms of MemGuardXt LKM. The hardware module is placed in front of a memory controller to differentiate among NoC sources with rate-constrained and best-effort traffic provisions and enhance support for mixed-criticality applications on MPSoCs. Similar to MemGuardXt, it supports a violation free-guaranteed operating mode for rate-constrained flows and dynamic adaptivity through EWMA prediction. Using NoC-independent, C++-based statistical simulation, we show improvements of our module over an equivalent hardware adaptation of genuine MemGuard without our extensions. Furthermore, using annotated SystemC modules of extended MemGuard module (based on synthesis), we evaluate hardware MemGuard at the memory controller of a NoC-based System-on-a-Chip (SoC) model using an MPEG4 traffic model and show similar improvements. The module hardware cost (obtained using synthesis from Xilinx Vivado HLS and Vivado) is comparable to an ARM AMBA AXI4 and much smaller than a 4x4 STNoC instance [260].

In this context, certain modern SDRAM controllers are highly predictable in terms of memory performance (bounding the execution time), since they operate based on pre-computed SDRAM patterns at design time (static schedules) [261, 262]. In general, these controllers suffer from poor performance and limited flexibility, which restricts them to a small set of applications [256, 263].

The introduced hardware MemGuard module is placed in front of the memory controller and aims at awareness of regulation decisions about concurrent RC and BE traffic sources, to prevent stealing of RC bandwidth, without removing reclaiming, including in VF mode. Furthermore, we detect overbooking by RC sources, to allow more BE traffic in its place and target more gradual bandwidth reclaiming to allow adaptive targeting of *maximum* bandwidth requirements of RC sources. Finally, we extend traffic prediction, by exploiting more general EWMA, which computes a weighted average of all past periods based on a parameter lamda (λ), which determines the impact of history. For each CPU core, when a new MemGuard period starts, EWMA is calculated using the following formula:

$$S_t = \lambda \cdot Y_t + (1 - \lambda) \cdot S_{t-1}, for \ t > 1, 0 \leq \lambda \leq 1 \ and \ S_1 = Y_1$$

where S_t is the consumed bandwidth moving average and Y_t is the actually consumed bandwidth, for some core in the tth regulation period. We use S_t as the prediction for period $t + 1$.

Focusing on the input interface of the hardware MemGuard module (at receive-side of a target device), we note that excessive traffic must be blocked from certain sources in order to prioritize other sources. Otherwise, if a source is requesting excessively more than its reserved memory bandwidth, its packets will appear with a disproportionately high frequency at the on-chip network input interface of the hardware component (i.e., the memory controller). MemGuard tries to postpone the majority of these packets, but it also needs to dequeue packets from the NoC, in order to advance RC traffic of other sources. Note that such "misbehaving" sources require large amounts of buffering within our bandwidth regulating component (holding pending requests); otherwise, it results in unbounded delay for other RC traffic.

To resolve this issue, we assume that all traffic sources in our system have a maximum number of outstanding accesses and correspondingly, the bandwidth regulating module provides per source buffering of that amount. This allows postponing packets of a traffic source that exceeds its reservation for as long as necessary. Such per-source buffering is required for a hardware MemGuard implementation at the receive-side of the bandwidth-regulated system, regardless of whether we use the genuine or the extended algorithm.

Usually, core caches employ a small number of Miss Status Holding Registers (MSHRs), e.g., see [264], which limits the maximum number of outstanding accesses. This implies an access protocol that requires either a response or an acknowledgment, to release initiator resources. On the flip side, core uncached accesses and device DMA (usually, to uncached memory) are seldom implemented with a maximum number of outstanding accesses, as initiator resources can be released immediately after the access is issued. Thus, the proposed hardware bandwidth regulating design will not work correctly in systems that support such memory traffic sources, without an outstanding access upper bound (see also Subsection 7.2.1). Nevertheless, the proposed design supports an independent and relatively long queue for traffic sources of zero reservation, which can tolerate a number of accesses from sources without a maximum outstanding access limit, without compromising bandwidth regulation.

7.2.1 Limitations of Hardware MemGuard

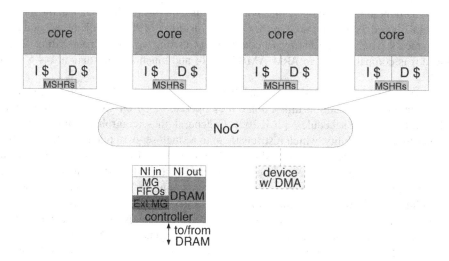

Figure 7.5: Hardware MemGuard Component System Placement

Figure 7.5 represents the location of a hardware MemGuard component in a quad-core system. MemGuard is placed after the input network interface of the DRAM controller to manage the order, in which packets are delivered to the controller. Although such a component can be used to regulate access to other types of memory (e.g., in Section 7.2.4 is used for SRAMs), it is designed for DRAM bandwidth management. Per-source buffering First In First Outs (FIFOs) are presented as an independent part of our extensions. Figure 7.5 also shows a potential DMA-capable device without an outstanding access upper bound, which our design can tolerate to a high degree.

In addition to such devices, there are some other DMA devices that potentially introduce limiting factors for the effectiveness of a hardware MemGuard component. Per-regulated-source buffering depends on the regulated sources having an independent NoC access point, so that they can be identified in the request packet header (to allow for an acknowledgment) and be classified separately. In case the access point of the NoC is shared between the cached cores and uncached cores, a special provisioning in the per-source buffering FIFOs component will be required to identify the protocol. It is also important that NoC access over shared caches or coherence directories with placement or protocols that mask, modify, or completely remove the regulated source information from DRAM accesses, will be more complicated and can even make the design of a hardware MemGuard module unattainable.

7.2.2 Architecture of the Hardware MemGuard

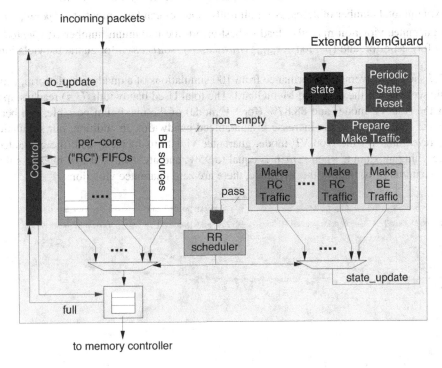

Figure 7.6: Block Diagram of the Extended MemGuard

Figure 7.6 represents the block diagram of the hardware module for the extended MemGuard. There is one FIFO per regulated source (serving an RC flow), plus a BE-traffic FIFO (8 and $8 \cdot number\_of\_sources$ packet-entries respectively in the implementation and evaluations of following sections). In each cycle of a regulation period, the algorithm accumulates per FIFO state data (in the "Prepare Make Traffic" block) and computes the conditions to pass a packet from each of the FIFOs (in the "Make RC/BE Traffic" blocks); for details on implementation of Make RC/BE Traffic and relative hardware cost (from Vivado HLS and Vivado tools), see [260]. A Round-Robin (RR) scheduler iterates over Make Traffic decisions to pass or postpone a FIFO's top packet, but only for non-empty FIFOs and selects the first positive one. State affected by the Make Traffic decision is then updated. At the beginning of a new regulation period, per FIFO state is reset and EWMA state is updated (by the "Periodic State Reset" block).

7.2.3 Synthetic Traffic Evaluation

We consider the relative performance of our C++ models of Genuine vs. Extended MemGuard hardware modules for different synthetic traffic configurations (uniformly random, Variable Bit Rate (VBR) and bursty) arriving from two or more "hypothetical" cores. We consider both Violation Free (VF) and Non-Violation Free (Non-VF) modes. Each core reserves $r_{min}/number\_of\_sources$, except for the one, which has a zero reservation.

For improved accuracy in our simulation experiments, we assume $n = 10240$ periods and repeat each experiment for $m = 100$ times. The regulation period duration is $period = 20$ units (i.e., up to

20 accesses can be forwarded to the memory controller during a period). Hence, the total number of simulated memory accesses by all cores is: $(number\_of\_sources \cdot period \cdot n \cdot m)$. All MemGuard performance metrics are normalized in a range from 0 to 100%, by taking into account:

- The maximum number of bandwidth reclaims: $period \cdot number\_of\_sources \cdot n$;
- The maximum total number of accesses for all traffic sources can be computed as: $period \cdot n$;
- Since a guarantee violation may also lead to best-effort, the maximum number is: $(period - r_{min}) \cdot n$ for best-effort and $(period - r_{min}) \cdot number\_of\_sources \cdot n$ for guarantee violations.

Figure 7.7 shows the average performance from 100 simulations of a quad-core platform, using a VBR traffic scenario for the extended MemGuard. The total Used bandwidth (U_i) reaches up to 99.02% (for the Non-VF mode) and 88.87% (for VF mode) of the maximum possible, and bears an almost linear relationship to the increase in guaranteed bandwidth. In addition, our simulation experiments reveal that for the Non-VF mode, guarantee violations increase with the guaranteed bandwidth reaching maximum when r_min is equal to 95%, and are always at low rates less than 10% of the maximum possible. For the VF mode, there are zero guarantee violations.

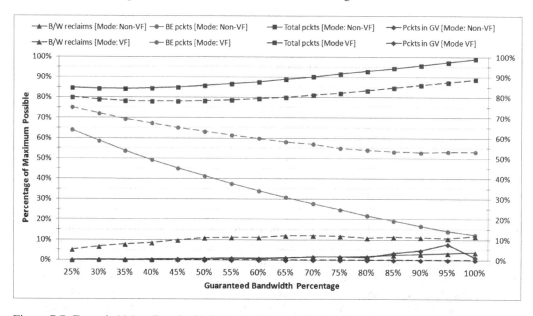

Figure 7.7: Extended MemGuard with Variable Bit Rate Traffic (Quad-core)

Similarly, for the genuine MemGuard (graph omitted due to space limitations) we observe the following:

a Used bandwidth is almost similar in the Non-VF mode, but much lower on average in the VF mode.

b Bandwidth reclaim requests are much higher (e.g., 35% vs. 12.03% in VF mode).

c Best-effort packets are less by 10% in Non-VF and 20% or more in the VF mode.

d The number of guarantee violations is similar.

Figure 7.8 considers bursty traffic and compares performance of extended versus genuine Mem-Guard. Results indicate that extended MemGuard continues to improve compared to genuine algorithm for a larger number of cores. As it manages a smaller number of bandwidth reclaims (15 to 35

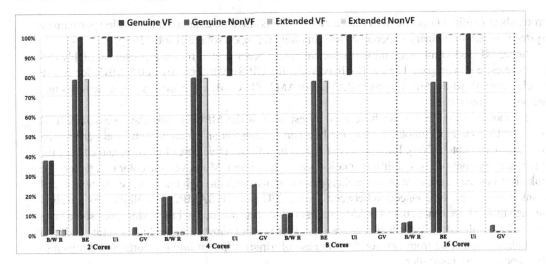

Figure 7.8: Max/Min Ranges for Bandwidth Reclaim (B/W R), Best-Effort (BE), Used Bandwidth (Ui) and Guarantee Violations (GV) in Violation Free and Non-Violation Free Mode for Genuine (Left Pair of Bars) and Extended (Right Pair of Bars) MemGuard for 2, 4, 8 and 16 Cores

times less), has higher best-effort traffic rate (up to one order of magnitude), utilizes almost 100% of the provided bandwidth, and generates a comparable number of violations for the Non-VF mode; notice that for simplicity in Figure 7.7, only minimum and maximum bounds are shown for the same r_min range. Simulations based on random and VBR traffic have drawn similar conclusions.

7.2.4 NoC-Based Evaluation

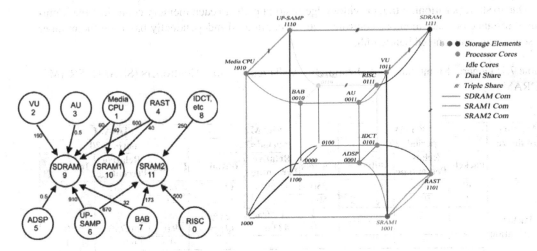

Figure 7.9: Mapping the MPEG4 Task Graph (Shown on the Left) on a Custom 16-Node Hypercube NoC, where a Separate MemGuard Instance is Located at each Memory Component (SRAM1, SRAM2, SDRAM)

By adapting the hardware MemGuard to a custom on-chip network traffic management, we are able to differentiate RC and BE traffic using NoC virtual channels, or a special tag in the packet, enabling fine granularity traffic classification to applications.

Our evaluation examines the performance of genuine versus extended MemGuard hardware protocols at multiple target devices (memory controllers) that are connected to a cycle-approximate SystemC model of a binary hypercube NoC topology (cf. open source HSoC [265]).

In the MPEG4 graph, nine initiators (active cores) generate and transmit packets at highly different rates, e.g., compare 1580 MB/s bandwidth for UP-SAMP (core 6) to 0.5 MB/s for ADSP blocks (core 5), while three targets (SRAM1, SRAM2, SDRAM) are passive storage elements that are only receiving data.

As shown in Figure 7.9, we have used a best map of the MPEG4 task graph resources (IPs) onto the 16-node NoC topology using off-line partitioning via an efficient high-level partitioning tool known as Scotch [266]. Idle cores are marked in yellow, active cores in brown, and each of the three memory controllers in a different color (green, red, blue). Moreover, the color of a hypercube link identifies the destination of the packets that are transferred through that link, i.e., if the color of a link is blue (e.g., the connection between Media CPU and UP-SAMP), then all packets traveling through this link are sent to the SDRAM memory device (i.e., the one in blue). Moreover, some links are marked with a double (or triple) *share sign*. This symbol shows that the link is actually shared by packets originating from two (or three in case of triple) different initiator cores and destined to the same memory controller.

In this optimized mapping, based on the underlying e-cube shortest-path routing algorithm, each memory controller receives packets through edge-disjoint paths. More specifically, paths directed to any memory controller are distinct from the paths to any other memory controller. This aspect provides static partitioning at NoC-level and allows for the following two important properties:

- Non-interaction of the traffic flows directed to different memory controllers.
- Consequently, based on the underlying MPEG4 traffic model, predictability of the traffic flow arriving at each memory controller. In this case, certain links are actually shared by packets originating from different initiator cores, which are destined to the same memory controller (cf. dual or triple shares in Figure 7.9). However, this buffering issue does not actually affect MemGuard performance.

Due to static partitioning that provides edge-disjoint paths to each memory controller, the Mem-Guard instance on each memory controller can be configured independently based on the number of sources and guaranteed bandwidth.

Table 7.1: Generic MemGuard Initialization for All Three Memory Controllers (SRAM1, SRAM2, SDRAM)

Memory	SRAM1			SRAM2			SDRAM		
Initialize	period=20			period=20			period=20		
	Packets	Relative Percentage	Initial Q_i	Packets	Relative Percentage	Initial Q_i	Packets	Relative Percentage	Initial Q_i
Packet Distribution	40	50%	$Q_1 = 2$	250	15.69%	$Q_1 = 1$	600	33.46%	3
	40	50%	$Q_2 = 2$	500	31.39%	$Q_2 = 3$	60	3.35%	1
				173	10.86%	$Q_3 = 1$	32	1.78%	1
				670	42.06%	$Q_4 = 3$	910	50.75%	6
							0.5	0.03%	1
							190	10.6%	1
							0.5	0.03%	1
Total	80	100%	$\sum Q_i = 4$ guaranteed	1593	100%	$\sum Q_i = 8$ guaranteed	1793	100%	$\sum Q_i = 14$ guaranteed

More specifically, assuming an EWMA parameter $\lambda = 0.2$ (to emphasize history) and a period of 20 quanta (time slots) for each of the memory controllers in MPEG4, we use reservations of

- 4 time quanta for SRAM1 IP (receives from 2 sources);
- 8 quanta for SRAM2 IP (receives from 4 sources); and
- 14 quanta for SDRAM IP (receives from 7 sources).

Notice that the number of sources corresponds to the number of "make_RC_traffic" blocks in each memory controller. Initial reservation of the quanta to different traffic flows at each memory controller are based on the MPEG4 rates as shown in Table 7.1. Thus, the initial reservations for the traffic sources of SDRAM IP are (3, 1, 1, 6, 1, 1, 1), with 6 corresponding to UP-SAMP and 3 to RAST (the two maximum rates). Notice that, since EWMA is in place, algorithm bandwidth reservations quickly adapt to the actual rates, especially since a small period (20 quanta) is used for all cases.

Table 7.2: Performance at Different Memory Controllers (MPEG4)

NonVF(Genuine, Extended)				
Memory	**B/W Reclaim**	**BE**	**GV**	**Delay (MG periods)**
SRAM1	(136, 62)	(21, 69)	(2, 4)	
SRAM2	(3661, 1167)	(1082, 2309)	(192, 153)	300, 205
SDRAM	(2691, 2490)	(210, 2444)	(66, 56)	
Total	(6488, 3719)	(1313, 4822)	(260, 213)	
VF(Genuine, Extended)				
Memory	**B/W Reclaim**	**BE**	**GV**	**Delay (MG periods)**
SRAM1	(441, 55)	(15, 31)	(0, 0)	
SRAM2	(22302, 2685)	(16, 28)	(0, 0)	1698, 1675
SDRAM	(14410, 3016)	(0, 8)	(0, 0)	
Total	(37153, 5756)	(31, 67)	(0, 0)	

More specifically, our SystemC simulation uses MPEG4 traffic with more than 60k packets; there is very small difference (1-2%) from large sets of experiments with up to 998k packets. Results shown in Table 7.2 shows that when using MPEG4 task graph to generate traffic, extended MemGuard generates a smaller number of bandwidth reclaims (so-called interrupts) and more best-effort traffic (in both Non-VF and VF mode) resulting in an improved bandwidth rate (routing all packets with a smaller delay), while also generating less guarantee violations in Non-VF mode than the genuine version. Finally, we note that the actual value of λ (set to 0.2 in all experiments) does not affect the performance since MemGuard is able to adjust to MPEG4 traffic rates very quickly.

We have also performed experiments involving distributed database transactions based on parallel equi-join operations. Equi-joins are implemented using parallel hashing of keys, which results to an all-to-all NoC communication pattern, whereas the length of each vector depends on the distribution of the keys and the hash function. In our model, eight initiators make parallel accesses to eight different target memories via the hypercube NoC to access 2x5k records, resulting in a total of 80.000 RC packets being transmitted to all memory controllers (with ideally 10.000 packets per memory controller). In this scenario, we use a period of 20 time quanta, for all MemGuard configurations and for guaranteed bandwidth all possible values correspond to a minimum "8" to the maximum of "20" quanta. Our simulation results show similar behavior characteristics as MPEG4, with Extended MemGuard producing more best-effort traffic and less bandwidth reclaims in Non-VF mode than the genuine version, while the number of guarantee violations is larger. Moreover, both genuine and extended MemGuard take almost the same simulation time to successfully schedule all packets.

An interesting open issue remains to evaluate the precise interplay between the proposed hardware mechanisms and existing QoS techniques at the network interface and router layer in NoC-based multi-core SoCs, such as ST Microelectronics' STNoC. Moreover, using extended MemGuard as the core algorithm, it is possible to experiment with bandwidth regulation at network interface- and router-level.

7.3   Hardware Support at Network-on-Chip Level

When considering a NoC-based MPSoC, bandwidth regulation policies interact with address interleaving and QoS policies at the network interface. Address interleaving balances transactions across memory nodes, allowing for efficient use of NoC bandwidth, improved scalability and reduced congestion. In particular, *address interleavng* alters the traffic pattern and thus also the performance of bandwidth regulation policies for all uncached accesses. Interleaved memory was initially proposed to compensate for the slow speed of Dynamic Random Access Memory Device (DRAM), by distributing memory addresses evenly across memory banks. In this way contiguous memory requests are headed towards each memory bank in turn, i.e., they are accessed sequentially, one at a time. This results in higher memory throughputs due to hiding the memory refresh time of DRAMs, i.e., reduced waiting time for the memory banks to become available for write or read operation.

Since the 1970s, preliminary interleaved schemes have been integrated in different generations of commercial systems, such as IBM 360/85, Ultrasparc III and Cray-Y MP. Nowadays, the industry attempts to incorporate a large number of low power DDR controllers - known as channels - [267] that will be accessed in burst by multiple initiators, such as CPUs, Direct Memory Access (DMAs), or other peripherals. A channel is accessed depending on the transaction address through a memory map. Having different channels allows balancing the network load amongst the channels. This is realized with an address interleaving scheme implemented in the Initiator's Network Interface (INI). An important parameter set during configuration time is the interleaving step, i.e., the amount of network packets sent per transaction by an initiator prior to changing destination memory node. This corresponds to the number of bytes to be written before changing memory node.

In fact, several commercial multi-core SoCs support bank interleaving. For example, Intel regularly issues best practices for optimized DDR memory performance in systems integrating e.g., Xeon processors and servers, which support interleaving. These specifications require an even number of banks per channel, an even number of ranks per bank and an odd/even number of channels depending on memory size. Moreover, the optimal number of ranks is related to the number of Dual In-line Memory Modules (DIMMs) and the operating frequency of the memory subsystem [268]. In addition, Arteris [269] and Sonics [270] integrate memory interleaving with their interconnect solutions, however they do not provide detailed experimental results. Sonics SonicsGN-3.0 NoC provides limited results, only for two channels using off-chip DDR3 memories.

7.3.1   STNoC Implementation of Address Interleaving

STMicroelectronics NoC (STNoC) is a configurable interconnect, whereas an integrated interactive EDA design flow (composed of I-NoC, Meta-NoC environments with GUI, low and intermediate-level device driver generators and validation subsystems) is used to design a custom cost-effective network-on-chip solution on a single die that may include optional hw/sw components at the Network Interface (NI) and possibly NoC layer. These components jointly implement advanced system/application services via a set of low-level primitives. For instance, optional STNoC services include end-to-end QoS via a fair bandwidth allocation protocol implemented at NoC layer and target NI, firewall mechanisms implemented at the initiator NI, or fault tolerant routing by implementing on-the-fly post-silicon routing function re-programmability at the NI. In this context, ST has recently implemented native support of address interleaving for QoS reasons.

STNoC internally uses source-based routing, i.e., the routing path depends only on the source address of the request transaction. Address interleaving is a QoS feature implemented in the shell (IP-specific subsystem) of the initiator network interfaces, on top of the NoC layer which provides the routing service, including packetization and size/frequency conversion. Address interleaving is able to modify the current STNoC memory map settings. If instantiated at STNoC configuration time, it can be switched on/off and reprogrammed at runtime through a set of registers. When address interleaving is disabled, then the current classical address decoding applies.

Notice that in order to guarantee flexibility and data integrity, all INIs must be able to route traffic consistently towards the same address space at runtime. Hence, safe reprogramming of address interleaving registers across all related INIs is supported via a coherent reprogramming system driver, which allows an external *secure code* to stop incoming traffic, reprogram all related INIs, wait for reprogramming to end and then restart NoC traffic.

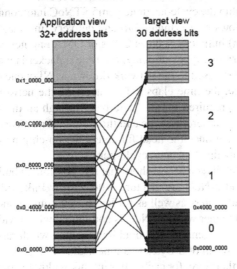

Figure 7.10: Address Interleaving in STNoC

While a classical routing function associates an address to a routing through a memory address range, STNoC address interleaving is based on defining at configuration time which memory region will be part of an Interleaved Memory Region (IMR); STNoC supports up to 16 IMRs. If the incoming transaction address falls in an IMR, the associated routing path is computed based on the current address, configured interleaving step and number of channels for the given IMR. As a result of this computation, an appropriate routing path is selected, leading to one of the channels constituting the interleaved memory set. Ordering issues related to request transactions heading to the same destination are assumed to be resolved at IP-level.

Figure 7.10 illustrates the operation of STNoC address interleaving. Different pages (corresponding to access requests from an initiator) are mapped towards different memory controllers connected to a different node (each color on the right side corresponds to a different node); in fact, this specific case corresponds to 1 IMR with 4 channels of $1GB$ each.

STNoC address interleaving is implemented by spreading the memory space into different memory nodes, with each node equipped with a different memory controller. The memory controllers are usually located in contiguous nodes. Notice that the number of channels and their memory map location are fixed parameters intrinsic to the system. Another solution available is that requests to addresses can change in turn towards different banks belonging to the same memory controller, located on a node; this feature comes with restrictions and has not yet been tried in post-synthesis.

7.3.2   Evaluation Framework: Performance and Power Consumption

This section gives an overview about the setup and environment, on which we evaluate the address interleaving scheme. We aim to achieve high performance without consuming unnecessarily silicon resources and power. Tailoring the system wisely could result in a well-balanced setup, e.g., a 32-byte link-width for the NoC might not be necessary if an 8-byte link-width is sufficient. We vary the link-width and keep fixed the number of Virtual Channels (VCs) across our experiments, equal to 2. The router's buffer size was set equal to the number of flits comprising a packet. On the other hand, we set large queues for the NIs in order to avoid potential overflows when new data arrive; this is unrealistic but it prohibits packet dropping and allows us to find the time needed to serve a packet in the worst case and also assess saturation cases by measuring the time a packet remains in the NI queue prior to releasing it into STNoC. We measure the NoC delay at flit- and packet- level, then derive the throughput and power consumption for different cases.

We perform experiments within the cycle-accurate gem5 STNoC interconnect simulation framework published in [271] that allows measuring end-to-end delivery times and power consumption. We connect to this framework, a) initiators, such as CPUs or DMA engines, and b) targets, such as memory controllers, and measure two types of delay; the time a packet is waiting in the NI queue before it is split up into flits and released to the network (known as NI queue delay), and then, after the head flit leaves the NI queue, the time elapsed for traversing the network until it reaches the destination. The latter is the time required for the flit to pass through all the routers in its path and reach the target NI. This way me measure the time and throughput for transferring the entire packet (read/write requests); we recall that once the head flit enters the on-chip network, the next flits of the packet are transmitted in a pipelined manner, cycle-per-cycle.

Our experiments (for more details see [272]) show that it is possible to build large-scale systems with even over 32 nodes, around a NoC architecture with narrow link width. This accounts for considerable savings in wires and routers as well as for less power consumption. We have observed that router's total power is much higher when STNoC is configured with a wide link. This is mainly due to the router's clock power; as a generic rule, doubling the link width results in a double clock power. Thus moving from a link width of 8 Bytes to a link width of 32 Bytes, results in a 4x power overhead at the router clock. Furthermore, for each different link width, the power increases slightly with the amount of channels. In terms of RTL cost, the typical configuration implementation adds an overhead of 5% when synthesized in 28nm FDSOI ST technology.

As a future extension, the gem5 STNoC simulation environment and related methodology can be used for high-level configuration of address interleaving parameters. For instance, in use cases with specific and well-defined communication patterns, this framework can be used to explore and auto-generate different address interleaving features, such as the number of IMR, the default IMR start/end addresses in the memory map, the offset, number of channels and default step value for each IMR. In this respect, our framework can extend the STNoC configuration toolsets, i.e., I-NoC and Meta-NoC. More specifically, it can interface with existing IP mapping (and partitioning) tools able to configure the connectivity memory map of IPs to STNoC network nodes and define the corresponding routing paths and NI look-up tables for embedded applications with well-defined access patterns.

The precise interplay of hardware MemGuard with SoC and NoC QoS policies and address interleaving is an interesting topic that remains unexplored. The gem5 STNoC cycle-accurate framework is very useful for exploring trade-offs across system- and network-on-chip policies.

7.4 Mixed-Criticality Network-on-Chip

This section introduces the DREAMS solutions for mixed-criticality NoCs. The first two subsections (i.e., 7.4.1 and 7.4.2) introduce the requirements of an on-chip interconnect in Mixed-Criticality Systems (MCSs). Then in Subsection 7.4.3 the overall architecture of the DREAMS NoC is described and the following subsections describe the internal building blocks and the operational specification of the NI. In particular, Subsection 7.4.6 describes how the NI establishes a segregation between different criticalities and supports different communication paradigms.

7.4.1 Support for Mixed-Criticality

In mixed-criticality SoCs, application components of different criticality levels interact and coexist on a shared chip. A fundamental prerequisite for this integration is that safety-critical components are free from interferences, with respect to other safety-critical as well as low-critical components. Otherwise, low-critical components have the potential to cause a failure of safety-critical ones and the entire system would have to be certified to the highest level of criticality, which is economically and technically infeasible for low-critical components. Isolation of interference at the chip-level is technically achieved by partitioning [273], both in temporal and spatial domains.

A temporal separation of resource allocations enables us to use the same resource for different components without affecting each other. In the DREAMS proposed solution, this is achieved by dividing the bandwidth network between the safety-critical and low-critical messages in a time-triggered manner. In addition, safety-critical messages are transmitted based on a predefined schedule to exhibit a determinism behavior.

In addition to temporal partitioning, a spatial separation of the resources allows to eliminate interferences, if more resources are available (e.g., routers, memory size). In the proposed solution, messages of different criticalities are separated at the NI level to achieve simpler router architecture. Hence, the concept of *source-based routing* [274] is employed, in which the sender of the message determines the path, through which the message traverses the routers to reach the destination. On-chip paths are a segment of the *virtual links* (cf. Section 2.1.3.2) and are defined at the design time.

7.4.2 Support for Heterogeneous Communication Paradigms

As discussed in Section 2.1, support for different timing models is demanded in mixed-criticality systems. Periodic injection of *Time-Triggered (TT)* messages is desired in safety-critical applications, as it offers predictability, determinism and inherent fault-isolation [275]. However, in this communication paradigm an efficient resource-utilization cannot be achieved, as the resources need to be reserved based on a *predefined schedule*.

As apposed to TT messages, *Best-Effort (BE)* messages are preferred for low-critical applications, in which a failed or delayed delivery of a message does not result in a catastrophe. As the best-effort messages require no resource reservation and no timing restrictions, this type of communication supports legacy-application and provides an efficient bandwidth utilization.

Rate-Constrained (RC) communication paradigm [276] has been introduced to address a reasonable trade-off between bounded worst-case latency and resource sharing by defining the *rate-constraints* as well as the *priorities*.

7.4.3 Overall Architecture

Figure 7.11 demonstrates a typical SoC in mixed-criticality systems, in which an NoC establishes a message-based communication for the connected *tiles*. From the architectural point of view, each tile

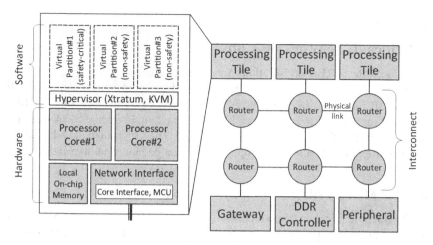

Figure 7.11: Example Network-on-Chip in MCSs, © IEEE 2016, Reprinted with Permission from [12]

can host a single-core or multi-core processor, an IP core, e.g., memory controller (to be shared by several other tiles) or a chip-to-cluster gateway, which is responsible for the redirection of messages between the NoC and the off-chip communication network.

In case several parallel application components run on a single processing tile, the computational power can be virtualized by a *hypervisor* [277] (cf. Chapter 6), in order to isolate the software components. Hypervisors are widely used in mixed-criticality systems to provide components of different criticality levels with isolated execution environments [224] termed as *virtual partitions* (dashed boxes on top of the hypervisor in Figure 7.11).

The first building block of the interconnect is the NI, which serves as an interface for the internal components of the tile (e.g., the processor core, the memory controller, etc.) by receiving and delivering the messages. In addition, the NI generates the message segments (known as packets and flits), in case a message needs to be injected into the interconnect. At the opposite direction, i.e., for the incoming messages, the NI assembles each message from the received segments and stores it to be read by the connected tile. In addition to the above functions, in mixed-criticality applications, the NI needs to address the challenges that are addressed previously (i.e., Subsections 7.4.1 and 7.4.2). This is performed by Mixed-Criticality Unit (MCU), which is placed inside the NI (highlighted box in Figure 7.11) and is described in this section.

The second building block of the interconnect is the on-chip router, which relays the messages from a source to a destination, each of which can be an NI or an adjacent router. The number of input and output units at each router and the connection pattern of these units represent the topology of the on-chip network (e.g., star, ring, spidergon). In the proposed solution, the *source-based control* approach has been used to keep a simple architecture for the routers. As a result, the routers do not need to care about the criticalities, but only the priorities. Physical links are the third building blocks of the interconnect, which act as a glue element among the NIs and the routers and realize the physical interconnection among them.

7.4.4   Network Interface

The building blocks of the proposed architecture for a mixed-criticality network interface are illustrated in Figure 7.12. On the left hand-side, virtual partitions show how the computation power of the processor core(s) is virtualized by the hypervisor. A hardware wrapper (e.g., AXI or OCP)

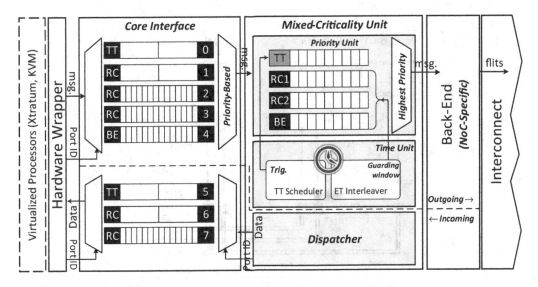

Figure 7.12: Building Blocks of the Proposed Architecture

at the core side provides the physical access to the memory-mapped ports that are inside the *core interface*.

The MCU contains the time unit, the priority unit and the dispatcher. The *time unit* is the only element of the NI that understands the concept of global time and triggers the time-triggered activities, such as the transmission of the periodic messages, based on the given schedules. In addition, this unit maintains the guarding windows, in case timely blocking is used. The *priority unit* contains *priority queues* and enables the NI to inject messages of the highest priority when permitted by the guarding windows.

The back-end has no understanding about the concept of temporal and spatial partitioning and performs solely the NoC-specific operations and is comparable with the NIs of existing NoCs.

The remainder of this section elaborates on each of these buildings blocks.

7.4.5 Core Interface Using Ports

The core interface provides the cores with an access to the NI by offering the *ports* and *registers*. Each port provides a memory-mapped interface between the processing cores and the NoC and stores the messages until they are delivered to the NoC or fetched by the core. Ports are isolated physical memories that establish spatial partitioning at transport layer. They are used to maintain the isolation that is established at the application layer by the hypervisor (cf. Figure 7.13).

Ports are classified as *input* or *output*, from the point of view of the application. Output ports store the messages that are written by the application running on the core. Based on the temporal constraints and the priorities, which are defined by the configuration of the port, the message is injected into the NoC and delivered to the NI of the destination tile. Once the message is delivered at the destination side, it is stored at an input port by the dispatcher to be read by the respective core.

Each port is composed of three internal units, i.e., data area, port configuration and port status units, which are briefly described below:

Data Area

The data area is implemented by a memory that stores the messages. Based on the direction of the port, the memory is either written to or read from by a driver at the application layer. Depending on

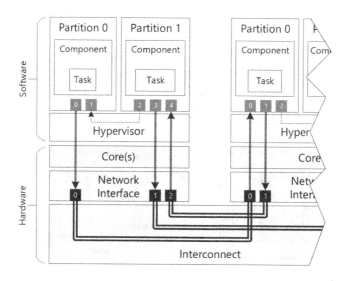

Figure 7.13: System-Wide Closed Encapsulated Communication Channels, © IEEE 2016, Reprinted with Permission from [12]

the semantics of the message, the data area can be implemented either as a single buffer or a queue. As the *state messages* need not be queued, a buffer is used for the data area. Since the buffer is accessible by the cores as well as the NI, a synchronization mechanism such as a semaphore, or non-blocking write [278] is employed, in order to avoid conflicts between the read and write operations. Unlike state messages, newer *event messages* must not overwrite the older ones, but rather they need to be accumulated. This is technically done by queue-ing event messages using FIFOs. Hence, the access to the data area can be performed without any synchronization mechanism.

Configuration Unit

Each port is configured by a set of parameters, a part of which can be updated at run-time. Configuration parameters are associated with individual ports and include the design configuration parameters such as the port ID, the direction (i.e., input or output), the semantics (state or event), the traffic type (i.e., TT, RC and BE), the timing parameters depending on the traffic type, the priority and the message size.

The configuration unit enables the application layer to have a limited access to a subset of the configuration parameters, e.g., the stored path to the destination and the Minimum Inter-arrival Time (MINT) value. However, the remaining configuration parameters (e.g., port direction, port size) are statically initialized and can only be read by the internal building blocks of the NI.

Status Unit

Status registers contain the statistical information about each port and are exposed to the application. Status flags provide the application with information, such as the number of available messages at the port, if the port is empty or full. Moreover, this unit collects the error logs, which reflect the faulty behavior of the application. For instance, in case the application enqueues an over-sized message or at a rate, which is higher than the allowed rate (based on the schedule or the rate-constraints).

7.4.6 Mixed-Criticality Unit

The challenge in MCSs is to establish safety along with efficient resource usage, where safety is established by segregation and efficient resource utilization is achieved by sharing the resources. In the proposed solution, the Mixed-Criticality Unit (MCU) establishes temporal partitioning by supporting a collision-free transmission of TT messages. Moreover, an efficient resources utilization is achieved by exploiting the remaining bandwidth for low-critical subsystems. As shown in Figure 7.12, the MCU contains three building blocks, the time unit, the priority unit and the dispatcher.

The time unit is the only unit, which is aware of the *global time base* and performs the run-time scheduling of the NoC resources based on a predefined *schedule*. This unit performs the periodic transmission of TT messages and traffic shaping of RC messages. If there is remaining bandwidth, it also forwards BE messages.

This unit is composed of two independent schedulers, the TT scheduler and the Event-Triggered (ET) interleaver. The TT scheduler ensures that there is no interferences for messages of the safety-critical subsystems by imposing temporal constraints. The ET interleaver establishes a temporal separation of time-triggered and event-triggered messages by preserving the guarding windows.

The priority unit aids the time unit by sorting the messages based on priorities and enqueue-ing the messages into the respective priority queue. The highest priority is dedicated to TT messages, as they shall not incur any delay. This is achieved by the guarding windows that are kept based on a predefined dedicated schedule.

The dispatcher is used for the incoming message at the destination NI to forward the message to the respective input port. This unit is triggered whenever a new message arrives at the NI. It retrieves the destination port from the message and enqueues it into the respective port. The message waits at the port until the respective application running on the core reads the port. In case of an event-triggered message (i.e., RC and BE message) an interrupt can be sent to the core in order to notify the core about the arrival of a new message. In case of a time-triggered message, there is no need for a notification from the NI, since the core will read the port at the predefined instant in its schedule.

7.4.6.1 Operational Description

The functionality of the MCU can be classified into the following three aspects:

Periodic Transmission of TT Messages

Transmission of TT messages adheres strictly to the principles of the TT communication paradigm [279]. This communication paradigm provides *a priori* known knowledge about the instant, at which the message arrives at the routers and the destination, thereby establishing a deterministic communication system for safety-critical subsystems.

These kind of operations are performed according to a predefined configuration in temporal and spatial aspects. Temporal in the sense that each operation is triggered at defined instants given by the *TT schedule*. The schedule is a circular linked list of entries, each of which contains the ID of the port that contains the message and the phase (i.e., the deviation from the beginning of the period), at which the message is injected into the interconnect. Spatial partitioning for safety-critical messages is achieved through a predefinition of the following spatial configuration, the port, from which the message is dequeued, the entire path, through which the message is relayed by the routers and the port, into which the message is enqueued at the destination.

In order to establish a chip-wide collision-free communication of TT messages, the operation of the schedulers of different NIs shall be harmonized by a *global time base* to have a common understanding of the time, despite different clock domains. Otherwise, two different messages might meet each other at the same resource, due to the clock drift at one of the sender nodes. The global time base is a low-frequency digital clock [280], which is based on the concept of sparse time base [192] and is generated through an internal or external clock synchronization (e.g., TTEthernet).

This clock provides a system-wide *notation of the time* between different components of the system and enables them to interpret the schedule and deadlines for events within the system.

Traffic Shaping of RC Messages

The RC communication paradigm aims at establishing a well-shaped flow of data. These messages are characterized by two parameters; the MINT (also know as BAG [10]) and the jitter. Successive RC messages are guaranteed to be offset by a configured MINT, which is imposed by the scheduler due to insufficient bandwidth [10]. Jitter is defined as the maximum admissible deviation from the MINT.

The functionality of the MCU for RC messages is inspired by the traffic shaping function described in ARINC Specification 664P7 [10]. The aim of the traffic shaper is to control the bandwidth according to the given MINT in the port configuration. Consecutive RC message at each port can not be injected into the respective priority queue if the MINT is not elapsed. Furthermore, sufficient bandwidth must be allocated such that the introduced delays and temporal deviations (jitter) meet the defined limits.

Relaying of BE Messages

Since BE messages possess the lowest priority within the system, there is no guarantee whether or when the BE messages can be delivered. Hence, they are injected into the interconnect only if there is a remaining unused bandwidth by TT and RC messages.

In contrary to TT and RC messages that the destination of all messages of each port is predefined at the design time by the parameters, individual BE messages can have their own destination address. As a result, the given destination by the application layer is decoded at the port and used by the back-end for the routing information.

7.4.6.2   Integration of Time-Triggered and Event-Triggered Traffics

Sharing the communication resources between TT and ET (i.e., RC or BE) traffics requires a particular segregation mechanism, as the ET messages can delay the TT traffic. This is due to the fact that each ET message blocks the resource for a limited time span, during which the message is traversing, while at the same time the scheduled injection of a TT message may arrive. Hence, in order to control the impact of ET messages (which are typically used for low-critical flow) on TT messages (which are used for safety-critical flow), one of the following mechanisms can be used.

The first approach, which is known as *timely blocking* [281] uses the concept of temporal segregation to eliminate this impact. The second approach, which is widely used in already existing NoCs, uses the concept of *priority* to bound the impact. In this approach, the message of higher priority has the precedence over lower-priority messages in acquiring the resource if a collision happens. In the following parts, these two approaches are elaborated.

Timely Blocking by Guarding Windows

Fault containment is achieved by temporal partitioning between safety-critical and low-critical messages using *guarding windows* (grayed area in Figure 7.14), through which the transmission of any ET messages is blocked. This eliminates the impact of ET messages on TT messages at the cost of lower bandwidth usage. More precisely, during each guarding window, the NI does not inject any ET messages into the interconnect to ensure that TT messages are not delayed. Guarding windows can be applied at different levels (cf. [281]).

The ET Interleaver (cf. Figure 7.12) enforces guarding windows in a TT manner, as they need to be harmonized with TT messages. Each guarding window is composed of a *clean-up* and a *white* slot (cf. Figure 7.14). During the clean-up slot, low-critical messages are not allowed to use the resource, in order to avoid preemption of those messages. This is due to the fact that if a low-critical

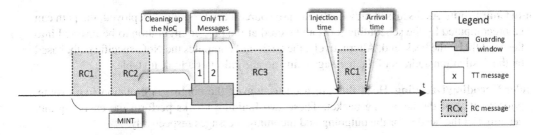

Figure 7.14: Integration of TT and ET Messages, © IEEE 2016, Reprinted with Permission from [12]

message starts using the resource, it takes time until the resource becomes free for the safety-critical message. The duration of clean-up slot is equal to the maximum transmission time of a low-critical message.

Shuffling by Prioritized Virtual Channels

In contrast to timely blocking, shuffling requires no temporal synchronization and guarding windows. Hence, the communication of low-critical messages is never blocked for the sake of safety-critical messages. However, this approach has the drawback that the safety-critical traffic might incur a bounded delay that is caused by the low-critical traffic.

Shuffling is technically achieved by prioritized VCs [282]. VCs employ the concept of virtualization and offer several channels out of a single physical link by using multiple buffers (each of which acts as a VC) at both terminals of each physical link. This provides an efficient utilization of the link and enables the flits to bypass a blocked router by a contention from a cross-cutting traffic, if the forthcoming path is not blocked.

In addition, VCs offer a shorter delay for the messages of higher priority. This is achieved by a mapping between the VCs and the priorities to let the flits of the higher priority acquire the switch prior to the lower priorities.

In MCSs, VCs can be mapped further to the traffic types. Three priority classes can be identified, some of which can have multiple sub-priorities. More precisely, the highest priority in the network belongs to the TT messages (as shown in Figure 7.12). Since periodic messages are sent according to a predefined schedule and it is assumed that the schedule induces no collisions, there are no sub-priorities needed for the periodic messages.

The second priority class is assigned to RC messages. However there can be different levels of sub-priorities for different sporadic messages (the number after the term RC in Figures 7.12 and 7.14 denotes the priority). In case two sporadic messages of different priorities compete for using a resource, the one of higher priority wins and the other waits for the resource.

BE messages possess the lowest priority class in the network. There is no guarantee whether and when these messages arrive at the destination. According to the implementation, further priorities for those messages might be defined.

7.4.7 Back-End

The back-end has no understanding of the concept of time and criticality and it performs all the NoC-specific operations. This unit provides the building blocks of the NI with an interface to the routers by generating the packets and consequently the flits and sending them to the routers at the sender tile. In addition, this unit disassembles the received packets and flits from the NoC and sends them to the destination input ports for incoming messages at the destination tile [283].

In the following, major contributions of this unit are elaborated [284].

Route Computing As discussed previously, in case source-based routing is deployed, the path can be precomputed by the scheduling tools and stored at the port configuration to be inserted into the head flit by the back-end. Alternatively, the back-end computes the NoC-specific path, based on the destination address of the message using an algorithm or a look-up table.

Header Encoding/Decoding Head flits are used to convey the control information (e.g., route, priority, etc.) of the belonging packet. These two building blocks perform the encoding and decoding of the header for the outgoing and incoming messages respectively.

Packetization/Depacketization For outgoing messages, after the head flit is prepared by the previous building block, the message needs to be divided into fixed-length packets. In case of incoming messages, this building block extracts the data (i.e., the payload) and the destination port from the body flits and the head flit respectively and constitutes the message to be forwarded to the respective port.

Virtual Channel Allocation The allocation of a VC is packet-based, i.e., after the packet is constructed by the back-end, this building block allocates a VC to the packet. After the allocation of the VC to a packet, the process of injection of the flits starts. This process is controlled by a credit-based flow control [284], by which the back-end tracks the number of vacant buffers in each VC at the next router.

8

Cluster-Level Communication Services

T. Koller

Universität Siegen

M. Abuteir

TTTech Computertechnik AG

A. Eckel

TTTech Computertechnik AG

A. Geven

TTTech Computertechnik AG

L. Kohútka

TTTech Computertechnik AG

L. Rubio

IK4-Ikerlan

C. Zubia

IK4-Ikerlan

8.1   Off-Chip Network 342
    8.1.1   Time-Triggered Ethernet 342
              8.1.1.1   Gateways 345
              8.1.1.2   Time-Triggered Ethernet Switch 351
    8.1.2   EtherCAT 354
              8.1.2.1   Safety Communication Layer 354
8.2   Security Services 360
    8.2.1   Risk Analysis 361
              8.2.1.1   Securing DREAMS Services 361
              8.2.1.2   Potential Attacks 362
    8.2.2   Security at Multiple Levels 363
    8.2.3   Security Classification 364
    8.2.4   Cluster-Level Security 365
    8.2.5   Secure Time Synchronization 367
    8.2.6   Application-Level Security 369
              8.2.6.1   Message Exchange Format 369
              8.2.6.2   Supported Algorithms 371
              8.2.6.3   Evaluation 371

This chapter describes the DREAMS cluster-level communication services, Safety Communication Layer (SCL) and secure communication services. As shown in Figure 8.1 the cluster level consists of the off-chip components and also includes the on-chip / off-chip gateway.

Section 8.1 describes the TTEthernet and EtherCAT extensions to support real-time and the mixed-criticality requirements on top of Ethernet communication. Both communication protocols are used in the development of demonstrators (see Section 11), EtherCAT for the wind-power demonstrator and TTEthernet for the safety-critical domain and health-care demonstrators.

In Section 8.2, the security services for the cluster and application levels are described. The cluster-level security services secure the communication between the off-chip components. The

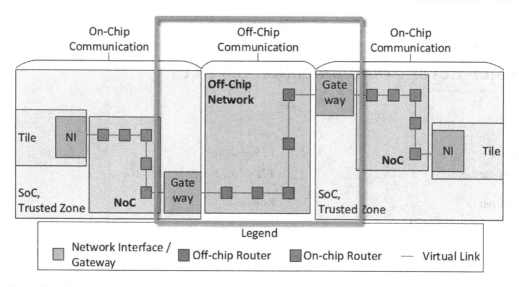

Figure 8.1: Cluster-Level Communication Components

application-level security services provide a secure end-to-end channel between the applications. Using the application-level security services, also the secure resource management communication is achieved. The exact usage is described after the resource management is introduced (see Section Section 9.6.1 for further details).

8.1   Off-Chip Network

As explained in the description of the DREAMS waistline structure of services (see Section 2.2), Ethernet provides a foundation for the deployment of higher level communication services (e.g., Internet Protocol (IP)). Ethernet has been extended with several approaches in order to support the required real-time and mixed-criticality requirements, e.g., audio video bridging based on an IEEE Standard, EtherCAT, ARINC 664 [10], TTEthernet [20]. In the DREAMS project, TTEthernet and EtherCAT are extended for this purpose.

8.1.1   Time-Triggered Ethernet

Time-Triggered Ethernet (TTEthernet) [20] is a real-time Ethernet extension standardized by the SAE. It supports message exchanges with bounded transmission delays, low jitter and high channel utilization. TTEthernet establishes a global time base through clock synchronization and provides fault containment for failures of switches, communication links and nodes. In particular, mixed-criticality applications are supported, where safety-critical subsystems (e.g., protection safety) and non-safety-critical subsystems (e.g., multimedia) are combined in a single system [285]. For these different types of subsystems, TTEthernet includes suitable communication mechanisms ranging from best-effort messaging with a high channel utilization to predictable real-time messaging based on a time-triggered communication schedule.

A TTEthernet network consists of a set of nodes and switches, which are interconnected using bi-directional communication links. TTEthernet combines different types of communication on the same network. A service layer is built on top of IEEE 802.3, thereby complementing layer two of the Open System Interconnection (OSI) model [18].

TTEthernet supports synchronous communication using so-called time-triggered frames. Each participant of the system is configured offline with pre-assigned time slots based on a global time base. This network access method based on TDMA offers a predictable transmission behavior without queuing in the switches and achieves low latency and low jitter.

The bandwidth that is either not assigned to time triggered frames or assigned but not used is free for asynchronous frame transmissions. TTEthernet defines two types of asynchronous frames: rate-constrained and best-effort frames. Rate-constrained frames are based on the AFDX protocol and intended for the transmission of data with less stringent real-time requirements [20]. Rate-constrained frames support bounded latencies but incur higher jitter compared to time-triggered frames. Best-effort frames are based on standard Ethernet and provide no real-time guarantees.

The different types of frames are associated with priorities in TTEthernet. Time-triggered frames have the highest priority, whereas best-effort frames are assigned the lowest priority. Using these priorities, TTEthernet supports three mechanisms to resolve collisions between the different types of frames [286, 287]:

- **Shuffling:** If a low priority frame is being transmitted while a high priority frame arrives, the high priority frame will wait until the low priority frame is finished. That means that the jitter for the high priority frame is increased by the maximum transmission delay of a low-priority frame. Shuffling is resource efficient but results in a degradation of the real-time quality.

- **Timely Block:** According to the time-triggered schedule, the switch knows in advance the transmission times of the time-triggered frames. Timely block means that the switch reserves so-called guarding windows before every transmission time of a time-triggered frame. This guarding window has a duration that is equal to the maximum transmission time of a lower priority frame. In the guarding window, the switch will not start the transmission of a lower priority frame to ensure that time-triggered frames are not delayed. The jitter for high priority frames will be close to zero. Timely block ensures high real-time quality with a near constant delay. However, resource inefficiency occurs when the maximum size of low-priority frames is high or unknown.

- **Preemption:** If a high priority frame arrives while a low priority frame is being relayed by a switch, the switch stops the transmission of the low priority frame and relays the high priority frame. That means that the switch introduces an almost constant and a priori known latency for high priority frames. However, the truncation of frames is resource inefficient and results in a low network utilization. Also, corrupt frames result from the truncation, which can be indistinguishable to the consequences of hardware faults. The consequence is a diagnostic deficiency.

The TTEthernet frame format is fully compliant with the Ethernet frame format. However, the destination address field in TTEthernet is interpreted differently depending on the traffic type. In best-effort traffic, the format for destination addresses as standardized in IEEE 802.3 is used. In time-triggered and rate-constrained traffic, the destination address is subdivided into a constant 32-bit field and a 16-bit field called the virtual-link identifier. TTEthernet communication is structured into virtual links, each of which offers a unidirectional connection from one node to one or more destination nodes. The constant field can be defined by the user but should be fixed for all time-triggered and rate-constrained traffic. This constant field is also denoted as the 'CT marker' [20]. The two least significant bits of the first octet of the constant field must be equal to one, since rate-constrained and time-triggered frames are multicast messages.

TTEthernet ensures a mixed-criticality strategy that respects the time-triggered and rate-constrained time characteristics in the form of the communication schedule; an integration strategy

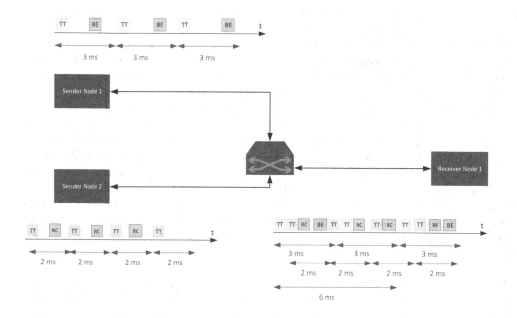

Figure 8.2: Example of Dataflow Integration

based on the above possibilities must be chosen. Figure 8.2 presents an example TTEthernet network consisting of three end systems and a single switch. The example consists of two end systems that send frames, a switch that integrates the frames from the two senders, and a receiver that receives the integrated dataflow from the switch. As depicted, sender 1 sends a time-triggered frame (TT) with a period of 3 milliseconds and best-effort frames (BE). Sender 2 sends a time-triggered frame (TT) with a period of 2 milliseconds, best-effort frames (BE), and rate-constrained frames (RC). The resulting integrated data flow is depicted on the right.

TT communication requires synchronicity between the time-triggered senders. In order to establish and maintain synchrony, a management function is executed to establish a global notion of time based on which messages can be exchanged. Each end system stores the message dispatch points in time for TT messages in a local table. The reception of TT messages is based on the message identifier. The schedule tables are generated off-line and need to be guaranteed to be collision-free. In order to deduce the mixed-criticality integration, we must thus first ensure synchrony at the highest priority. After this, the next level of priorities is given to TT messages, then to RC messages and finally BE messages can be sent in remaining bandwidth gaps. The problem of generating the scheduling is generally known as the scheduling problem and is aggravated when latency requirements are tight and system utilization is high. An efficient utilization requirement of hardware often requires combined synthesis of processor schedules and bus schedules (e.g., task and network schedules). When we consider hierarchical networks, the efficient scheduling of such networks also becomes increasingly complex.

A typical DREAMS mixed-criticality network may look like the one depicted in Figure 8.3. The figure highlights a network where applications of different criticality levels have communication needs that are subdivided in non-critical (e.g., infotainment), time-critical (e.g., navigation) and safety-critical (e.g., control system) data paths. All components suffer from the 'lift-up' effect, i.e., the fact that when it handles only a marginal amount of critical data, it requires the full spectrum of certification that is in line with the criticality of the data. In the example, this is highlighted by the colors green, yellow and red. Although only few applications require the highest level of criticality,

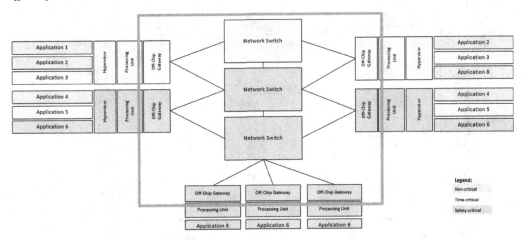

Figure 8.3: Example of a Mixed-Criticality Network

a large number of components in the system must support this. The off-chip network is highlighted in green.

8.1.1.1 Gateways

In DREAMS, two types of gateways are presented based on network connection: on-chip / off-chip gateway and wireless gateway.

On-Chip / Off-Chip Gateway

In order to establish the end-to-end communication over heterogeneous and mixed-criticality networks in DREAMS, gateways are used. The connection between off-chip networks, as well as between off-chip and on-chip networks is established through gateways that fulfill the services described in Section 2.2.4.3. Within the DREAMS architectural style, the gateway services are part of the core communication services that are to be offered by any DREAMS-based platform and are at the heart of the waistline architecture. An on-chip / off-chip gateway relays selected messages from the NoC to an off-chip network and vice versa, while performing the necessary protocol transformations. At off-chip level, the (1) off-chip networks and (2) off-chip gateways belong to the core communication services. Each cluster has a corresponding off-chip network, where the networks of different clusters can be connected through an off-chip gateway.

The gateway core is responsible for redirecting incoming messages based on timely redirection, protocol conversion, monitoring, and configuration services. The network interfacing provides the interface between the Media Access Control (MAC) and the gateway core. Furthermore, classification and serialization of the packets is performed in the network interfacing. In order to realize fault-tolerance, the gateway can include multiple network MACs. Each network MAC connects the gateway to either an off-chip network (e.g., TTEthernet) or an on-chip network (e.g., STNoC). In case of network redundancy, multiple network MACs are required. Thus, the network interfacing is responsible for merging identical incoming messages and duplicating outgoing messages to be sent to different MACs.

The generic gateway functionality is depicted in Figure 8.4. In the figure, the actual gateway functionality is implemented in the so-called 'Message Relay Entity'. The left and right legs of the figure represent the two sides of the gateway.

Gateway Services

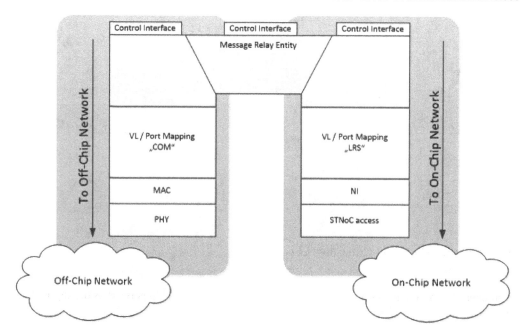

Figure 8.4: Generic On-Chip / Off-Chip Gateway

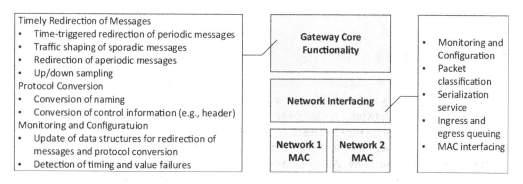

Figure 8.5: Gateway Services [16]

The Gateway Abstraction Layer (GAL) encapsulates the off-chip gateway interface as a set of communication ports. The gateway services have been defined in the architectural style [21] and are shown in Figure 8.5.

The services are subdivided into four groups that together make up the gateway services as in the architectural style. This chapter shortly revisits them in their respective logical groups:

- Group 1 - Timely message redirection services

    - GWS1: Time-triggered redirection of periodic messages
    - GWS2: Traffic shaping of sporadic messages (RC)
    - GWS3: Redirection of aperiodic messages (BE)
    - GWS4: Up / down sampling

- Group 2 - Protocol Conversion services

    - GWS5: Conversion of naming

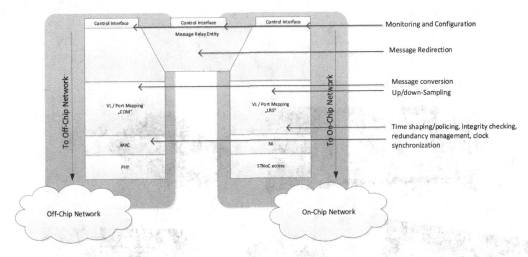

Figure 8.6: Gateway Elements

   – GWS6: Conversion of control information

• Group 3 - Monitoring and Configuration services

   – GWS7: Update of data structures for redirection of messages and protocol conversion

   – GWS8: Detection of timing and value failures

• Group 4 - Global Time services

   – GWS9: Provision and propagation of a global distributed timebase

These services are implemented on the gateway in the different gateway layers. Note that the physical implementation in the layers and the logical grouping as described in the previous deliverable is not always the same. For functional, complexity and efficiency reasons, some services are implemented close to the physical connections, whereas other services are implemented at the message relay entity, as depicted in Figure 8.6.

Platforms for Off-Chip Gateway Services

Within the DREAMS project, the off-chip gateway has been implemented in different hardware platforms as shown in Figure 8.7: TTTech PCIe and XMC cards, Xilinx Zynq ZC-706, ARM Juno Board and Freescale QoriQ T4240QDS. The ARM Juno Board and Freescale QoriQ T4240QDS both utilize PCIe to interface with the DREAMS TTEthernet device depicted Figure 8.7, which serves as a carrier board for the TTTech TTE XMC card shown in the top right. The FPGA on this device (Altera STRATIX IV) is used to instantiate the DREAMS services on the device and connect the ARM Juno board and the Freescale QoriQ board to the mixed-criticality network. The backbone of the network is provided by the networking platform based on the 'TTEthernet A664 Lab Switch' (Figure 8.8), which instantiates required DREAMS communication services.

Wireless Gateway

The goal of deterministic wireless communication is the reduction of jitter and latency in the context of distributed DREAMS nodes that are connected through a wireless network.

Several fundamental restrictions limit the transferability of the integral set of guaranteed properties inherent to time-triggered communication into the wireless domain. Some of these restrictions

Figure 8.7: Platforms for Off-Chip Mixed-Criticality Gateway Services

Figure 8.8: TTEthernet A664 Lab Switch

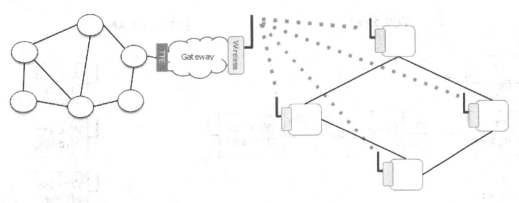

Figure 8.9: Network Gateway Connecting TTEthernet with a Wireless Sub-Network

are: limited bandwidth and unreliability of communication channels; larger latencies due to wireless specific access control protocol (e.g., 802.11 DCF/EDCA) and, above all, the fact that existing wireless protocols are designed for the co-existence of multiple contention-based communications potentially sharing the same media (i.e., same band on the radio spectrum).

Nevertheless, a specifically designed gateway may enable the establishment of a unified notion of time across the wired and wireless domains allowing a certain degree of deterministic communication. Figure 8.9 depicts one such scenario in which the gateway interfaces between a wired TTEthernet network and a set of wireless nodes.

Wireless Implementation

The contrast between inherent properties of the two sides of the gateway introduces a number of constraints to the services provided by a generic gateway, as depicted in Figure 8.4. These services can be scaled as needed to accommodate specific wireless constraints.

In order to establish a baseline for communication across the wireless channel for DREAMS, an experimental setup was designed with the focus on the implementation of collision-free communication on the basis of IEEE 802.11 enabling synchronized wireless nodes as an extension to a regular TTEthernet network. The goal was to estimate the boundaries of what can be achieved and is reasonably feasible with existing technologies. In this respect, 802.11 is a perfect candidate and a well-established wireless standard with plenty of available off-the-shelf components. Our choice for the 802.11 wireless standard is justified by a number of reasons, namely:

- It is a widely used technology for which an extensive set of low priced COTS exist.

- It provides an easy translation of Ethernet frames due to the similar frame format.

- The network can be arranged in a managed mode in which nodes are connected to a single access point, which suits our model for the gateway.

- A software based solution is possible thanks to the extended support for 802.11 clients as well as access point drivers are available for common platforms for which TTEthernet support is also available (e.g., Linux).

- The different flavors of 802.11 allow for a different range of options in terms of throughput and bandwidth.

The following components are depicted in Figure 8.10, where red arrows indicate wired traffic (Ethernet domain) and green arrows indicate wireless traffic (802.11 domain).

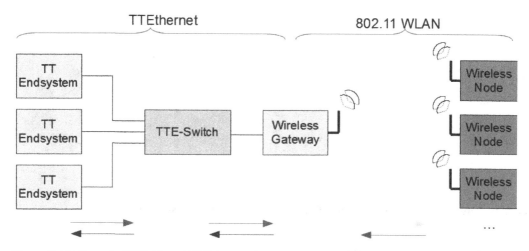

Figure 8.10: Extended 802.11 and TTEthernet Gateways Use-Case

- **TT Endsystem:** Wired TTEthernet end-system with scheduled critical traffic.

- **TTE-Switch:** TTEthernet switch acting as TTEthernet compression master (i.e., clock 'master').

- **Wireless Gateway:** Wired TTEthernet node and 802.11 wireless access point, acting as TTEthernet-WLAN gateway. Also has the role of a clock synchronization gateway for the WLAN network taking as reference the TTEthernet clock.

- **Wireless Node:** 802.11 Wireless node sending and receiving scheduled frames to / from wireless bridge. It also maintains time synchronization with the wireless gateway.

Network Synchronization

Note that in the setups considered, we are utilizing the Access Point (AP) as master for the synchronization process of the wireless devices. That is, the transmission of the beacon, by the AP (or the node emulating the AP), and the posterior reception of the beacon at the nodes, marks the beginning of the inter-beacon-interval. Hence, from this event, each device sets its own clock as the initial moment for the inter-beacon-interval, and it is from this reference point, that the transmissions scheduled by the different wireless devices are carried out. Summarizing, we are using the transmission of the beacon as synchronization means, however, this provides just a relative time reference, not an absolute one. This approach is in line with the 'gateway as time master' concept. A further level of detailed synchronization can be implemented by transmitting synchronization information in the slots scheduled to synchronize the different wireless devices on the basis of e.g., SAE AS6802 synchronization [288].

Schedule Adaptation

Due to the translation between the wired and the wireless domain, it is possible that the schedule in both domains should be modified. Some factors to be considered, and some clarifications regarding our prototype, are described in what follows.

- Maximum frame size for non-fragmented wireless packets: If the maximum frame size for non-fragmented wireless packet is too low, the information from the wired network would be fragmented, pushing the requirement for a schedule adaptation in the wireless domain. In our case, we considered only the case in which there are no fragmented packets and all the packets have the same size.

- Additional mandatory frames inherent to the wireless protocol (e.g., ACK, RTS / CTS): In the current implementation, for the transceiver mode setup, additional frames used by WiFi were handled in the following way. The ACK is not required, because we do not implement a retransmission scheme. On the other hand, regarding the RTS / CTS frames, we assume that the environment is relatively clean (in an electromagnetic sense) and there are no hidden nodes, thus, we do not implement any RTS / CTS step. Again, this is a consequence of the predefined schedule, under the assumption that the interference is low. Other message types were ignored for the current implementation framework.

- Fixed vs variable slot sizes in the wireless domain: In our prototype, for simplicity, we used slots with the same size in the wireless domain. If it turns out that the slots should have different duration, or the duration in the wireless domain would be different from the duration in the wired domain, a schedule adaptation should be carried out. This goes beyond the scope of the current prototype implementation.

- Management frames (e.g., beacon frames): The beacon frame can be present in a wireless network, but not necessarily must be mapped in a wired network. Therefore, its use can bring advantages like the centralized synchronization, but on the other hand could require a schedule adaptation in the wireless medium, with respect to the wired one.

Bounded Latency

The wireless implementation described in this chapter is the first step towards the development of a wired-wireless gateway. However, in a wired-wireless gateway there will be certain latency, while the information is transferred from the wired card to the wireless one, or vice versa. In the current implementation activities, this latency is not considered (yet), as the main focus is on the wireless transmission and compatibility to 802.11. In a next step, the latency required in the gateway implementation itself will be investigated.

8.1.1.2 Time-Triggered Ethernet Switch

Standard Ethernet does not guarantee either the real-time capability or bounded transmission delays. Therefore, it is not suitable for the real-time and safety aspects in mixed-criticality systems. In the architectural model of the TTEthernet switch, we assume temporal and spatial partitioning that guarantees predictable timing of periodic time-triggered messages and bounded latencies for sporadic rate-constrained messages. The TTEthernet switch extends the standard Ethernet switch by adding the time deterministic message transfer capabilities, while retaining full compatibility with the requirements of IEEE 802.3.

Such a system simplifies the design of complex distributed systems and applications as well as processing of critical (periodic, sporadic) and non-critical Ethernet traffic (aperiodic). High-priority periodic messages are routed through the switch according to a predefined schedule with fixed latency and a transmission jitter in the sub-microsecond range. Sporadic messages are routed based on the AFDX protocol. All other messages are forwarded based on standard Ethernet when bandwidth is available.

As shown in Figure 8.11, the switch consists of multiple ports and a bridge. Each port contains a physical layer and a MAC layer. The physical layer is built according to IEEE 802.3. The MAC layer is based on IEEE 803.2 with the extensions explained in the following. The MAC layer checks the validity of the destination address for an incoming message and distinguishes between traffic types

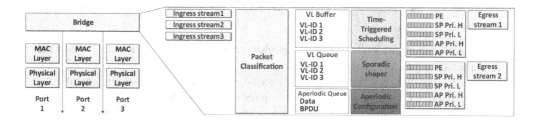

Figure 8.11: System Model of a Switch [16]

based on connection-oriented and connectionless communication. The connection-oriented communication is used for the periodic and sporadic messages. Aperiodic messages use the connectionless communication. We regard a message as a tuple with the following elements:

- Message in connection-oriented communication:
  $< type, VLID, data >$

- Message in connectionless communication:
  $< type, destination address, data >$

In TTEthernet switch, the constant field is extracted from the destination address using the bit mask $0xfffffff0000$. In case the constant field has the predefined value of the 'CT marker', this message is either time-triggered or rate-constrained. Otherwise, the message will be regarded as aperiodic. The switch distinguishes between periodic and sporadic messages using the 'EtherType' value. The IEEE Standardization Authority has assigned the values 0x88d7 [20] and 0x0800 [10] for the 'EtherType' fields of time-triggered and rate-constrained messages.

Figure 8.11 shows the block diagram of the bridge. The task of the bridge is to handle and forward ingress messages to the egress ports depending on the traffic type. The bridge model contains five layers: the bridge classification layer that determines the traffic type of an ingress message, the TT scheduling layer that processes the periodic messages, the sporadic shaper layer that handles the rate-constrained messages, the aperiodic configuration layer that handles the aperiodic messages, and the egress port layer for realizing shuffling and timely block.

Processing of Periodic Messages

For each periodic message, a static communication schedule defines three parameters: period, phase and maximum frame size. Periodic messages are scheduled to be transmitted periodically, where the phase parameter defines the exact start time relative to the start time of the period. In each period, multiple periodic messages with different phases are supported. In addition, the static communication schedule defines the ingress and egress ports of each time-triggered message, as well as the buffer size in the bridge.

When a periodic message arrives at the bridge from the MAC layer, the bridge classification layer checks the integrity and validity of the message. The integrity checking verifies that the message has the correct size and arrives from the correct ingress port as defined by the communication schedule for the virtual link of the message. Valid messages are put into the corresponding virtual-link buffer, which provides buffer space for exactly one message. In case this buffer is full and another message arrives with the same virtual-link identifier, the newer message replaces the old one.

The TT scheduling layer is responsible for relaying the periodic messages from the virtual-link buffer to the queue for periodic messages at the correct egress port according to the communication schedule. The communication schedule also determines the point in time when the periodic message is relayed, thereby ensuring the deterministic communication behavior.

For each egress port, the bridge has five egress queues with decreasing priorities: one queue for periodic messages, two queues for sporadic messages (each one for a different priority class) and two queues for aperiodic messages (also for two different priority classes).

The egress port layer forwards the messages from the egress queues to the MAC layer according to the priority. The highest priority is assigned to periodic messages, whereas aperiodic messages have the lowest priority.

Also, the shuffling and timely block mechanisms are realized in this layer. The timely block mechanism disables the sending of other Ethernet messages in the bridge during a guarding window prior to the transmission of a periodic message. For the shuffling mechanism, no guarding window is needed. In the worst-case, the bridge delays a periodic message for the duration of an Ethernet message of maximum size (i.e., 123 μs in case of 100 Mbps).

The virtual-link buffer and the static communication schedule ensure fault isolation for failures of nodes. For example, a babbling idiot failure of a node involving the transmission of untimely messages cannot affect the timing of messages from other nodes. In addition, masquerading failures are detected by the bridge classification layer. A masquerading failure occurs if an erroneous node assumes the identity of another node.

Processing of Sporadic Messages

Sporadic messages are encapsulated by a specific bandwidth allocation. The message flow of sporadic messages is associated with two main parameters for each virtual link: the Bandwidth Allocation Gap (BAG) and the jitter. The BAG value defines the minimum time between two Ethernet messages that are transmitted on the same virtual link. Jitter may be introduced by multiplexing all virtual links into shared egress queues.

In addition to these parameters, the bridge needs for every virtual link information about the ingress and egress ports, as well as the required queue size. When a sporadic message arrives at the bridge, the message is checked in the filtering unit of the bridge classification layer. The size of the message must be below the maximum message size and the ingress port must comply with the configuration parameters of the virtual link. Valid messages are enqueued into the corresponding virtual-link queue.

The RC shaper layer realizes the traffic policing for the sporadic message by implementing an algorithm known as token bucket. This layer checks the time interval between consecutive messages on the same virtual link and moves sporadic messages from the virtual-link queue to one of the sporadic egress queues. We distinguish between two priority classes of sporadic messages with two corresponding sporadic egress queues. The egress port layer is responsible for forwarding the messages from the egress queue to the MAC layer.

Processing of Aperiodic Messages

The Spanning Tree Protocol (STP) as defined by IEEE 802.1D is used to establish a loop-free topology for communication of aperiodic messages [289]. The supported periodic messages include Bridge Protocol Data Units (BPDU) and best-effort data messages. BPDU messages are exchanged between switches to determine the network topology, e.g., after a topology change has been observed.

The switch needs two queues (called aperiodic queues) for these two aperiodic message types. The bridge classification layer puts incoming aperiodic messages into the BPDU queue or the aperiodic data queue.

The aperiodic configuration layer consists of two processes: a process that handles aperiodic data messages and another one handling BPDU messages according to STP. Each process relays aperiodic messages from an aperiodic queue to one of the aperiodic egress queues. TTEthernet supports two classes of best-effort messages with each priority class having one aperiodic egress queue. The egress port layer forwards the messages from the egress queue to the MAC layer as explained previously for sporadic messages.

8.1.2 EtherCAT

Ethernet for Control Automation Technology (EtherCAT) [290] is an Ethernet based fieldbus solution used in the DREAMS wind-power demonstrator (see Section 11.1). EtherCAT utilizes standard frames and the physical layer defined in the Ethernet Standard IEEE 802.3, with short cycle times ($\leq 100\ \mu$ s) and low jitter for accurate synchronization ($\leq 1\ \mu$ s), which makes it suitable for hard and soft real-time requirements in automation technology.

EtherCAT supports multiple network architectures and topology variations. For example, EtherCAT can be used by a master device to exchange information with distributed Input / Output (I/O) modules within a redundant EtherCAT ring. A basic EtherCAT network is mainly composed by a master and one or more slaves that exchange information cyclically between them; thus, master sends outputs and receives inputs from slaves. These outputs and inputs variables are allocated in datagrams, and these datagrams shape the frames that will travel around the EtherCAT network.

The EtherCAT Network Information (ENI) [291] provides the required information to accomplish the task of linking the cyclic commands with the variables. So, during start up, the master device configures and maps the process data on the slave devices.

EtherCAT technology uses a dual-port memory so the application delivers the data to one side of the memory, while the data received from the medium is allocated at the other side. The cyclic commands, which are mainly a piece of this dual-port memory, contain a certain number of variables, each with its offset and length. Thus, once it is identified the offset and the length of the piece of the memory that forms the datagram, the relation with the variables can be obtained.

Modern communication systems not only realize the deterministic transfer of control data, they also enable the transfer of safety-critical control data through the same medium. EtherCAT utilizes the protocol Fail Safe over EtherCAT (FSoE) for this very purpose. The EtherCAT safety technology was developed according to IEC 61508, it is certified for safety applications and it is standardized in IEC 61784-3 [292]. The protocol is suitable for safety applications with a Safety Integrity Level up to SIL3.

8.1.2.1 Safety Communication Layer

Safety Communication Layers, which are implemented as parts of safety-related systems according to IEC 61508 series, provide the necessary confidence in the transportation of messages (information) between two or more participants on a communication channel in a safety-related system, or sufficient confidence of safe behavior in the event of channel communications errors or failures.

The resulting Safety Integrity Level (SIL) claim of a system depends on the implementation of the selected functional safety communication layer within the system, as the implementation of a functional safety communication layer in a standard device is not sufficient to qualify it as a safety device.

The transmission system forms an integral part of the safety related system, which must be protected to guarantee the end-to-end communication integrity. In a transmission system according to IEC 61508-2 [132] two options exist:

- *White Channel:* The entire channel is designed, implemented and validated according to IEC 61508 and IEC 61784-3 [292] or IEC 62280 [293] series.

- *Black Channel:* The channel is not designed, implemented or validated according to IEC 61508, and the measures shall be implemented in the interface with the communication channel in accordance with the IEC 61784-3 or IEC 62280 series as appropriate. This means that all the measures necessary to implement transmission of safety data in accordance with the requirements of IEC 61508 shall be performed by an additional safety communication layer.

While IEC 61508 is not restricting the use of communication technologies, IEC 61784-3 [292] focuses on the use of fieldbus based functional safety communication systems and IEC 62280 [293] safety related communication in transmission systems on railway applications for communication, signaling and processing systems.

DREAMS architecture follows the black channel approach; this means that all the measures necessary to implement transmission of safety data in accordance with the requirements of IEC 61508 shall be performed by an additional SCL. According to the IEC 61784-3, the SCL is an application level service on top of a non-safety related communication stack.

- It enables 'safe' data exchange between applications

- It must be developed with a life cycle equivalent to the highest safety level (SIL) in the application.

- It requires at least the detection of the following types of errors (FMEA): e.g., corruption, message incorrect order, message outside temporal requirements, message lost, message duplicated.

Communication errors and measures to detect them

The main tasks of a SCL are the following: (i) deliver correct data and (ii) send the information to the right destination. Various errors may occur when messages are transferred in complex network topologies, such as, hardware failures, electromagnetic interference, or other influences. A message can be lost, occur repeatedly, be inserted from somewhere else, appear delayed or in an incorrect sequence, and / or show corrupted data. In the case of safety communications, there may also be incorrect addressing: a standard message erroneously appears at a device and pretends to be a safety message. Different transmission rates may additionally cause bus component storage effects to occur. The following communication error descriptions can be found in the literature [292–294]:

- **Corruption:** Messages may be corrupted due to errors within communication channel participant, the transmission medium or message interference.

- **Unintended repetition:** Due to an error, old messages are repeated at an incorrect point in time.

- **Incorrect sequence:** Due to an error, the predefined sequence (e.g., natural numbers, time references) associated with messages from a particular source is incorrect.

- **Loss:** Due to an error, a message is not received or not acknowledged.

- **Unacceptable delay:** Due to an error, messages may be delayed beyond their permitted arrival time window.

- **Insertion:** Due to an error, a message is inserted that relates to an unexpected or unknown source entity.

- **Masquerading:** Due to an error, a message is inserted that relates to an apparently valid source entity, so a non-safety relevant message may be received by a safety relevant participant, which then treats it as safety relevant.

- **Addressing:** Due to an error, a safety relevant message is sent to the wrong safety relevant participant, which then treats reception as correct.

As shown in Table 8.1, different safety measures provide protection against one or more of the communication errors previously described. It shall be demonstrated that there is at least one corresponding safety measure or combination of safety measures for the defined possible errors. Measures commonly used to detect deterministic errors and failures of a communication system are:

- **Sequence number:** A sequence number is integrated into messages exchanged between message source and message sink. It may be realized as an additional data field with a number that changes from one message to the next in a predetermined way [292]. Sequence numbering consists of adding a running number (called sequence number) to each message exchanged between a transmitter and a receiver. This allows the receiver to check the sequence of messages provided by the transmitter [293].

- **Time stamp:** In most cases the content of a message is only valid at a particular point in time or time slot. The time stamp may be a time, or time and date, included in a message by the sender [292]. When an entity receives information, the meaning of the information is often time-related. The degree of dependence between information and time can differ between applications. In certain cases old information can be useless and harmless and in other cases the information could be a potential danger for the user. Depending on the behavior in time of the processes which interchange information (e.g., cyclic, event controlled) the solution may differ. One solution which covers time-information relationships is to add time stamps to the information. This kind of information can be used in place of or combined with sequence numbers depending on application requirements [293].

- **Time expectation / time-out:** During the transmission of a message, the message sink checks whether the delay between two consecutively received messages exceeds a predetermined value. In this case, an error has to be assumed [292]. In transmission (typically cyclic) the receiver can check if the delay between two messages exceeds a predefined allowed maximum time. If this is the case, an error shall be assumed [293].

- **Connection authentication / source and destination identifiers:** Messages may have a unique source and / or destination identifier that describes the logical address of the safety relevant participant [292]. Multi-party communication processes (e.g., unidirectional, broadcast) need adequate means for checking the source of all information received, before it is used. Messages shall include additional data to permit this. Messages may contain a unique 'source identifier', or a 'unique destination' identifier, or both. The choice is made according to the safety related application. These identifiers are added in the safety related transmission functions for the application [293].

- **Feedback message:** The message sink returns a feedback message to the source to confirm reception of the original message. This feedback message has to be processed by the SCL [292]. Where an appropriate return transmission channel is available, a feedback message may be sent from the receiver of safety-critical information to the sender. The contents of this feedback message may include data derived from the original message, data added by the receiver or additional data for safety / security purposes [293].

- **Different data integrity assurance systems:** If safety and non safety related data are transmitted via the same communication channel, different data integrity assurance systems or encoding principles may be used (different hash functions), to make sure that non safety related messages cannot influence any safety function in an safety receiver [292].

- **Identification procedure:** Open transmission systems can additionally introduce the risk of messages from other (unknown) users being confused with information originating from an intended source (a form of masquerade). A suitably designed identification procedure within the safety related process can provide a defense against this threat. Two types of identification procedure can be distinguished [293]: bi-directional identification (e.g., return communication channel with exchange of entity identifiers between senders and receivers of information) and dynamic identification procedures (e.g., dynamic exchange of information between senders and receivers).

- **Safety code:** In transmission systems, in general, transmission codes are used to detect random and / or burst errors, and / or to improve the transmission quality by error-correction techniques. Even though these transmission codes can be very efficient, they can fail because of hardware faults, external influences or systematic errors [293]. The safety related process shall not trust those transmission codes from the point of view of safety. Therefore a safety code under the control of the safety related process is required additionally to detect message corruption. The safety case shall demonstrate the appropriateness, in relation to the required safety integrity and the nature of the safety related functions, of the following: the capability for detection of expected systematic types of message corruption and the probability of detection of random types of message corruption.

Table 8.1: Effectiveness of Measures on Errors According to IEC 61784-3 [292]

Communication Error	Measures							
	Sequence Number	Time Stamp	Time Expectation	Connection authentication	Feedback message	Data integrity assurance	Redundancy with cross checking	Different data assurance systems
Corruption						X	X	X
Unintended Repetition	X	X					X	
Incorrect Sequence	X	X					X	
Loss	X				X		X	
Unacceptable Delay		X	X					
Insertion	X			X	X			
Masquerade				X	X			
Addressing				X				

DREAMS SCL implementation

IEC 61784-3 lists and specifies most relevant industrial and automation technology fieldbuses (e.g., Foundation Fieldbus, PROFINET, Interbus, Ethernet/IP, PowerLink). IEC 61784-3-3 PROFIsafe has been chosen as the foundation for DREAMS SCL for safety systems up to SIL3. This part of the IEC 61784-1 series specifies a safety communication layer (service and protocol) based on IEC 61784-1, IEC 61784-2 and IEC 61784-3. PROFIsafe uses the 'Master-Slave' mechanism for transmission of safety telegrams. The master, which is typically called the 'F-Host', cyclically exchanges safety-relevant data with all its configured slaves called 'F-Devices'. The DREAMS SCL implements multiple safety measures specified by the standards mentioned before (see also Section 10.3.1): e.g., consecutive numbering to ensure completeness and the right order of transmitted safety Protocol Data Units (PDUs)s, time expectation with acknowledgment, data integrity check (CRC).

The developed SCL piece of software provides an Application Programming Interface (API) for both the communication protocol stack and the safety application software that uses it:

- High API (App-SCL): API provided by the SCL to the application (from Device application to 'F-Device' driver).

- Low API (SCL-Medium): API provided by the SCL for the integration with the communication protocol stack (e.g., EtherCAT). The SCL can be considered a 'universal' application level service on top of any non-safety related communication stack.

According to the standard [292], the message Protocol Data Unit (PDU) has the format described in Table 8.2, where

- The F-Input / Output Data is the payload that is interchanged between the 'F-Host' and the 'F-Device'.

- The status and control bytes (see Tables 8.3 and 8.4) are used by the 'F-Device' / 'F-Host' to provide information about the communication state to its counterpart. These two bytes are different between them; basically, the 'F-Host' uses the control byte to send safety-related commands to the 'F-Device' and the status byte is used by the 'F-Device' to answer 'F-Host' commands and to report any detected failure that may lead to a safety state. The format of both bytes is detailed below:

  - Status Byte (see Table 8.3)
    * Bit 0 is set when the 'F-Device' (its technology firmware) has new parameter values assigned. In this project it is not being used.
    * Bit 1 shall be set by the specific device technology, if there is a malfunction in the 'F-Device'.
    * Bit 2 is set if the 'F-Device' is recognizing an 'F' communication failure, i.e., if the consecutive number is wrong or the data integrity is violated. This bit information enables the 'F-Host' to count all erroneous messages within a defined time period T and to trigger a configured safe state of the system if the number exceeds a certain limit (maximum residual error rate).
    * Bit 3 is set if the 'F-Device' is recognizing an F communication failure, i.e., if the watchdog time in the 'F-Device' is exceeded.
    * Bit 4 is set by the Functional Safety Communication Profile (FSCP) 3/1 protocol layer during start-up and in cases of any communication error. In addition the 'F' part of the specific device application can set this bit also.
    * Bit 5 is a device-based toggle bit indicating a trigger to increment the virtual consecutive number within the 'F-Host'.
    * Bit 6 is set when the F-Device has reset its consecutive number counter.
    * Bit 7 is reserved for future FSCP 3/1 releases. It is reserved and still not used.

- Control Byte (see Table 8.4)

  - Bit 0 is set by the application within an 'F-Host' in case of a parameterization request ('F-Device' needs new parameters). In this project it is not used.

  - Bit 1 is set by the 'F-Host' driver corresponding to the variable "OA_Req_S". This signal is not safety related and should be used by the 'F-Device' to indicate locally the request for an operator acknowledgment ('OA_C').

- Bit 2 is set when the 'F-Host' detects a communication error, either by the status byte or by itself. As a consequence thereof the counter of the virtual consecutive number within the 'F-Device' will be set to '0'. Bit 2 shall be reset again after an error has gone. Thereafter the consecutive numbering resumes.

- Bit 3 is reserved for future FSCP 3/1 releases.

- Bit 4 can be set to force the outputs of an 'F-Device' to configured or built-in fail-safe values.

- Bit 5 is a host-based toggle bit indicating a trigger to increment the virtual consecutive number within the 'F-Device'.

- Bits 6-7 are reserved for future FSCP 3/1 releases. They are not still used in this project.

- The Cyclic Redundancy Check (CRC) is the procedure to calculate redundant data to be added to the message in order to detect errors which may arise during the transmission from the influence of physical data corruptions. The CRC part of the safety PDU can be of 3 or 4 bytes (it is selected through the F-Parameters). Although only one CRC is sent (the CRC2), the SCL calculates two CRCs. The first one is CRC1 and it is calculated with the F-Parameters. However, this CRC in not sent. After doing this, the SCL calculates the second one, using the first CRC1 as initial value. There are 2 options for the CRC2 that is going to be sent in the PDU. The first one is formed by 3 bytes and the second one, by 4 bytes. CRC1 is needed to calculate both options. The standard provides 3 different polynomials for the CRC to be calculated.

Table 8.2: Safety PDU Format [292]

F-Input / Output Data	Status / Control Byte	CRC2
Maximum 123 bytes	1 byte	3 or 4 bytes

Table 8.3: Safety PDU Status Byte [292]

Bit7	Bit6	Bit5	Bit4	Bit3	Bit2	Bit1	Bit0
Reserved	Consecutive Number Counter	Toggle	FV activated	Watchdog timeout	CE CRC	Device fault	Parameter

The developed DREAMS SCL has 3 main states as described below. Any unexpected error will trigger the 'safety communication state'. In that case, the 'F-Host' will command the 'F-Device' to switch to safety state through the 'Activate FV' flag (control byte) and the 'F-Device' will acknowledge it through the 'FV activated' flag (status byte). In case of getting into 'safety communication state', the 'F-Host' uses two flags prior to switch to regular communication: 'OA_e' and 'OA_C'. After an error, and once the first exchange of messages without any issue takes place, the 'F-Host' raises the 'OA_e' flag. After a second exchange without errors, and if the 'OA_e' flag is raised, the 'F-Host' raises the 'OA_C' flag. In order to reestablish the regular state both flags have to be raised.

Table 8.4: Safety PDU Control Byte [292]

Bit7	Bit6	Bit5	Bit4	Bit3	Bit2	Bit1	Bit0
Reserved	Reserved	Toggle	Activate FV	Reserved	Consecutive Number Counter	OA Request S	Parameter

- **Initial state:** The SCL starts in a safety state and depending on its behavior during the start up, it moves into the 'safety communication state' or the 'regular communication state'. After the first 4 iterations, and if no error has been detected, the system enters into 'regular communication state', which is the desired state (scenario without errors reported). Nevertheless, should an error be detected, the system moves into the 'safety communication state'.

- **Safety Communication State:** The system moves into this safety state due to an error in the communication and will be able to recover to 'regular communication state' after 4 correct iterations.

- **Regular Communication State:** The SCL stays in this state when the communication is operating correctly without reported errors.

8.2   Security Services

In DREAMS, security services are provided for the protection of on-chip / off-chip communication, time synchronization, resource management and execution services. These services are provided at different layers and levels. These include chip level security, cluster level security and application level security. These topics are explained in detail in their respective subsections. A threat analysis of the complete system was performed to identify the potential threats to the most important assets that need to be protected. This includes identification of the potential threats to the assets, the potential attackers, examples of potential attack scenarios and means to protect against the attacks. It is identified that the communication services, both on the on-chip and off-chip levels, are vulnerable to different kinds of attacks and therefore they need to be protected in different ways, such as for confidentiality, authenticity, integrity, etc.

Resource management is at the core of the DREAMS project (see Chapter 9). In resource management, the complete (distributed) system is monitored, configured and reconfigured for resource usage when and as needed. Resources are initially allocated, and in case of failures, the resources are reallocated from another part of the system or applications are moved to other parts of the system. It is important that the resource management is performed uninterrupted and therefore the resource management components need to be secured in order to protect them and ensure that the system is running smoothly. Additionally, time synchronization is also a very important aspect of the DREAMS project, and it needs to be ensured that the time synchronization services are executed uninterrupted and without any impact on their security. Last but not the least, the execution services also need to be protected against security threats.

8.2.1 Risk Analysis

In this section the risk analysis of the whole system is performed. The security risks, potential vulnerabilities in the system and the ways in which these vulnerabilities can be exploited are listed. The potential attackers are also classified and the attack targets are listed. The most important factors considered in this section are: types of attackers, location of the attackers and attack targets.

The first and foremost question is identifying the potential attackers. It is important to consider who a potential attacker might be. It is noteworthy here that the attacker might not only be a human but it could also be a rogue executable / application.

In DREAMS, the attackers are classified as internal attackers and external attackers. Internal attackers are the ones who have access to the NoC, whereas external attackers have no physical access to the NoC but they can still attack the communication being done over the network, i.e., cluster level communication. An attacker might target different components of the system, such as resource management components, user applications, time synchronization components and / or messages exchanged between them.

As the DREAMS architecture is structured into physical and logical views, security has to be considered for both views.

8.2.1.1 Securing DREAMS Services

The most important DREAMS architectural services that require security are briefly discussed below:

- **Communications Services:** The communications in DREAMS is performed over the on-chip network (NoC communication) or over the off-chip network (cluster-level communication). The attacker having access to the NoC is termed as internal attacker. By having access to the NoC, the attacker can read the memory for secret information such as keys and can generate authentic messages. An external attacker on the other hand has no access to the NoC, but he can still intercept and try to manipulate off-chip communication. Such an attacker can capture messages to analyze them offline, replay messages such as time synchronization messages and / or try to generate fake messages on behalf of a system component. Figure 8.12 shows which kind of communication is accessible to which kind of attacker. The on-chip as well as off-chip communication needs to be protected from manipulations through appropriate security services.

- **Resource Management Services:** Resource management services ensure the adaptability of applications of mixed criticality. They ensure system reconfiguration upon changes in the system's environment. In DREAMS the most important resource management services are the Global Resource Manager (GRM) (see Section 9.5) and the Local Resource Manager (LRM) (see Section 9.4). The GRM makes decisions based on the information from the LRMs. LRMs are tasked by the GRM to commit the decisions made by the GRM. Thus sending wrong decisions on behalf of the GRM to the LRMs can cause havoc in the system. On the other hand, sending false / masqueraded information on behalf of one or more LRMs to the GRM will influence the GRM to make wrong decisions and thereby force the corresponding LRMs to implement wrong decisions. This will also have a devastating influence on the correct state of the system. Therefore the communication between the GRM and the LRMs has to be appropriately protected. Further description about security in the resource management is given in Section 9.6.1.

- **Time Synchronization Services:** The global time synchronization services of DREAMS ensure that the local clocks in the system have the same value at the same points in real-time. This

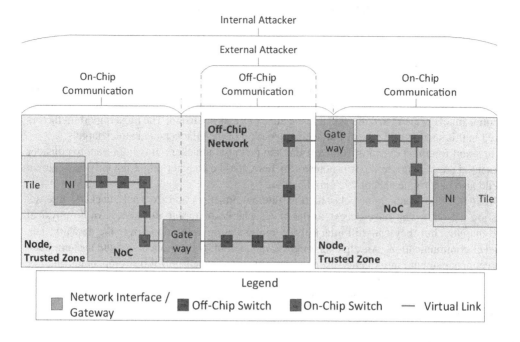

Figure 8.12: Differentiation of Internal and External Attackers

is important for the functioning of the time triggered network. However, these services can be targeted by attackers in mainly two different ways. The attacker might aim at the communication processes, thus compromising the time synchronization. The attackers might also target the time values of the individual clocks, disrupting the global time view at the attacked node.

- **Execution Services:** The execution services provide the ability to run applications in the DREAMS system. Execution services are provided through the hypervisor and ensure a transparent on-chip / off-chip communication. The virtualization layer of the hypervisor abstracts the underlying hardware and ensures spatial and temporal partitioning. Spatial partitioning ensures that the memory of the partitions is isolated, thus a failure in a partition does not impact the other partitions in the node and the overall system. As one application is assigned to one partition, the internal attackers can only access and manipulate the messages of their allocated partition, whereas an external attacker cannot gain access to any partition.

8.2.1.2   Potential Attacks

In this section, a list of the potential attacks on the system is given and their consequences are discussed.

- **Sniffing Attacks:** An attacker might sniff the packets, to see which kind of communications are taking place over the system. The spoofed packets are normally analyzed offline by an attacker. An example would be to sniff the packets belonging to the patient health information, exchanged by the patient monitoring system. Although this is not an active attack, the confidentiality of communication is compromised by this attack.

- **Man-in-the-Middle Attack:** An attacker might insert himself between two real communication parties and make them believe that they are talking to the legitimate communication partners directly. However, in reality they would be talking through the attacker, who can monitor and or disrupt the communication. The attacker might simply suppress some messages, so that the legitimate recipient never receives them or the attacker might alter or delay the messages to achieve certain objectives as described below.

- **Suppressed Status Messages:** An attacker might suppress certain messages. This can be achieved by specifically targeting certain message types or suppressing messages at random. In the first case, the attacker needs to be able to interpret the message types, whereas the second case is relatively easier to perform. As an example, the resource management components, such as GRM and LRM are continuously exchanging status messages. If such messages are suppressed, they have different effects. If the messages from LRM cannot reach GRM, the GRM will assume that the subsystem has failed. This will result in GRM issuing reconfiguration commands. Other messages that can be potentially dropped include the time synchronization messages from master clock to the slave clocks, thereby disrupting the correct functioning of the overall time triggered communication.

- **Delayed Status Messages:** An attacker might be able to delay certain messages to achieve certain unexpected results. This is related to the above mentioned attack, but here the attacker is not dropping the messages, rather delaying them by a certain amount. As an example, the reconfiguration messages from GRM to LRM or the status messages from LRM to GRM can be delayed. Other messages that can be potentially delayed include the time synchronization messages from master clock to the slave clocks.

- **Denial of Service (DoS) Attacks:** An attacker might be able to insert useless message of his own in the network (or replaying an existing packet a large number of times) to choke a switch or an end node. This will result in choking the switch, the end node, or the switch and all the end nodes connected to the switch and thereby disrupting critical communications which should otherwise take place timely. An example includes disrupting the communication involving a patient monitoring system, or the flight management system.

- **Masquerading Attacks:** Masquerading attacks can be performed in many ways. An attacker might masquerade as GRM, as LRM or as any other application. If an attacker masquerades as GRM, he might be able to send false configuration / reconfiguration messages to one or more LRMs. By sending false status message to GRM on behalf of an LRM, or by sending wrong information about the failure of components, the GRM might be forced to choose inappropriate configuration.

8.2.2   Security at Multiple Levels

The risk analysis and the potential attacks described in the previous section show the requirement of protecting the system on different levels. The security services have to be provided on the data link layer and the application layer. Figure 8.13 shows the different layers of the security services. Whereas the data link layer security services secure the connection on layer 2, the application layer security services secure the connection between the applications on layer 7. Two main levels of security services provided in DREAMS are:

- **Cluster-Level Security Services:** Cluster-level security services cover the off-chip communication. Thus, a message originating from a node till it reaches the end node is secured using the cluster level security services. In DREAMS, these services are provided through Media Access Control Security (MACsec) protocol, which is defined in the IEEE 802.1AE standard [295].

- **Application-Level Security Services:** Application-level security services provide end-to-end security and protect against certain attacks on the chip level. This is provided as an optional set of services in the form of a security sub layer. Applications have the ability to bypass it if end-to-end security is not needed, e.g., by in-flight entertainment system.

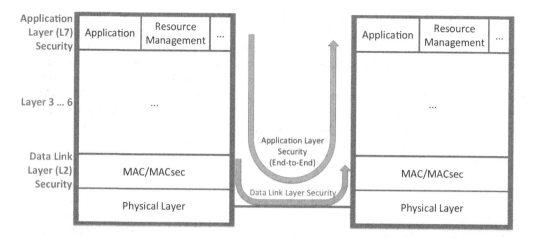

Figure 8.13: Security in the Data Link Layer and Application Layer

This classification leads to different understandings of the term end-to-end. End-to-end might refer to the applications. In this case, security services have to be provided at the application level. End-to-end might also refer to the cluster-level communication, i.e., the communication between two nodes. Under this interpretation of end-to-end communication, security services have to be provided at the data link layer. Figure 8.14 shows the different understandings of the term end-to-end.

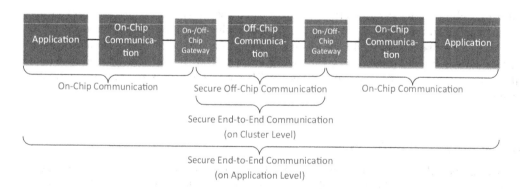

Figure 8.14: Secure End-to-End Communication

8.2.3 Security Classification

The following basic security services are provided in the context of DREAMS project at the application and / or cluster level:

- **Confidentiality:** Confidentiality ensures the privacy of information, such as the privacy of the exchanges messages.

Table 8.5: Classification of Security Services

Security Level	Supported Security Service
0	No Security
1	Integrity
2	Integrity & Authentication
3	Integrity & Authentication & Confidentiality

- **Integrity:** Integrity ensures that if the data or messages are modified, the modification will be detected by the intended recipient.

- **Authentication:** Authentication ensures the authenticity of data origin as well as the authenticity of the communication partner.

- **Availability:** Availability services ensure that the system and / or its components are available when needed.

- **Access Control:** Access control services protect against unauthorized access to system components.

In DREAMS, four different levels of security services are defined as shown in Table 8.5 [296].

- Security level 0 is used by those applications that do not need any security service, e.g., when there is no confidential data to transmit and authenticity and integrity are also not required, such as in-flight entertainment system.

- Security level 1 is used by applications, which require the transfer of data to be protected against modifications, and in case if modifications occur, they should be able to detect them.

- Security level 2 is an extension of security level 1, where not only integrity is provided but also the authenticity of the communication partner is ensured. An example is the exchange of time synchronization message, which might not be treated as confidential, but the authenticity of sender as well as integrity of the time value is important.

- Security level 3 is used by most of the critical applications that need privacy of communication in addition to security level 2. Such applications include resource management components, such as GRM and LRM.

To prevent replay attacks, e.g., to resend an old reconfiguration request, security level 1, 2 and 3 use time-varying parameters. Checking these time-varying parameters, the receiver can detect, if he already received this message.

8.2.4   Cluster-Level Security

A cluster is an interconnection of nodes using the on-chip / off-chip gateways and the off-chip switches. In DREAMS, the nodes are connected using the TTEthernet technology. MACsec is used to protect the cluster level communications. MACsec protects against many of the above discussed potential attacks and provides the following guarantees [295]: data origin authentication, data integrity, confidentiality, replay protection, bounded receive delay and DoS attack mitigation.

However, non-repudiation and protection against traffic analysis is not covered by MACsec. MACsec is a layer 2 protocol and it supports connectionless communication, which is required by real-time systems. MACsec establishes two Secure Channels (SCs) on each link. One SC is established in each direction. The pair of SCs over a link forms a Secure Connectivity Association

(CA). Each CA represents groups of nodes or switches connected via unidirectional links. During the flow of a frame from source node to destination nodes, each switch decrypts the incoming data and re-encrypts it before forwarding it. Due to different SCs between the switches, the data might be encrypted using different keys, and other parameters. Key management as well as the establishment of CAs is not covered by 802.1AE, but is specified separately by 802.1X-2010 [297].

MACsec defines a default cipher suite of Galois / Counter Mode (GCM) [298] of Advanced Encryption Standard (AES) [299] with a 128 or 256 bit key, i.e., GCM-AES-128 or GCM-AES-256, to protect the frames. GCM supports and enables authenticated encryption operation [300], i.e., authentication and encryption in one step. MACsec frame format is mostly the same as Ethernet frame format, with the addition of a security tag and a Message Authentication Code or Integrity Check Value (ICV) field.

Security features of MACsec are implemented directly in the MAC layer of TTEthernet IP core. Each port of the core consists of one TX module and one RX module, where TX is used for frame transmission and RX is used for frame reception. MACsec is implemented as an extension of existing RX and TX modules and these extended versions are called RX_{MACsec} and TX_{MACsec}. Thus, each RX and TX of every port represents one MACsec hardware module with its own GCM-AES module that is used for data encryption, decryption and authentication. This means that the amount of MACsec modules used in TTEthernet IP core can be described as shown in Equation (8.1).

$$M = 2 * P \tag{8.1}$$

M is the number of MACsec modules and P is the number of ports. Switches that have higher amount of ports may require too much logic resources for implementation of MACsec features for all ports. For this reason, the amount of MACsec supported ports may be limited and some ports may be chosen to omit MACsec support. Thus, a trade off between FPGA resource costs and amount of secured ports is recommended for switches with many ports.

Figure 8.15 represents a diagram depicting MACsec processing, i.e., composition and decomposition of MACsec frames for cases when encryption is used. For frame transmission, the communication layer provides 'destination address', 'source address', 'Ethertype' and 'payload' to the TX_{MACsec} module, where the MACsec frame is created. Since the MACsec frame is a standard Ethernet frame, 'preamble', SoF (Start of Frame) and CRC are generated in ordinal way. 'Destination address' and 'source address' are copied to the MACsec frame without any changes, but they are also simultaneously used for generation of the ICV (Integrity Check Value). Then, the MACsec header is generated, which consists of MACsec 'EtherType', 'control bits', Packet Number (PN) and an optional SCI. The MACsec header is used for the ICV generation at the same time as well. The original 'EtherType' and 'payload data' provided from the communication layer are encrypted by the 128-bit version of AES that operates in the GCM (Galois / Counter Mode). The encrypted data are then inserted behind the MACsec header and simultaneously used for the ICV generation. Then, the ICV is generated and inserted behind the encrypted data. Finally, the CRC is created the same way as for any other type of an Ethernet frame.

For frame reception, the processing starts with ordinal Ethernet frame processing, i.e., the 'preamble' and SoF are identified. Then, 'destination address' with 'source address' are captured and simultaneously provided to the GCM-AES module for the ICV generation. Then the 'Ether-Type' is analyzed to check whether the incoming frame is a MACsec frame. If it is not a MACsec frame, the ICV generation is cancelled and frame is processed in the ordinary way. If it is a MACsec frame, the MACsec header is further analyzed and simultaneously used for the ICV generation. After that, the encrypted 'EtherType' and the encrypted payload arrive. These encrypted data are used for decryption and ICV generation as well. Then, the ICV is received, which is compared to the generated ICV. If these values are equal, then the authentication was successful. If not, then the authentication fails and an error is sent to the communication layer, and thus the received frame is

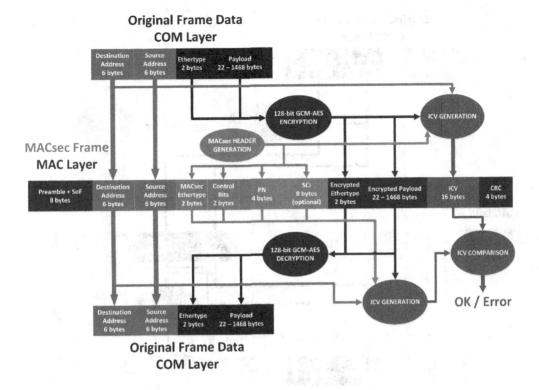

Figure 8.15: MACsec Frame Encapsulation and Processing Diagram Using Encryption and Decryption

dropped. The last step is the ordinary CRC check, which may also produce an error in case that the frame was corrupted.

The encryption itself can be turned off, while the remaining MACsec functionality responsible for message authentication is still active. This situation is shown in Figure 8.16. It is important to mention, that the ICV generation requires the encryption of the data due to the GCM algorithm. Thus, the encryption process must be used regardless of whether the frame must be encrypted or not. Therefore, the time required for MACsec processing without frame encryption is the same as the time required for MACsec processing with frame encryption.

The implemented MACsec modules are using 128-bit version GCM-AES algorithm only. Other options are not available in the current version of the TTEthernet IP core. This algorithm was selected because MACsec standard defines it as a default algorithm to be used. It is necessary to ensure that the throughput of the MACsec processing is not lower than the speed of frame transmitting, or receiving, respectively. In our case a speed of $1~Gbit/s$ is used in a combination of $125~MHz$ clock frequency that is used for the logic implemented in a FPGA. According to these parameters, one byte per clock cycle is transmitted / received and thus, the throughput of MACsec processing must be at least 1 byte per clock cycle. If this timing requirement is not met, then the latency caused by MACsec cannot be constant, which is very important for minimizing the jitter in distributed real-time systems.

8.2.5 Secure Time Synchronization

Time synchronization ensures that the time at the local clocks is the same at the same time in real time. Cluster-level communication services are used in DREAMS by the time synchronization ser-

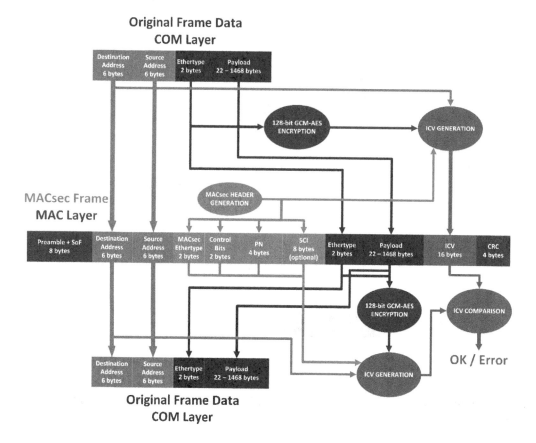

Figure 8.16: MACsec Frame Encapsulation and Processing Diagram without Encryption and Decryption

vices. Time is synchronized using the TTEthernet synchronization mechanism. The time synchronization service is secured in DREAMS to protect against the attacks on time synchronization as discussed already. Exchange of messages during time synchronization does not require confidentiality as privacy of the synchronization messages is not important. However, integrity and authenticity of the time synchronization messages have to be ensured. This in turn ensures that the time values have not been changed during transit and that the messages come from trustworthy sender.

Protocol Control Frames (PCFs) are used by TTEthernet to exchange time synchronization messages. MACsec is used for securing the synchronization PCFs. The ICV field of MACsec provides integrity and authentication services. Since encryption and decryption might increase the processing overhead, omitting them might help in reducing the overall processing time. However, still the security services of integrity and authentication increases the jitter as compared to the situation when MACsec is not used. This is due to the fact that CAs between the switches have to be established before MACsec can provide the required security services. As due to the used algorithms, omitting confidentiality does not lower the overhead in the presented implementation. Hence, confidentiality, integrity and authentication are used for the secure time synchronization.

PCF frames use their own specific 'EtherType' value. These frames are protected the same way as the frames types by MACsec. In the TX_{MACsec} modules of devices that are configured as synchronization master, standard PCF frames are periodically created, then encapsulated into MACsec frame and then sent to the network. In the RX_{MACsec} modules of all devices, these frames are translated into standard PCF frames (i.e., authenticated and decrypted if they were encrypted), then they are identified by PCF 'EtherType' and processed in a standard way.

8.2.6 Application-Level Security

For end-to-end security, application level security services are provided in DREAMS. These services are provided in the form of a security sub-layer on top of the communication layer of the underlying hypervisor, i.e., XtratuM [198]. Two options were considered for providing the security services, i.e., either as a security sublayer, running in its own partition, which each application can communicate to using standard communication services. Thus, an application can send its data to the security sublayer for encryption or decryption. However, this still means that the data is transferred without any security between the partitions. There is also a security overhead in doing so. The other option considered and chosen was to provide security services in the form of a sub-layer over the communications layer of the hypervisor. Thus, each application can transfer its data to the security sub-layer and specify the required security level. The security layer provides the appropriate level of security and hands over the data to the communication layer for transmission to the other end. This is similar to the standard way of data transmission in the absence of a security sub-layer.

The application level security services are split into four levels as stated before. Some applications, such as in-flight entertainment system, might not need any end-to-end security. In such a case, they can opt for no security service, and thus, the communication will bypass the security sub-layer. This is possible by choosing the provided security level 0.

8.2.6.1 Message Exchange Format

Each of the security levels require the definition of an appropriate message format [210]. The message format defines a header and a trailer (on security level 0, there is no trailer). The message format of security level 0 is shown in Figure 8.17. The data fields for security level 0 are defined as follows:

- *frame.length* (4 bytes): specifies the length of the message.

- *message* (variable length): contains the message itself. The message has a variable length. Its length is defined in the *frame.length* field. It is not encrypted. Integrity and authentication checks are not possible.

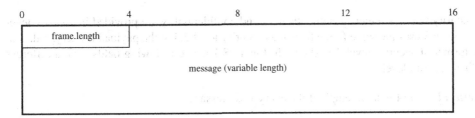

Figure 8.17: Security Level 0 Frame

Security level 1 provides integrity. There is no separate frame for security level 1 as using symmetric cryptographic mechanisms, the separation between integrity and authentication is not possible. Internally, security level 2 is used, if an application requests security level 1.

Integrity and authentication are provided by security level 2. The payload is not encrypted, but protected against manipulation. The message format of security level 2 is shown in Figure 8.18. The data fields for security level 2 are defined as follows:

- *aad.length* (4 bytes): length of the Additional Authenticated Data (AAD), i.e., the length of the header. The length is not fixed because of the *flags* field.

- *plaintext.length* (4 bytes): length of the message.

- *nonce* (12 bytes): Initialization Vector (IV) for the security mechanisms.

- *sourceID* (4 bytes): XtratuM source ID.

- *destID* (4 bytes): XtratuM destination ID.

- *portDesc* (4 bytes): XtratuM port descriptor.

- *portType* (4 bytes): XtratuM port type.

- *flags* (variable length): if required, different additional flags can be defined. For now, the flags are not used.

- *plaintext* (variable length): message itself. The message has a variable length. Its length is defined in the *frame.length* field. The field is not encrypted, but the integrity and authentication can be checked using the authentication tag.

- *tag* (16 bytes): authentication tag to ensure integrity and authentication. The authentication tag secures both the header, starting at *aad.length* and ending at *flags*, and the plaintext.

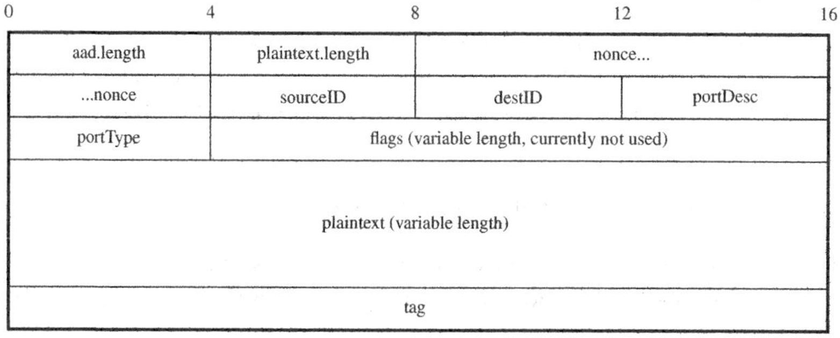

Figure 8.18: Security Level 2 Frame

The security services integrity, authentication and confidentiality are provided in security level 3. Security level 3 uses the same frame format as security level 2, but the payload is encrypted. The message format of security level 2 is shown in Figure 8.19. The following fields have a different content than security level 2:

- *ciphertext.length* (4 bytes): length of the encrypted message.

- *ciphertext* (variable length): encrypted message. The message has a variable length. Its length is defined in the *frame.length* field. Integrity and authentication can be checked using the authentication tag.

- *tag* (16 bytes): contains the authentication tag to ensure integrity and authentication. The authentication tag secures both the header, starting at *aad.length* and ending at *flags*, and the ciphertext.

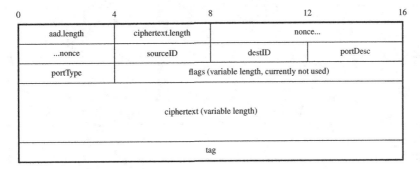

0	4	8	12	16

aad.length	ciphertext.length	nonce...	
...nonce	sourceID	destID	portDesc
portType	flags (variable length, currently not used)		
ciphertext (variable length)			
tag			

Figure 8.19: Security Level 3 Frame [210]

8.2.6.2 Supported Algorithms

A choice of the following lightweight cryptographic algorithms [301, 302] are provided for securing the end-to-end communication:

- **CLEFIA-OCB:** CLEFIA is standardized in [302]. Together with the Offset Codebook (OCB) mode [303], it can be used to provide authentication. CLEFIA-OCB provides a ciphertext and an authentication tag when a plaintext, a key and a nonce / Initialization Vector (IV) is provided as input. The lengths of nonce and the input block for CLEFIA is 128 bits. CLEFIA is based on Feistel Network [304].

- **ChaCha20-Poly1305:** ChaCha20 is a stream cipher, which is provided as an alternative cipher to the block cipher CLEFIA. Poly1305 is a cryptographic Message Authentication Code (MAC), which is used for providing data integrity and authentication. Together, the combination ChaCha20-Poly1305 is standardized as an authenticated encryption cipher in RFC 7539 [305] and RFC 7905 [306]. The combination is used in many protocols such as Transport Layer Security (TLS), and used by Google to secure the communication between their servers and Android phones.

8.2.6.3 Evaluation

In this section, the overhead of the security library is evaluated. The clock cycles are measured for the complete function call including XtratuM inter-partition communication functions. There is no separate measurement for security level 1 as using symmetric cryptographic mechanisms, the separation between integrity and authentication is not possible. Internally, security level 2 is used, if an application requests security level 1.

In Figure 8.20 the impact of the security sublayer using no security services is compared to the direct usage of the XtratuM communication services without using the security sublayer. As neither CLEFIA-OCB nor ChaCha20-Poly1305 are used on security level 0, these measurements are independent from the selected algorithm on the channel. Using small message sizes, the overhead is insignificant. Starting with 128 byte messages, the required clock cycles increase because of additional copy operations of the data.

CLEFIA-OCB Measurements

Figure 8.21 shows the clock cycles required to send a message using the respective security level in the CLEFIA-OCB mode. Level -1 means, that the security sublayer is not used, i.e., the XtratuM communication functions are used directly. The measurements for security level -1 and security level 0 show about the same values as in Figure 8.20. This is because the security algorithms are

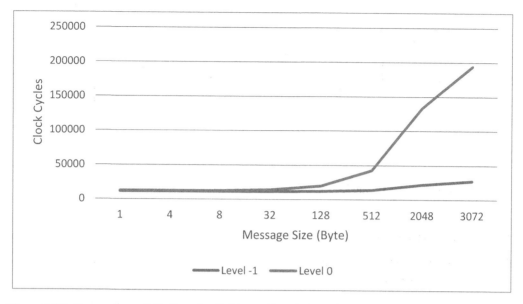

Figure 8.20: Comparison of No Security Sublayer (Level -1) and the Security Sublayer on Level 0

not involved. Security level 2 and 3 require about the same amount of clock cycles due to the construction of the OCB mode of operation.

The measurements of the required clock cycles receiving and checking (and decrypting, on level 3) the message show similar values as sending the message. Figure 8.22 shows the measurements for receiving a message. The required clock cycles are slightly higher as the message authentication code has to be checked.

ChaCha20-Poly1305 Measurements

Figure 8.23 shows the clock cycles required to send a message using the respective security level in the ChaCha20-Poly1305 mode. Level -1 means, that the security sublayer is not used, i.e., the XtratuM communication functions are used directly. Also here, the measurements for security level -1 and security level 0 show about the same values as in Figure 8.20, because the security algorithms are not involved. Due to the construction of the ChaCha20-Poly1305 combination, the security level 2 is faster than the security level 3 where all three security services are provided, i.e., confidentiality, integrity and authenticity.

Receiving a message using ChaCha20-Poly1305, the clock cycles are slightly higher than when sending a message. Figure 8.24 shows the measurements for receiving a message using ChaCha20-Poly1305. Also here, the additional check of the message authentication code leads to the increase.

Comparison

In this section, the two algorithms CLEFIA-OCB and ChaCha20-Poly1305 are compared regarding their required clock cycles.

- Figure 8.25 shows the required clock cycles on security level 2. Especially when sending large messages, there is a major advantage of ChaCha20-Poly1305. The same applies for receiving messages as shown in Figure 8.26.

- Figure 8.27 shows the required clock cycles on security level 3. Here, the advantage of ChaCha20-Poly1305 compared to CLEFIA-OCB is smaller, but it is still significant. Again, the same applies for receiving messages as shown in Figure 8.28.

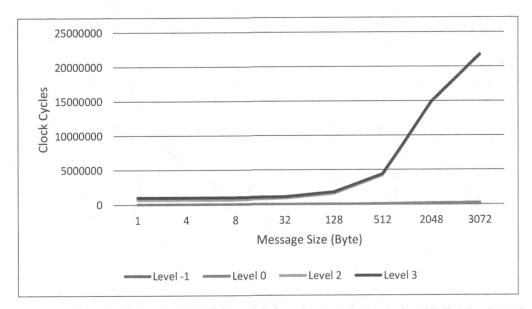

Figure 8.21: CLEFIA-OCB – Clock Cycles – Send, No Security Sublayer (Level -1) to Level 3

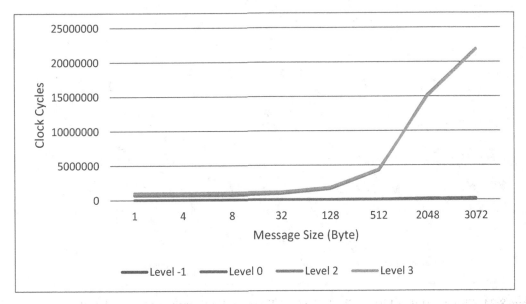

Figure 8.22: CLEFIA-OCB – Clock Cycles – Receive, No Security Sublayer (Level -1) to Level 3

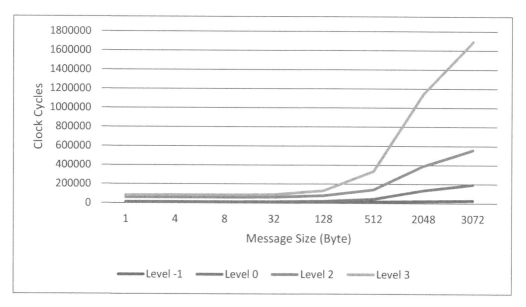

Figure 8.23: ChaCha20-Poly1305 – Clock Cycles – Send, No Security Sublayer (Level -1) to Level 3

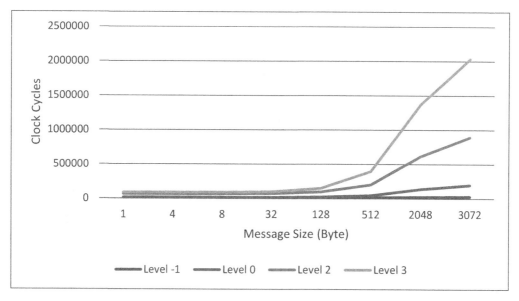

Figure 8.24: ChaCha20-Poly1305 – Clock Cycles – Receive, No Security Sublayer (Level -1) to Level 3

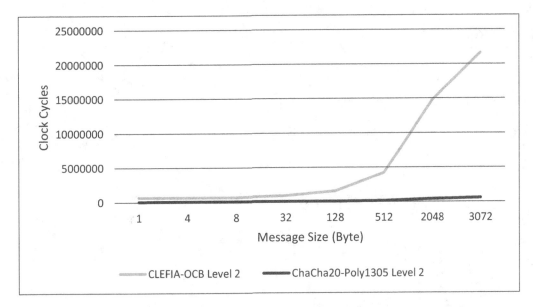

Figure 8.25: Comparison of CLEFIA-OCB and ChaCha20-Poly1305 on Level 2 (Send)

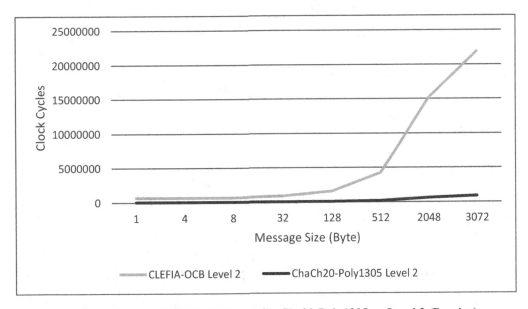

Figure 8.26: Comparison of CLEFIA-OCB and ChaCha20-Poly1305 on Level 2 (Receive)

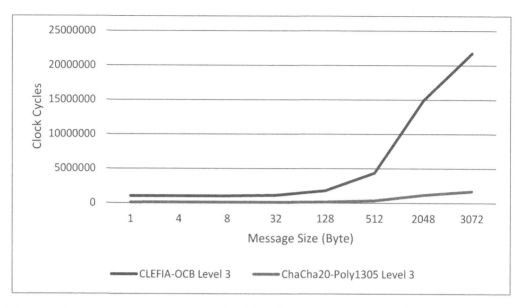

Figure 8.27: Comparison of CLEFIA-OCB and ChaCha20-Poly1305 on Level 3 (Send)

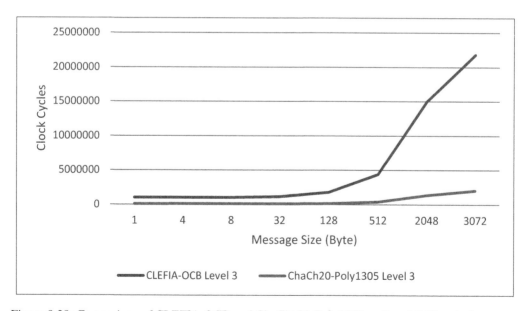

Figure 8.28: Comparison of CLEFIA-OCB and ChaCha20-Poly1305 on Level 3 (Receive)

9

Resource Management Services

G. Gala
Technische Universität Kaiserslautern

D. Gracia Pérez
THALES Research & Technology

G. Fohler
Technische Universität Kaiserslautern

C. Pagetti
ONERA

9.1	Overview of DREAMS Resource Management		378
9.2	Local Resource Monitor or MON		379
	9.2.1	MON for Core Failure	380
		9.2.1.1 General Approach	380
		9.2.1.2 Implementation	380
	9.2.2	MON for Deadline Overrun	381
		9.2.2.1 General Approach	381
		9.2.2.2 Implementation	382
	9.2.3	MON for Quality of Service	382
		9.2.3.1 General Approach	382
		9.2.3.2 Implementation	383
9.3	Local Resource Scheduler or LRS		384
	9.3.1	General Approach	384
	9.3.2	Implementation	385
	9.3.3	Requirements on Applications	386
9.4	Local Resource Manager or LRM		387
	9.4.1	Core Failure Management	387
		9.4.1.1 Implementation	388
	9.4.2	Deadline Overrun Management	388
		9.4.2.1 General Approach	388
		9.4.2.2 Implementation	390
	9.4.3	QoS Management	390
		9.4.3.1 Implementation and Choices	392
		9.4.3.2 Improving QoS with Monitors Information	392
		9.4.3.3 Applications with QoS Support	393
		9.4.3.4 LRM QoS and Deadline Overrun Managers	393
9.5	Global Resource Manager or GRM		394
	9.5.1	Implementation	394
	9.5.2	Global Reconfiguration Graph	395
9.6	Resource Management Communication		397
	9.6.1	Secure Resource Management Communication	400
		9.6.1.1 Security Services and Levels	400

This chapter is dedicated to the DREAMS Resource Management Services (RMS), which are composed of the Global Resource Manager (GRM), Local Resource Managers (LRMs), Local Resource

Schedulers (LRSs) and Resource Monitors (MONs) services. Section 9.1 provides an overview of the Resource Management (RM) services by giving a short definition of each RM component. Sections 9.2, 9.3, 9.4 and 9.5 elaborate on the MON, LRS, LRM and GRM respectively by describing the general approach and exhibiting an implementation of each. At the end, in Section 9.6 the communication between the aforementioned building blocks, including the security services are described.

9.1   Overview of DREAMS Resource Management

DREAMS provides a set of services for system-wide adaptability of mixed-criticality applications consuming several resources via global integrated RM. The resource management in DREAMS is in charge of maintaining the execution of applications even in the presence of failures. The resource management services are realized by Local Resource Managers on each multi-core and a Global Resource Manager running on one of the multi-core platforms. The Global Resource Manager knows at any time the global view of the overall system and is able to construct the local configurations of each multi-core.

The resource management aims at supporting two types of failures:

- permanent core failures leading to local and global reconfiguration;

- temporal overload situations, e.g., when best-effort applications have a too consuming access to shared resources, leading to adaptation.

The DREAMS middle-ware relies on the Time and Space Partitioning (TSP) principles compliant with Integrated Modular Avionics (IMA) [197, 307], which is the de-facto standard for current aircraft design. TSP is ensured by the XtratuM hypervisor [198] and an additional layer, named DRAL for DREAMS Abstraction Layer, provides specific interface to the application to access the DREAMS services. Among those services, DRAL offers the capabilities to reconfigure the application ports accessing the TTEthernet network.

The fault-tolerant services, referred as the DRM run-time library, are implemented on top of XtratuM and DRAL. They consist of:

1. MON service, which monitors the health of cores and critical applications,

2. LRS service, which performs the run-time scheduling based on the configuration set by the LRM, and

3. LRM service, which either adopts the initial configuration, or a configuration requested by the GRM, or selects a new configuration from the ones available and reports current state to the GRM.

4. GRM service is the central component, which has comprehensive knowledge about the system and it communicates with and co-ordinates all the LRMs. It makes global decisions based on the information received from LRMs; it obtains new configurations from an offline-computed set of configurations and send the reconfiguration orders to the LRMs. It also manages an external input to manually trigger a system-wide reconfiguration.

The four types of RM building blocks (GRM, LRM, MON, LRS) can be arranged across the DREAMS platform in many different configurations as explained in Section 2.2.9.4. All the potential local and global reconfigurations are computed off-line by the DREAMS toolchain. Thus, when a failure occurs, the resource managers simply apply the transition decided off-line.

Application model

We consider mixed-critical systems, where we differentiate the following two types of application.

Definition 9.1 (Application model). *An application can be a:*

- critical application. *Such an application must respect its timing constraints and in particular the WCET must fit in the allocated slots. Moreover, it cannot be stopped apart if the application encounters an internal error or if the executive layer fails. A critical application app is defined as a set of periodic or sporadic tasks $app = \{\tau_i = (C_i, AET_i, T_i)\}$ where C_i is the WCET, AET_i is the average execution time and T_i is the period or minimal inter-arrival time;*

- best-effort application. *Such an application has less strong constraints. We accept to interrupt them as long as a minimal QoS (quality of service) is ensured. A best-effort application is defined as $app = (U_i, AU_i)$ where U_i is the worst-case asked utilization and AU_i is the average utilization.*

System model

A configuration consists in defining temporal slots on the multi-core and mapping the applications in the slots.

Definition 9.2 (Configuration). *A configuration (also denoted* plan *in the hypervisor terminology) consists of:*

- *a* major cycle *(MaC), the length of which is denoted as* MaC_length;

- *a set of slots sl_i distributed over the cores and the MaC. A slot is defined as $sl_i = ([s_i, e_i], n_i)$, where s_i is the start time, e_i is the end time and n_i is the number of cores where the slot is allocated;*

- *a mapping of the jobs of critical applications in the slots. Jobs are unrolled on the MaC and we know for each job $\tau_{i,j}$ to which slot sl_k it belongs to. We know moreover in which order the jobs inside a slot are executed;*

- *a mapping of best-effort applications in the slots. For instance, app_i is executed in the slots $sl_{j_1}, \ldots, sl_{j_p}$.*

Definition of the Notions of Reconfiguration and Adaptation

A reconfiguration consists in moving from one configuration to another and this happens when a core has failed. An adaptation consists in degrading a configuration and this occurs when a temporal overload situation happens. Adaptations are handled locally by the LRM, whereas reconfigurations are performed locally by the LRM or globally by the GRM.

9.2 Local Resource Monitor or MON

The MON makes the LRM enable to take reconfiguration or adaptation actions due to the safety and performance requirements by collecting information from the system and the applications. It is in charge to detect three different failures:

1. MON for core failures;

2. MON for deadline overrun;

3. MON for quality of service (QoS).

9.2.1   MON for Core Failure

The MON is regularly executed on each core to monitor the operation of the core. In case the core is operational, this service updates a shared structure that is in nominal mode. Otherwise, if the core is failed, the service is consequently not executed and is not able to modify the shared structure, thereby the failure can be detected. For an efficient detection, this operation must be accurately scheduled and ordered. Hence, each core performs this action at a pre-defined known time.

9.2.1.1   General Approach

The MON updates the shared structure in an asynchronous manner, but at distinct pre-defined instants. The exchange between the MON calls and the LRM is based on shared memory. At the end of the MaC, all the (still alive) LRMs detect the cores that have failed since the last MaC.

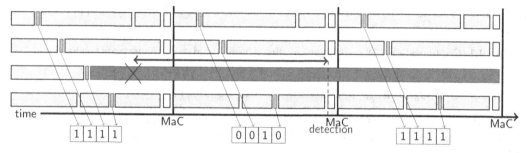

Figure 9.1: Example of Core Failure Detection on a Quad-Core - Green is Used for Tasks, Violet for MON, Yellow for LRM and Gray for Failed Core

Figure 9.1 depicts an example of core failure detection on a quad-core processor. Different processing activities run on individual cores (shown by wide green slots) and the MON is executed once on each processing core (shown by narrow violet slots). During each MaC, each MON toggles the respective entry of the shared structure and at the end of the MaC, the LRM (shown by narrow yellow slots) checks the value of the shared structure. In this example, a core failure on the third core during the second MaC is detected, as the respective value in the shared structure is not updated.

9.2.1.2   Implementation

The following source codes exhibit how the main functionalities of the MON can be implemented. The represented function in Code 1 initializes the shared data structures. The second function in Code 2 updates the structure each time MON is called.

Code 1 (For Declaration of Shared Memory and Alive Buffers).

```
void mon_core_failure_init() {
  struct dlrm_mon_core_failure *data =
    (struct dlrm_mon_core_failure *)
    dlrm_lrm_malloc(sizeof(struct dlrm_mon_core_failure));
  dlrm_config.lrm_desc->shared_mem->mon_core_failure_data = data;
  int32_t *alive = data->alive_buffer;
  for(int i=0; i < DLRM_MAX_CPU; i++) {
    alive[i] = 0;
  }}
```

Code 2 (For Checking and Updating the Sampling Port).

```
void mon_core_failure_run(int32_t partid) {
    uint32_t myid = dlrm_config.part_desc->part[partid].hwcpuid;
    struct dlrm_lrm_desc *lrm_desc = dlrm_config.lrm_desc;
    struct dlrm_lrm_shared *shared = lrm_desc->shared_mem;
    int32_t *alive =&shared->mon_core_failure_data->alive_buffer[myid];

    *alive = 1- *alive;
}
```

9.2.2 MON for Deadline Overrun

In case of deadline overrun, the LRM interrupts best-effort applications to allow the critical ones to respect the deadlines. The MON for deadline overrun service extends the deadline warning detection method that is described in [87].

9.2.2.1 General Approach

In many cases, under-provision of the platform can lead to problematic situations, where critical applications may overrun their deadlines. To avoid any timing failures for critical applications, a detection mechanism analyzes intermediate deadlines and adapts the processor demands by interrupting the best-effort applications. Hence, an adaptation consists in interrupting the best-effort applications. Each critical application monitors its execution and checks if the application is in danger of overrunning the deadline. In this case, the MON service notifies the LRM that a deadline overrun might occur.

The partition slots for critical applications contain internal *observation points* that are defined off-line and correspond to the moments, where the MON is executed. In addition, the temporal behavior between the tasks is monitored by the main MON slots to avoid modification of the partition code.

Figure 9.2 illustrates an example of adaptation to avoid internal deadline failure in a quad-core processor. In this example, both critical and best-effort tasks are executed. As shown in the figure, only the two upper cores contain the observation points (shown by small violet slots) and the LRM (shown by small yellow slots) inside the critical tasks. In this example scenario, the execution of the best-effort tasks is suspended by the LRM of the critical tasks (shown by blue arrows), based on the information that is received from the MON (shown by red arrows). The red slots represented that the best-effort task is suspended by the LRM.

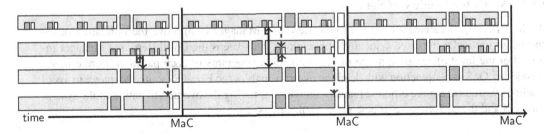

Figure 9.2: Example of Internal Deadline Failure Adaptation by the LRM Inside Critical Tasks - Green is Used for Tasks, Violet for MON, Yellow for LRM and Red for Suspended Processes

The evaluation of whether the interferences caused by the best-effort tasks can be tolerated by critical tasks is performed by a safety condition. Thanks to the positioning of observation points between tasks, the safety condition is given in Eq. 9.1.

$$\mathsf{ET}(x) \leq internal\ deadline(x) \tag{9.1}$$

where $\mathsf{ET}(x)$ is the monitored execution time of *Part* until point x and *internal deadline(x)* is a pre-computed constant giving the maximal possible internal deadline.

9.2.2.2 Implementation

The implementation of the MON for deadline overrun is hardware dependent and the timing is computed with local core registers. Code 3 represents the source code of an implemented deadline overrun check.

Code 3 (Source-Code of Deadline Overrun Check).

```
int dlrm_mon_deadline_overrun_check(uint32_t tidx,
             struct dlrm_slot_desc_item *slot) {
  static int iso_mode = 0;
  uint32_t elapsed;
  uint32_t limit;
  struct dlrm_mon *data = dlrm_config.lrm_desc->shared_mem->mon_data;
  struct dlrm_qos_job *qos_job = &data->jobs[slot->jobidx_base + tidx].qos;
  uint32_t cycles_per_ms = get_core_freq() / 1000;

  elapsed = dlrm_mon_slot_get_elapsed_cycles();
  limit = slot->obsdate[tidx] * cycles_per_ms;

  // end of slot return to parallel mode
  if (tidx == slot->num_tasks - 1) {
    iso_mode = 0;
    return 0;
  }

  // keep isolation mode
  if (iso_mode)
    return 0;

  // check the safety condition
  if (elapsed > limit) {
    iso_mode = 1;
    return 1;    }
  return 0;
}
```

9.2.3 MON for Quality of Service

The MON for QoS service allows the Local Resource Manager to improve the utilization of resources, in particular in mixed-criticality applications. This keeps the Local Resource Manager informed of the level of resource usage of the application programs of each partition. Furthermore, considering QoS in conjunction with deadline overrun enables the Local Resource Manager to avoid deadline overruns in a more efficient manner, thereby improving the overall utilization of the system resources and the performance of best-effort applications.

9.2.3.1 General Approach

The execution of the MON for QoS service takes the advantage of deadline overrun monitoring actions to capture the behavior of critical applications. In order to avoid any additional modification of existing applications, the MON for QoS is executed at most once, each time the MON for deadline overrun is executed (wider violet slots in Figure 9.2). The reason for reducing the number of measurements is that depending on the targeted architecture and the system requirements, too frequent measurements might degrade the performance of the system. The minimal granularity corresponds to the measurement of a complete slot.

Best-effort applications (and thus partitions) are not necessarily required to be monitored by the MON for QoS services, as the QoS actions in the LRM do not rely on but can be improved by the behavior of best-effort applications. Depending on the hypervisor support, this monitoring can be done without modifying the applications at regular intervals (e.g., through a regular time interruption at partition level), or at the beginning and the end of the best-effort partitions slots. Otherwise the best-effort applications can be modified to regularly monitor its execution and inform the LRM through a dedicated API.

Monitoring for QoS takes advantage of hardware monitors to retrieve more detailed information on the resource usage. These are typically continuous counters implemented as registers that are incremented each time an action in the system occurs (e.g., load/store access from the core, cache miss, etc.). In order to use them, they need to be initialized at the beginning of the code section that requires measurement and collected at the end. Initialization of the counter depends on capabilities of the targeted system, but typically it consists of (1) resetting the counter (i.e., setting it to 0) or (2) reading its value and keeping it. For collection, the counter registers are simply read.

The actual measured metric depends on the action during the initialization phase: (1) if the counters were reset, the read value corresponds to the measured metric, and (2) if the value of the counter was simply stored during the initialization then the measured metric is equal to the counter value that is read during the collection phase minus the value that is read during the initialization phase. While the second option can be used even when the targeted system supports resetting the counters, it is preferred to use the first, where possible to avoid overflowing the register counters. It is important to note that when MON for QoS runs before the first measured section in a slot, it only runs the initialization phase.

9.2.3.2 Implementation

The number of performance monitors and their types depend on the targeted embedded system and the architecture. For instance, a system without an L2 cache will not have an L2 miss counter register, whereas a system with an L2 cache might have one. As part of the DREAMS project, support for two embedded systems has been developed: DREAMS Harmonized Platform (cf. Section 2.6, with a dual-core ARMv7) and the Freescale QorIQ T4240QDS (with a 12-core Freescale PowerPC 6500).

Code 4 shows an example of how most of the performance monitoring registers are read and reset on the DREAMS Harmonized Platform. It uses assembly instructions to read and write the relevant ARMv7 co-processor registers.

Code 4 (Monitoring QoS PMU Register Code).

```
@ Resets the counters
.global reset_pmn
reset_pmn:
    MRC     p15, 0, r0, c9, c12, 0
    ORR     r0, r0, #0x02
    MCR     p15, 0, r0, c9, c12, 0
    BX      lr

@ Returns the value of the Nth counter (N supplied as an argument in r0)
.global read_pmn
read_pmn:
    AND     r0, r0, #0x1F
    MCR     p15, 0, r0, c9, c12, 5
    MRC     p15, 0, r0, c9, c13, 2
    BX      lr
```

As this assembly code is strongly platform-dependent, we provide a high-level C programming interface to interact with the registers (Code 5). This interface masks the hardware-specific operations with simple initialization, reset, and read functions.

Code 5 (Monitoring QoS Interface Code).

```
// initializes the performance monitoring unit and configures the set of
// performance monitors to be used
void dlrm_mon_init()
{
  enable_pmu();
  enable_ccnt();

  pmn_config(0, 0x03); // Data cache miss
  pmn_config(1, 0x06); // Data read
  pmn_config(2, 0x07); // Data writes
}

// resets the PMU and cycle counters to zero
void dlrm_mon_reset()
{
  reset_pmn();
  reset_ccnt();
}

// reads the PMU and cycle counters into the supplied struct
void dlrm_mon_read(dlrm_mon_status_t *status)
{
  status->cycles = read_ccnt();
  status->data_cache_misses = read_pmn(0);
  status->data_reads = read_pmn(1);
  status->data_writes = read_pmn(2);
}
```

9.3 Local Resource Scheduler or LRS

The LRS completes the DLRM services and schedules the execution of the tasks from critical and best-effort applications.

9.3.1 General Approach

The LRS starts its execution once a partition slot starts. The first time that the LRS is executed (typically during plan 0 schedule of the hypervisor), it launches the application initialization to set up its internal state for execution. Afterward, it initializes the schedule for application tasks for different slots and planned configurations. The precomputed schedules for the tasks can be retrieved from a configuration file and/or from the corresponding application.

After the initialization of the application and the schedule, the LRS launches a predefined list of application tasks for that slot in a sequential manner. The execution of the LRS is stopped after all of the tasks are executed, even if there is remaining time in the current partition slot. The execution of the LRS can conceptually span in multiple partition slots. However, for simplicity it is assumed that this starts and ends in the same partition slot.

time ⟶

Figure 9.3: Example of Critical Partition Slot Execution - Green is Used for Partition Slot, Orange for the Application Task, Purple for LRS, Yellow for MON and Violet for LRM

Figure 9.3 illustrates an example of a critical partition slot that is controlled by the LRS. In this figure, different colors are used:

Green represents the partition slot.

Purple represents the execution of the LRS.

Orange represents the execution of the application task.

Yellow represents the execution of the MON.

Violet represents the execution of the LRM.

As shown in the figure, the execution of the LRS starts at the beginning of the partition slot and lasts until the execution of the last task and the corresponding MON and LRM slots are terminated. At the beginning of the slot and before the execution of the first task, the LRS executes the LRM to send possible partition stop signals required by QoS management of the LRM. In addition, it runs the MON services to initialize the MON monitors that will be used to collect the performance statistics of the upcoming task.

Between the execution of two consecutive tasks, again a sequence of MON and LRM are executed. The MON is executed to collect the performance monitors of the just executed task and the current execution time of the slot. The LRM is executed to use the just collected execution time for checking potential deadline overrun and taking preventive actions, like stopping the best-effort concurrent partitions. In addition, the LRM sends possible partition resume or stop signals that are required by the QoS management. The second execution of the MON initializes the MON monitors for the upcoming task.

Finally, after the execution of the last task within the slot, the MON services are executed to collect the performance monitors of the last task. In addition, the LRM needs to be executed to send resume signals to the best-effort partitions that were stopped for QoS management or deadline overrun avoidance.

9.3.2 Implementation

Code 6 represents a simplified implementation of the LRM being executed at each slot of critical partitions. The code follows the steps that were discussed above.

Code 6 (LRS for Critical Partitions Slots).

```
void lrs(int partition_id, int slot_id,
    int num_tasks, task_t *task_list) {
  int task_id = 0; // task ids go from 0 to (num_tasks - 1)
  // first task (task_id = 0)
  lrm_qos(partition_id, slot_id, task_id);
  mon_init(partition_id, slot_id, task_id);
  execute_task(task_list[task_id]);

  // middle and last tasks
  for (task_id = 1; task_id < num_tasks; task_id++) {
    time_t last_task_exec_time =
      mon_collect(partition_id, slot_id, task_id - 1);
    lrm_deadline(partition_id, slot_id, task_id - 1, last_task_exec_time);
    lrm_qos(partition_id, slot_id, task_id);
    mon_init(partition_id, slot_id, task_id);
    execute_task(task_list[task_id]);
  }

  // after the last task
  mon_collect(partition_id, slot_id, task_id - 1);
  lrm_deadline(partition_id, slot_id, task_id - 1);
}
```

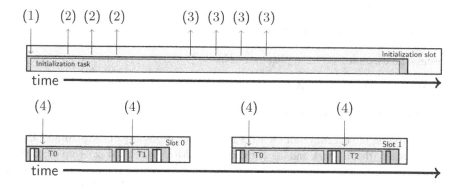

Figure 9.4: Example of Interactions between the LRS and an Application during Initialization and a Major Cycle - Green is Used for Partition Slot, Orange for the Application Task, Purple for LRS, Yellow for MON and Violet for LRM

9.3.3   Requirements on Applications

The usage of the LRS requires a certain applications architecture. The LRS applies the application schedule to the application tasks by providing the following interfaces:

1. Application callback interface for the application initialization and the tasks and their precomputed schedule declaration to the LRS

2. Interface for the applications to declare the tasks

3. Interface for the applications to define their task schedule

4. Task callback interface that the LRS uses to launch the execution of the tasks

The callback interface 1 is only called at the first time the partition is executed, e.g., when the system is initialized. During its execution, the application initializes its internal structures and uses the interface calls 2 and 3 to declare its tasks and its precomputed tasks schedule according to the different scheduling plans the LRM can choose from. Once all the applications of the system have been initialized, the LRS uses the task callback interface 4 to execute the tasks according to the current LRM schedule and the task schedule defined by the application with the interfaces 2 and 3.

While the initialization might seem dynamic, the task schedule declared by the application during the initialization is typically precomputed, i.e., computed off-line.

Figure 9.4 represents an example, in which the interactions between the LRS and an application takes place. In this example, the LRS calls the application through the callback interface 1. During the application initialization task, the application declares three tasks it contains (T0, T1, T2) using interface 2 and using interface 3 their intra-partition schedule:

- Task T0 is to be executed in slot 0

- Task T1 is to be executed in slot 0, and as it is being declared after T0 it will be executed after the previous T0

- Task T0 is to be executed in slot 1

- Task T2 is to be executed in slot 1, and again as it is being declared after T0 in slot 1 it will be executed after the execution of T0 in slot 1

With this information the LRS can launch the application tasks during regular major cycles using the callback interface 4, as depicted in the bottom time-line of Figure 9.4.

Some observations from the example in Figure 9.4:

- During the initialization, the monitor services do not need to be called, but they could be called to keep initialization statistics.

- The initialization can span multiple time slots.

- For simplification, this example declared only one schedule, but the final system might require multiple, e.g., one schedule when all the cores work and another when one core has failed. The LRS interface 3 allows the application to declare its different schedules by passing a schedule id parameter in addition to the slot and task ones.

9.4 Local Resource Manager or LRM

As described earlier, the LRM either adopts the configuration from the GRM to particular resources (e.g., processor core, memory, I/O) or selects a new configuration from the ones available and reports state of the resource (from MON) to the GRM.

The LRM is considered to provide the following services, each of which is described in the following subsections:

- Core failure management

- Deadline overrun management

- QoS management for non-critical task

9.4.1 Core Failure Management

In case of core failure, the LRM is in charge of reconfiguring the multi-core in order to enable the local applications to continue the execution. The reconfiguration strategy, in case not all applications could be locally hosted after the failure(s), follows the following two rules:

1. Critical applications are locally reconfigured in priority.

2. Applications must be moved entirely, i.e., an application cannot run on two multi-core chips at the same time.

Each core is associated with an LRM. All LRMs operate synchronously and in parallel, but only one is the *master*, which applies the reconfiguration. If the master LRM was hosted by a failed core, a new master is selected. The reconfiguration strategy of masters is known off-line.

The LRM takes decisions for local reconfiguration at the end of the MaC by collecting all failed cores. This entails that several failures may happen during a MaC and decisions could consider multiple failures. To avoid non deterministic decisions, we impose reconfiguration graphs to be symmetric.

Example 1. *Figure 9.5 illustrates an execution on the DREAMS Harmonized Platform (DHP) [308] composed of a dual-core ARM processor. In this example, during the first MaC, the system operates in configuration C_1, where two user partitions (P0 on core 0 and P1 on core 1), two LRMs (in yellow) and two MON (in violet) execute.*

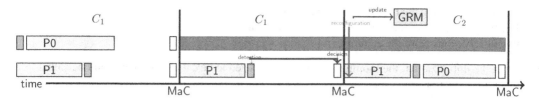

Figure 9.5: Execution on ARM in Case of Permanent Core Failure

*During the second MaC, core 0 fails and the failure is detected by the MON. The LRM takes a local reconfiguration decision that consists in moving to C_2 at the third MaC. The LRM informs the GRM of the current configuration via an **update** message. The reconfiguration graph that is shown on the left-hand side of Figure 9.6 illustrates how the LRM takes the decision about the new configuration. The numbers on the arrows represent the ID of the core, whose failure triggers the move to the new configuration shown by the tip of arrow. For instance, the transition from C_1 to C_2 takes place in case core 0 fails and the LRM selects C_3 in case core 1 fails. The transition of $C_1 \rightarrow C_4$ corresponds a global reconfiguration, where the ARM will host another application in case a core of another multi-core fails. In this case, the GRM sends an **order** to the LRM to switch to the new configuration.*

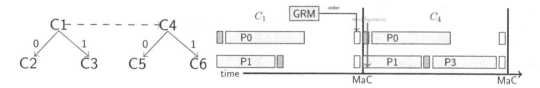

Figure 9.6: Local and Global Reconfiguration Graph and Timeline

9.4.1.1   Implementation

Code 7 represents an implemented example of the LRM for the reconfiguration in case of a core failure.

9.4.2   Deadline Overrun Management

As described earlier, the MON is in charge of monitoring internal deadlines of critical applications and if there is an overrun, the LRM immediately stops the best-effort applications.

9.4.2.1   General Approach

Computing an upper bound for application Worst Case Execution Times (WCETs) on a multi-core chip and reserving this amount of time for all tasks leads to an over-provisioning of the platform. However, the WCET is rarely reached and most of the time, the Average Execution Time (AET) is far below the capacity of the platform. This is the reason why it is accepted that a multi-core be over-utilized by the applications. The proposed model is detailed in the below definition.

Definition 9.3 (Under-Provisioned Platform)**.** *Any multi-core can be over-utilized in the following way:*

- $\sum_i \frac{C_i}{T_i} + \sum_j U_j >$ *number of cores: the overall utilization exceeds the multi-core capacity;*

Code 7 (LRM Core Failure Code).

```
void lrm_core_failure_run(struct lrm_state *state) {
    static xm_u8_t core_status[DLRM_MAX_CPU] = {0};
    struct dlrm_reconf_line*
        reconf_table = dlrm_config.system_desc->reconf_table;
    xm_u8_t i;
    int32_t nbcpu= dlrm_config.system_desc->num_cores;
    int32_t *alive =dlrm_config.lrm_desc->shared_mem
                        ->mon_core_failure_data->alive_buffer;
    xmPlanStatus_t plan;
    xm_s32_t err = XM_get_plan_status(&plan);
    int32_t current_configuration = plan.current;

    // if global reconf asked
    xm_s32_t flags;
    struct order_message msg;
    flags=receive_order(&msg,OrdersPort);
    if(flags == DLRM_COMM_SUCCESS) {
        current_configuration = msg.mode;
    }

    // is there a core failure
    int32_t next_conf ;
    int needed_reconf=0;
    static xm_u8_t expected_value = 0;
    expected_value = 1- expected_value;
    static int nb_master_failed=0;
    uint32_t num_master = state->current_lrm_master_part;
    next_conf = current_configuration;
    // computation of all failing cores
    for (i = 0; i < nbcpu ; i++) {
        if (alive[i] != expected_value && core_status[i] == 0) {
            // core value has failed
            core_status[i] = 1;
            needed_reconf=1;
            if (next_conf != -1) {
                next_conf = reconf_table[next_conf].next_plan[i];
            }}}

    // update LRM master
    uint32_t
        cpu_master = dlrm_config.part_desc->part[num_master].hwcpuid;
    while(core_status[cpu_master]==1){
        nb_master_failed++;
        cpu_master = dlrm_config.lrm_desc->master_order[nb_master_failed];
    }
    if(XM_PARTITION_SELF==
            dlrm_config.lrm_desc->master_order[nb_master_failed]) {
        state->current_lrm_master_part =
            dlrm_config.lrm_desc->master_order[nb_master_failed];
    }

    // send update to GRM
    lrm_core_failure_send_update(next_conf);
}
```

- $\sum_i \frac{AET_i}{T_i} + \sum AU_j \ll$ *number of cores: the overall average utilization is far below the multi-core capacity;*

- $\sum_i \frac{C_i}{T_i} <$ *number of cores: the overall utilization for the critical applications fits the multi-core capacity.*

In other words, the best-effort applications are those leading to the overtaking of the provisioning. This situation will be handled as proposed in [87], which means that we will monitor the critical applications regularly; If an internal deadline is exceeded, the best-effort applications will be interrupted and resumed once the critical applications are not any longer endangered.

Example 2. *Let us consider again the Example 1 and assume that partition P0 is safety-critical, while P1 is best-effort. During the first* MaC *of Figure 9.7, no deadline overrun occurs, while in the second, a problem is detected by the MON, leading to the interruption of P1 by LRM. In the next* MaC, *all partitions execute and no problem is encountered.*

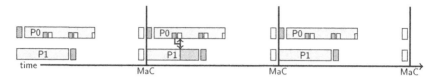

Figure 9.7: Adaptation in Case of Deadline Overrun

9.4.2.2   Implementation

Code 8 represents an implemented example of the LRM for a reconfiguration in case of a deadline overrun.

Code 8 (LRM Deadline Overrun Code).

```
void lrm_deadline_overrun_run(int toswitch) {
  static unsigned shared = 0;
  int i;
  xm_s32_t err;
  switch (toswitch) {
  case 1: // request for isolation
    shared++;
    if (shared == 1) {
      for (i = 0; i < dlrm_config.part_desc->num_partitions; i++) {
        if (is_non_critical(dlrm_config.part_desc->part[i])) {
          err = XM_suspend_partition(i);
        }}}
    break;
  case 2: // request for ending isolation
    if (shared == 0)
      break;
    shared--;
    if (shared == 0) {
      for (i = 0; i < dlrm_config.part_desc->num_partitions; i++) {
        if (is_non_critical(dlrm_config.part_desc->part[i])) {
          err = XM_resume_imm_partition(i);
        }}}
    break;
  }}
```

9.4.3   QoS Management

The objective of the LRM QoS services is to enhance the system usage when a potential deadline overrun is detected, to maximize the best-effort applications requested usage (i.e., the actual partition schedule) and to improve QoS for best-effort applications.

Figure 9.8 represents an example, in which the LRM manages the QoS (Note: MON between slots have been removed to simplify the mechanism description). The figure shows a quad-core processor, on which the first core (C0) is executing a critical application, while the other three cores

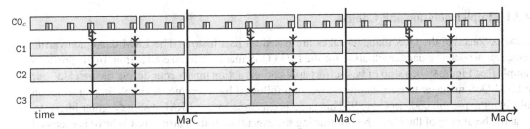

Figure 9.8: Example of Unused System Usage due to a Potential Deadline Overrun Detection

(C1 – C3) are only running best-effort applications. At some point during the execution of the first MaC the deadline overrun management detects a potential failure and consequently sends a suspend signal to the best-effort applications running on other cores; Later, when the critical application slot finishes, the best-effort applications are resumed (shown by dashed blue lines in Figure 9.8). Using this approach, in this example the best-effort applications are granted less than 75% of the requested core time (and obviously the other processor and system resources).

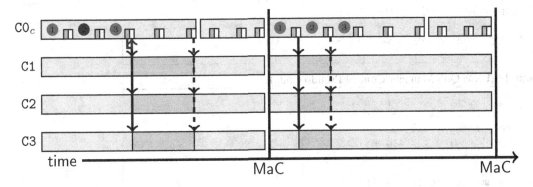

Figure 9.9: LRM QoS Management with Deadline Monitoring Information

To improve the usage of system resources and satisfy the demands of best-effort application (usage of allocated slots), the LRM QoS management uses the deadline overrun management data and the monitoring information. Using the deadline overrun data, the QoS management can determine which of the monitored segments had endangered the critical application slot deadline (see the Section 9.4.3.1 for a technique to determine which segments endangered the deadline). Following the previous example, in Figure 9.9, we observe in the first MaC that among the three segments preceding the suspend signal submission, the second segment is the segment that mostly endangered the deadline. This is depicted by the bullets (1, 2, 3) with different colors: green for no effect, orange for medium effect and red for significant effect on WCET. With this information, the QoS manager during the second MaC can force the best-effort applications to stop after the execution of the first prove and release them (dashed blue arrows in Figure 9.9) after the execution of the second deadline overrun prove, thus halting the execution of best-effort applications during the second segment. Using this action, the best-effort applications can use more than 85% of the allocated core time, thus gaining more than 10% of available execution time when compared with the case of only using deadline overrun management (see Figure 9.8). The LRM might reconsider the prevention action in future MaC if the best-effort applications resource usage is reduced.

9.4.3.1 Implementation and Choices

All the actions that the QoS manager takes are based on past history. The LRM deadline overrun manager must detect a danger situation for the LRM QoS manager to take an action. In the previous example (see Figure 9.9), a stop of best-effort applications action must occur during the first MaC for the LRM QoS manager to take its preventive stop followed by a resume. A simple implementation looks like the pseudo code in Code 9. This code is executed by the LRM at some regular intervals, i.e., at the beginning of the MaCs before starting the execution of the partitions. Code 10 is executed at the beginning of each segment to control if a stop or resume system call to the low critical applications is issued.

Code 9 (LRM QoS Manager Management Pseudo Code).

```
// code run by the LRM at the beginning of the MaC
if (deadline_overrun_detected) {
  int id = get_offending_segment();

  // set the segment to stop the non critical applications
  //   and resume their execution once the segment is finished
  // by default all stop segment entries are set to false
  stop_segment[id] = true;
  // by default all resume segment entries are set to false
  resume_segment[id + 1] = true;
}
```

Code 10 (LRM QoS Manager Control Pseudo Code).

```
// code run by the LRM preceding the segment
// it receives as input the current segment id (cur_id)
if (stop_segment[cur_id]) {
  stop_offending_partitions(); // offending partitions are know
                               //   at design time
}
if (resume_segment[cur_id]) {
  resume_offending_partitions();
}
```

To determine which segment has endangered the application slot deadline, we need to determine which of the executed segments suffered at most from the other applications running in parallel on the other cores. For that purpose we divide the execution time of each segment by its computed WCET, effectively determining which of the segments deviated more from its WCET.

9.4.3.2 Improving QoS with Monitors Information

The Local Resource Manager QoS manager can further improve the management decisions by taking into account performance monitor measurements from the Monitoring service. With the performance measurements of the critical application and the measurements of each of the best-effort applications, the QoS manager can check which of the best-effort applications was slowing down the critical one, and just send the stop signal to that application. Proceeding with the previous example, Figure 9.10 shows a scenario, in which the application on core 2 (C2) causes the WCET degradation of segment 2, and the QoS manager stops only C2. Thus, non-critical applications in C1 and C3 are granted 100% of their allocated time, and only the application in C2 is degraded by less than 15%, i.e., less than 5% of the global system usage.

The quality of the action the LRM QoS can take depends on the available monitors on the target architecture. For example, more accurate actions can be taken if the target architecture provides monitors at the core bus interface and system memory (DRAM) levels than if the architecture just provides monitors at the core bus interface level.

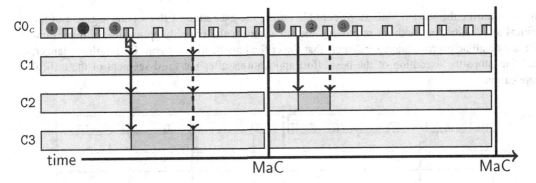

Figure 9.10: LRM QoS Management with Deadline and Performance Monitoring Information

9.4.3.3 Applications with QoS Support

Best-effort applications may provide QoS support. These applications should have various discrete resource utilization levels, e.g., low, medium, high, each of which corresponds to a different QoS. For instance, *low* resource level uses less resources but also gives lower QoS as compared to *medium* or *high* resource levels. The LRM can improve performance of critical task without having to stop the best-effort tasks. Hence, the best-effort task that is slowing down the critical tasks can now be switched to a lower resource level instead of being stopped completely. The new resource level can be determined based on a feedback from performance monitoring measurements. Continuing the previous example (Figure 9.10), the application on core 2 (C2) that causes the WCET degradation of segment 2, is set to *low* resource level. The QoS manager can also set multiple best-effort tasks to lower resource levels to improve the performance of the critical task, thus allowing all applications to continue executing.

Code 11 (LRM QoS Manager for Application with QoS Support Pseudo Code).

```
// code run by the LRM at the beginning of the MaC
if (deadline_overrun_detected) {
        int id = get_offending_segment();

        // set the segment to stop the non critical applications
        //
and resume their execution once the segment is finished
        // by default all stop segment entries are set to false
        decrease_qos_segment[id] = true;
        // by default all resume segment entries are set to false
        increase_qos_segment[id + 1] = true;
}

// code run by the LRM preceding the segment
// it receives as input the current segment id (cur_id)
if (decrease_qos_segment[cur_id]) {
        decrease_qos_offending_partitions();
}
if (increase_qos_segment[cur_id]) {
        increase_qos_offending_partitions();
}
```

9.4.3.4 LRM QoS and Deadline Overrun Managers

With the presence of the deadline overrun manager, the critical applications will never be endangered, as the deadline overrun manager stops the best-effort applications even if the QoS manager has applied some QoS actions. Figure 9.11 takes the example in Figure 9.10, on which the QoS manager during the second MaC stops core C2 during the execution of the second segment of the critical application, but without improving the situation for the critical application. In this example, the deadline overrun manager detects that the critical application on core C0 is still on danger, and thus stops the execution of the best-effort applications after the third segment of the critical application.

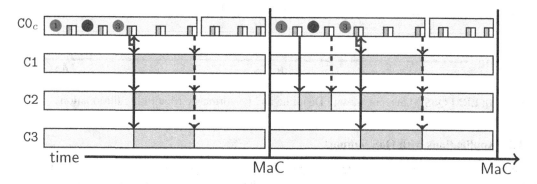

Figure 9.11: LRM Deadline Overrun and QoS Managers Combined Behavior

9.5   Global Resource Manager or GRM

The GRM provides services for a system-wide management of resources based on predetermined global off-line configuration tables. Fulfilling or recognizing system-wide constraints is not possible by viewing a single resource in isolation, but by enforcing a system-wide view, which may require global decisions to be made. When the conditions, on which the off-line configuration is based change significantly at runtime (e.g., core failure), then this information is communicated by the LRM (via the MON) to the GRM. The GRM may select a different off-line configuration from a predefined set of configurations and sends it to the LRMs that are involved in the reconfiguration. The LRMs then apply the new configuration on their respective nodes via the LRS.

9.5.1   Implementation

The implementation of the GRM can be performed in several different manners. The first option is to implement it as a software module in a separate node with no other tasks running alongside it on the same core. On one hand, this will make development of the GRM itself, and the reconfiguration and optimization logic easier. On the other hand, there will be some cost associated to the exclusive dedication of one node to the GRM. Another option is to integrate the GRM into an existing node in the system, e.g., running it as a critical application alongside other applications. This would mean less cost overhead, and the communication with the LRMs would be easier using the DREAMS Abstraction Layer ports. Nevertheless, this will make developing the resource manager harder, and will decrease its flexibility as it is constrained by other applications running in that node. Finally, a hardware implementation is possible, but in the context of some use-cases it could be too costly and not justifiable.

The GRM matches LRM updates to off-line configuration tables. Hence, it does not have a large workload and a simple finite state machine can perform it. This is because in DREAMS resource management solution, complexity of the decision making is broken down by several aspects of the resource management architecture.

Conceptually, one GRM exists in the DREAMS system, although distribution is possible for fault-tolerance and scalability considerations. Therefore, the GRM can be realized either by a single node or a set of nodes. However, we regard the GRM as a single entity in context of this book.

9.5.2 Global Reconfiguration Graph

As described earlier, the LRM manages the global reconfiguration requests from the GRM. The operation of the LRM is constrained by a complete symmetric local reconfiguration graph. In addition, the LRM sends an *update* message to the GRM every MaC for *membership* checking and local reconfiguration change information. This is because the global reconfiguration needs to be stored at the GRM.

Example 3. *Consider a DREAMS platform composed of three multi-core chips N_0, N_1 and N_2, such that N_0 and N_1 have four cores, while N_2 has two cores. Assume that N_2 hosts only the GRM (and no user application) and that N_2 is not subject to any failure (DREAMS assumption). There are four user-applications A_1, A_2, A_3 and A_4 to be mapped on the distributed platforms.*

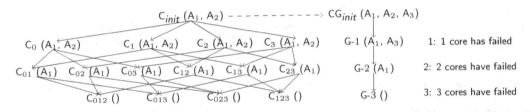

Figure 9.12: Local Reconfiguration Graph for N_0

The local reconfiguration graphs computed by GRec (cf. Section 5.3.1) are shown in Figures 9.12 and 9.13. Each state of the graph is a tuple ⟨ID of the configuration (or XtratuM plan), the applications hosted on the multi-core⟩. According to the requirements, the graphs must be symmetric and complete. This entails that the number of states is equal to the number of cores failure combination -1 (because the case, in which all cores have failed is not stored). In the example, since a node has four cores, there are at least $2^4 - 1 = 15$ states just for taking into account the failures. In Figure 9.12, these cases are shown on the left-hand side, for instance C_{012} is the configuration, where cores 0, 1 and 2 have failed. Among the represented configurations, some may address a global reconfiguration. In each graph, there is only one configuration which reaches CG_{init}. In order to be complete, GRec computes all possible combinations of cores failure from this state. In Figure 9.12, a short version is shown on the right-hand side, where G-i represents all states, in which i cores have failed. Note that the reconfiguration graph has $15 + 15 = 30$ states. The graph 9.13 is a full-compact version, since all states are of form i or G-i.

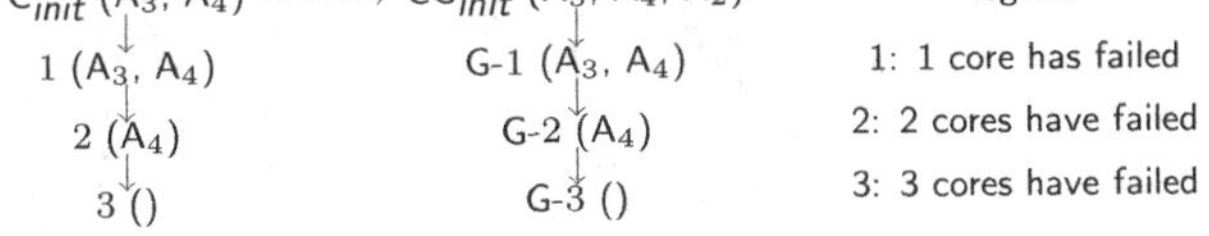

$$C_{init} (A_3, A_4) \dashrightarrow CG_{init} (A_3, A_4, A_2)$$

	legend
$1 (A_3, A_4)$ $\quad$ G-1 (A_3, A_4)	1: 1 core has failed
$2 (A_4)$ $\quad\quad$ G-2 (A_4)	2: 2 cores have failed
$3 ()$ $\quad\quad\quad$ G-3 $()$	3: 3 cores have failed

Figure 9.13: Local Reconfiguration Graph for N_1

The global reconfiguration graph is the Cartesian product of the two reconfiguration graphs. Thus it has potentially up to $30 \times 30 = 900$ states. However, the example of Figure 9.14 represents *a compact version which has less states, as some local transitions required a global transition as well. For instance, if N_0 moves from C_0 to C_{01}, the application A_2 will be relocated if possible on N_1. The minimal information to be stored in GRM are shown in Figure 9.15 (non-compact version).*

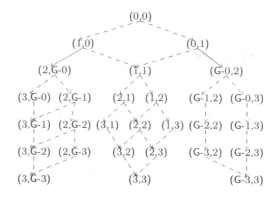

Figure 9.14: Global Reconfiguration Graph

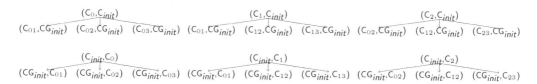

Figure 9.15: Implementation

Since the number of the plain arrows in the global reconfiguration can vary from very low to very high (depending on the possible local reconfiguration changes), instead of applying a C static-structure generation (as is done for the local reconfiguration graph) a C function can be generated. The message value sent by the LRM (m) corresponds to the configuration of the LRM node (C) and the Node ID (id).

$$m = C + (O * (id + 1)) \qquad (9.2)$$

where O is a fixed offset (e.g., 10000). For a total of 'n' nodes in the system, message values 0,1,2...n-1 are reserved for the failure of the last core of the node, as no message is received from the node but some global reconfiguration from the GRM may be needed. The global reconfiguration graph function for the example is partially represented in code 12. The reconfigurations to be applied are only illustrated, when there are some failures in node N_0.

Code 12 (GRM Reconfiguration Graph in C).

```
static int current_configuration_LRM_Node[2]={1,1};
// configuration of each LRM
// Cinit = 1, C0 = 2, C01 = 3, C02 = 4, C03 = 5, C012 = 6,
// C013 = 7, C023 = 8, C1 = 9, C12 = 10, C13 = 11, C123 = 12,
// C2 = 13, C23 = 14, , C3 = 15, CGinit = 16, CG0 = 17,
// CG01 =18 , CG02 = 19, CG03 = 20, CG012 = 21, CG013 = 22,
// CG023 = 23, CG1 = 24, CG12 = 25, CG13 = 26, CG123 = 27,
//CG2 = 28, CG23 = 29, CG3 = 30
// We consider failure on node N0 only for this example
if(receive_update(&msg_update, LRM_Node[0])<0)
        msg_update.message = 0; //no message received from LRM
// has LRM 0 sent an update?
// msg corresponds to reached configuration
switch case (msg_update.message){
        case 0: //last core of N0 has failed
        //take action if a configuration corresponding to this exists
        case 10002: case 10009: case 10013: case 10015:
                    break;
        // just update the configuration
        case 10003: case 10004: case 10005: case 10010:
        case 10011: case 10014:
                    // unique failure in N0
            if (current_configuration_LRM_Node[1]==1){
            // A2 must be reallocated on N1
      msg_order.mode=6;
      send_order(msg_order,LRM_Node[1]);
                }
                break;
        case 10006: case 10007: case 10008: case 100012:
                    if (current_configuration_LRM_Node[1]==1
                    && current_configuration_LRM_Node[0]==1 ){
                    // double failure in N0 during 1 MaC
                        msg_order.mode=6;
                        send_order(msg_order,LRM_Node[1]);
                }
                    break;
        case 10016: case 10017: case 10018: case 10019:
        case 10020: case 10021: case 10022: case 10023:
        case 10024: case 10025: case 10026: case 10027:
        case 10028: case 10029:
            break; //A2 cannot be reallocated
        default:
          break;
        }
current_configuration_LRM_Node[0]=msg ;
```

9.6   Resource Management Communication

The LRMs communicate regularly the status of local resources to the GRM or to the supervisor LRM (in case of hierarchical architecture), and the GRM or the supervisor LRM can send back reconfiguration orders, when required. The communication between the RM components is performed by *sampling* and *queuing* communication channels of the XtratuM hypervisor (cf. AR-INC653 [197]).

A message can be sent from a source RM component to other RM components via these channels. The RM components can access the communication channels via the corresponding sampling or queuing ports. Sampling ports are used for *state* messages, as the receiver requires only the most updated message. The message remains at the port until the message is read by the receiver, or the sender overwrites the message at the port by a new one. Queuing ports are used for *event* messages, as all of the messages need to be processed by the receiver. If a new message is sent before an old one is read by the receiver, the messages are buffered. The delivery of the message at the receiver is performed in an FIFO order.

Channels, ports, maximum message sizes and maximum number of messages for queuing ports are defined in the XtratuM configuration files. Since XtratuM does not ensure security of the message exchanged between RM components, all messages are sent via a security library as described in Section 9.6.1.

Figure 9.16 represents a hierarchical architecture of the RM components and the communication channels between the components. As shown in figure, two different communication channels can be used for communicating between higher-level and lower-level RM components (e.g., between the GRM and the top-most LRM).

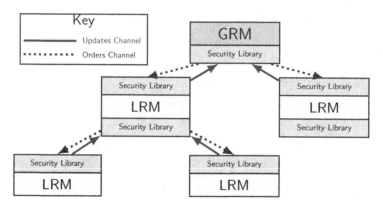

Figure 9.16: Hierarchical Architecture of Resource Management Components and Communication Channels

Update Channel

This channel is used by an LRM to send status updates or global reconfiguration requests to the GRM or a supervisor LRM. The channel is a queuing channel with a buffer length of 32, so that all the update messages from the LRM can be delivered sequentially. One update channel exists between each pair of LRM- GRM. Each update message is 64 bit long (without security headers) and consists of the current configuration being executed by the node corresponding to the LRM, message value, and the type of the update that is being transmitted, i.e., the reconfiguration request or status update. The contained value in the update message is explained in Section 9.5.1.

Figure 9.17 represents the format of update messages, which must be sent by each LRM to the GRM or supervisor LRM at least once during each MaC. Otherwise, the GRM (or supervisor LRM) considers the corresponding resource as failed (e.g., all cores have failed in case of multi-core chip), as no update message has been received from that LRM. Thus, the update message is used as a membership sign as well.

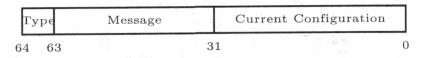

Figure 9.17: Resource Management Updates Message Format

Order Channel

This channel is used by the GRM to send status reconfiguration orders to the LRMs. The order channel is a sampling channel, so that the LRMs only receive the most updated order that is sent by the GRM. Each order message is 64 bit long (without security headers) and consists of a new configuration to be applied by the LRMs. The reconfiguration orders need to be applied immediately or at the end of MaC. Figure 9.18 represents the format of orders messages.

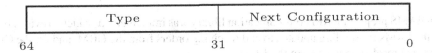

Figure 9.18: Resource Management Orders Message Format

In case the RM components reside on different nodes, the update and order messages are sent in a time-triggered manner, which is summarized in Table 9.1. As described earlier, to achieve redundancy, one LRM is instantiated on each core of the DREAMS nodes. In this case, each LRM has its own update (queuing) channel to the GRM, as there is no multi-cast queuing channels. In contrary, there is only one multi-cast order (sampling) channel from GRM to all LRMs on a single node. Figure 9.19 represents an example for the communication of the RM components in a system with three nodes - the DREAMS Harmonized Platform and two other multi-cores.

Table 9.1: Summary of the Properties for Resource Management Communication Channels

Communication Channel	Port Type	Off-chip Communication	Source	Destination
Updates	Queuing	Time-Triggered	LRMs	GRM or supervisor LRM
Orders	sampling	Time-Triggered	GRM or Supervisor LRM	LRMs

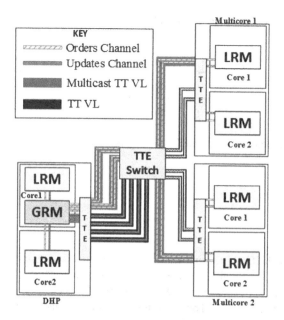

Figure 9.19: Resource Management Communication with Three Nodes

In the DREAMS project, an RM communication library was implemented, which provides functionality for the LRMs for sending updates to and receiving orders from the GRM, and for the GRM to send orders and receive updates from the LRMs.

9.6.1 Secure Resource Management Communication

Security services are essential for the communication of RM components, as those components accomplish critical tasks in the system. The resource management is an interesting target for an attacker, as it manages the system and attacking the resource management allows the attacker to manipulate this management. Attacked resource management components can lead to wrong decisions and application of wrong orders [210, 309].

The GRM and the LRM are the most important components that have to be secured. If an attacker intends to attack the GRM, he can masquerade himself as an LRM and sends false availability, energy or error information to the GRM. As a consequence, the GRM can take a wrong decision (e.g., selection of a wrong configuration), based on the fake information received from the attacker. If an attacker wants to attack an LRM, the attacker can masquerade himself as a GRM and send false instructions to one or more of the LRMs, on behalf of the GRM. In this case, the LRM or the LRMs apply inappropriate local configurations. As all of the attacks might lead to an unpredictable operation of the system, it is important to secure the RM components.

9.6.1.1 Security Services and Levels

Figure 9.20 represents the data flow between a GRM and an LRM. If one of the components, e.g., the GRM sends a message to the LRM, the security sublayer encrypts the data and relays it to the XtratuM hypervisor. Afterwards, XtratuM forwards the secured message to the receiver partition to be delivered to the security sublayer. Afterwards, the security sublayer checks and decrypts the message and sends the message to the LRM.

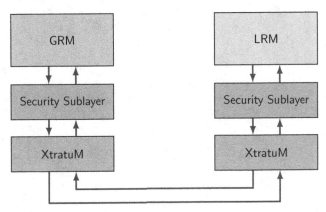

Figure 9.20: Secure Resource Management Communication

As described earlier in this chapter, there are two different types of channels between the RM components, i.e., the update channel and the order channel. Below we analyze the required security level (cf. Section 8.2) for each channel.

1. Order Channel: Reconfiguration messages are sent from a higher-level RM component to a lower-level one. To ensure the correctness of the reconfiguration orders and to make sure that the messages are not faked or manipulated, integrity and authentication need to be checked. In addition, as the orders might contain sensitive information, confidentiality needs to be fulfilled. The combination of confidentiality, integrity and authentication requires the usage of security level 3.

2. Update Channel: Status messages are sent from a lower-level RM component to a higher-level one. This includes update information and reconfiguration requests. Like the order channel, integrity and authentication are required for this channel to provide a protection against manipulation and faked messages. As the status messages may include information that shall not be exposed to the attacker, confidentiality is also required for this channel. Hence, for this channel, the third level of security is used, as confidentiality, integrity and authentication are required.

To prevent replay attacks, e.g., to resend an old reconfiguration request, the security sublayer uses time-varying parameters, thereby the receiver can detect if it has already received this message.

10

Safety Certification of Mixed-Criticality Systems

I. Martinez

IK4-Ikerlan

G. Bouwer

TÜV Rheinland Industrie Service GmbH

F. Chauvel

SINTEF

Ø. Haugen

SINTEF

R. Heinen

TÜV Rheinland Industrie Service GmbH

G. Klaes

TÜV Rheinland Industrie Service GmbH

A. Larrucea Ortube

IK4-Ikerlan

C. F. Nicolas

IK4-Ikerlan

P. Onaindia

IK4-Ikerlan

K. Pankhania

TÜV Rheinland Industrie Service GmbH

J. Perez

IK4-Ikerlan

A. Vasilevskiy

SINTEF

10.1   DREAMS Safety Certification Strategy 405
10.2   Certification and Compliant Items 406
     10.2.1   Need and Importance of Certifications 407
     10.2.2   Accreditation 408
     10.2.3   EC Type-Examinations 408
     10.2.4   Certification Requirements According to IEC 61508 409
     10.2.5   Compliant Items According to IEC 61508 409
10.3   Modular Safety Cases 409
     10.3.1   Modular Safety Case for Cluster-Level Mixed-Criticality Networks 410
         10.3.1.1   Linking Analysis 413
10.4   Mixed-Criticality Patterns 414
     10.4.1   Hypervisors 414
         10.4.1.1   NoC Accessible Memory Area Diagnosis Pattern 415
         10.4.1.2   Critical Partition Diagnosis Pattern 415
         10.4.1.3   Communication Input / Output Server Pattern 416
         10.4.1.4   Digital Input / Output Server Pattern 417
     10.4.2   COTS Multi-Core Device 418
         10.4.2.1   Shared Memory Diagnosis Pattern 418
         10.4.2.2   Cache Coherency Management Diagnosis Pattern 420
         10.4.2.3   Inter-Connection Management Unit Diagnosis Pattern 421
         10.4.2.4   Interrupt Controller Diagnosis Pattern 422
     10.4.3   Mixed-Criticality Network 423
         10.4.3.1   Network-on-Chip Diagnosis Pattern 424

10.5   Functional Safety Management Process for DREAMS Architecture   425
         10.5.1   IEC 61508 Functional Safety Management   425
                  10.5.1.1   Phase 10 'Realization': V-model   425
         10.5.2   Tools   427
10.6   Certification of Mixed-Criticality Product Lines   428
         10.6.1   Families of Systems and Product Lines   428
                  10.6.1.1   A Brief Introduction to Variability Modeling   428
                  10.6.1.2   The Base Variability Resolution Model   430
                  10.6.1.3   The Benefit of Product Lines   430
         10.6.2   Piecewise Certification   431
                  10.6.2.1   Product Lines and Modularity   431
                  10.6.2.2   What Is a Piece?   432
                  10.6.2.3   Piecewise Verification   433
                  10.6.2.4   What Do Standards Say about Piecewise Certification?   433
                  10.6.2.5   Tool Certification and Its Relationship to Piecewise Certification   434
                  10.6.2.6   Piecewise Certification and Product Lines   434
         10.6.3   IEC 61508 Certification   435
10.7   Method for Certifying Mixed-Criticality Product Lines   437
         10.7.1   Certification Support in DREAMS   437
         10.7.2   Certification Arguments   440
         10.7.3   Database of Argument Models: Mixed-Criticality System   440
                  10.7.3.1   Adapting Mixed-Criticality Systems for Reuse as Argument
                             Models   440
                  10.7.3.2   Supporting Certification for Multiple Safety Standards   441
         10.7.4   Arguments of Compliance to Safety-Standards   441
                  10.7.4.1   Safety-Compliant Use of the DREAMS Tool Inventory   441
                  10.7.4.2   Mapping Argument Models to the FSM Documents   444
         10.7.5   Arguments Based on Verification, Validation and Testing   444
         10.7.6   Summary   445

As explained in the introduction (see Chapter 1), in federated architectures where each major functionality is provided by a dedicated embedded system, the ever increasing demand for additional functionalities leads to an increase in the number of ECUs, connectors and wires. This leads to an increase in the overall cost, size, weight, power consumption and complexity [310], which in some cases limits the scalability of current federated architectures. For example [50]:

- "Wind power: A modern off-shore wind-turbine dependable system manages up to three thousand inputs / outputs, several hundreds of functions are distributed over several hundred nodes grouped into eight subsystems interconnected with a field-bus and the distributed software contains several hundred thousand lines of code [93, 311]."

- "Automotive: The software component in high-end cars currently totals around 20 million lines of code, deployed on as many as 70 Electronic Control Units (ECUs) that accounts for 30% of overall production costs [312]. The Volkswagen Phaeton has 61 ECUs, 11.136 electrical parts, 2.110 cables and 3.860 meters of cables with a weight of 64 kg [313]."

- "Railway: The ever increasing request for safety, better performance, energy efficiency and cost reduction in modern railway trains have forced the introduction of sophisticated dependable embedded systems [314]. The number of ECUs within a train system is in the order of a few hundred [315, 316]."

An integrated approach could reduce the amount of ECUs, wires and connectors with the associated improvements in terms of cost, size, weight, power, scalability and reliability [1,2]. However, certification of mixed-criticality systems according to safety standards such as IEC 61508 [56] becomes a challenge because sufficient evidence must be provided to demonstrate that the resulting integrated system is safe for its purpose [50].

As explained in previous sections, multi-core and virtualization technology can support the development of mixed-criticality systems by means of software partitions. However, as stated by IEC 61508, whenever a system integrates safety functions of different criticality, sufficient independence of implementation must be shown among these functions. If there is not sufficient evidence, all integrated functions will need to meet the highest integrity level. Sufficient independence of implementation is established showing that the probability of a dependent failure between the higher and lower integrity parts is sufficiently low in comparison with the highest safety integrity level [132]. For this purpose, as stated in IEC 61508-2 Annex F, spatial and temporal independence are key to ensure the independence of execution ("that elements will not adversely interfere with each other's execution behavior such that a dangerous failure would occur") [50]. Different safety standards define similar needs for spatial and temporal independence, partition or isolation. For example, as described in Section 3.1, the avionics domain DO-178/ED-12 requires spatial and temporal partitioning.

However, widely available Commercial Off-The-Shelf (COTS) multi-core processors were not designed with a focus on hard-real time applications but towards the maximal average performance instead. This is the source for multiple temporal independence, partition or isolation issues and a challenge for the certification of mixed-criticality systems [50,311,317–325].

This chapter develops the overall DREAMS safety certification strategy (see Section 10.1). Section 10.2 describes basic certification concepts and their applicability to mixed-criticality multi-core systems. Sections 10.3, 10.4, 10.5, 10.6 and 10.7 develop the safety certification strategy with the description of modular safety cases, mixed-criticality solution patterns, DREAMS Functional Safety Management (FSM), product lines and a methodology for the certification of mixed-criticality product lines.

10.1 DREAMS Safety Certification Strategy

The DREAMS safety certification strategy builds on top of previous successful research projects (e.g., GENESYS [326, 327], TERESA [328–330], MultiPARTES [224, 311, 331], PROXIMA [332–334]) and the positive assessment of several safety concepts for dependable mixed-criticality embedded systems that meet IEC 61508, ISO 13849, EN 5012X and ISO 26262 safety standards [50,311,333,334]. Different lessons were learned from those research projects, which have been used to define the certification strategy developed in DREAMS:

- Feasibility: It is technically feasible to develop and certify partitioned multi-core mixed-criticality systems with current versions of safety standards (e.g., IEC 61508, ISO 13849, EN 5012X, ISO 26262), but the effort is high.

- Temporal independence and isolation: With respect to IEC 61508, temporal isolation support (e.g., HW and SW support) simplifies the safety argumentation, but temporal independence does not necessarily require temporal isolation guarantees. The lack of temporal isolation guarantees and rare (undocumented) temporal events could reduce the availability of the system but should not jeopardize safety if appropriate fault avoidance and control mechanisms are implemented (e.g., watchdog), which could lead the system to a safe-state [50,311].

- Cost competitive development and certification: Diagnosis strategies, modularity and mixed-criticality solution patterns are required among others, to enable cost efficient certification. In addition to this, if variability of product families are not considered from the beginning, minor variations of the system require a complete revision and update of the safety-concept and certification.

The DREAMS certification strategy aims to pave the way towards the cost competitive development and certification of mixed-criticality systems, tackling several challenges and lessons learned, by means of:

- Modular safety cases (see Section 10.3) to limit the impact of changes to specific modules of the system, enable reusability / modularity and reduce the complexity of the system (simplification strategy).

- Mixed-criticality patterns (see Section 10.4) that provide reusable solutions for the development of mixed-criticality systems based on partitions, multi-core devices and networks.

- DREAMS FSM (see Section 10.5) that maps safety related tools defined in Chapter 5 with development phases defined in IEC 61508.

- Definition of a methodology for the certification of mixed-criticality product families (see Sections 10.6 and 10.7).

This strategy takes into consideration among others a subset of the DREAMS architectural style described in Chapter 2, hypervisor based execution environments described in Chapter 6, a subset of the safety related tools described in Chapter 5 and a subset of the modeling and development processes described in Chapter 4.

And this certification strategy is used for the development of safety-concepts for different domain specific case studies: IEC 61508 SIL3 wind-power [50, 311] (see Section 11.1), EN 5012X SIL4 railway signaling [333] and ISO 26262 ASIL D automotive cruise control [334].

10.2   Certification and Compliant Items

As explained in Section 2.4.1, IEC 61508 [56] is a generic international safety standard considered as a reference by multiple domain specific standards: e.g., IEC 62061 and ISO 13849 for safety of machinery, ISO 26262 for automotive, EN 5012X for railway, IEC 60880 for nuclear power plants, IEC 61511 for safety instrumented systems.

IEC 61508 [56] defines requirements that safety-related items, components, devices and systems must fulfill in order to reduce risks and to assure their safety functions. Essentially it focuses on risk consideration where items are assessed in qualitative and quantitative terms in relation to probability of failure of a 'function' or in terms of systematic integrity. There are four different levels of 'Safety Integrity' covering a qualitative Systematic Capability (SC) 1 to 4 rating and a quantitative Safety Integrity Level (SIL) 1 to 4 rating. These ratings actually classify the safety level / risk reduction of a product or system. As a rule of thumb, the highest the Safety Integrity Level (SIL) the highest the certification cost

Today the industry typically requires that such items (either complex or low complex hardware or software) are subject to 'independent assessment / certification' by an accredited certification body of the European Union (EU) community. As part of an independent assessment, the accredited organization's experts would perform a detailed assessment according to the applicable clauses of the IEC 61508 requirements. Based on actual evidence as proof of applied calculations, measures or solutions gathered during the assessment with a positive result, compliance to IEC 61508 can be confirmed in form of an assessment report and consequently a certificate can be issued.

It should be noted, that the IEC 61508 'as technical standard' considers the 'world' as not 100% perfect and consequently risks of failure within a function are finite. Therefore one cannot assume that all items are 100% perfect or free of failure/fault(s)/error(s). Hence 'safety integrity' is classified in terms of 'probability of failure'. The main objective is to reduce risks as low as reasonably practicable.

10.2.1 Need and Importance of Certifications

For the industry certification of products is important, as a qualified and neutral approval by third parties confirms that a product meets technical requirements and thus expectations of customers.

In this context product liability becomes important. Legally product liability means that manufacturers, distributors, suppliers, retailers, and others who make products available to the public are held responsible to injuries those products might cause [335]. The Product Liability Directive 85/374/EEC is a document released by the Council of the European Union defining a regime of strict liability for defective products [336]. According to this EC Directive 85/374/EEC amongst others the liability of the producer in relation to the injured person arising from the Directive cannot be limited by other provisions [336].

This leads to certain obligations of manufacturers, distributors, suppliers, retailers, and others who make products available to the public [337]. The placing of products to the market at large has to conform to the following EC regulations:

- EC Regulation 765/2008/EC [338] defines the requirements for accreditation and market surveillance in connection with the marketing of products,

- EC Regulation 764/2008/EC [339] defines procedures with regard to the application of certain national technical regulations for products which were placed on the market in another Member State of the EC,

- EC Directive 2001/95/EC [340] on general product safety for consumer products,

- EC harmonization rules for certain sectors (e.g., Machinery Directive, Toys Directive, Low Voltage Directive, Medical Device Directive),

- Decision No. 768/2008/EC [341] of the European Parliament and the Council of July 9th, 2008 on a common legal framework [342] for marketing of products (relevant to the drafting of legislation).

Products - within the scope of the EC harmonization rules - shall meet the essential requirements defined therein. As a rule this means before products are placed on the market: a conformity assessment shall have to be carried out, the technical documentation must be drawn up for proof of conformity, an EC declaration of conformity shall be issued and the CE marking must be affixed.

The EC declaration of conformity is a specific certificate which can only be issued for products based on defined product types and following the relevant EC directives e.g., the Machinery Directive 2006/42/EC [343]. But for compliant items which are not complete products, only basic compliance certificates can be issued - not an EC type-examination certificate.

Conformity assessment, also known as compliance assessment, is any activity to determine, directly or indirectly, that a process, product, or service meets relevant technical standards and fulfils relevant requirements (see ISO/IEC 17000, Conformity Assessment - Vocabulary and General Principles [344]).

A conformity assessment of a product can be carried out by a conformity assessment body and depending on the scope by a certification body, inspection body or laboratory.

In case of positive assessment a certificate as defined document is issued. The process of assessment and issue of certificate is commonly called 'certification'. A product certification body confirms the presumption of conformity for a product for the future as long as the respective certificate is valid and the product has not been changed. It may be necessary that surveillance assessments are carried out to check that the product is manufactured unchanged as expected.

10.2.2   Accreditation

Accreditation is an approval by a national accreditation body following ISO/IEC 17011 [345] (e.g., in Germany the DAkkS "Deutsche Akkreditierungsstelle GmbH") that a conformity assessment body meets the requirements defined in harmonized standards (see EN ISO/IEC 17020 for inspection bodies [346], EN ISO/IEC 17025 for laboratories [347] and EN ISO/IEC 17065 [348] for certification bodies) which are based on international standards.

The accreditation of a conformity assessment body substantiates trust and traceability within the results of its conformity assessments.

The accreditation body assesses whether the conformity assessment body / certification body performs its activities professionally, competently and in compliance with legal and normative requirements and on an internationally comparable level.

10.2.3   EC Type-Examinations

Products sold in the European Economic Area (EEA) must meet requirements of applicable EC directives. The CE marking is a mandatory marking for products which are sold within the European Economic Area and indicates that the product conforms to the applicable EC Directives [349].

Three European regional standardization bodies, European Committee for Standardization (CEN), European Committee for Electrotechnical Standardization (CENELEC) and European Telecommunications Standards Institute (ETSI), work on behalf of the European Commission on harmonized European Standards (EN's). Manufacturers, other economic operators, or conformity assessment bodies can use harmonized standards to demonstrate that products, services, or processes comply with relevant EU legislation [350]. The harmonized standards must be listed in the "Official Journal of the European Union". The use of these standards remains voluntary.

The CE mark is a declaration by the manufacturer that a product complies with the respective EC directive and hence the harmonized standard(s) referenced by the directive(s) [351].

If harmonized European standards exist and the product complies with these standards, it is presumed that the product is in conformity with the requirements defined in the relevant EC Directives.

For some products (these are defined by the European Commission, see [352]), special conformity assessment bodies ('Notified Bodies') must verify that the product meets specific technical requirements. This is not obligatory for all products. If the product is not verified by an independent body, then it is the manufacturer's decision to give proof if it complies with the technical requirements. This includes estimating and documenting the possible risks when using the product. The associated technical dossier should include all documents proving that the product complies with the applicable technical requirements. Finally the CE marking can be affixed on the product. The marking must be visible, legible and indelible. An EC declaration of conformity stating that the product meets all legal requirements must also be drafted, signed and published [349].

Importers must make sure that the products they place on the market comply with the applicable requirements and do not present a risk to the European public. Distributors must also be able to demonstrate to national authorities that they have acted diligently and with due care and have affirmation from the manufacturer or the importer that the necessary measures have been taken [349].

10.2.4 Certification Requirements According to IEC 61508

The EN 61508 derived from the IEC 61508 is a harmonized standard, but not listed in the "Official Journal of the European Union". Requirements of the EN 61508 are product independent. This means, that for a product certification 'like for the EC-Type-Examination' further product specific standards must be considered and applied.

For certification bodies - like TÜV Rheinland - the EN ISO/IEC 17065 [348], describes the comprehensive certification process basically in the following steps: application, application review, evaluation, review, certification decision, surveillance (incl. prolongation) and termination, reduction, suspension or withdrawal of certification.

'Evaluation' is the same activity which is referred to in IEC 61508 as 'assessment' or 'functional safety assessment' since it is a standard for Functional Safety. IEC 61508 does not request a certification but recommends that independent accredited certification bodies provide conformity assessment services. IEC 61508 requires different levels of independence (independent person, independent department or independent organization) for the functional safety assessment depending on the consequences of malfunction or non-compliance starting from minor injury up to many people killed. Certification will provide the highest level of independence for functional safety assessments because the assessments will be done by an independent organization.

10.2.5 Compliant Items According to IEC 61508

A 'compliant item' is an item that meets individual parts of the requirements of IEC 61508. For example compliant software items may be qualified in three ways according to IEC 61508-2 [132]: compliant development, proven-in-use argumentation and subsequent qualification of a software item.

A safety manual shall be provided by the manufacturer for a compliant item. The safety manual for a compliant software item must provide amongst others basically the following information according to IEC 61508-3 [353]: functional specification, configuration information, constraints, systematic capability, assumptions on use, implementation instructions, release information and known anomalies, compatibility information and evidences.

In summary, it can be concluded that a certification of a compliant item according to IEC 61508 by an independent accredited certification body provides the highest possible certainty that there will not be any problems with a later needed EC type examination certification of the respective product where the compliant item is embedded in.

10.3 Modular Safety Cases

As explained in Section 2.4.1, a safety case is a documented body of arguments and evidences, intended to justify that a system is acceptably safe for a given set of constraints (e.g., application, operating environment, hypothesis of usage). A safety case can be developed as a composition of Modular Safety Cases (MSCs). This modular approach can be used to limit the impact of changes to specific modules of the system, enable re-usability and reduce the complexity of the system (simplification strategy). There are three predominant safety case notation languages:

- Goal Structuring Notation (GSN) [120] is a graphical argumentation notation language that explicitly represents the individual elements of a safety argument (requirements, claims, evidence and context) and their relationships.

- Claim, Arguments and Evidence (CAE) [354] is a notation language composed of claims, arguments and evidence elements.

- Structured Assurance Case Metamodel (SACM) [355] combines multiple argument specification models and harmonizes the common elements of the GSN and CAE notation languages.

As listed below, several generic and reusable DREAMS IEC 61508 compliant MSCs have been defined using GSN and CAE notation languages [356–359]. For simplification purposes, this section summarizes only one of the developed MSCs (see referenced documents for further details).

- Execution environment: Generic hypervisor with linking analysis to the XtratuM hypervisor [356] and COTS multi-core device with linking analysis to the Zynq 7000 [357]

- Chip-level communication services: Generic Network On Chip (NoC) [358]

- Cluster-Level communication services: Generic black-channel communication protocol with linking analysis to Ethernet for Control Automation Technology (EtherCAT) and associated Fail Safe over EtherCAT (FSoE) [359]

10.3.1 Modular Safety Case for Cluster-Level Mixed-Criticality Networks

Figure 10.1 shows a top level representation of a MSC that defines the safety-related arguments that a mixed-criticality network shall fulfill to be compliant with IEC 61508. This safety standard considers networks provided by the manufacturer *as they are* (*black channel* networks) and networks developed in accordance with a safety standard (e.g., IEC 61508, IEC 61784-3) (*white channel* networks) (see subsection 7.4.11.2 of IEC 61508-2 [132]) [359].

As explained in Section 8.1.2.1, *white channel* networks shall be designed, implemented and validated according to IEC 61508 and IEC 61784-3 or IEC 62280. In the case of *black channel* networks, it is assumed that parts of the communication channel cannot be designed, implemented and validated according to a safety standard. Therefore, additional measures and diagnosis techniques are required to ensure that the failure performance of the communication process is compliant with the IEC 61508 and the IEC 61784-3 safety standards.

The following points summarize the safety arguments defined in the MSC for an IEC 61508 compliant mixed-criticality network (see also Section 8.1.2.1 and [323, 358]):

- White channel networks and the safety-related parts of the black channel network (e.g., Safety Communication Layer (SCL)) should be *developed in compliance with the IEC 61508 and IEC 61784-3 standards*, against the required SIL and residual bit error rate values. The residual bit error rate is the number of bit errors per time unit based on the total number of bits received during a time interval. For instance, as stated in Table 1 of IEC 61784-3, for a SIL of 3, the probability of dangerous failures per hour and the maximum permissible residual error rate of the functional safety communication system shall be lower than $10^{-9}/h$.

- A safety network shall guarantee that the reception time interval of two consecutive messages is below a predefined time value defined by the system integrator. Otherwise, a communication error shall be assumed. This requirement is usually referred to as the *idle current, closed-circuit or de-energized to trip principle*.

- Common *communication errors* in mixed-criticality networks include the corruption, unintended repetitions, incorrect sequence, loss, unacceptable delay, insertion, masquerade and addressing. A safety mixed-criticality network shall consider those errors and shall implement *deterministic remedial measures* recommended in IEC 61784-3 to detect and mitigate common communication errors. For example: sequence number, time stamps, expectation time, connection authentication, feedback messages, data integrity assurance, redundancy with cross checking and different data integrity assurance techniques may be implemented to that end.

- In addition to the defined methods to estimate residual errors, *further fault cases* defined in IEC 62280 shall be considered and controlled by white channel networks. IEC 62280 recommends the measures and safety services to reduce the risk associated with the threats of communication systems. For instance, it suggests the message authentication, integrity, timeliness and sequence checking, source and destination identify, feedback message and cryptography techniques to reduce associated risks with the communication threats (e.g., repetition, deletion, insertion).

- *Error reaction mechanisms* shall be implemented by an IEC 61508 compliant network to achieve a safe state in case that a communication error is detected. For example, it can stop the communication between specific components of the system. In the case of black channel networks, the reactions to errors mechanisms shall be defined in the documentation provided by the network developer.

- The *reaction and response times* of a safety net should not exceed the time values specified by the network manufacturer, even in the presence of a failure. To that end, a safety communication system shall provide a predictable and a deterministic communication between the components of different criticalities (e.g., safety and non-safety).

- A *combination of the measures* quoted before (e.g., sequence numbers, time stamps, expectation time, connection authentication) shall be implemented by the safety communication network to detect communication errors. The quantity and timing errors to be considered are provided by the network manufacturer or the Functional Safety Communication Profile (FSCP). The utilization of a common function within the fieldbus communication by the specific groups of participants is called a profile. For example, communication profile family 12, commonly known as EtherCAT is based on the IEC 61158 standard [290] and the safety communication layer specification is defined in IEC 61784-3-12 [360]. FSCP 12 describes a protocol for transferring safety data up to SIL3 between FSCP 12 devices.

- The transmission / reception among safety and non safety-related messages through white channel networks may be affected by interferences, leading to the failure of the communication between the components connected to the network. White channel networks shall guarantee the *non-interference of non safety-related communication* to ensure the compliance with the temporal and spatial independence requirements. For instance, Time-Triggered (TT) communication systems may be used to ensure the temporal independence. These networks use a priori knowledge about the permitted component behavior to block faulty messages.

- A safety communication network can be tested in two ways: In the first instance, the components of the network may be tested together, thus obtaining an *exhaustive diagnosis* of the network. In the second scenario, the components of the network are tested independently by test beds or simulators. Nevertheless, both scenarios should follow the parameters set in the FSCP. An FSCP defines suitable conformance testing to assess services and measures of IEC 61784-3. A safety network shall support an adequate diagnostic strategy to avoid configuration related issues that can jeopardize the safety of the communication system and associated subsystems. For that reason, qualified tools of classes T2 (e.g., verification tools) and T3 (e.g., compilers) shall be used for designing, developing and configuring the safety network.

- Failure mode and effects analysis (e.g., FMEA, FMECA, FMEDA).

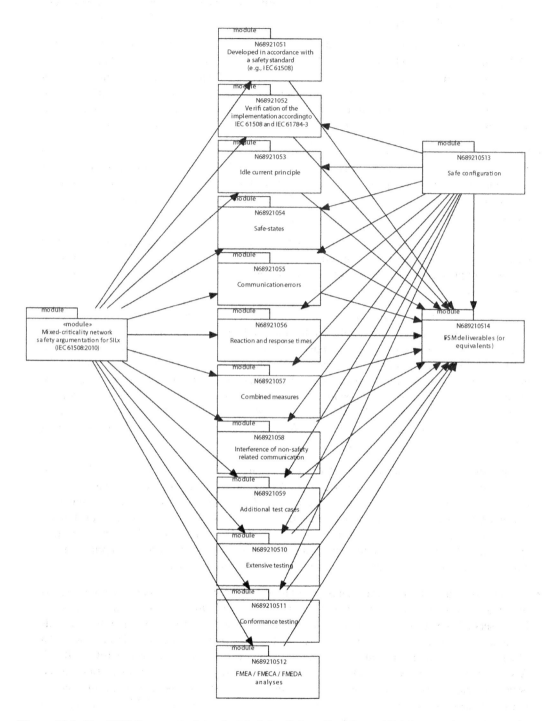

Figure 10.1: Top GSN Representation of a Modular Safety Case for an IEC 61508 Compliant Mixed-Criticality Network [323]

10.3.1.1 Linking Analysis

A linking analysis document defines the way in which commercial or custom components meet the safety arguments stated in the MSCs. This section summarizes the linking analysis of the Safety over EtherCAT network with respect to the MSC for an IEC 61508 compliant mixed-criticality network previously described.

Section 8.1.2 described EtherCAT, associated Fail Safe over EtherCAT (FSoE) and the developed DREAMS Safety Communication Layer (SCL). FSoE defines a safety single-channel communication layer (SCL) for transferring messages with different criticality. The transport medium (EtherCAT) is referred to as a *black channel* network, which is not included in the safety considerations. Furthermore, it is assumed that the SCL (custom or commercial) on top of EtherCAT network applies to *fail-safe* systems and that follows an *IEC 61508 compliant development process* with a residual error rate probability. For example, the FSoE is developed with a residual error rate lower than $10^{-9}/h$.

The following items present the way in which FSoE fulfills the safety requirements stated in the MSC.

- *Idle current principle*: The SCL diagnoses the EtherCAT *black channel* network by separate watchdog timers, detecting delays on the communication paths and the reception of EtherCAT frames. The *system integrator* provides the parametrization and the documentation of the measures and diagnostic techniques which are implemented by the communication network.

- *Communication errors and Combined measures:* The SCL supports a set of measures and diagnostic techniques for controlling the common communication errors (see Section 8.1.2).

- The *reaction time* of the EtherCAT SCL depends on the arriving time of the order from the master node (calculated) plus the order's propagation time required to achieve the upper communication layer. The processing time of the application on top of the slave depends on the application itself, and therefore, it is out of the scope of this analysis.

- The *response time* of the SCL on top of a *black channel* network is determined, among other things, by the network topology implemented in the system architecture.

- The custom or commercial SCL provides *non-interference of non safety-related communications*, thus ensuring the temporal and spatial independences.

    - The *temporal independence* is guaranteed using TDMA based protocols. Although the average performance of those protocols is not as good as for randomly accessed networks, it is one way to assure the temporal independence. Also, data networks such as the Avionics Full Duplex Ethernet (AFDX) with rate constraints can be implemented for the same purpose. This SCL supports the time multiplexed concept, where the access to the medium is given via a token. The token is represented by the communication frame where each network slave writes and reads the transmitted information.

    - The *spatial independence* is achieved through the isolation of the applications from the Hardware (HW) using a HW network controller. The controller passes the information to the application layer when the received information is pointed as correct. The spatial independence between *black channel* slaves is inherent to the network definition and the communication generation. The HW network controller provides random and systematic failure diagnosis in order to detect failures during design time and execution time (see IEC 61508-2 and IEC 61508-3 [132, 353]).

- The SCL on top of the *black channel* network is tested using an *extensive testing* where the worst-case scenarios are taken into consideration. Those scenarios are defined by the *network manufacturer* or the FSCP 12 in IEC 61784-3 [360].

- *Conformance testing* is mandatory to verify that the device complies with the communication requirements of IEC 61508-2 [132] and IEC 61784-3 [292]. Those tests are required for checking the communication devices that should work with other communicating devices which are provided by other manufacturers. They are implemented when the integration tests cannot be carried out. The conformance test is also required for the *black channel* network and the SCL. An example of the conformance testing of a black channel system and an SCL can be found within the EtherCAT conformance test tools [361].

- The EtherCAT SCL follows an IEC 61508 compliant development process where *qualified tools* are used for its design, development and configuration [362].

- The SCL shall provide the Failure Mode and Effects Analysis (FMEA) / Failure Mode, Effects and Criticality Analysis (FMECA) / Failure Modes, Effects, and Diagnostic Analysis (FMEDA) analysis to assess random and systematic failures.

10.4 Mixed-Criticality Patterns

Dependable design patterns provide reusable design solutions, artifacts and knowledge that support engineers in the development of dependable embedded systems [330, 363, 364]. This concept can be extended to support mixed-criticality patterns with reusable solutions to recurrent challenges that need to be addressed in the development of mixed-criticality systems [325]. For example, according to IEC 61508-3 (see Annex F) the common causes of execution interferences are the shared use of random access, peripherals and processor time, the communication between the elements necessary to achieve the overall design and fault-propagation. Those interferences are common in COTS multi-core and many-core devices where resources are shared among cores.

Based on this, several reusable mixed-criticality design patterns have been defined [365] for:

- Hypervisor and partitions (see Section 10.4.1): NoC accessible memory area diagnosis, critical partition diagnosis, communication input / output server and digital input / output server [365]

- COTS multi-core device (see Section 10.4.2): shared memory diagnosis, cache coherency management, inter-connection management unit diagnosis and interrupt controller diagnosis [365]

- Mixed-criticality network (see Section 10.4.3): NoC diagnosis [365]

10.4.1 Hypervisors

A hypervisor (see Section 6.1) is a layer of Software (SW) or a combination of SW and HW that allows running several independent execution environments, also called partitions, in a single embedded computing platform. Partitions are logical divisions of memory with static or dynamic cycle and execution time. They can have assigned one or more peripherals and can be developed for different levels of criticality. Among products in this category, XtratuM [216], PikeOS [222], Wind River [366] and QNX [367] hypervisors can be commercially distinguished.

Partitioning a system using a hypervisor can give rise to spatial independence, temporal independence and real-time constraints. The broad trend of partitioning multi-core and many-core systems requires inter-partition communication mechanisms such as shared memories or NoCs. NoCs are commonly used communication systems to avoid the problems associated with the use of shared memories [368]. Those memories imply temporal and spatial interferences due to memory inconsistencies and cache coherency problems. However, the use of NoCs may increase the complexity of the system and may involve challenges to certification such as guaranteeing that the critical memory assigned to the network is not accessed by the hypervisor or the partitions.

The following paragraphs present further failure scenarios where solutions to tackle spatial and temporal interferences and manage the resources in multi-core mixed-criticality systems are presented.

10.4.1.1 NoC Accessible Memory Area Diagnosis Pattern

NoCs are widely implemented communication systems to avoid Point To Point (P2P) individual communication paths between the components of mixed-criticality systems. They enable the creation of logic paths to interchange data. However, on-chip networks can access critical memory areas in use by other components, causing errors that can jeopardize the safety of the system. The most significant impact caused by the memory access of a NoC is the breaking of the temporal isolation, which can also be endangered due to delays caused by a high amount of traffic in the NoC.

This cross-domain pattern defines the following 3 solutions to detect, manage and avoid failures in the critical memory areas which are accessible by the NoC.

Different patterns are proposed for the detection, management and failure avoidance in critical memory areas that are accessible by the NoC, as listed below and described with more detail in [365]:

▶ | Solution 1: | HW isolation of NoC

Dedicated memories or memory areas may be assigned to the NoC for incoming / outgoing message buffers and its internal operations. In addition, the NoC may be attached to a different bus as the processing cores, which should provide a complete isolation from the processing cores. This solution scheme can be implemented using a dual-port Random-Access Memory (RAM) where one port is accessible by the NoC and the processing cores use the other port.

▶ | Solution 2: | Input / Output Memory Management Unit

Memory Management Units (MMUs) allow controlling the access of Direct Memory Access (DMA) transfers programmed by the bus-master capable Input / Output (I/O) devices. Consequently, the DMA transfers cannot overwrite and read from the restricted memory addresses. An I/O MMU enforces the spatial isolation and avoids the overwriting of the safety-sensitive memory regions by the NoC.

▶ | Solution 3: | Additional monitoring mechanisms

See the following design pattern *Critical partition diagnosis pattern*.

10.4.1.2 Critical Partition Diagnosis Pattern

When dealing with partitioned mixed-criticality systems, failures caused by the exchange of information are quite probable. The lower criticality functionalities can interfere on the higher criticality functionalities. This pattern analyzes the occurrence of temporal interferences generated by multiple accesses in parallel to the shared memory and the occurrence of spatial interferences caused by failures of the hypervisor.

Different patterns are proposed for the measurement and detection of interferences caused by non-critical partitions on critical partitions and attest the system's temporal and spatial independence, as listed below and described with more detail in [365]:

▶ | Solution 1: | Limit the concurrency

Critical tasks may be executed without concurrency in partitioned mixed-criticality systems. When a critical task is running on a certain core of a multi-core processor, the other cores do *idle* only for the duration of the task. This consideration enables avoiding contention.

The limitation of the concurrency can be achieved by appropriately configuring the partitions execution window at design time. The loss in performance can be leveraged by tuning the amount of time that a core (running a critical task) executes without concurrency.

The maximum amount of interferences that can be suffered by one core due to accesses to shared memory and the bus bandwidth used by the other cores can be calculated using off-line analysis. Those analyses are characterized by non-automate sampling with discontinuous sample measure and evaluation.

Concurrency can be guaranteed up to a certain safe time limit based on the temporal constraints of the safety-critical tasks and the maximum amount of interferences.

▶ Solution 2: **Assess the spatial isolation**

This solution presents a diagnostic partition that periodically checks the data of the critical memory areas, including the hypervisor's code and the partitions' code and data. Checksum and similar mechanism recommended by the IEC 61508 safety standard are perfect candidates to ensure that no accidental modification of code or data takes place. Partitions can individually implement measures and diagnostic techniques (e.g., Error-Correcting Code (ECC), checksum, parity bit) to check the code and data.

▶ Solution 3: **Assess the temporal isolation**

The measures and techniques recommended by the IEC 61508 safety standard can also be implemented to assess the temporal isolation. For instance, the *Program temporal sequence monitoring* technique may be implemented to monitor the execution of safety-critical tasks regarding their temporal response and to ensure that the temporal isolation is not compromised due to a failure of the hypervisor.

10.4.1.3 Communication Input / Output Server Pattern

The functionalities (e.g., safety-related, non safety-related) executed on multi-core mixed-criticality systems usually require communication. Shared memories, on-chip buses (e.g., TTNoC), off-chip buses (e.g., EtherCAT) and local buses (e.g., PCIe, digital I/Os and RS485) may be implemented to that end.

These communication systems have their pros and cons. For instance, shared memories are prompt to interferences in architectures with more than one core [369]. To overcome issues related to the shared memories, on-chip and off-chip communication networks are one of the alternatives for providing internal and external communications. For example, the communication between partitions with different criticality levels may be carried out through an STNoC on-chip network and a Time Triggered Ethernet (TTE) network with or without real-time capabilities. However, the use of NoCs implies additional interferences in general with challenges for certification. The challenges regarding NoCs are discussed in Section 10.4.3.1.

On the other hand, the number of functions integrated into mixed-criticality systems which require communication tends to increase, mixing the communication requirements of different criticality (e.g., safety, real-time, security) and hampering the development and certification of mixed-criticality networks. As a result, the underlying mixed-criticality communication systems can require an adaptation procedure or shall be modified to cover new requirements. Those processes may incur a higher engineering and certification cost.

This pattern defines a communication server that simplifies the system design and development of mixed-criticality systems [325]. The communication server centralizes and manages the communication between partitions and device external I/Os. This server is logically abstracted from the processor control of the communication network (e.g., using partition ports) and manages the assignment of the peripherals to the partitions implementing the exclusive access to peripherals technique that enables the interference freeness among different criticality functionalities that manage non-exclusive device resources.

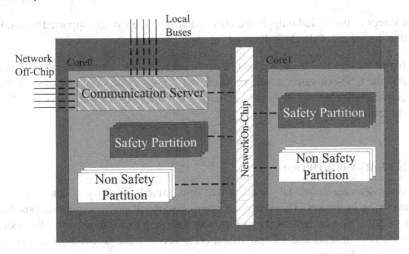

Figure 10.2: Communication I/O Server Representation [323]

10.4.1.4 Digital Input / Output Server Pattern

Digital Input Outputs (DIOs) are used in multi-core mixed-criticality systems to communicate external components such as sensors and actuators. In single core architectures, DIOs can be managed without significant difficulty, as there is a single core. Instead, in multi-core and many-core architectures where hundreds of thousands of functionalities with different criticality levels can be implemented (e.g., safety, non safety, real-time), one DIO may be requested at the same time by more than one functionality. Therefore, the management of I/O interfaces is hindered in multi-core and many-core devices [325], which, at the same time may cause temporal interferences.

Besides, from a product line perspective, the number of DIOs requested by a product might change. Those upgrades may lead to scalability problems and an adaptation process to fit with the new specifications.

This cross-domain pattern proposes the implementation of a Digital Input Output System (DIOS) partition to centralize the management of the DIOs of mixed-criticality systems [325]. The DIOS is a consistent concurrent manager of DIOs which is abstracted from platform and hypervisor details to assure reusability. Furthermore, it periodically updates the values of the inputs and refreshes the information of the partitions where the inputs are required. This server can be reused on different system architectures, thus simplifying the system design.

Afterward, the measures and diagnosis techniques defined in the following paragraphs are executed, so that, if a failure is detected, the outputs would be refreshed with the safe value instead of with the value provided by the partitions. In the case that different partitions try to update the same output with different values, the partitions would reach a safe state, and the outputs would be updated to the default value.

- The *Cyclic Redundancy Checks (CRCs)* of the register's values associated to DIOs are periodically compared against the values which are already stored by DIOS. The comparison period is determined by the minimum refreshing period of the digital outputs.

- It is checked that the partitions in charge for updating the digital outputs refresh the values of the DIOS partition. To that end, a *token* is implemented, which is updated every time that the communication is performed, always agreeing with the expected values in the DIOS. This solution can also be applied to the remaining partitions of the system, but with the digital inputs, for assuring the communication between the partitions and the DIOS.

- The register values of the digital outputs are checked against the values supported by the DIOS every time that the values of the digital outputs change.

- The digital inputs are checked under a pre-configured timeout to detect whether their values can be modified. If the time-out value is not specified, the default value would be used (a month), and the developer shall integrate an output to change the values of the inputs in a controlled non safety way for testing purposes.

10.4.2   COTS Multi-Core Device

COTS multi-core devices are commonly used in real-time mixed-criticality systems due to their low-cost and short time to market. Those devices provide sophisticated components (such as the P4080 CoreNet [370]) which increment the complexity and may cause drawbacks. For example, the simultaneous running of tasks and the resource sharing between the processor cores can lead to temporal and spatial interferences.

Different research studies propose techniques to improve the performance of the multi-core devices by reducing the memory interferences of the applications [256, 371, 372] and to control the mapping of application's data to memory channels [373]. Some of those techniques focus on scheduling policies which provide request prioritization and reduce the inter-partition interferences.

The coherency of the processing cores, the memories and the programmable logic can also be prompt to interferences in general. In single-core systems, the coherency of the cores and the memory is not a problem such as there is only one processing unit that can read/write from/to the memory. Conversely, in multi-core and many-core systems, there can be two or more processing units executing at the same time. So, it may be the case that they need to access the same memory location at the same time. If no processor changes the data of the accessed memory location, they could share data indefinitely.

However, the coherency protocol may fail in the case of memory data is changed by some processing core, leading to data inconsistencies and possibly causing a failure of the system. Consequently, we identified a need for extending the current measures and diagnostic techniques recommended by the IEC 61508 safety standard for multi-core and many-core systems.

The following patterns aim to solve, detect and control the failures of remarkable, challenging components of multi-core systems, including shared memories, cache coherency units, interconnection management units and interrupt controllers. Those patterns can be reused as the basis to extend the current measures and diagnosis techniques recommended by IEC 61508.

10.4.2.1   Shared Memory Diagnosis Pattern

The sharing of resources is a habitual implementation in multi-core mixed-criticality devices to improve the performance. Those multi-core devices integrate cache memories for private use (e.g., L1 cache) or to communicate the components of the device (e.g., L2 cache). Secondary cache memory is commonly used to improve the performance of the system when the processor generates significant data traffic.

The IEC 61508 safety standard covers the failures caused by memory sharing such as the causal factors of the execution interference between components of a single computer platform (see Annex F of IEC 61508-3 [353] *"Techniques for achieving non-interference between SW components on a single computer"*). However, in systems with more than one core, the measures and techniques recommended by that standard are not applicable at all due to the shared memory can be accessed by the processing cores at the same time, giving rise to data corruption. Measures and diagnostic techniques recommended by the IEC 61508 safety standard should be reviewed and extended for covering the issues caused by the use of shared memories in multi-core architectures.

This cross-domain pattern [323,374] proposes the following reusable generic solutions to detect, evict and manage the failures related to the shared memories in HW architecture with more than one processing core.

▶ ☐ Solution 1: ☐ **Limit the use of shared memories [323, 365, 374]**

As described in [323, 374], this solution proposes to limit as much as possible the use of the shared memory and in the case that it is implemented to control its access for avoiding parallel accesses. For instance, the shared memory of the Zynq-7000 zc706 multi-core device can be disabled for avoiding interferences [375].

Another possibility is to replace the shared memory by an on-chip network, thus avoiding interferences caused by the use of shared memory. NoC systems provide benefits regarding spatial and temporal segregation. However, as analyzed in subsection 10.4.3, the management of NoCs for communicating components with different criticality implies fundamental challenges to certification [376].

▶ ☐ Solution 2: ☐ **Cyclic Redundancy Check with comparison [323, 365, 374]**

As described in [323, 374], this solution proposes a CRC based diagnosis with comparison technique to detect failures in the shared memory, which is cyclically executed (e.g., 50 ms). It considers two implementation scenarios where the first scenario builds on a COTS multi-core device which allows integrating functionalities with different criticality levels into its processing cores and the second scenario that consists of a partitioned system where functionalities with different criticalities are executed on top of partitions (e.g., safety, security, real-time).

Figure 10.3 shows a simplified representation of a multi-core device that consists of two cores, a shared memory and two independent memories (Memory A and Memory B) for storing and reading the application data and the CRCs.

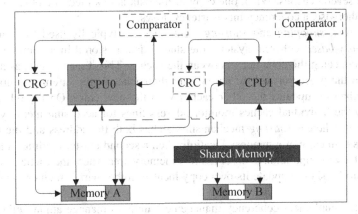

Figure 10.3: Shared Memory Diagnosis Pattern - Block Diagram. © Elsevier 2017. Reprinted, with permission, from [365].

For reasons of simplification and understanding, a scenario where both cores execute the same functionality is taken as a basis for describing the steps of the CRC and comparison-based diagnosis pattern.

- Step 0: The first time that this solution is executed, both cores calculate the CRC (called old_CRC) of the data of the processing cores. Those CRCs are stored in different memory locations of *Memory A*.

- Step 1: In the first step, the data of the processing units are saved, through the shared memory, in different memory locations of *Memory B*.

- Step 2: In this step, the old_CRCs, the CRCs which are stored in Memory A at the first time, are read by both cores.

- Step 3: In this step, the data stored in Memory B is read by both cores through the shared memory. Then, the CRC of the read data is calculated (one CRC per core).

- Step 4: In this step, the comparison of the gold_CRCs and new CRCs is executed. If the CRCs match, the execution continues. Otherwise, a fault-tolerance technique or a safe-state should be executed.

- Step 5: In the last step, once the CRCs are compared, and it is checked that the shared memory does not corrupt the data, the current CRCs (one CRC per core) are stored in Memory A. Those CRC would be called old_CRCs in the next execution cycle.

The CRC and comparison technique may be extended to improve the diagnosis coverage, for example, implementing partition redundancy. In addition, a diagnosis scenario with redundant HW can also be implemented. This scenario executes the comparison of the CRCs at partition level and system level. This comparison can be realized internally by a SW comparator or externally by a HW comparator. In the case that a SW based comparator is implemented, a communication network should be used to spread of the resulting CRCs through the entire system. This approach enables to detect whether the shared memory fails due to a failure of some component of the device or the device itself.

10.4.2.2 Cache Coherency Management Diagnosis Pattern

As described in [323, 374], cache coherency is the consistency of shared resource data that ends up stored in multiple local caches (e.g., L1 and L2 cache). A coherency mechanism stores the copies of the data saved in several caches. When one copy of the data is modified, the other copy should also be modified. Otherwise, a coherency inconsistency arises.

The *directory-based*, *snooping* and *snarfing* techniques are typically used to guarantee coherency. In a *directory-based* coherent system, the shared data is stored in a common directory that is used for guaranteeing the coherency between the caches. This directory acts as a filter that grants permission to the processor to load an entry to the cache. When an entry is updated, the directory-based mechanism updates the other caches with the new entry. On the other hand, in *snooping* coherency the individual caches monitor address lines for accessing memory locations that they have cached, whereas *snarfing* mechanism watches both the address and the data in an attempt to update its own copy of a memory location when a second master modifies a location in the main memory. If a write operation is observed to a memory area where the cache has a copy of data, the cache controller should update its own copy located in the snarfed memory with the new data.

Multi-core devices implement coherency management units to manage among others the coherency of the processing cores, the memory and the programmable logic. For instance, the Zynq-7000 zc706 multi-core device implements a Snoop Control Unit (SCU) to manage the coherency [375]. However, the coherency techniques do not fully guarantee freeness off data propagation inconsistencies. Here is where this cross domain pattern focuses, ensuring that changes to the data are propagated through the device and if not, detecting whether a coherency failure occurs. To that end, this pattern defines the following three solutions.

▶ | Solution 1: | **Configuration Check**

The coherency management unit shall be configured reasonably to minimize the interferences. The wrong or incorrect configuration of the coherency management unit may lead to the loss of coherency and the resultant failure of the system. Therefore, this first solution proposes to periodically check the configuration of the coherency management unit, comparing it with the expected configuration or the last valid configuration set. In addition, this solution assumes that the chosen configuration is free of systematic faults and that, therefore, a set of measures and diagnostic techniques for ensuring that it is protected against unexpected configuration changes should be implemented.

For instance, the periodic read-back (see Table A.10 of IEC 61508-2), modification protection (see Table A.17 of IEC 61508-2) and the failure detection by online monitoring (see Table A.15 of IEC 61508-2) techniques may be implemented.

▶ Solution 2: Diagnose random failures

The software can manage the memory regions shared among certain sets of coherent masters. In addition, it can ensure that the share-ability mappings between the masters are consistent to avoid unexpected behaviors and inconsistencies. For instance, a protection mechanism such as a MMU can be used to control the memory, manage permissions to blocks of the memory and translate the virtual addresses to physical addresses. This solution assumes that the coherency management unit implements a set of measures and diagnosis techniques to detect and control random faults such as the wrong addressing, partial update or single bit errors faults. For instance, watchdog timers may be used for detecting the temporal deviations, a CRC with comparison technique may be implemented for detecting unexpected data modifications, (see pattern 'shared memory diagnosis' in Subsection 10.4.2.1) and an ECC and / or a parity bit technique can be implemented to detect data consistency violations, including partial update or single bit error failures.

▶ Solution 3: Diagnose systematic failures

Systematic faults can also affect the coherency management unit. These faults can be sourced from the HW design, the environmental stress, external influences and operational failures. This third solution considers the implementation of the measures and diagnostic techniques recommended in Tables A.15 to A.17 of IEC 61508-2 [132] for managing the systematic faults in the coherency unit. The selection of the measures and techniques depends on the HW available and the SW supported by the system architecture, giving rise to different combinations of measures and diagnostic techniques.

In addition to the solutions defined, this pattern considers that fault avoidance and fault control measures should be implemented in systems with cache coherency. For instance, the use of the shared memory may be limited to an absolute minimum required for operation, the use of multiple threads and tasks for one safety function can be restricted to a minimum and write accesses to the memory can be assigned statically to the tasks. The automatic invalidation of cache lines after a defined period is required to ensure that caches are flushed periodically and that they keep coherent. The faults can be controlled implementing communication protocols with additional messaging between sender and receiver of the information. For example, flags to indicate whether the information is updated and received can be applied. This procedure enables detecting the order violation fault.

Extra coding information may be integrated with the data to detect data consistency violations. To that end techniques such as CRC, ECC or parity information may be used. It is safe to assume that a HW that implements the ECC technique or the parity technique may have bugs (e.g., ARM: 751475—Parity error may not be reported on full cache line access (eviction / coherent data transfer / 'cp15' clean operations [377]). Further techniques implement data structures that match the cache architecture (e.g., the maximum size of one cache line for optimal performance) and implement additional measures and diagnosis techniques. Examples are ECC techniques, scrubbing, the implementation of timing expectations and error detection for the shared memory communication or common communication error related measures such as sequence numbers.

10.4.2.3 Inter-Connection Management Unit Diagnosis Pattern

COTS multi-core devices implement interconnection buses to communicate the components with different criticality levels integrated on them (e.g., cores, memories). Interconnection buses may apply different protocols to switch the traffic through the device's components such as AMBA Advanced eXtensible Interface (AXI) and AXI Coherency Extension (ACE). For instance, the Zynq-7000 zc706 multi-core device implements an interconnection manager that consists of a set of interconnecting blocks or switches. Those blocks manage, among others, the communication between the cores of the processor, the memories (e.g., On Chip Memory (OCM), DDR, L2 cache), the peripherals (e.g., Input Output Processor (IOP)), the Performance Level (PL) (if applicable). The interconnection units, in general, are prone to uncertainties related to their behavior (e.g., lack of information). The interconnection scheme of each COTS multi-core device is unique. For instance, the Zynq zc706 multi-core device provides an interconnection scheme that includes several ports and interconnection blocks such as the AXI_HP and AXI_GP ports, going through several interconnect blocks such as the *central interconnect*, the *OCM*, the *SCU*, the *memory interconnect* or the *IOP*.

This cross-domain pattern provides a set of generic solutions to measure and detect faults in the interconnection management units of multi-core devices. To that end, it assumes that the core, the cache memories (e.g., L1 and L2 memories), the OCM memory, the PL (if applicable) and its associated components (e.g., Block RAM (BRAM)), the timers and the interrupt controller are checked in advance. This pattern considers the following three solutions for testing the interconnection management units.

► | Solution 1: | Check the configuration of the interconnect management unit

The interconnect management unit shall be configured reasonably to provide minimum possible interferences. The blocks that compose the interconnect manager are set up using registers. The configuration of their registers should be used to manage their behavior, thus leading to an incorrect or partial behavior. Therefore, to detect whether the configuration of the interconnection management unit changes, this solution proposes the implementation of the periodic read-back check with a comparison of the interconnects manager's configuration registers.

► | Solution 2: | Diagnose random failures

In the case that maximum latency is required by a multi-core device, the Quality of Service (QoS) modules may be used. Those modules ensure expected throughput and latency in the system design. They regulate the masters that do not guarantee maximum latency (e.g., core, DMA, IOP).

On the other hand, QoS modules can be used to resolve issues related to contention. They implement a two-level arbitration abstraction scheme. The first scheme is based on the priority indicated by the QoS register. The highest QoS value has top priority. The second scheme builds on a least recently granted scheme. It is used when multiple requests are pending with the same QoS signal value.

An interconnect manager shall also consider the measures and diagnosis techniques for typical faults such as wrong addressing or wrong data forwarding, including partial transmissions or single bit errors. For instance, watchdog timers may be implemented to detect temporal deviations and the CRC with comparison (see *Shared memory diagnosis pattern*), the ECC and the parity bit diagnosis techniques may also be implemented to detect data consistency violations, considering partial updating or single bit error failures.

► | Solution 3: | Diagnose systematic failures

The interconnect management units may be affected by systematic faults. Those faults can be caused by the HW design, environmental stress or operational failures. This solution considers the implementation of a set of the measures and diagnosis techniques recommended in Tables A.15 to A.17 of IEC 61508-2 for detecting and controlling the systematic faults. It is assumed that the selection of the measures and diagnosis techniques depends on the HW platform and/or the SW implemented by the system architecture. Consequently, the selection of the measures and diagnosis techniques may vary. In addition, the possibility of systematic errors in the configuration of the interconnection management unit shall also be addressed by those techniques.

10.4.2.4 Interrupt Controller Diagnosis Pattern

Interrupt controllers manage the prioritization of the tasks in multi-core devices. For instance, the interrupt controller of the Zynq-7000 multi-core device, which is referred to as the Generic Interrupt

Controller (GIC), is internally composed of one or more distributed blocks depending on the number of cores and one or more core interface blocks. The interrupt distributor centralizes the sources of interrupts before dispatching the one with the highest priority level to an individual core. This controller ensures that one core can only take an interrupt targeted to several cores at a time. The sources of interrupts are identified by a unique interrupt ID number, a configurable priority and a list of the cores which are targeted. The interrupts that are handled by the interrupt controller can originate in cores, Private Peripheral Interrupts (PPIs), PL, Processing System (PS), Shared Peripheral Interrupts (SPIs) and Software Generated Interrupts (SGIs).

On the other hand, the core interfaces should perform the interrupt priority masking and preemption handling for the cores of the device. Each core interface block provides an interface for each processor that operates within interrupt controller.

An interrupt controller can fail, or the request or the assignment of interrupts may fail, thus affecting to the execution of the tasks of the cores. Table A.1 of IEC 61508-2 [132] defines the requirements for faults that shall be detected and measured to guarantee the safety of the interrupt handling. However, this standard is intended for single computing systems where a resource is not shared between more than one component, and therefore, the measures and diagnosis techniques recommended by this standard (see Annex F of IEC 61508-2) are not at all applicable to the interrupt controllers used for managing the execution of tasks with different criticality into the same device.

This cross-domain pattern defines the following design solutions to diagnose the interrupt controllers. It is assumed that the cores, the L1 and L2 cache memories and the OCM memory, the PL and associated components (e.g., BRAM), the timers, the interconnection management units and the coherency management units are checked in advance.

▶ | Solution 1: | Check the configuration

The interrupt distributor and the core interfaces of the interrupt controller can be configured independently using registers. This first solution proposes to periodically check the configuration registers of the interrupt controller to detect whether the configuration is modified.

▶ | Solution 2: | Diagnose random failures

The interrupt controller component can be the subject to unexpected internal failures which can be caused by direct-current (DC) faults, drift and oscillations and reset-related faults. In Table A.1 of IEC 61508-2 [132] techniques and measures for diagnostics and recommended maximum levels of diagnostic coverage for an interrupt controller are defined.

▶ | Solution 3: | Diagnose systematic failures

Tables A.15 to A.17 of IEC 61508-2 [132] recommend techniques and measures for controlling systematic failures, including techniques and measures to control systematic failures caused by the HW design, environmental stress or operational failures. These techniques shall be implemented to detect systematic faults that can occur in the interrupt controller. For example, the possibility of systematic errors in the configuration of the interconnection management unit shall be addressed using these techniques.

10.4.3 Mixed-Criticality Network

NoCs embed the solutions associated with traditional off-chip networks into the chip. They can be implemented for safety-critical applications, providing support for TT and Event-Triggered (ET) (Rate-Constrained (RC) and Best-Effort (BE)) traffic. TT NoCs are predictable communication systems with inherent non-interferences among their components (such as processing cores, peripherals and memories). Instead, ET NoCs are non-predictable networks.

The shift towards NoCs leads to challenges such as supporting multiple types of communication as well as supporting applications with different criticality levels. For instance, the Time Triggered Network On Chip (TTNoC) communication does not support the transmission of ET messages system [258] and AEtheral NoC does not support the transmission of RC messages [378]. However, the integration of functionalities with different criticality on a single computing platform communicating through a NoC can lead to communication errors.

This cross-domain pattern aims to provide a generic and reusable fault-tolerant on-chip communication network with support for different computation and communication models as well as mixed-criticality systems.

10.4.3.1 Network-on-Chip Diagnosis Pattern

This pattern manages the prioritization of the communication of different criticalities with scheduling, routing, traffic shaping and error detection.

In accordance with the IEC 61508 safety standard, this pattern can be considered as a SCL network implemented on top of a *black channel* network (see Figure 10.4). Therefore, it is assumed that parts of the communication channel, i.e., the NoC cannot be designed, implemented and validated according to a safety standard.

The SCL shall be compliant with a safety standard (such as IEC 61508 and IEC 61784-3). The SCL must fulfill the safety-related requirements stated in the safety standard, including a safety life-cycle development process. A linking analysis of this NoC pattern is presented in [358] where the way in which this SCL fulfills the safety-related requirements defined in the MSC for an IEC 61508 compliant mixed-criticality network is analyzed.

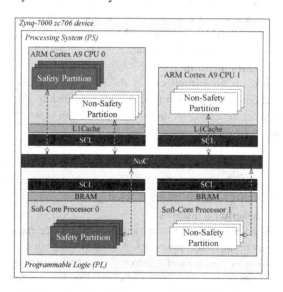

Figure 10.4: NoC Diagnosis Pattern - Block Diagram. © IEEE 2016, Reprinted with Permission from [358]

This pattern supports the following requirements to schedule, route, shape traffic and detects the communication errors:

- *Support for TT and ET (BE and RC) timing models:* TT messages are periodically transmitted for achieving predictable timing with minimal latency and no jitter. Instead, BE messages do not have timing restrictions and do not fulfill requirements of non safety applications. RC messages offer a reasonable trade-off between resource reservation and latency.

- *Provide fault-isolation:* This NoC pattern provides fault-isolation at SW level using a hypervisor virtualization mechanism and establishes partitioning at HW level using encapsulated communication channels.

- *Compatibility to a wide range of NoCs:* This pattern shall be integrable on a broad variety of NoCs, enabling the system to support both TT and ET communications, despite only ET transmissions are backed by the underlying network.

- *Support of hard real-time applications:* This pattern shall ensure that messages of the system meet the pre-specified deadlines in all situations defined in [41]. For this purpose, this pattern shall provide a schedule that enables to achieve deterministic communication.

- *Support of mixed-criticality systems:* The communication of applications with different criticality levels that interact and coexist on a shared computing platform requires protection mechanisms that establish chip-wide segregation. Virtualization mechanisms such as hypervisors are not considered enough to establish the segregation because non safety partitions can influence the safety-related ones. Therefore, this approach shall provide rigid temporal and spatial partitioning by setting up a chip-wide partitioning.

- *Establish temporal and spatial segregation:* It establishes temporal segregation among TT messages assigning different time slots to the TT messages of each tile and guaranteeing that no other tile can inject messages within the slots of other tiles. In addition, this NoC pattern ensures the temporal segregation between TT and ET messages. ET messages can be injected into the network only if there is no ongoing or upcoming TT communication.

On the other hand, this SCL ensures spatial segregation. To that end, it assigns separate memory areas to the ports of the network for storing the messages and establishes that the messages with different criticality levels are routed through separate paths (source based routing).

10.5 Functional Safety Management Process for DREAMS Architecture

10.5.1 IEC 61508 Functional Safety Management

According to IEC 61508, a FSM specifies the way in which the functional safety is achieved throughout the development process. The FSM must specify the roles and responsibilities of the participants throughout the phases of the development process (e.g., XtratuM hypervisor roles 'system architect', 'system integrator' and 'application supplier' defined in Section 6.3.8), including the measures and diagnostic techniques to avoid systematic faults and to ensure that the system achieves the targeted SIL. The IEC 61508 safety standard defines a life-cycle development process with 16 phases including the analysis, realization and operation phases. The DREAMS based realization phase is described in this section. The remaining phases are outside the scope of this specific project.

10.5.1.1 Phase 10 'Realization': V-model

A V-model is a suitable methodology for the realization of the mixed-criticality system. Figure 10.5 presents a V-model development process, based on IK4-Ikerlan's certified IEC 61508 SIL3 FSM [379], where continuous arrows specify the dependencies between the phases of the development process. At the end of each phase, a verification process against the results of the next step (broken vertical arrows) is executed to check the consistency of the phases inputs and outputs. In addition, this development process is split into a design branch and a testing branch (green arrows). The design branch covers the development phases related to the definition and design of the HW and / or the SW. Instead, the testing branch covers the phases related to the integration, Verification and Validation (V&V) activities of the HW and the SW. The design and testing branches are linked to the test plans, which can differ depending on the implementation in HW or SW. For instance, the development of HW requires applying measures and diagnostic techniques recommended by IEC 61508-2, whereas the development of SW requires applying measures and diagnostic techniques recommended in IEC 61508-3. The solid arrows on the left side of each test plan indicate the origin of the test plans while the solid arrows on the right side denote the actions which are required to be performed during the testing. The dashed lines to/from the test plans indicate the verification activities between the phases.

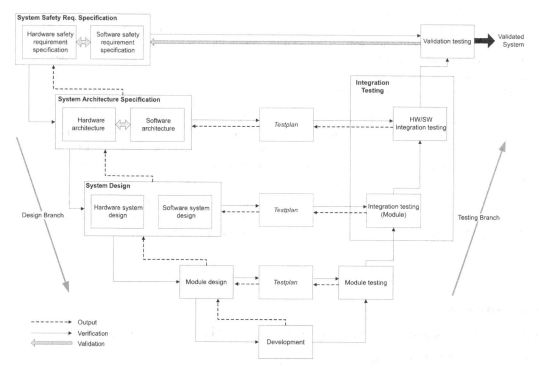

Figure 10.5: V-model Realization Phase Diagram [380]

The 'system safety-related requirement specifications' block specifies the safety-related demands of the HW and the SW, and it defines the validation plan that shall be followed in sub-phase 'validation testing' for validating the HW or/and SW system(s). It helps in the system requirements gathering and classifying. Sub-phase 'system architecture specification' defines the HW and SW system modeling and architecture definition. This sub-phase also defines the integration tests that shall be executed in the HW or / and SW 'integration testing' sub-phase. Sub-phase 'system design' defines the system architecture design and specifies a verification plan for the modules integration tests. Sub-phase 'module design' sets out the way in which each HW and SW component (module) should be implemented into the system. Furthermore, it defines the communication protocols to be used for communicating the components, configures the SW components related to the HW architecture and defines the verification plan for testing the modules in sub-phase 'module testing'.

These HW and SW modules and attached interfaces (Application Programming Interfaces (APIs)) are developed / implemented in sub-phase 'integration testing'. Once the HW and/or SW modules are developed/implemented, the testing branch starts. In this branch, the modules developed/implemented are tested using the test plans defined in sub-phase linked. In addition, the integration of the modules and the HW and/or the SW is tested in the 'integration testing' sub-phases using the verification plans defined in sub-phases linked. Finally, sub-phase 'validation testing', the whole system is validated according to the validation plan specified in the first phase of the development process to check that system's functionality.

10.5.2   Tools

SW tools are commonly used nowadays for all of the process steps, ranging from a simple text editor to the implementation of sophisticated verification algorithms. DREAMS has developed / extended models, design, verification, configuration and fault injection toolsets for mixed-critical systems (see Section 4.4).

According to the IEC 61508 safety standard, these tools can be classified as on-line and off-line tools. On-line tools can directly influence the safety-related embedded system. Instead, off-line tools support a phase of the software development process and cannot directly influence the safety-related embedded systems during run-time. Off-line tools can be classified depending on their usage in classes T1, T2 and T3. T1 tools generate no outputs which can directly or indirectly contribute to the executable code of the safety-related system (e.g., documentation tools). T2 tools support the testing and verification of the design or executable code, and they cannot directly influence in the executable software (e.g., diagnostic tools and V&V tools). On the other hand, T3 tools generate outputs that can directly or indirectly contribute to the executable code of the safety-related system (e.g., engineering tools, translation tools and configuration tools).

In our case, the tools that are used to edit the model of the DREAMS system are tools of class T3. These tools enable the end user to manually create or edit the descriptions of the system under design, share the descriptions with other tools of the same toolchain and visualize the data generated by other tools. Among others, the tools developed/extended in this project are the AutoFOCUS 3 (AF3) / System model editor, timing model editor, safety model editor and the Base Variability Resolution (BVR) variability editor.

Design tools are used for automating certain aspects of the design and checking the consistency of the systems. To that purpose, they apply procedures such as the heuristic, exploration and optimization algorithms. The design toolset developed in the project DREAMS is composed of the AF3 / Design Space Exploration (DSE) (see Sections 5.1.2 and 4.7.3), BVR Product Generator (see Section 4.7.1), RTaW / Timing Decomposition, RTaW – Timing (T2 tool) / On-Chip Time-Triggered Scheduler (see Section 5.2.1), TTEthernet Plan (T1 tool), Xoncrete, GRec and MCOSF tools.

On the other hand, verification tools are also used for verifying the design choices made at design time. The verification toolset includes the RTaW – Timing / Evaluation (T2 tool) and the safety constraint checker tools.

In relation to the configuration-related tools, they are used for generating the configuration files of the mixed-criticality system. The configuration files are based on verified models, thus avoiding the occurrence of errors caused by a manual edition of configuration files. This toolset is composed of tools of class T3 which are used for configuring a XtratuM hypervisor, a TTEthernet off-chip network, an on-chip network interface layer, a resource management and a virtual platform.

The last toolset is composed of fault injection tools which are used for testing the system and watching and validating its behavior in the presence of faults. Tools of this category are the STNoC fault injection and IK4-Ikerlan's EtherCAT fault injection frameworks. Since the safety constraint checker and fault injection tool help to identify flaws in the design and implementation of safety-related systems without any contribution or effect to the executable code, they are considered tools of category T2.

The tools presented in preceding paragraphs are intended to be used in the phases of the safety development process presented before, thus generating the input / output documents required to certificate a mixed-criticality system. Note that except the tools for injecting faults, all other tools are used during the left branch (design branch) of the development process. Extended descriptions of these tools can be found in Chapter 5.

Those cross-domain patterns provide reusable generic solutions and diagnostic techniques which may be implemented in the integration (verification) phase of a safety development process. Thus, the use of those patterns may provide the documentation set defined before in this section. They may be listed in the verification plan and may provide evidences for the verification report.

10.6 Certification of Mixed-Criticality Product Lines

A product line is set of product samples that share and manage a common set of features satisfying the needs of particular market areas. Those systems may be developed from a common set of reusable core. As an example, we can consider the scenario shown in Figure 10.6 where different levels of detail (granularity) are used for representing a product line. Those levels of abstraction include proper life-cycle development processes which shall be followed to develop compliant items and enable re-usability.

In Figure 10.6, vertically, we can distinguish a product line that consists of systems (e.g., wind turbines). Systems can be, at the same time, composed of a set of subsystems (e.g., 'supervision and control unit'), which can be composed of additional subsystems or components (e.g., platforms, sensors, actuators, COTS multi-core devices, hypervisors, operating systems). This hierarchical structure can occur recursively at many levels of abstraction for developing product lines of different criticality domains (e.g., railway, automotive, lift, avionics).

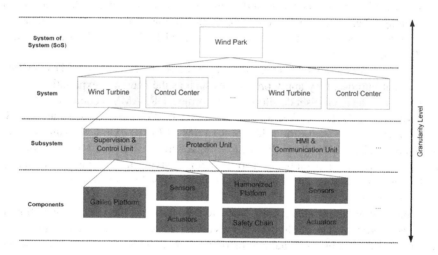

Figure 10.6: Product Line Abstraction [323]

10.6.1 Families of Systems and Product Lines

10.6.1.1 A Brief Introduction to Variability Modeling

Modeling is a core practice in science and engineering. In short, modeling aims at easing the resolution of a particular problem while preserving our ability to gain hindsight about it. Regardless of the approach, any modeling activity is always strongly coupled to the problem it aims at. Modeling is inherently difficult because it requires striking the right balance between discarding enough of the reality to tame its complexity and still retaining enough to maintain significance. This paradox known in engineering as the principle of incompatibility [381] directly impedes our ability to reuse models.

Modeling experts tackle complexity with to two main leverages: abstraction and separation of concerns. Abstraction discards any aspect of the reality that is irrelevant to the problem of interest whereas separation of concerns divides the problem into separate sub problems of smaller complexity. When exploiting the resulting models, hindsight often comes from our ability to distinguish between what varies and what remains. Mathematics and especially geometry, topology and algebra produced many models where the hindsight comes from the duality between what changes and what remains. Differential equations, to name only one, capture invariant relationships, which permit to understand how complex dynamics varies.

From the modeling perspective, Product Line Engineering (PLE) explicitly captures variability in order to maximize reuse throughout the development and maintenance process, in order to improve productivity while reducing risk and cost. PLE therefore distinguishes between what remains and what varies in a system. The 'things that remain' often reflect hard constraints in the application domain of interest whereas parts opened to variations reflect potential areas for business and innovation. The main objective is therefore to automate the derivation of a new product from the prescription of its features.

Modeling the variability among a family of products relies on the notion of feature. A feature stands for "a unit of functionality of a software system that satisfies a requirement, represents a design decision, and provides a potential configuration option" [382]. Products are thus characterized by the features they provide, and in turn, product lines are characterized by the union of all features provided by their products. The simplest way to capture the commonalities (respect to the variability) among a set of products is using a feature table. The table relates the set of products with the set of possible features: specifying for each product the features it provides. This approach is commonly used to provide customers with a comparison of a range of products such as mobile phones, TVs, etc.

A better way to capture variability within a set of products is through a feature model [383,384]. A feature model gathers the commonalities within a product line into a tree of features, where some features are mandatory (the commonalities) whereas others may be optional or exclusive (the variability). Figure 10.7 illustrates the graphical notation associated with such feature models on a subset of what could be the DREAMS platform. This simplified DREAMS platform encompasses hardware components, hypervisors and Operating Systems (OSs). The platform may or may not include hypervisors, which are either based on OS virtualization or on hardware virtualization, or both. By contrast the platform will always include an OS, which will be one of the three possible alternatives listed in Figure 10.7.

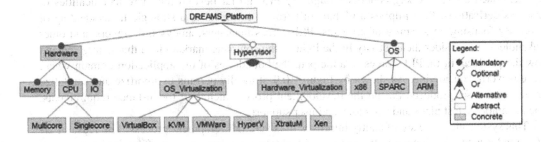

Figure 10.7: A Simplified Feature Model Capturing the Variability of Hardware Platforms

The proposed notation is not expressive enough to capture all the additional constraints that may exist and restrict the set of products which can be derived from a given product line. In our example, it may be that some OS virtualization technologies are not available for all operating systems, and the choice of one restricts the possible OSs.

Orthogonal Variability Models (OVM) [385] and its successor the Common Variability Lan-

guage (CVL) [143] attempted to overcome the inherent tight coupling between the product derivation procedure and the underlying technologies. They resolve variation points and variants, not
anymore in the technological space, but on a domain-specific model. The resulting model of the
product can thus be finalized using the associated domain-specific tooling.

10.6.1.2   The Base Variability Resolution Model

Base Variability Resolution (BVR) is a language built on CVL technology, but enhanced due to
needs of the industrial partners of the VARIES project (http://www.varies.eu), in particular
Autronica [142]. BVR is built on CVL, but CVL is not a subset of BVR. In BVR we have removed
some of the mechanisms of CVL that we are not using in our industrial demo cases that apply BVR.
We have also made improvements to what CVL had originally. For the purpose of DREAMS we
may just say that BVR is a continuation of the CVL language with associated tooling. Figure 10.8
below illustrates the variability model of the simplified DREAMS platform in the BVR tool.

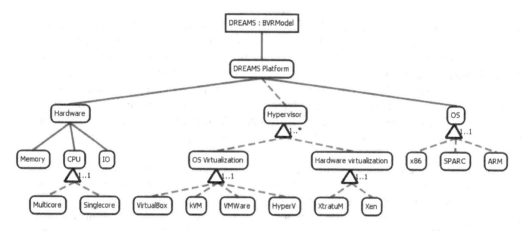

Figure 10.8: The Variability Model of the DREAMS Platform Modeled Using the BVR Tool

10.6.1.3   The Benefit of Product Lines

As mentioned above, the key benefits brought by PLE go far beyond the mere technicalities of
products derivations: PLE implies a global shift from a technical to a strategic understanding of
reuse. By focusing on a family of products, PLE forces managers, analysts, designers, and other
stakeholders to consider their activity in the light of the complete market niche that is targeted and
how the products fit in. PLE forces to anticipate the boundaries of the application domain where
reuse is worth considering. As shown by Figure 10.9, it ideally permits to amortize and capitalize
on any core asset produced during the development process including for instance requirements,
documentations, test plans and test cases and user support.

    This systematic reuse significantly improves the overall development process. The reuse of ar-
tifacts reduces the development effort as it avoids duplicating development effort. It also increases
artifacts' internal quality as the probability to find and correct defects increases with the reuse rate.
Reuse eventually ensures the internal spread and consolidation of the domain-specific knowledge
accumulated throughout successive products developments. Market agility also significantly gains
from PLE adoption. PLE enables faster responses to new customers needs and more generally to
new market trends. Existing products can be quickly extended with other features already available
in other products and new products can be built from new and yet unforeseen combinations of features requested by customers. PLE adoption may thus bring a competitive edge as it empowers the
user with the ability to build the product that fits her very needs.

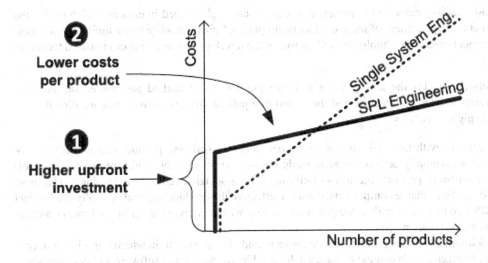

Figure 10.9: The Cost of Software Product Lines

10.6.2 Piecewise Certification

As product line technologies build upon the re-usability of the underlying software technologies (e.g., objects, components, services), they form a promising approach to reduce the cost of certification and re-certification, by maximizing reuse of certified units, here-after called pieces. Yet, reuse is always challenging as it is difficult to ensure that third-party components will perform correctly in an environment for which they were not explicitly designed. Among others, the USD 500 million crash of Ariane 5 in 1996 [36], remains a strong evidence of the opposition between reuse and verification and in turn, certification. Below, we review the key techniques available to verify reusable piece of software and we discuss how they could be extended to certification.

10.6.2.1 Product Lines and Modularity

Building highly reusable software pieces has been the major impetus for several breakthroughs in Software Engineering: routines, modules, objects, components, services, aspects, etc. While re-usability lacks a precise and well-accepted definition, it is worth to note the distinction that exists between reused and reusable software: Reused components are not necessarily reusable 'per se' (they may be the only available alternative) and reusable components are not necessarily reused (e.g., poor visibility, wrong time-line). Yet, the technologies successively proposed to develop reusable software pieces shed some lights on their key characteristics: effectiveness, generality, cohesion, substitutability and visibility, to name only a few.

- Effectiveness is the 'sine qua none' condition for re-usability: software that fails doing what they are supposed to do will certainly not be reused as is.

- Generality reflects the extent to which the problem solved by the software is common and directly impacts its re-usability. The use of software pieces that only solve linear differential equations is for instance restricted to linear mathematical models.

- Cohesion characterizes pieces of software that have a single and well-defined responsibility. Cohesion requires some level modularity, in order for different concerns to be isolated into different units (i.e., modules). The resulting highly cohesive units are more easily understood, and in turn, more easily reused. Object-oriented technologies significantly contribute to the definition of general and cohesive abstractions.

- Substitutability calls for the presence of explicit and well defined interfaces, which enable the (dynamic) replacement of individual software pieces. Substitutability was a limitation of classical object-oriented technologies and motivates the development of components-based technologies.

- Visibility reflects the ability for a software piece to be identified and reused on the spot. Service-oriented architecture and the underlying publish-discover-invoke scheme directly promote higher visibility.

By contrast with the technologies mentioned above, a software product line is a framework which helps eventually deliver products made out of a set of reusable software pieces. To be effective, a software product line has to build upon an associated reusable technology: The more reusable are the available components, the more effective the product line will be. Software Product Line (SPL) contribute to realize the potential existing within an existing set of application specific and reusable software pieces.

In addition, a software product line defines a bounded context, within which reuse is worth considering. In practice, this context becomes delineated by the associated software architecture, which reduces the need for re-usability. Software components only need to meet the re-usability requirements within this architecture, and need not be generally reusable. Assumptions regard for instance communication protocols, middleware technologies, data encoding, etc. Building a software product line is thus tightly coupled to key architectural choices and eases the identification and development of the missing reusable software components.

Finally, as mentioned previously, reuse as understood in SPL goes beyond the sole underlying technology. SPL aims at reusing as much as possible any assets produced during the development cycle: requirements, analysis, tests, and possibly certification evidences.

10.6.2.2   What Is a Piece?

As mentioned before, various software units of reuse have been proposed in the literature ranging from simple routines to high-level services. From the standpoint of certification and re-certification, a key aspect is the substitutability of a part: the ability to replace a unit by another one, while still guaranteeing that the whole is operational (e.g., safe, correct). The need for substitution of individual parts drove the development of component-based system, where "a software component is a unit of composition with contractually specified interfaces and explicit context dependencies only. A software component can be deployed independently and is subject to third-party composition" [386]. This general definition encompasses any piece whose dependencies with its environment are well specified so as to enable its substitution by an equivalent.

By analogy with the way relationships between people are specified by contracts constraining the rights and duties of each party, relationships between software pieces follow contracts specifying the assumption a piece makes and the guarantees it provides. Ultimately, this so-called assume / guarantee paradigm aims at building the specification of a complete system in a bottom-up manner: by assembling the individual specifications of its parts.

A textbook example of contract characterizes the reuse of a square root function, called 'sqrt'. As described below, this contract expresses under which conditions one can calculate the square root of a real number.

sqrt(in x: Real, out r: Real) assume: x$>=$0 guarantee: $x = r2$

It is worth to note that the contract is completely disconnected from the algorithm actually used to compute the square root. The assume / guarantee paradigm is indeed strongly linked with information hiding as it requires pieces to expose the necessary and sufficient information to enable reuse. Ideally, this contract could also include extra-functional concerns, such as memory consumption, energy consumption or execution time, although such extensions are generally difficult to leverage for formal verification.

Design by contract [387] is a direct application of the assume / guarantee ideas in Software Engineering practices. Following ideas of the Floyd-Hoare logic [388], each routine is equipped with pre and post conditions capturing its semantics. Pre-conditions explicit the assumptions made by the routine to perform correctly, whereas the post-conditions specify the properties guaranteed after completion. Such contracts are fragments of specification embedded at runtime into assertions in order to detect discrepancies between the specification and the actual behavior of the routines. Coupled with testing techniques, contracts have shown to be an effective verification and diagnostic tool. As we shall see in the next section, contracts can also be used to verify various types of properties on software assemblies. Contracts are actively researched as a means to capture various software interactions, especially synchronization and quality of service [389].

10.6.2.3 Piecewise Verification

Verification in Software Engineering is driven by two main approaches: automated testing and formal verification. Testing takes products at the end of development iterations and checks their adherence to specification (e.g., correctness, performance, usability). By contrast, formal verification builds mathematical models of systems and proves their correctness, performance, usability, etc. We review below the main characteristics of the two approaches.

A variety of assume / guarantee specifications and other contract-based specification have been developed as a means to compose (resp. decompose) formal system specifications. We shall restrict ourselves to an overview of the main ideas, but interested readers may find a comprehensive review of contract-based models in [390]. As explained assume/guarantee specification are couples (A,G) where A models the assumption a component has on its own environment, and G models the guarantees it offers to its environment. A and G are generally formal processes, whose composition may be subject to safety and liveness issues. The composition of such processes is yet not as straightforward as it may seem: mutually dependent components lead to circular reasoning which must handle carefully. To this end, various composition operators, have been developed for specific models. Communicating Sequential Processes (CSP) [391], the Temporal Logic of Actions (TLA) [392, 393], Focus [394] or BIP [395] to name a few are examples of formalisms using such composition rule.

More recently, there has been an attempt to combine such formal verification techniques with architectural description languages (ADL). Allen [396] for instance defined the formalization of components and their assemblies into well-defined software architectures. Various Architecture Description Languages (ADLs) have been then proposed such as ACME, ArchJava, Unified Modeling Language (UML) 2.x [45, 46], Systems Modeling Language (SysML) [144] or Modelica [397].

10.6.2.4 What Do Standards Say about Piecewise Certification?

Ideally, any change in the product or in the related development process shall trigger the re-certification of the newly derived products. As stated in IEC 61508-1, modification requests in the element, process or associated standard / legislation that apply to the element trigger the re-certification of the element (e.g., product).

Certification and re-certification thus induce a significant increase of the development process costs. Collected data [398] showed for instance that certification expenses with respect DO-178B increase by a factor of 3 to 5, depending on the associated criticality level. Standards and especially safety standards such as IEC 61508 or ISO 26262 recognize indeed the need for reuse in both system development and certification.

By analogy with advances in software architecture which enhanced re-usability, modular certification aims at taming the high cost of certification by offering reusable certification pieces. If a system is made out of independently certified components, replacing a component should only trigger the re-certification of that very component together with the overall assembly architecture.

Recalling the techniques surveyed in this section, modular certification should be able to leverage existing piecewise verification techniques.

10.6.2.5   Tool Certification and Its Relationship to Piecewise Certification

Certification is inherently about collecting evidence that a given product (in the case of safety) adhere to some given requirements. Eventually, there is never any absolute guarantee that the requirement hold and certification is only about consolidating the confidence one may have. Evidence can be collected from the product (functional safety), but also from the development process that was used to deliver the product. Knowledge about the process indirectly supports the product-based evidence, as a sound and generally well-established development process is more likely to yield a well understood product.

Although various software development processes are possible, a widely accepted one is the V-model (see Section 10.5), which requires that every design step (i.e., requirement analysis, system design and implementation) be secured by appropriate Verification and Validation (V&V) procedures. Traceability is needed for certification purposes and all V&V activities shall be properly documented.

Process certification also covers the tools that are used throughout the development process such as Computer-Aided Software Engineering (CASE) tool, compiler, code generators, etc. For instance, the IEC 61508 distinguishes between three categories of tools depending on their impact on the final product. Category T1 includes all tools that do not directly impact the safety of the final product, such as text editors for instance. Category T2 includes tools that contribute to improve safety, but which do not contribute to the running code such as static code analyzers or model-checkers. Finally, Category T3 includes tools that directly contribute to the running code of the product such as code generators or compilers. While tools in the first category need not fulfill any specific requirements, the risk induced by using tools from Category T2 and T3 shall be assessed separately. In addition, tools from category T3 shall provide evidence that they meet their specifications.

10.6.2.6   Piecewise Certification and Product Lines

Various techniques and challenges related to the certification of safety-critical software systems are detailed in [399]. Product lines appeared as a promising approach for industries that produce families of such systems. The literature also reports about several attempts to apply product line ideas to safety critical systems in automotive or avionic, especially. Dordowsky *et al.* [400] described for instance how they built the NH90 product line of medium weight multi-role military helicopters, which resulted in 23 variants. While this work focused on using product lines to assemble software components, certification of modular components (w.r.t. DO-178B and AC20-148) is not addressed but is clearly identified as 'a major cost block'. Alternatively, Hutchesson and McDermid [401] showed how to leverage recent advances in model-driven engineering and component-based systems to foster to verification of high-integrity product lines. This approach contributes to provide the evidence needed to achieve modular certification. Habli [402] made a first attempt to apply product line techniques to safety analysis. He proposed a specific metamodel to capture safety concerns in a product-line so that product-line safety and development artifacts can be jointly reused in a traceable and justifiable manner. Braga *et al.* [403] described a product line of Unmanned Aerial Vehicles (UAV) named Tiriba. This work is a first attempt to leverage feature models to foster certification, as they capture the impact that each feature has on certification and thus can contribute to the impact analysis required when deriving new products (see Section 10.6.2.4). Conmy and Bate [404] discussed the challenges of using product line and component-based design to ease the assembly of safety arguments.

Although product line engineering is perceived as a promising technique to do modular certification, research is still in its infancy. As identified above, product lines are a potential enabler for impact analysis, when products have to be re-certified. In practice re-certification will address the modification itself as well as the supporting evidence. All not affected parts of the safety related system as well as all affected evidence should be reused and need not to be re-certified. A second potential is the ability to easily identify compliant component or solutions that could meet the requirements of systems compliant with existing standards. Building blocks can be certified independently and yet be available as COTS, ready for assembling new safety-critical systems. Only the resulting assemblies would therefore require certification.

10.6.3 IEC 61508 Certification

In the case of DREAMS project, we take the IEC 61508 safety standard as the reference standard for developing the safety argumentation of a product line. IEC 61508 is a reference safety standard for several domain-specific safety standards such as ISO 26262 (automotive), EN 5012X (railway) and ISO 13849 (machinery). Therefore, the safety argumentation defined for developing an IEC 61508 compliant product line can be extended to accomplish domain specific safety standards and support different levels of criticality (e.g., SIL1 to 4 according to the IEC 61508 safety standard, Automotive Safety Integrity Level (ASIL) A to D according to ISO 26262). Nevertheless, it is noteworthy that the adaptation process that is required to extend the IEC 61508 compliant product line may include further requirements, measures and diagnostic techniques recommended by those standards that shall be met to accomplish safety certification.

The variability and safety-related arguments of a product line can be represented using the GSN [120], CAE [354] and SACM [355] safety case notation languages. For instance, the generic MSCs for IEC 61508 compliant components, including a hypervisor, a safety partition, a COTS multi-core device and a mixed-criticality network, are defined in the DREAMS project using the CAE notation. These MSCs define the basic safety requirements that shall be met by those components to be IEC 61508 compliant items. However, the MSCs generated using CAE have been ported to GSN language due to limitations of CAE and extended elements provided by GSN, which enable representing modular systems using the 'module' and 'contract' elements. The module extension is a package of arguments to provide a general view of the argument structure. It is rendered as a rectangle with a second smaller rectangle that represents the reference to a module containing an argument.

GSN also provides the contract element for specifying the interrelationships between the safety arguments of each abstraction layer, linking the goals to be supported with the supporting goals and minimizing the impact of changes between interrelated layers. In this particular case, the contract element details how a commercial/custom component fulfills the safety-related requirements defined in the generic safety argumentation for heterogeneous safety architectures.

Based on the representation scheme shown in Figure 10.10 and the module and contract elements provided by GSN, we define the following four abstraction layers for developing a safe product line.

- The generic safety-related arguments for a safety standard (e.g., IEC 61508) compliant component: e.g., a hypervisor, a safety partition, a COTS multi-core device.

- The safety-related arguments for application independent safety standard compliant commercial or custom components: e.g., a Zynq-7000 multi-core device, the XtratuM hypervisor.

- The safety arguments for a specific product sample. e.g., a wind turbine product sample based on the DREAMS architecture style (see Section 2).

- The safety-related arguments for a product line which should be common to all possible product samples of a product line. e.g., a wind turbine product line.

The representation of a product line may be automated for achieving an optimum product configuration. This may depend on the safety requirements for a product sample. Automation can be

Figure 10.10: Product Line Abstraction Layers Representation Using GSN

accomplished using tools, which can also automate the generation of the safety case reports. Automation tools can also be used to complete the information on the product sample safety contract, linking the requirements of the product sample to the safety requirements of the optimum application independent component (see Figure 10.10). For instance, DSE toolset resolves variability models, selecting alternative candidate deployments of the logical components of the HW target and automatically assembling the safety argumentation models per each product sample. The outputs of DSE include a partial argumentation model that is mapped to a set of semi-automatically generated evidence documents.

Continuing with the representation style presented in the previous paragraphs, different product lines based on the DREAMS architecture style can be generated. A product line representation exposes the safety arguments related to the product line, identifying the way in which a product sample meets those requirements and presenting the linkage between the product sample and the commercial or custom components that make it up. Figure 10.11 represents an overview of a product line where a safety contract regarding the qualified tools selects the best combination of components for a particular product sample. For instance, the product sample can choose from amongst different multi-core processors (e.g., Zynq-7000, P4080, Hercules).

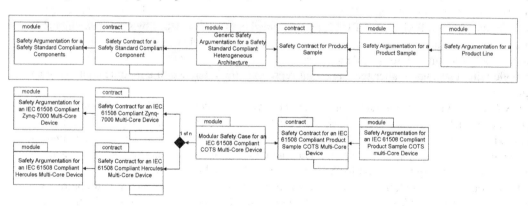

Figure 10.11: Product Line GSN Representation - Choose Best Combination of Components

The representation hierarchy presented in the previous paragraphs can be extended to develop product lines of different application domains (e.g., railway, automotive) as the presented safety argumentation scheme enables reusing the safety argumentation blocks of the commercial and custom components. Each domain specific safety standard defines additional requirements and measures and diagnostic techniques that shall be met to accomplish safety certification. In addition, this representation hierarchy can be extended to develop product samples with different levels of criticality (e.g., SIL1 to 4 according to the IEC 61508 safety standard). This approach is extended in [405] using the wind-turbine demonstrator as an example.

The harmonization of the underlying requirements from different safety standards is an ongoing trend [406], although currently, no cross-domain solution copes with the differences between the safety standards.

10.7 Method for Certifying Mixed-Criticality Product Lines

This section describes the proposed method for the certification of mixed-criticality product lines.

10.7.1 Certification Support in DREAMS

The mixed-criticality systems considered in DREAMS have some special properties that affect certification and approvals, namely:

- The need for total separation between critical and non-critical applications.

- The high degree of configurability and re-configurability.

The first of these properties, on the need for total separation, is covered in Chapter 6, where detailed argumentation chains are described both generically and in particular for the XtratuM hypervisor to assert how virtualization will keep applications fully separated.

To achieve the second property, on the high variability and configurability, we shall demonstrate how explicitly modeling product-lines of mixed-criticality systems makes it possible to generate detailed and product specific arguments for any individual product up for approval. DREAMS aims at the cost-effective development process for mixed-criticality product lines, where product samples may share common features, subsystems or components. The overall DREAMS development approach relies on composition of modules to design the products, thus enabling the final users to reuse safety assurance artifacts either at subsystem or component levels when analyzing the safety properties of the samples (e.g., safety manuals). Likewise, the DREAMS certification approach also consists of a modular composition of arguments that links the safety claims, starting from the safety requirements, to the final evidences that the developers have to provide to the certification authorities. This composition of arguments shall reflect the design rationale, ending in a number of proofs such as test results analysis and validation, analysis results, formal proofs, referred documents or available certificates for pre-certified items.

The DREAMS toolset supports the semi-automated Design Space Exploration (DSE), seeking for the best possible product configuration for a given set of requirements, also including safety requirements. During the design stage, these tools automatically select alternative candidate deployments of the logical components on the target hardware, resolving the variability models and going through a number of verifications in a safety evaluator component: the Safety Compliance and Rules Checker. As for the certification, this toolset also automatically assembles an argumentation model for each product sample. The outcome after completion of the DSE is a partial argumentation model, which will be mapped to a set of certification documents according to the Functional Safety Management process chosen by the DREAMS user. These documents are generated in a semi-automated fashion, taking into account that part of the final evidences would come from V&V activities to be carried out at later stages of the project, and some of these could not be available at DSE phase.

The modular certification approach in DREAMS is supported by a database of argument models for each pre-certified or certifiable component. The DREAMS argumentation database eases tackling with safety standards from different application domains requiring specific argumentation patterns, as it can handle multiple argumentation variants tailored to the domain certification requirements. For instance, for a part having an IEC 61508 argumentation template besides a DO-178C argumentation model the DREAMS toolset could assemble differentiated argumentations for certifying products containing this component, according to either safety standards.

A substantial burden of the certification process comes from the compilation of final evidences resulting from V&V. Note that safety standards impose process redundancy in the safety product development to prevent systematic errors, stated as a requirement for strict separation between activities related to product design and implementation and the quality assurance activities, including V&V tasks. In order to comply with the isolation between development and verification, the DREAMS certification support shall support interoperability with V&V information repositories. The DREAMS database-based argumentation provides such an extensible framework to relate diverse information sources into a single argumentation model. In the event of development iterations, the DREAMS approach facilitates updating the whole set of information, enabling improvements at the coherence between arguments. In the near future, automated argumentation model checkers may be integrated in the DREAMS toolset and help the certification process by automatically unveiling the weak or missing points in an argumentation, therefore reducing iterations and lowering the overall certification costs. The DREAMS certification approach would also yield benefits during other phases of the product life-cycle, as the same rework on the argumentation models is applicable to the product evolution (e.g., substitution of parts, retrofitting, enhancements to the safety product line), requiring a re-assessment of the resulting configuration.

Figure 10.12 presents the equivalences between DREAMS argumentations and the generic argumentation templates as defined in the Open Platform for EvolutioNary Certification of Safety-critical Systems (OPENCOSS) project [407].

The DREAMS support to the certification of mixed-criticality systems is twofold:

- Product-oriented Certification Support: DREAMS Harmonized Platform (DHP) provides safety components that are re-usable across different application domains. These components can be also combined in different deployments, as to satisfy the requirements specific to each product variant of a mixed-criticality product line.

- Process-oriented Certification Support: DREAMS toolset provides automated analysis and configuration features that would ease the staged development of the mixed-criticality product line, and eventually provide evidences to support the safety claims.

When the DREAMS toolset explores the DSE, it relies on oracles that compute the performance of a candidate product for different properties, including safety. The safety oracle analyzes the suitability of the product architecture and the safety manuals of the intended components and subsystems to establish which safety integrity level can be claimed. This evaluation is based on a rationale elicited from the safety standard, and the information provided from the safety manuals. A preliminary Safety Concept (SC) is then generated, based on safety-relevant information from the DREAMS platform components and the system composition analysis, i.e., a given mixed-criticality product configuration. This preliminary SC can be either completed at later stage with additional arguments and evidences (e.g., integrating information gathered from V&V), or may refer to documents that shall be available prior to the certification to support the safety claim.

We identify the following challenging issues for assembling preliminary SCs for Mixed-Criticality Systems (MCSs):

1. The integration of fragments of diverse information, described in diverse formats and possibly stored on distributed information sources (e.g., FTAs, HAZOPs, requirements databases, natural language documents, system models);

2. The preservation of traceability and coherence between information cross-references;

3. The description of the safety claim argument in a human-friendly format.

The integration of the safety-relevant information is one of the main concerns of projects OPENCOSS and AMASS-ECSEL—we suggest the reader seeking for an in-depth discussion on this topic to review [408, 409]. In DREAMS, we relate the information sources by using textual links (e.g., a URL or a bibliography reference entry) to tackle issue 1, and the toolset will partially address the issues 2 and 3 by relying on compositional argument models to build the MCSs

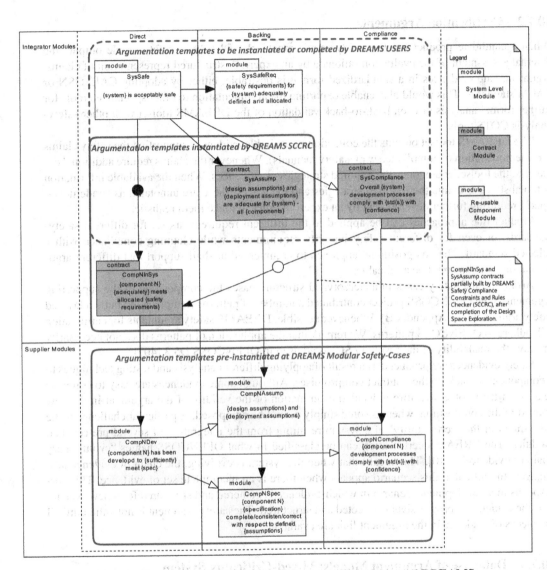

Figure 10.12: Mapping between OPENCOSS Argumentation Templates and DREAMS
Certification Arguments [324]

safety claim argument. To this end, we require re-usable and coherent information about the modular safety-cases, and this should be ideally arranged in some argument database. In order to simplify the composition of arguments in the AF3 environment, we selected the GSN notation to represent the arguments. Already developed MCSs for DREAMS components are translated to GSN and stored in the argument database.

10.7.2   Certification Arguments

When a candidate product fits the safety requirements, the DREAMS safety oracle outputs the external assessment of the evaluation rationale by an expert. A structured representation of claims, arguments and evidences in a standardized format is desirable, either by adopting CAE, GSN or SACM standards. This would also enable exporting the argumentation to other specialized tools for further formal analysis or even back-to-back validation of the DREAMS toolset with other safety-analysis COTS tools.

The DREAMS toolset outputs the collection of available evidences supporting the safety claims (i.e., demonstration of compliance w.r.t. safety manuals). Whenever the claims require additional evidences, the toolset enumerates the required strategies or evidences. When the available information is too abstract to define the required strategies and evidences, these are annotated as "underdeveloped", requiring further development by an expert in order to make them realistic.

Product line approaches can be applied to the different requirements set for different safety-standards, or even for different safety integrity requirements while claiming compliance with a selected standard. The compositional approach to argumentation shall support that different argumentation patterns could be applicable.

Arranging the safety cases in a stereotyped structure eases the composition of the supporting arguments. The OPENCOSS project contributed a number of patterns (templates) to build a layered safety argument (see Appendix B). Whenever possible, DREAMS safety arguments for certification will adhere to OPENCOSS patterns. We remark that the application of patterns does not necessarily improve the readability of the resulting SC, as advised in OPENCOSS D5.3 [407].

Safety evidences are produced as a result of applying different analysis and testing techniques to a component to back up the contract compromises. Although some evidences are easy to compose (e.g., fault trees) others require a detailed examination of the validity of the argument in the new context of the composition, whereas some simply cannot be composed. A particular challenge is the assessment of the 'certifiability' of a system resulting from the integration of several pre-certified modules. The DREAMS certification can be classified in what OPENCOSS names "evolutionary chain of evidence" [410]. This means that when the system is evolving, the chain of evidence also changes. In particular, this scenario appears when there is an incomplete set of evidence. This corresponds to a development scenario in which evidence is gathered and structured for a new system, thus the evidence is progressively collected and structured. The safety argument is not valid until all the pieces of evidence in the argument link are available.

10.7.3   Database of Argument Models: Mixed-Criticality System

This section describes the proposed approach towards the reuse of Mixed-Criticality Systems as argument models and required support for different safety standards.

10.7.3.1   Adapting Mixed-Criticality Systems for Reuse as Argument Models

The argument database contains interrelated GSN graphs that represent the same arguments written in CAE. We should remark that each of these argumentation models consist of two layers: a generic safety case argumentation for an abstraction of the considered component, and a specific safety argumentation for a particular component choice. Figure 10.13 depicts this structure: the abstract argumentation modules (folders with light green background on the left) are supported by a corresponding safety argument for a concrete component (folders at the rightmost part). The situation where several alternative components can be used as replacements is represented by the GSN Choice elements, and the justification about why the real component supports the abstract argument is represented by the contract arguments.

10.7.3.2 Supporting Certification for Multiple Safety Standards

Product line approaches can be applied to different requirements sets for different safety-standards, or even for different safety integrity requirements while claiming compliance with a selected standard. The compositional approach to argumentation shall support the application of different argumentation patterns. The OPENCOSS project defined a safety argumentation metamodel, hybridizing the graphical depictions of GSN with the extended expressiveness of SACM, bringing to the Common Certification Language (CCL). CCL argument models can be agnostic to the specific safety standard, and by using mapping components, one can adapt an argumentation model to the argument structure required by a particular safety standard. OPENCOSS claims that this effectively provides cross-domain translatability of the safety arguments. Unfortunately, CCS is not yet standardized, neither are the required tools available to the public. In DREAMS, a simple alternative is to include variants of the argument modules for cross-domain safety components, where each of these variants is tailored to the argumentation structure required by the applicable safety standard (see Figure 10.14). These variants can also be stored in the argument database, and would be recalled by an extended Safety Compliance Constraints & Rules Checker (SCCARC) when needed.

10.7.4 Arguments of Compliance to Safety-Standards

Current safety standards require a demonstrable compliance to development process paradigms. The IEC 61508 safety standard defines the general requirements for each phase of the development process. Part 7 of IEC 61508 provides an application guideline and a collection of recommended techniques for the design, development and verification of safety critical systems. Unless specifically addressed, the project team shall define the actual selection and combination of techniques considering the required systematic capability level, but there is no guarantee of positive assessment from the certification body.

10.7.4.1 Safety-Compliant Use of the DREAMS Tool Inventory

The DREAMS tool inventory provides a number of features and use cases intended to ease the development of mixed-criticality product-lines. Some of the tools implement analysis capabilities that help assure the safety of each feasible system variant. The following list recalls the toolset, and summarizes the potential evidences generated by each tool. Once available, these evidences can be referred from the argumentation model to support the safety claims. Noticeably most of the tools work with models of the product variant, and therefore the satisfaction of the safety claims by the final system implementation shall be demonstrated by complementary means (additional verification, validation and testing).

- AF3: The following outputs are generated: logical model of the system, system platform architecture model and mapping of application components on execution units.

- Timing Model Editor: The following outputs are generated: model of the logical architecture, extended with timing constraints (repetition and end-to-end latencies).

- Safety Model Editor: The following outputs are generated: system safety functions model and library of HW / SW components models annotated with safety properties.

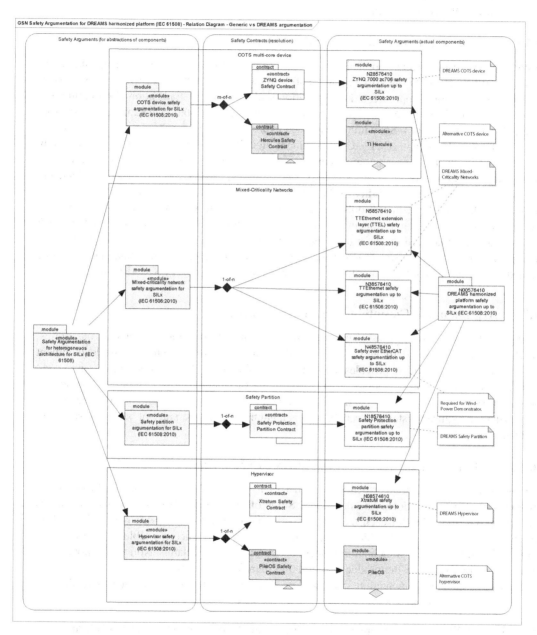

Figure 10.13: Layered Structure of the DREAMS Database of Argumentation Models for Reusable Components [324]

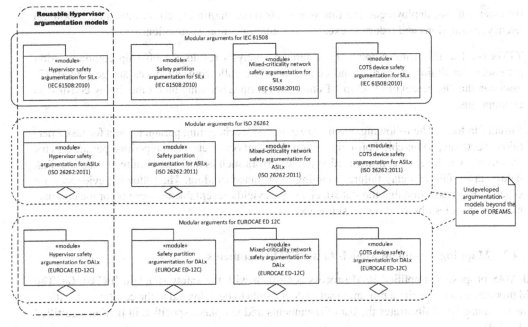

Figure 10.14: Layered Structure of the DREAMS Database of Argumentation Models for Re-usable Components [324]

- RTaW-Timing decomposition: The following outputs are generated: analysis results for the sub-latency constraints for each scheduling domain.

- TTE-Plan: The following outputs are generated: off-chip network scheduling parameters. The following evidences are generated: (1) Software to hardware allocation deployment model and (2) software components redundancy model.

- Xconcrete: The following outputs are generated: partition task scheduling parameters for the nominal mode.

- RTaW-OnChip-TT-Sched: The following outputs are generated: on-chip network scheduling parameters. The following evidences are generated: (3) Software to hardware allocation deployment model and (4) software components redundancy model.

- GRec: The following outputs are generated: partition / task scheduling parameters for a set of core failure related modes and core failure related mode transitions for the configuration of the GRM.

- MCOSF: The following outputs are generated: partition task scheduling parameters for transition modes and transition mode parameters for the configuration of the GRM.

- Safety Constraint Checker: The following outputs are generated: analysis results for the safety properties of the system architecture and preliminary GSN argument model for preparation of safety case report. The following evidences are generated: (5) documented assertion that deployment model (SW components into SW partitions, SW hypervisors, and tiles) satisfies the system safety requirements.

- RTaW-Timing Evaluation: The following outputs are generated: analysis results for the timing properties of the system deployment and timing model. The following evidences are generated:

(6) assertion that deployment and timing models meet the timing constraints, i.e., maximum latency assumptions and recurrent execution /communication assumptions.

- TTVerify: The following outputs are generated: analysis results for off-chip communication parameter set timing properties and constraints. The following evidences are generated: (7) assertion that the model of off-chip TT communication scheduling meets the timing constraints assumptions.

- Virtual Platform: The following outputs are generated: scheduling parameter sets for task / partition execution and on-chip / off-chip communication; occurrence and repetition parameters for the injection of different communication related errors such as omission failure, corruption, link failure, crash failure, delay failure, babbling idiot, masquerading. The following evidences are generated: (8) assertion that the statistical analysis yields acceptable test results for the communication latencies observed on the simulator.

10.7.4.2 Mapping Argument Models to the FSM Documents

DREAMS propose a simplified FSM process compliant with the safety standard IEC 61508. The FSM procedures will specify a minimum set of documents to be delivered to the certification authority (e.g., Figure 10.15 illustrates the list of documents and templates specified in Ikerlan's certified FSM [379]).

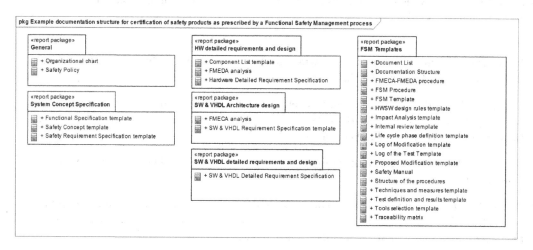

Figure 10.15: Example FSM Documentation Structure [379]

The DREAMS certification approach based on argumentation models helps in arranging the information supporting the safety claims, while improving the overall consistence. To reduce manual rework on the documentation, a mapping between the argument models and the actual certification documents is required. In addition a report generator to translate the argumentation models into a human readable document (using user defined standardized document templates) is also required.

10.7.5 Arguments Based on Verification, Validation and Testing

The DREAMS toolset allows the developer to generate a number of evidences supporting the safety claims. Nevertheless, to credit a mixed-criticality configuration for certification, the claiming organization shall elaborate many other complementary arguments and their corresponding evidences.

Process redundancy and qualification of tools as required by safety standards imply that these additional pieces of information should be generated by an independent team, and possibly using different tools and formats. The integration of this information into a single argumentation falls beyond DREAMS's objectives. However, the underlying argument database constitutes a viable integration infrastructure: it could be used to store and refer the arguments and evidences generated in project tasks, executed concurrently with the design and analysis of the mixed-criticality product line and later completed with the verification and testing results.

10.7.6 Summary

The certification of safety-critical systems requires a considerable effort to provide coherent information about the system properties and its development, in an understandable and reviewable form. Preserving the coherence of the information with the information split across multiple sources is challenging. Argumentation models are valuable tools to integrate diverse information sources into a global overview of the safety claims and the rationale and evidences to support these claims. While mixed-criticality product lines define alternative designs satisfying the safety requirements, safety remains an emergent property specific to each particular product configuration. Thus for certification purposes, one shall bundle the safety-relevant information on a per-product configuration. In DREAMS, a number of pre-built safety components form the basis to build the different variants, using two levels of abstraction to ease the replacement of component. These foundation arguments are stored in an argument database, which could also support the development of complementary supporting arguments (e.g., collection of V&V arguments and evidences). This database will be accessed from the SCCARC component to assemble a preliminary SC for each mixed-criticality configuration, generated in a semi-automated fashion, and containing the required document references to support the overall safety claim.

The DREAMS compositional approach to certification may be enhanced in a number of ways:

- Translation of the modular argumentation models to a neutral argumentation language—as proposed in the OPENCOSS project. Thus, a single set of argumentations can be composed and suited to the requirements of a domain-specific safety standard, simplifying the maintenance of the argumentation database.

- Adoption of modern standardized argumentation languages (e.g., SACM) that enhance the expressiveness of the argument models.

11

Evaluation

J. Perez
IK4-Ikerlan

M. Coppola
STMicroelectronics

M. Faugère
THALES Research & Technology

D. Gracia Pérez
THALES Research & Technology

M. Grammatikakis
TEI of Crete

A. Larrucea Ortube
IK4-Ikerlan

A. Mouzakitis
Virtual Open Systems

A. Papagrigoriou
TEI of Crete

P. Petrakis
TEI of Crete

V. Piperaki
TEI of Crete

I. Sarasola
IK4-Ikerlan

G. Tsamis
TEI of Crete

11.1   Wind-Power Domain 448
    11.1.1   Introduction 448
    11.1.2   Demonstrator Description 448
    11.1.3   Results and Conclusions 452
11.2   Safety-Critical Domain 453
    11.2.1   Mixed-Criticality and Multi-Cores 455
    11.2.2   Fault Management 458
11.3   Healthcare Domain 459
    11.3.1   Out-of-Hospital Use-Case: Security-Performance Tradeoffs 460
        11.3.1.1   On-chip Security and Data Privacy 462
        11.3.1.2   Network On Chip Firewall: Definition, Setup and Access Requests 463
        11.3.1.3   Hierarchical Driver and Security Services 464
        11.3.1.4   Security-Performance Tradeoffs 465
    11.3.2   In-Hospital Use-Case: Hospital Media Gateway 469
        11.3.2.1   Infotainment Functionality 469
        11.3.2.2   Single and Multi-Room Scenario 469
        11.3.2.3   Soft Real Time ECG Analysis and Visualization 470
        11.3.2.4   Experimental Framework and Results 471

This chapter provides an evaluation of the technologies developed in the DREAMS project, using three evaluation demonstrators representative of the domains targeted in the project (avionics, wind-power and healthcare as described in Chapter 3). Each demonstrator combines different contributions of the DREAMS project and analyzes the results from different perspectives:

1. The wind-power demonstrator (see Section 11.1) integrates the building blocks, tools and methods listed in Table 11.1. The analysis is based on Key Performance Indicators (KPIs) with respect to the wind-power specific challenges and objectives to be solved (see Section 3.2).

2. The safety-critical domain demonstrator (see Section 11.2) integrates the building blocks, tools and methods listed in Table 11.1. The analysis focuses on the temporal interference between safety-critical and non safety critical partitions deployed in different combinations of hardware platforms, Stressing Benchmarks (SBs) and configurations.

3. The healthcare demonstrator (see Section 11.3) integrates the building blocks, tools and methods listed in Table 11.1. The analysis focuses on the implementation of the NoC, memory / network bandwidth regulation, real-time performance and security performance trade-offs.

11.1   Wind-Power Domain

This section summarizes the wind-power domain demonstrator and analyzes a set of Key Performance Indicators (KPIs) used to measure the impact of DREAMS contributions with respect to the previously described challenges (see Section 3.2). The demonstrator aims to achieve a higher degree of integration between the 'control and supervision' and 'safety protection' functions, thus making the overall solution more robust, maintainable and flexible, taking into account the integration of mixed-criticality requirements.

A detailed description and analysis of the demonstrator is provided by a set of references of interest [38, 50, 115, 311, 323, 324, 358, 405, 411].

11.1.1   Introduction

Section 3.2 provides an introduction to the wind-power demonstrator describing the wind-turbine 'Human Machine Interface (HMI) and communication', 'control and supervision' and 'safety protection' functions (see Figure 3.2), the Galileo platform, EtherCAT communication and required safety certification.

11.1.2   Demonstrator Description

As shown in Figure 11.1, the demonstrator is based on the DREAMS Harmonized Platform (DHP) and Galileo platform. The DHP is a COTS Zynq 7000 board [375], which provides two ARM Cortex A9 cores and an FPGA where DREAMS technologies and services are implemented (e.g., STNoC, XtratuM hypervisor). The DHP and Galileo platforms are interconnected via a PCIe interface.

Figure 11.2 sumarizes both the current simplified solution and the proposed solution based on DREAMS contributions:

- Current solution: The wind-turbine 'HMI and communication' and 'control and supervision' functions are integrated in the Galileo platform. The Galileo platform is connected via EtherCAT communication to a set of distributed Input / Outputs (I/Os) and the 'safety protection' functions are implemented as external subsystems.

- Proposed solution: 'Safety protection' functions are integrated in the DHP instead of external

Table 11.1: DREAMS Building Blocks, Tools and Methods Integrated into each Demonstrator

Group	Element	Safety-Critical	Wind-Power	Healthcare
Modeling (Section 4)	AF3 (Section 4.2.1.3)	X	X	
	Safety Model Editor (Section 4.4.1)		X	
Algorithms and Tools (Section 5)	AF3 - DSE (Section 5.1.2)		X	
	RTaW / Timing Decomposition (Section 5.2.1.7)			
	RTaW-Pegase / On-chip TT Scheduler (Section 5.2.3.5)			X
	TTE-Plan (Section 5.2.4)	X		
	XONCRETE (Section 5.2.2)	X	X	
	GRec (Section 5.3.1)	X		
	MCOSF (Section 5.3.2)			
	RTaW-Pegase / Evaluation (Section 5.4.6)			X
	Safety Constraint Checker (Section 4.4.3)		X	
	XtratuM Hypervisor	X	X	
	TTEthernet off-chip network	X		
	On-chip Network Interface			X
	Resource Management	X		
Execution Environment (Section 6)	XtratuM (Section 6.3)	X	X	
	Linux-KVM (Section 6.4)			X
On-Chip Network (Section 7)	Chip-Level Communication Services (Section 7)			X
Off-Chip Network (Section 8)	TTEthernet (Section 8.1.1)	X		
	EtherCAT + SCL (Section 8.1.2)		X	
	Security Services (Section 8.2)	X		X
Resource Management (Section 9)	MON (Section 9.2)	X		
	LRS (Section 9.3)	X		
	LRM (Section 9.4)	X		
	GRM (Section 9.5)	X		
Safety Certification (Section 10)	MSCs (Section 10.3)		X	
	Patterns (Section 10.4)		X	
	Product Line (Section 10.7)		X	
Platform	DHP	X	X	
	Freescale T4240QDS	X		
	Galileo		X	
	Healthcare platform			X

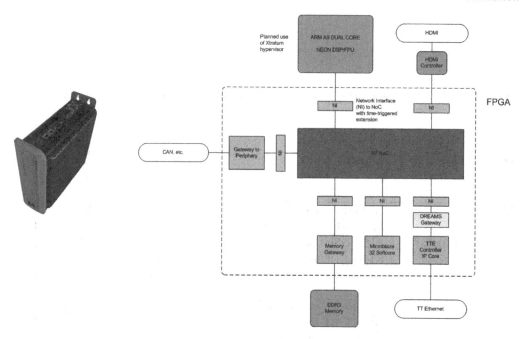

Figure 11.1: Galileo and the DREAMS Harmonized Platform

protection subsystems, and the DHP is interconnected with Galileo via PCIe interface. This solution, where protection subsystems are implemented in the DHP, may be also used to achieve heterogeneous redundancy, thus providing a Hardware Fault Tolerance (HFT) of 1. This increase of HFT is theoretical and the requirements for on-chip redundancy detailed in IEC 61508-2 Annex E [132] must be met in order to achieve HFT of 1.

As previously explained, the wind-power demonstrator is described and analyzed in the following set of (public) references of interest [38, 50, 115, 311, 323, 324, 358, 405, 411]:

- The PhD thesis "Development and Certification of Dependable Mixed-Criticality Embedded Systems" [323] describes the wind-turbine demonstrator system architecture, mixed-criticality product line, linking analysis of re-used Modular Safety Cases (MSCs) (see Section 10.3), usage of mixed-criticality patterns (see Section 10.4) and detailed FMEA / FMECA / FMEDA analysis of the used Zynq 7000 COTS multi-core device.

- The PhD thesis "Re-use of Tests and Arguments for Assessing Dependable Mixed-criticality Systems" [324] describes the wind-turbine demonstrator system architecture, AutoFOCUS 3 (AF3) based system model, safety modeling assessment and GSN based product line assessment using MSCs.

- The publication "Building Product-lines of Mixed-Criticality Systems" [115] describes a simplified software architecture model, product line variability and modeling, Design Space Exploration (DSE) and safety model analysis of the wind-power demonstrator.

- The publications "DREAMS Toolchain: Model-Driven Engineering of Mixed-Criticality Systems" [38] and "Building Product-lines of Mixed-Criticality Systems" [115], briefly describe the usage of DREAMS tool-chain for the development of a simplified wind-power demonstrator.

- The publication "GSN Support of Mixed-Criticality Systems Certification" [405] describes the

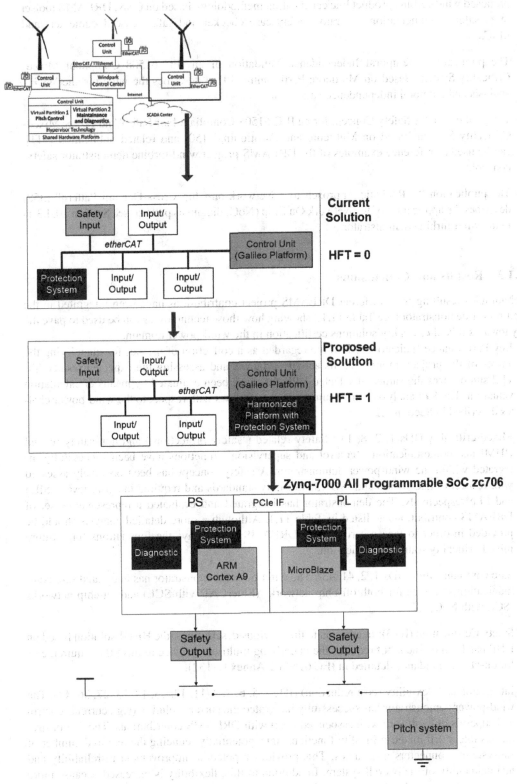

Figure 11.2: Wind Turbine - Current and Proposed Solutions

proposed wind-turbine product line certification methodology based on GSN, DREAMS toolset (e.g., safety argumentation generator, safety case checker and safety case documenter) and MSCs.

- The publication "Temporal Independence Validation of an IEC 61508 Compliant Mixed-Criticality System Based on Multicore Partitioning" [411] describes the wind-turbine WCET analysis and temporal independence validation.

- The publication "A Safety Concept for an IEC 61508 Compliant Fail-Safe Wind Power Mixed-Criticality System Based on Multicore and Partitioning" [50] and related publication [311], can be used as reference examples of the DREAMS project wind-turbine demonstrator safety-concept.

- The publication "A Realistic Approach to a Network-on-Chip Cross-Domain Pattern" [358] describes the application of the Network On Chip (NoC) diagnosis pattern (see Section 10.4.3.1) to the wind-turbine demonstrator.

11.1.3 Results and Conclusions

Technologies resulting from different DREAMS project contributions have been integrated in the wind-power demonstrator (see Table 11.1), showing how those technologies can be used to pave the way towards mixed-criticality solutions certification in the wind-power domain.

Key Performance Indicators (KPIs) are regarded as a collection of metrics for quantifying the objectives of the project, monitoring its activity progress and assessing the expected results. Table 11.2 summarizes the values of selected KPIs that have been evaluated by means of calculation or evaluation. The KPI analysis can be summarized as follows with respect to the wind-power challenges described in Section 3.2:

- Mixed-criticality (IDs 1, 2, 8, 14): Safety related ('safety protection') and non-safety related ('HMI and communication', 'control and supervision') functions have been successfully integrated within the wind-power demonstrator. A safety-concept has been positively assessed with respect to IEC 61508 and ISO 13849 safety standards and required integrity levels, SIL3 and PLe respectively. The demonstrator has integrated and evaluated a representative set of DREAMS contributions as listed in Table 11.1. Although a more detailed analysis should be provided in order to face a certification, DREAMS project lays the foundations for a future mixed-criticality solution certification.

- Safe communication (IDs 1, 2, 41, 43): The wind-power demonstrator has integrated safe communication solutions for both off-chip networks (EtherCAT with SCL) and on-chip networks (SCL with NoC).

- Safety Certification (ID 3): In addition to the previous descriptions, the HFT 1 solution based on DHP has been defined, but requires the underlying multi-core device to meet the requirements for on-chip redundancy detailed in IEC 61508-2 Annex E [132].

- Integration and flexibility (cost reduction) (IDs 4, 5, 6, 7, 9, 11, 12, 13, 19, 23, 27, 36, 42): The wind-power demonstrator has successfully integrated the current solution (e.g., current 'control and supervision' software with associated OSs) with DREAMS contributions. The demonstrator has integrated mixed-criticality functions, thus potentially reducing the required number of subsystems, connectors and cables. This provides a potential improvement in reliability and maintainability of the overall system. In addition to this, flexibility is increased because more flexible and advanced safety logic can be implemented in the device where safety related decisions are taken.

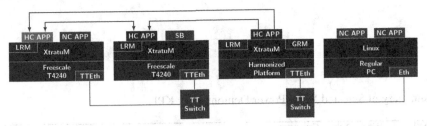

Figure 11.3: Use Case Target Deployment and Secured Communications between the Critical Applications (Green Arrows) [210]

Overall, it is concluded that the DREAMS wind-power demonstrator has contributed to advance in the field of mixed-criticality solutions for the wind-power domain. The technological development has been carried out in a demonstrator basis and should be used accordingly. However, this demonstrator establishes a starting point for the certification of mixed-criticality solutions that could be widely used to overcome the associated upcoming challenges in the wind-power industry.

11.2 Safety-Critical Domain

A safety-critical demonstrator was developed to highlight the applicability of the DREAMS architecture and the resource management capabilities. The demonstrator combines critical applications ('HC App') with non-critical applications ('NC App') using heterogeneous multi-core platforms, connected using a wired network.

Five different applications are deployed in the demonstrator, three critical and two non-critical. Additional non-critical applications (SBs) are introduced to stress the system. All the applications, sit on top of the DREAMS middleware and the XtratuM hypervisor, and they are deployed in two different hardware platforms: the Freescale T4240QDS and the DHP. Communications between the different hardware platforms are ensured by a TTEthernet network. The demonstrator mixes different types of traffic supported by TTEthernet:

- Rate-Constrained (RC) for the communication between the critical applications,

- and Time-Triggered (TT) for the communication between DREAMS services and the communication between Global Resource Manager (GRM) and Local Resource Managers (LRMs).

Figure 11.3 shows one of the applications target deployments based on two T4240s and one DHP platform. The GRM executes only on the DHP, while the LRMs execute on all nodes. Secure communication is exploited by the LRM and GRM instances in the system as described in the previous sections. The same secure communication library is used for the communications between the critical applications, displayed as green arrows in Figure 11.3.

In addition, critical applications are enhanced with the deadline overrun Resource Monitor (MON), to ensure they don't miss a deadline. The QoS MONs are used to improve performance of the non-critical applications running on the demonstrator. The core-failure MON executes on all cores of each node, and reports back the core status to the corresponding LRM. The Local Resource Scheduler (LRS) is used to schedule tasks on each node.

The whole demonstrator was developed using the AF3 tool to model it. Schedules for the nominal plans of each node were generated with XONCRETE, and extended reconfigurations generated with Graph of Reconfiguration tool (GRec). Configurations were produced for the three nodes and

Table 11.2: Summary of Selected Wind-Power Demonstrator KPIs

ID	KPI	Goal	Value	Comments
1	Achievable Performance Level (PL)	PLe	PLe	Heterogeneous on-chip redundancy of safety relevant logic. Category 3 requirements according to ISO 13849-1 [49] section 6.2.6. $MTTF_d$ high and DC_{avg} medium to high.
2	Achievable Safety Integrity Level (SIL)	SIL3	SIL3	Heterogeneous on-chip redundancy of safety relevant logic. HFT of 1 and SFF medium ($90\% < SFF < 99\%$). Techniques and measures to control systematic and random failures.
3	Achievable HFT	1	1	Heterogeneous on-chip redundancy of safety relevant logic in ARM and uBlaze. This HFT is theoretical, and the requirements for on-chip redundancy detailed in IEC 61508-2 [132] Annex E must be met in order to achieve certification.
4	Validated support for required OSs	Yes	Yes	The support for Windows Embedded CE 6.0, which is the key real-time OS for wind-power demonstrator is preliminary validated.
5	Minimum closed-loop cycle time	$1ms$	$1ms$	In the Galileo side, the Board Support Package (BSP) of Windows Embedded CE 6.0 has been modified so that 1ms interrupts are generated, thus allowing a time base of $1ms$ to schedule tasks. In the DHP, shorter times are possible since lighter operating systems are to be used, or even no operating system at all.
6	Minimum field bus cycle time	1ms	1ms	The minimum achievable cycle time with the EtherCAT master software stack is $1\,ms$.
7	Maximum jitter	$10\mu s$	$5\mu s$	Isolated executions of critical partition guarantee not to exceed this value. Need to follow XtratuM recommendations.
8	Fault containment by construction	Yes	Yes	Specific documentation for each building block and mainly, the DREAMS deliverable D2.3.1 [412] could represent enough evidences to consider fault containment by construction.
9	Percentage of integrated core services	50%	75%	Based on core services listed in D1.2.1 [21], three out of four core services have already been implemented in the DHP. The one missing is the core service defined as "Integrated resource management for time and space partitioning".
11	Percentage of system architecture/design modeled	70%	75%	The following metamodel elements from D1.4.1 [413] have been used: SW components (application components, virtual ports and virtual channels), HW elements (cluster, nodes, tiles, cores, RAM / ROM, communication networks, watchdog and clocks), system SW (hypervisors and partitions) and deployment components. Real-time and energy consumption models have not been used so far.
12	Percentage of software application modeled	50%	0%	Functional modeling of the application software is outside the scope of toolset. However, all entities of DREAMS software components defined in D1.4.1 [413] (e.g., component, ports, channels) have been used in the corresponding phase of wind-power demonstrator.
13	Model complexity	Yes	Yes	With the toolset, the expert has an integrated view of all models.
14	Temporal and spatial isolation by construction	Yes	Yes	Spatial isolation is guaranteed by the hypervisor and these evidences can be extracted from specific documentation of the virtualization layer and D2.3.1 [412]. Temporal independence could also be supported by the hypervisor.
19	Percentage of out of the box gateways	50%	0%	The only gateway required is PCIe IP in the FPGA, and it has been specifically integrated for the wind-power demonstrator.
23	Development steps covered by tools	60%	50%	Almost all development steps completed so far (50% of the V-cycle) have been supported by the DREAMS toolset, except application software design and development that are outside the scope of the DREAMS project.
27	Effort reduction for replacement of components	30%	0%	This KPI cannot be properly estimated at the moment. The use of models will clearly help to replace components and calculate the impact in the rest of the system, but this effort reduction can hardly be estimated with currently available information.
36	Percentage of support for relevant OS (RTOS and GPOS)	75%	75%	The following relevant operating systems are supported: Windows Embedded CE 6.0 (as RTOS and GPOS), Partikle (RTOS) and XAL. However, Windows Embedded Standard 7 is not supported, which would be the preferred option for the communications partition. Therefore, 3 out of 4 relevant operating systems are supported.
41	Safe data	> 10 bytes	64 bytes	Up to 64 bytes of safety relevant data is encapsulated in every frame. The maximum number of safety relevant variables to be transmitted depends on the number of bytes consumed by the variable types, and this could be increased if necessary.
42	Safe algorithm programming flexibility	Yes	Yes	There is no limitation regarding algorithm complexity in the programming of ARM and uBlaze safety partitions.
43	Network flexibility and scalability	Yes	Yes	The off-chip mixed-criticality network allows adding or removing elements and scaling the number of nodes. However, the on-chip network is not that flexible since a new FPGA design needs to be provided.

the TTEthernet switch (see Figure 8.8). The following sections show the capabilities of the generated solutions to manage the interferences in the targeted multi-cores and to keep the availability of the distributed system when core failure events occur. Simplified deployments were also developed:

1. A reduced deployment with two nodes (one T4240 and one DHP) connected through the real-time switch

2. Single-node deployments using only a T4240 platform over which all the critical applications are deployed. These deployments are used in the following sections, to simplify the evaluation description.

11.2.1 Mixed-Criticality and Multi-Cores

To evaluate the capabilities of the runtime solutions to integrate applications with different criticalities in a multi-core device, a study on a single T4240 platform was created, using only a cluster of the T4240 processor (i.e., four cores out of the twelve available).

The cores in the cluster share a second level cache and the whole memory subsystem. Three applications issued from a safety-critical industry domain and simple communication sink application (which we consider critical too) are deployed and scheduled in core 0 of the T4240 node: APP1—APP4. The rest of the cores, and the time slots not used by the critical applications scheduling, are used to execute non-critical applications. In our evaluation we used stressing benchmarks in order to create interferences in the target shared resources (L2 cache and memory subsystem). The Stressing Benchmark (SB) are:

- rr: an application performing random reads over a large (8MB) array.

- sr: an application performing sequential reads over a large array.

- rw: an application performing random writes over a large array.

- sw: an application performing sequential writes over a large array.

Critical and non-critical applications are combined in different deployment setups:

- Isolation (I): Only critical applications are executed in core 0. The other cores remain unused.

- Separate (S): Identical non-critical applications are scheduled in all the cores to execute at time windows not overlapping the schedule of the critical applications.

- Interference (F): Identical non-critical applications are scheduled in all the cores. The non-critical application executing in core 0 executes at time windows not used by the critical applications. Non-critical applications executing in the other cores use all the time available in the MAjor Cycle (MaC).

- Deadline overrun (D): Similar to the interference deployment setup but with the DREAMS resource management deadline overrun mechanism activated.

- QoS (Q): Similar to the interference deployment setup but with the DREAMS resource management deadline overrun mechanism extended with the QoS enhancements activated.

When combining the stressing benchmarks with the deployments setups we obtain the following experiments: I, Srr, Ssr, Srw, Srw, Frr, Fsr, Frw, Frw, Drr, Dsr, Drw, Drw, Qrr, Qsr, Qrw, and Qrw.

Tables 11.3 and 11.4 respectively show per application the MaC allocation time and the measured execution time per MaC (worst and median) in our experiments. APP4 is not shown in the tables because as previously described it is a simple sink application, not representative of the studied safety-critical domain.

- Table 11.3 shows the MaC allocation time per application on each deployment setup. All values are shown as the deviation percentage respective to MaC allocation time of the application when executing in the isolation deployment setup (I).

- Table 11.4 shows the critical applications execution time (worst and median, respectively 'Wt' and 'Med' in table) per MaC summary. All values are shown as the deviation respective to MaC allocation time of the application when executing in the isolation deployment setup (I). For deployment setup columns F, D and Q, cells are colored to indicate for which of the three the observed worst and median measurements were the worst (red), best (green) or in between (orange) among the three.

Table 11.3: MaC Allocation Time per Application on Each Deployment Setup

	Separate (S)	Interference (F)	Deadline Overrun/QoS (D/Q)
APP1	118	476.3	292
APP2	253	1021.2	608
APP3	107	431.9	277

Table 11.4: Critical Applications Execution Time (Worst 'Wt' and Median 'Med') per MaC Summary

	Separate (S)		Interference (F)		Deadline Overrun (D)		QoS (Q)	
	Wt	Med	Wt	Med	Wt	Med	Wt	Med
APP1								
sw	101	94.1	403.6	289.9	122.8	117.5	117	106.1
rw	101.5	94.1	206.6	194.9	114.1	106.7	110.5	103.2
sr	115.2	114	122.8	121.3	126.5	125.1	126	124.6
rr	115.3	114.1	123.3	122.1	126.5	125.2	126.2	124.6
APP2								
sw	100.2	95.8	200.5	174.9	190.5	173.3	180.5	147
rw	100.7	95.8	148.5	141.4	147.9	140.8	145.3	127.6
sr	129.3	128.8	139.8	138.6	147.9	146.6	150.2	146
rr	129.3	128.9	139.7	138.8	147.8	146.6	150.8	145.7
APP3								
sw	100.2	99.5	162.8	139.7	105.4	104.1	104.5	101.8
rw	100.3	99.5	120.9	110.7	105.2	102.2	103.4	101.5
sr	104	103.9	109.3	109.1	105.6	105.2	105.5	105.2
rr	103.9	103.9	109.7	109.4	105.3	105.2	105.5	105.1

For deployment setups I, S and F, MaC allocation time is computed as the sum of the application slots deadlines in the MaC. Slots deadlines are computed by XONCRETE. For a slot the deadline (DL) is computed as the sum of the jobs tasks worst observed execution time. For example, for a slot s, of partition p, with jobs sequence $j1, j2, j3$ corresponding to the execution of tasks $t1, t2, t1$ (i.e., jobs $j1$ and $j3$ execute task $t1$, and $j2$ executes $t2$) where the worst observed execution time of $t1$ in any slot of p is $w1$, and $w2$ for $t2$; the slot s deadline when running in isolation ($DL(s)$) is computed as $w1 + w2 + w1$. As the execution time of the tasks varies depending on the deployment setup I, S and F (see Table 11.4), the computed MaC allocation time for these deployment setups varies as can be observed in Table 11.3.

For deployment setups D and Q (i.e., when using the deadline overrun management mechanism), MaC allocation time is also computed as the sum of the application slots deadlines in the MaC. However, when using the deadline overrun management mechanism, slots deadlines (DL_{do}) are computed as the slot deadline computed when running in the S deployment setup plus the biggest slot's task worst observed execution time multiplied by the slowdown factor computed for the target (the T4240 in our study) minus one. Following the previous example, with a slowdown factor F if $w1 > w2$, then the deadline of slot s when using the deadline overrun mechanism ($DL_{do}(s)$) is $DL_{iso}(s) + (w1_{iso} \times (F - 1))$, where DL_{iso} and $w1_{iso}$ are respectively the slot deadline and the worst observed execution time for the task $t1$ when executing in the isolation deployment setup (I). The slowdown factor F is computed as *the worst observed execution time* slowdown observed when executing any critical application in the F deployment setup experiments (Fsr, Frr, Fsw, and Frw) compared to their execution in the isolation deployment setup.

In our evaluation using the T4240 processor with only one cluster, the worst slowdown measurement appears when using application APP1 on the Fsw experiment, with a slowdown on the worst observed execution time of 403.6% (see Table 11.4). This measurement sets the slowdown factor F for the T4240 processor with only one cluster used to ×4.306.

From Table 11.3 we can observe the first benefits of using the deadline overrun mechanism (deployment setups D and Q) on mixed-criticality systems. The MaC allocation time required by critical applications when running non-critical applications in parallel (deployment setups F, D and Q) is reduced when using the mechanism: from 476.3% to 292% for APP1, from 1021.2% to 608% for APP2, and from 431.9% to 277% for APP3; thus respectively for APP1, APP2, and APP3 ×0.61, ×0.6 and ×0.64 MaC allocation time of the F deployment setup is required when using the deadline overrun mechanism. This allows to create deployments with other critical applications that can use the freed time in the MaC.

As can be observed in Table 11.4, critical applications execution time is generally also reduced, for both, the worst observed execution time and median. The most significant gain is observed for APP1 when combined with the sw stressing benchmarks.

The worst observed execution time is reduced from 403.6% on the interference deployment setup to 122.6% (×0.30 reduction) when the deadline overrun mechanism is used and 117% (×0.29% reduction) when it is extended with the QoS enhancements. Likewise, significant reductions can be observed for the median execution time: ×0.42 with deadline overrun and ×0.38. The deadline overrun mechanism (with and without the QoS enhancements) achieves it by stopping the non-critical applications selectively to ensure the critical applications deadlines.

In some cases, we observe an execution time augmentation when using these solutions: for APP1 and APP2 when combined with the sr and rr stressing benchmarks, the most significant (×1.08) being the worst observed execution time of application APP2 on the QoS deployment when using the rr stressing benchmark. However, even in those cases, when using the DREAMS developed solutions deadlines of the critical applications deadlines were ensured.

Table 11.5 compares the usage of the T4240 multi-core processor by the non-critical applications when using the deadline overrun mechanism with and without the QoS enhancements, compared to the performance in the interference deployment. Only the MaC sections with the critical applications running were considered, i.e., performance is not collected when all the cores are being used to execute non-critical applications. For this measurement the critical applications slots deadlines (i.e., slot length) on I, D, Q deployment setups, were set to the same value: the deadline computed on the I deployment setup. The deadline overrun mechanism without the QoS enhancements was able to extract in the worst case 87% (when using the sw stressing benchmark) of the performance observed on the interference deployment. When using the sr stressing benchmark the performance observed was almost the same than when not using the mechanism. Furthermore, when using the QoS enhancements the non-critical applications performance is still improved by 2% and 3%.

Table 11.5: T4240 Processor Usage with Deadline Overrun Mechanism [414]

	Deadline overrun (D)	QoS (Q)
sw	87%	90%
rw	93%	95%
sr	99%	99%
rr	92%	95%

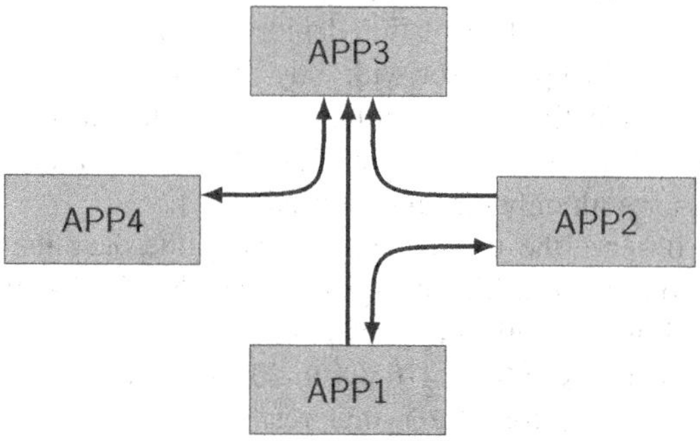

Figure 11.4: Critical Applications Communications in the Reduced Deployment for the Fault Management Evaluation [414]

11.2.2   Fault Management

A reconfiguration graph for the reduced deployment (one T4240 and one DHP) was created with the GRec tool to generate the nodes configurations (i.e., hypervisor plans) in the event of core faults. To reduce the number of configurations, only four cores of the T4240 node are considered (core0—core3). The two cores of the DHP node are considered (core0 and core1). A total of four critical applications issued from a safety-critical industry domain are deployed in the system: APP1-APP4. These applications are the same as in the previous section (Section 11.2.1). The applications communicate as described in Figure 11.4. In the initial plan only APP3 is deployed in the DHP node, while all the others are deployed in the T4240 node. Furthermore, two non-critical applications are also deployed in the system: NAPP1-NAPP2. One of them (NAPP1) is deployed in the T4240 node while the other one is deployed in the DHP node.

Multiple core fault sequences were simulated to evaluate the capability of the DREAMS resource management solutions to handle core faults. In all the scenarios, the resource management solution managed to keep the critical applications active, while there were enough cores available to execute them.

Table 11.6 shows how the resource management solution ensured system availability as a sequence of core failures was simulated in the above described reduced deployment. As can be observed critical applications (Crit.Apps. in Table 11.6) remain active with only two cores available in the system, core 0 of respectively the T4240 and the DHP nodes. The non-critical applications (NCrit.Apps. in Table 11.6) are removed from the system as core failures events occur, NAPP1 being stopped when 4 cores are still available, and NAPP2 when 3 cores are still available. Removing an additional core from the DHP core would stop reconfigurations, as that would stop the GRM running in that node. An additional core failure in the T4240 node would also render any possible reconfiguration of the critical applications deployment impossible. Effectively, the four critical ap-

Table 11.6: Example of Reconfigurations Managed by the Resource Management Solution in the Event of Core Faults

MAF	Core fault event	Cores available T4240	Cores available DHP	Crit.Apps. APP1	Crit.Apps. APP2	Crit.Apps. APP3	Crit.Apps. APP4	NCrit.Apps. NAPP1	NCrit.Apps. NAPP2
1	—	0,1,2,3	0,1	T4240 c0	T4240 c1	DHP c0	T4240 c2	T4240 c3	DHP c1
5	T4240(c3)	0,1,2	0,1	T4240 c0	T4240 c0	DHP c0	T4240 c2	T4240 c1	DHP c1
8	T4240(c2)	0,1	0,1	T4240 c0	T4240 c0	DHP c0	T4240 c1	—	DHP c1
12	DHP(c1)	0,1	0	T4240 c0	T4240 c0	DHP c0	T4240 c1	—	—
20	T4240(c1)	0	0	T4240 c0	T4240 c0	DHP c0	DHP c0	—	—

plications can not be scheduled in the DHP node, and even less in one single core of that node. All the reconfigurations were performed by the GRM in the MaC following the core failure detection.

The communications between the critical applications remained functional even when the deployment of APP4 is changed from the T4240 node to the DHP node. Effectively, the communication between APP3 and APP4 was done through the network on the initial deployment, while at cycle 20 after the failure detected in the T4240 core 1 the two applications are deployed at the DHP node, and their communications become local (i.e., in the same node). This did not require any special handling in the applications code, i.e., the applications are not aware of their deployment and thus if their communications are local or through the network. Applications perform their communication regularly through the DRAL interface. This demonstrates the capability of the DRAL hypervisor interface library to reconfigure communications.

11.3 Healthcare Domain

This section presents an open-source soft real-time 'eHealth' application based on extensions of PhysioNet WaveForm DataBase (WFDB), Open Source ECG Analysis (OSEA) and WAVE modules. This integrated framework focuses on real-time diagnosis of non-fatal pathogenic cardiac events, including Electrocardiogram (ECG) generation, processing and visualization phases.

We validate our telemedicine application using an initial prototype based on ARMv7 technologies for two separate use-cases: in-hospital ECG analysis and out-of-hospital soft real-time smartphone ECG processing. While for the in-hospital use case, we focus on the effects of memory / network bandwidth regulation technologies ('MemGuardXt', 'NetGuardXt') to real-time ECG processing and scalability, especially at the hospital media server, for the out-of-hospital use case we concentrate on fundamental security-performance trade-offs. Notice that in both cases, we have used the ST Microelectronics 'BodyGateway' device (BGW) sensor, with a custom Linux / Android driver.

- The out-of-hospital use-case focuses on real-time smartphone ECG processing and examines security-performance trade-offs. In this use-case, we consider transmission security via AES, patient anonymity using SHA-3 (keccak algorithm), and data privacy protection using Linux group ID; data privacy is supported in hardware by storing patient data and keys in BRAM and implementing a NoC Firewall architecture with a hierarchical Linux driver.

- In another use case intended for real-time ECG processing within the hospital, doctors use smart

Android devices to connect to a server for accessing ECG data of their patients. In this case-study, we examine interactions between MemGuard and NetGuard Linux kernel modules and real-time performance.

Both use-cases extend the current state-of-the-art, since real-time visualization of an annotated ECG signal has not been considered before. Moreover, our in-hospital use-case has already been ported on ARMv8 (ARM Juno board) with TTEthernet technology and will be validated in an actual clinical environment in the near future. However, currently the full application can only run on ARMv7 due to portability issues: visualization on ARMv8 is not possible due to issues with 32-bit compatibility libraries.

11.3.1   Out-of-Hospital Use-Case: Security-Performance Tradeoffs

Mobile health monitoring technology has the potential to bring a doctor's office to the patient's smartphone. In this section, we consider an end-to-end soft real-time out-of-hospital use-case that concerns transmission of patient ECG data from BGW pulse sensor via an Android device ('Patient App') to a cloud server for ECG analysis and annotation. The annotated ECG signal is subsequently transmitted on demand to another Android device ('Doctor App') for visualization.

Using a prototype featuring ARMv7 technology (two Odroid-XU4s running the 'Patient Apps' and 'Doctor Apps', and one Zedboard FPGA board as server), we evaluate soft real-time application performance requirements, as well as security overheads for supporting confidentiality, integrity and patient anonymity. Our real-time ECG analysis extends WFDB and OSEA open source software from offline towards real-time processing; the extension is currently available for download, except for BGW driver [415]. While AES-CCM incurs $\sim 30\%$, $\sim 20\%$, and $\sim 67\%$ of the end-to-end signal delay for 'Patient App', server, and 'Doctor App', anonymity cost (via SHA-3 hash) is marginal.

Our healthcare scenario considers soft real-time smartphone monitoring of patients out of the hospital environment in an attempt to evaluate real-time performance and security overheads for supporting transmission security and anonymity. As shown in Figure 11.5 patients wearing the BGW and carrying an Android device (usually a smartphone) with a Bluetooth connection to the device and a simultaneous WiFi connection to a cloud environment (receiving data and performing ECG analysis) can help medical personnel monitor annotated patient signals on their smartphone. Annotations are added asynchronously on top of the ECG signal to indicate specific chronic cardiac diseases.

'Patient App' Mapped to Odroid XU4 Running Android

The 'Patient App' allows the patient to configure connections to Bluetooth device (pairing) and to the cloud server (WiFi). Before transmitting ECG data to the cloud, the 'Patient App' writes a unique patient ID and current time of Android patient device when the device connects to the BGW.

- The patient ID provides a way to support anonymity [417], i.e., it uniquely determines (more specifically, hashes to) the MAC address of the BGW that the patient is wearing, as well as other private patient data. Anonymity permits post-processing analysis of aggregated ECG data on the cloud server, without endangering privacy of the patients.

- The time-stamp is used to provide metrics related to real-time operation. This can help identify issues, e.g., when BGW enters storing mode (saving data instead of transmitting due to Bluetooth connection delay) or if the WiFi connection from 'Patient App' to the cloud or from the cloud to 'Doctor App' has failed or is inactive.

Cloud Server for ECG Processing Mapped to Zedboard

In our prototype for the distributed server implementation, we use a Xilinx Zedboard platform.

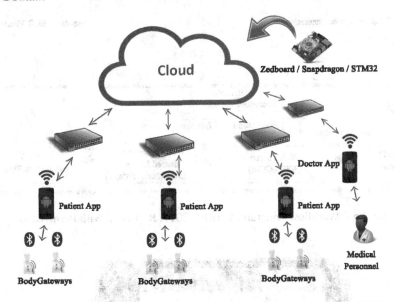

Figure 11.5: End-to-End System Architecture for Out-of-Hospital Use-Case. © IEEE 2017, Reprinted with Permission from [416]

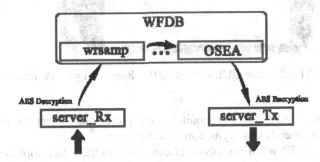

Figure 11.6: ECG Processing at Server. © IEEE 2017, Reprinted with Permission from [416]

Multiple devices running Linux may be mapped to a number of smartphone devices via a DNS service.

As shown in Figure 11.6, Zedboard can receive via WiFi, decrypt, store, convert to a standard format and process the ECG signal using WFDB and OSEA for detecting ventricular cardiac events. The `server_rx process` that runs on Zedboard receives ECG data for online cardiac heartbeat detection and classification using our WFDB and OSEA library extensions. In addition, the `server_tx` process can communicate asynchronously with the 'Doctor App' to retrieve a given patient's annotated ECG data, as explained in Section 11.3.1. The transmitted patient ID is used by the `server_tx` child process which inherits group access privileges of the connecting doctor (set by an administrator) to access patient data by hashing the patient ID via SHA-3 hash using keccak algorithm. Patient data includes private info as well as a key for accessing the patient's annotated ECG file for processing, encryption and transmission to 'Doctor App'.

'Doctor App' Mapped to Odroid XU4 Running Android

As shown in Figure 11.7, the 'Doctor App' implements an authentication mechanism based on a medical personnel ID to control access to sensitive medical data stored in the cloud. Then, upon

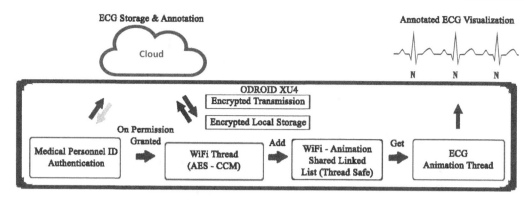

Figure 11.7: 'Doctor App' Workflow Diagram. © IEEE 2017, Reprinted with Permission from [416]

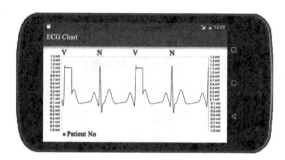

Figure 11.8: ECG Signal Visualization GUI. © IEEE 2017, Reprinted with Permission from [416]

successful login, the doctor must select to monitor a specific unique patient ID, thus providing anonymized access to the patient's vital signs stored on the server.

After sending the patient ID, a WiFi connection thread receives ECG data (triplets containing an incremental sample ID, the ECG signal, and a possibly asynchronous annotation) from cloud server and appends it into a shared buffer (linked list) and eventually a file. Another ECG animation thread also accesses the same list in order to decrypt info, plots ECG time series for the corresponding points, eventually emptying the list. Asynchronous posting of annotations on the chart appears as character "N" for normal and "V" for ventricular; see Figure 11.8. This graph shows the ECG signal with units (mV), and provides calibration, scroll and zoom functions.

11.3.1.1 On-chip Security and Data Privacy

Potential benefits from the adoption of Mobile Healthcare technologies include improved care, cost savings and reduction of errors. However, there are also several major legal, sociological and technical risks related to privacy and security.

Besides cryptography, NoC firewall security based on access control (deny rules) provides hardware multi-compartment isolation, preventing a process from accessing shared memory that has not been granted access. Hence a NoC firewall mechanism can be configured with a set of deny rules to protect sensitive data and shield applications from unauthorized access to physical memory regions containing critical data. This also prevents a malicious (or corrupt) device or driver from affecting other processes, or the OS itself.

NoC firewall technology has been developed to provide fine-grain page-level security. In this case, decisions are made on whether to accept or reject a specific request based on rules hidden within a page-level memory descriptor or via an independent memory management unit [418, 419].

In addition, notice that while [420] uses dedicated virtual channels to pass specialized rule-checking information processor and thread identifiers, our proposed coarse-grain, segment-level protection is based on the physical address of the NoC transaction request (excluding offset), similar to [271].

Unlike all previous works, we develop a generic hierarchical Linux driver infrastructure on top of our hardware NoC firewall architecture that supports modularity across different on-chip memory use cases related to access control in healthcare technologies and beyond. In addition, our framework does not use interrupts in order to avoid excessive delays; see [271].

The main purpose of our improved NoC firewall design is to examine the relative cost of supporting high-level security services, such as anonymity and key management for supporting privacy and security in a realistic soft real-time scenario related to mobile healthcare ('mHealth') technologies. None of these studies that involve NoC firewall have considered a realistic user application. More specifically, assuming that each patient has a unique 'eHealth' record (patient ID) that could be distributed to his doctor (simple user) freely or upon demand, we ensure a) that each doctor can access only patient data that he has permission to, and b) patient identity (including wearable pulse sensor MAC address) are hidden by using patient ID (and possibly a key) for validation. Thus, we attempt to design and implement modular firmware security solutions that:

- support anonymity of patient data in FPGA memory by hashing to an appropriate location in FPGA memory (BRAM) using the unique patient ID.

- protect patient data (hashed in system memory) when it must be exchanged with physicians for processing and visualization by providing hardware-based access control mechanisms. We prevent access from malicious or unauthorized physicians by setting firewall rules (based on group ID) in a hardware-based NoC firewall that protects all BRAMs. Therefore, malicious or unauthorized logical processes cannot read or write patient data in the 'eHealth' system by performing a surface attack.

Within this scope, we show that performance overheads for supporting software application security, confidentiality and integrity of healthcare data and anonymity of patient data using the NoC Firewall (and Linux group IDs) are small.

11.3.1.2 Network On Chip Firewall: Definition, Setup and Access Requests

As shown in Figure 11.9, we have developed a NoC firewall module with four sets of registers (one set per each input port) in front of the input FIFOs of a 4x4 Butterfly NoC; the number of sets is scalable and depends only on the size of the FPGA. The Butterfly NoC itself is configured with smaller 2x2 internal switching nodes using input-output buffering and each of its output ports is attached to a BRAM. Although a debug AMBA AXI4 interface (mapped to 0x80000000 to 0x80003FFF) provides direct access to the BRAMs, in the following subsection we focus on normal operation, i.e., setup of the firewall rules and normal access requests that travel through the firewall.

Each set of firewall registers can be configured independently to operate in two operating modes: 'Extended' and 'Simple'. Due to space limitations, we detail only 'Extended Mode', in which the firewall operates as a true NoC Firewall. More specifically, firewall rules in 'Extended Mode' are specified by providing both the input port of the access request, as well as the BRAM (destination) output port, i.e., out1 to out4 in Figure11.9. Hence, a firewall register set in 'Extended Mode' is configured to protect data in each of the BRAM memory-mapped address space from illegitimate access requests from specific input ports. In fact, provided BRAM ranges are relative to 0x40000000 which corresponds to the base address of BRAM1.

Figure 11.10 shows the NoC packet structure for an access request (read or write) via the firewall to a BRAM specified by a NoC output port via a NoC input port. The BRAM Offset is relative to the specified BRAM base address.

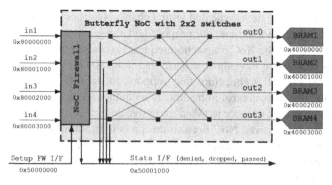

Figure 11.9: The NoC Firewall

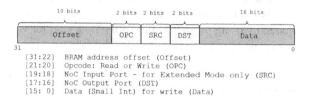

Figure 11.10: An Access Request in 'Extended Mode'

11.3.1.3   Hierarchical Driver and Security Services

The NoC Firewall driver is organized hierarchically into 3 layers in order to enhance modularity and re-usability. As shown in Figure 11.11, the proposed three layers includes a low-level I/O memory interface, as well as mid-level kernel system interface and user-level functions, with callbacks defined only between adjacent protocol layers.

The Low-Level Driver API (LLD) defines I/O memory methods to write in kernel space using `iowrite32()` method and read from kernel space using `ioread32()` method. These methods are called by the Mid-Level Driver API (MLD). For example, LLD is used to perform firewall setup by writing to memory-mapped setup registers using `iowrite32` method, and is also used to read firewall setup information from memory setup registers using `ioread32` method. LLD also calls `ioread32` to access to memory-mapped statistics counters. Finally, LLD uses `iowrite32` and `ioread32` methods to support both access via the firewall and direct access to BRAM.

The Mid-Level Driver API (MLD) defines firewall setup in specific output and input ports ranges, supports read / write access to input and output ports via NoC Firewall, provides direct access to different output ports (BRAMs), and enables different types of statistics for total passed / dropped packets. More specifically, MLD performs firewall setup for both 'Simple mode' and 'Extended mode'.

Finally, MLD supports direct access, in addition to access via the firewall. Direct access to BRAM is carried out via normal calls to functions `ioread32` / `iowrite32`. Access via firewall is much more complicated. It involves a) packetization of the write access request (see Figure11.10) at the NoC interface, as well as transfer of the NoC packet for execution at the BRAM b) both packetization of the read access request and response depacketization for read access.

In addition, notice that for security reasons access to all LLD and MLD functions are protected. Most can be called only from privileged users, i.e., administrators who have been subscribed to root group. In all others kernel panic is caused and no stack information is made available. Locking is provided to avoid consistency issues with simultaneous accesses from different user-space applications (e.g., using 'pthreads') or kernel-space programs (e.g., using 'kthreads').

The High-Level Driver API (HLD) completes the driver hierarchy by providing a configurable

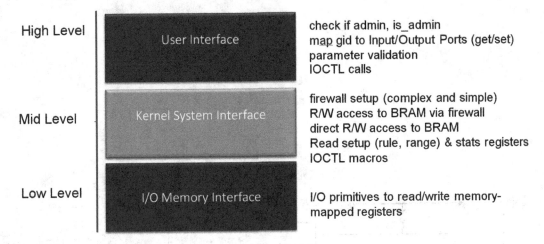

High Level	User Interface	check if admin, is_admin map gid to Input/Output Ports (get/set) parameter validation IOCTL calls
Mid Level	Kernel System Interface	firewall setup (complex and simple) R/W access to BRAM via firewall direct R/W access to BRAM Read setup (rule, range) & stats registers IOCTL macros
Low Level	I/O Memory Interface	I/O primitives to read/write memory- mapped registers

Figure 11.11: Hierarchical Linux Driver for the Firewall

way to map a particular use-case or programming scenario to mid-level firewall functionality. In particular, the HLD interface of the hierarchical Linux driver of the NoC firewall is used to support data privacy and anonymity in our in-hospital healthcare scenario, whereas doctors belong to clinics (see Table 11.7) and can access patients depending on whether patients are 'registered' in specific clinics.

In the HLD, we are able to support access control to all BRAMs by allowing a privileged user (e.g., system administrator) to write group tables in BRAM. Initially, firewall is temporarily set up to allow administrator access for writing these tables. Later after group tables are written, the firewall setup is modified to limit user access to these tables. Notice that group tables are programmed to limit access from specific user groups to the preset NoC firewall input / output ports, thereby limiting access to BRAMs via the NoC firewall. In our setup, each group corresponding to a different clinic (or department) would have access to a single BRAM containing data for all its patients; group tables are placed in the same BRAM, or alternatively in another BRAM shared across different groups.

For this reason, as shown in Figure 11.12, functions `setGidPerInport` / `setGidPerOutport` are used by the administrator, and symmetrically functions `getInportPerGid` / `getOutportPerGid` are used by user before each access. In case of an error, our driver causes kernel panic and no stack information becomes available.

In addition to user access control, we use an elaborate hashing mechanism (`hashPatientNo`) to identify offsets with which different healthcare patient data (e.g., `patient no, key, patient name`) is stored in BRAM (typically by root) for providing anonymity. This data can be subsequently read by users (physicians) via `readPatientData` function. Notice that implicit function `writePatientData` can only be called from root, otherwise it will fail. In fact, in our healthcare scenario, we allow only privileged users to perform write access via firewall. Such failures (and reasons behind them) can be easily detected by calling `testStats` that provides access to statistics logs.

11.3.1.4 Security-Performance Tradeoffs

In all figures, execution time for each process is calibrated to reflect processing of 256 ECG samples/sec for evaluating real-time.

In Figure 11.13 and Figure 11.14, we show performance of 'Patient App' and 'Doctor App'. For the 'Patient App', the Bluetooth thread (acting as BGW driver) takes ~63% of total execution time,

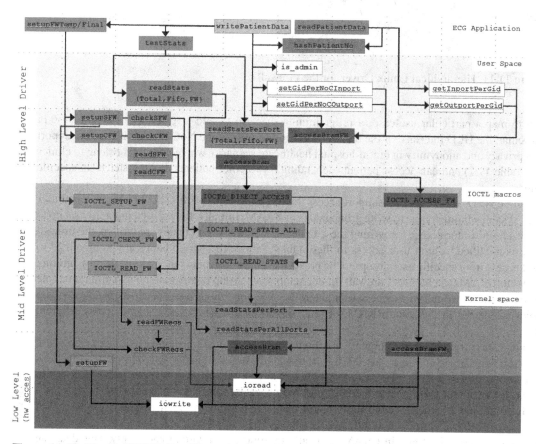

Figure 11.12: The NoCFW Driver Hierarchy

Table 11.7: Groups and Users in the Healthcare Scenario

Group	Users (Doctors)
root	system administrator
fwgroup	clinic
fwgroup1	clinic1
fwgroup2	clinic2

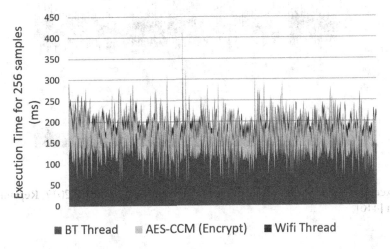

Figure 11.13: Execution Time Variations in a Trace of 'Patient App'. © IEEE 2017, Reprinted with Permission from [416]

while AES-CCM encryption for confidentiality and integrity protection takes ∼30%. However, for Doctor App, AES-CCM decryption takes 68%, while ECG Animation (easytest and preparation / locking mechanisms for next iteration) takes ∼21% of total time.

For the execution of the server, we have set affinity so that server_rx and server_tx processes share CPU0 of the ARM Cortex-A9 core, while ECG processing phases run on CPU1. As shown in Figure 11.15, more than 76% of the total execution time on Zedboard is spent for ECG processing; 20% in wrsamp function and 56% in easytest mainly for data filtering, but also for acquiring file locks (below 10ms with rare spikes), data conversions to EC-13 standard, merging of annotations with ECG data and preparation for data transmission. Finally, while AES takes ∼20%, server_rx and server_tx together take only 3% of the total execution time. Unlike AES, the cost for supporting anonymity via SHA-3 hash (keccak algorithm) is small (17 ms) compared to total time, while also finding if a doctor process group ID is legitimate is very small (<< 1 ms).

The healthcare application requires further optimization and porting to a parallel ARMv8 server to support efficiently more BGW devices at increased sampling frequency (256 samples/sec) without missing deadlines. With just one BGW device running at the maximum 256 samples/sec rate, our experiments reveal that we can support marginally real-time.

Finally, the administrator cost for involving the Linux kernel-space driver to setup anonymity service relates to:

- mapping group ID to input / output ports of the NoC Firewall; this takes on average 0.72ms and 1.45ms, respectively,

- setting up the firewall (initially for administrator setup, and later for user); this takes on average 4.42ms and 4.76ms.

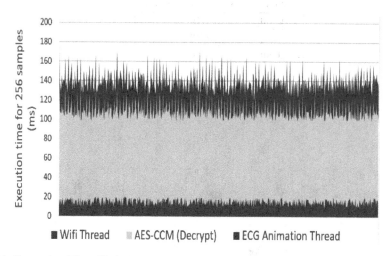

Figure 11.14: Execution Time Variations in a Trace of 'Doctor App'. © IEEE 2017, Reprinted with Permission from [416]

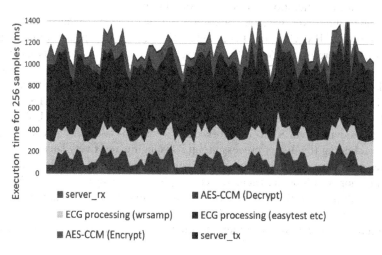

Figure 11.15: Execution Time Variations in a Trace of Server. © IEEE 2017, Reprinted with Permission from [416]

- storing the patient key in BRAM by invoking SHA-3 hash (keccak algorithm) takes on average 17.10ms.

Retrieving the anonymization key info in order to start streaming takes an average of 10.08ms. Unlike AES, anonymity costs are small compared to total cost of soft real-time healthcare application. The healthcare application requires optimization and porting to a parallel ARMv8 server to support efficiently more BGW devices without missing deadlines. With just one device, our experiments reveal that we can support marginally real-time.

11.3.2   In-Hospital Use-Case: Hospital Media Gateway

We consider memory and network bandwidth regulation policies ('MemGuardXt' and 'NetGuardXt') extending the current state-of-the-art (genuine MemGuard). Our algorithms support a) dynamic adaptivity by using exponentially weighted moving averages prediction and b) a guarantee violation free operating mode for rate-constrained traffic which is important for supporting mixed criticality on distributed embedded systems. The proposed algorithms implemented as Linux kernel modules (in x86 and ARM v7/v8) differentiate rate-constrained from best effort traffic and provide a mechanism for initializing (before the first period or later asynchronously) and dynamically adapting (at periodic intervals) guaranteed memory bandwidth per core or network bandwidth per connected (incoming or outgoing) network IP.

By examining a mixed-criticality scenario with video traffic and soft real-time electrocardiogram analysis on a hospital media gateway (Zedboard with two ARM v7 cores), we show that simultaneous use of 'MemGuardXt' / 'NetGuardXt' enables fine-grain bandwidth regulation for improved quality-of-service, when 'MemGuardXt' operates in violation free mode. To evaluate our kernel modules, we consider a realistic healthcare use case mixing non-critical processes related to patient entertainment with critical medical tasks associated with ECG analysis.

This solution combines different types of processes sharing the same system and network infrastructure.

11.3.2.1   Infotainment Functionality

Our Hospital Media Gateway (HMG) solution addresses end-user needs for infotainment by evolving the traditional hospital entertainment system (called linear TV) located in each room to an on-demand distributed system. Thus, in addition to critical medical services related to healthcare data acquisition, analysis, privacy protection and continuous physiological monitoring of the overall health and well-being, HMG also involves as a core function transmission of non-critical premium content for eventual consumption by patients located in different rooms. With this new feature, a number of smart devices can act as video clients (using a wired network) with regard to content services, eliminating the need for a dedicated high-end set-top-box per each end device.

11.3.2.2   Single and Multi-Room Scenario

In the single room scenario, medical patient data from up to 8 patients collocated in the same room who wear the BGW is transmitted via Bluetooth to local room gateways. Patients may move around hospital rooms as long as the Bluetooth signal-to-noise ratio is acceptable (usually 100m range). Figure 11.16 demonstrates the platform architecture that maps each local gateway to a 32-bit ARM v7 Odroid XU4 board.

A multiple room scenario is defined as follows. First, a wireless Bluetooth connection is used to transmit sensor data from multiple BGW devices in each room to a local room gateway. Then, this data is transmitted from each room gateway (32-bit Odroid XU4) to the HMG (32-bit Zedboard and eventually 64-bit ARM v8 Juno board) via the hospital's wired network. In order to guarantee soft real-time of the critical healthcare application, we must consider end-user delays, i.e., from

BGW device/driver to hospital media server for ECG analysis and visualization. Another important requirement is scalability of the healthcare infrastructure without affecting real-time. For example, the number of BGW devices (and resp. software processes) must scale with the number of monitored patients while the system is running.

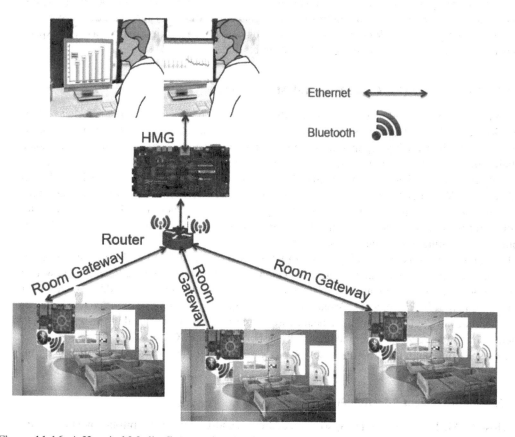

Figure 11.16: A Hospital Media Gateway in a Multi-Room Scenario

11.3.2.3 Soft Real Time ECG Analysis and Visualization

Our soft real-time ECG analysis is based on open source software, in particular an automated medical decision support system which diagnoses health issues, providing information to physicians. This system is able to detect and classify the heart beat signal to identify alarming situations, annotating such critical cardiac events graphically in soft real-time along with the ECG signal.

ECG analysis is initiated on the Zedboard (server) upon acquisition of the heart beat signal transmitted by the Odroid XU4 board. This signal is extracted by the BGW driver from raw data sent by the BGW sensor via Bluetooth and transmitted to the server at a maximum rate of 256 samples per second. As shown in Figure 11.17, our ECG analysis invokes our custom real-time extensions [415] of open source WFDB and OSEA software libraries whose main tasks are normalization of the input signal according to EC-13 standard, heart beat detection, and QRS classification. Finally, visualization uses WAVE, a fast XWindows application using the XView toolkit. WAVE supports fast, high resolution display of ECG signals at different scales, with asynchronous display of annotations via wave-remote function. It also handles remote access by Web browser, but this feature is not used in our tests.

Conversion to an EC-13 compliant format uses WFDB's `wrsamp` function which outputs two

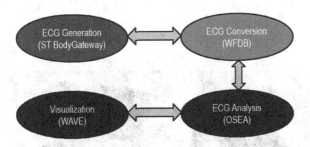

Figure 11.17: The ECG Analysis Process

files: a standardized ECG signal `synth.dat` and an ASCII "synth.hea" file which contains info about ECG data stored in the previous file e.g., the total samples. Then, our custom real-time version of `easytest` algorithm (part of OSEA) a) dynamically changes the sampling rate to 200 samples/sec and b) performs automated on-the-fly EC13-compliant heart beat detection and classification as Normal or Ventricular using different types of filters (e.g., noise reduction, QRS, SQRS) and related computations; these computations especially focus on variability of the R-R interval in the QRS complex compared to a 3 min training signal (chosen depending on the patient's age and sex). Annotations saved in "synth.atest" file, cf. [421, 422], indicate a very high positive predictivity (tested with MIT/BIH and AHA arrhythmia databases).

11.3.2.4 Experimental Framework and Results

We now study the effect of simultaneously performing memory / network bandwidth regulation (using 'MemGuardXt' and 'NetGuardXt') on a hospital media gateway (Zedboard with two ARMv7 Cortex-A9 cores). We apply NetguardXt to regulate two types of incoming network traffic on the server:

- Video-on-demand traffic arriving to Zedboard from an external server via an Ethernet router. Notice that data are saved locally, and eventually distributed to clients via streaming; this case is also similar to video transcoding. Since video streaming creates insignificant memory traffic (1-2 MB/sec), we focus on incoming traffic. However, the same prototype has been used to control quality-of-delivery of outbound video traffic using 'NetGuardXt'; see Figure 11.18.

- ECG network traffic arriving to Zedboard via an Ethernet router from two BGW sensors connected to an Odroid XU4.

ECG processing at the Zedboard involves running three processes on CPU0: a) a server process that opens independent TCP connections to receive ECG data from BGW devices connected to Odroid XU4, b) an initial consumer process that starts WAVE application for each connected client, and c) an animator process that uses our real-time extensions to ECG analysis (repositioning the initial training signal) to compute and asynchronously transmit the new annotated signal to WAVE (via wave-remote function). In addition, video-on-demand service runs on CPU1. Both CPUs are considered rate-constrained, and the following setup is used.

- 'MemGuardXt' configuration uses a fixed $period = 1\ msec$, $i = 2$, $\lambda = 0.2$, $r\_min = Q_0 + Q_1 = 90MB/sec$, $Q_{min} = 50KB/sec$; the relative rate Q_0/Q_1 (for cores 0 and 1) is set by one of three scripts; since Zedboard does not provide a counter for last level cache (L2) misses, we disabled L2 cache.

- Similarly 'NetGuardXt' configuration uses a fixed $period = 1\ sec$, $i = 2$, $\lambda = 0.2$, $r\_min = Q_0 + Q_1 = 70KB/sec$, $Q_{min} = 1000KB/sec$, while the relative Q_0/Q_1 ratio (for cores 0 and 1) is controlled dynamically by the same three scripts described below.

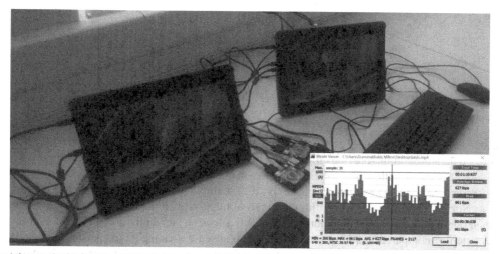

```
root@linaro-ubuntu-desktop:~/netguard_driver# echo "1500 50 300 1200" > /sys/kernel/debug/netguard/netguard_config
root@linaro-ubuntu-desktop:~/netguard_driver# echo "1500 50 1200 300" > /sys/kernel/debug/netguard/netguard_config
root@linaro-ubuntu-desktop:~/netguard_driver# ■
```

Figure 11.18: Regulating Video Streaming Quality-of-Delivery Using 'NetGuardXt'

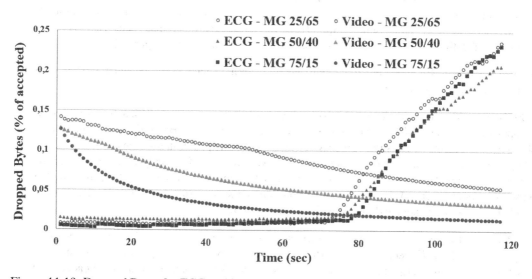

Figure 11.19: Dropped Bytes for ECG and Video as Percent of Accepted Ones

- The three scripts (MG 25/65, MG 50/40, and MG 75/15) form the driving force behind our experiment. They first fix 'MemGuardXt' Q_0/Q_1 ratio as 25/65 (first), 50/40 (second), or 75/15 (third script), and then periodically, every 20 sec, reconfigure the 'NetGuardXt' Q_0/Q_1 ratio (for cores 0 and 1) always with the same sequence: {18/72, 16/74, 14/76, 12/78, 10/80, 8/82}. Thus, each script runs for 2 minutes, gradually decreasing assigned network budget for ECG (and increasing that of Video), while keeping a fixed memory bandwidth ratio for both applications. VF mode is used by default for both guards; only in Figure 11.22, 'MemGuardXt' is used in Non-VF mode.

Selection of Q_0/Q_1 rates for 'MemGuardXt' / 'NetGuardXt' was based on initial experiments that evaluated performance of ECG processing and video-on-demand in isolation, in order to locate regions in the system parameter space where mixed criticality effects are interesting. For example, 'MemGuardXt' condition $Q_0 + Q_1 = 90MB/sec$ was based on experimenting with memory bandwidth requirements when both ECG and video application run simultaneously without restrictions.

Dropped Bytes for ECG and Video (as Percent of Accepted Ones) for the corresponding configuration script, i.e., ECG rate decreases every 20 sec.

Figure 11.19 shows the dropped bytes for ECG and video (as percent of accepted ones) for the corresponding configuration script, i.e., ECG rate decreases every 20 sec. This data extracted from kernel logs, shows that gradually decreasing ECG network bandwidth via 'NetGuardXt' from 18kB/sec to 8KB/sec (in 20 sec intervals), results in an increasing cumulative ECG drop rate and decreases the drop rate of video traffic.

In Figure 11.20 we show corresponding 'MemGuardXt' performance for MG 75/15 case. Notice that in this case all MemGuard figures continue to scale well despite the decreased network bandwidth, i.e., TCP retransmissions due to drops at the incoming network interface (see Figure 11.19) appear to be manageable in real-time. Configuring TCP retransmission timeout options is interesting for ECG [110]. Figure 11.21 shows an execution trace at the server for MG 75/15. While ECG server, consumer and animation processes (involving `wrsamp` and `easytest`) share CPU0, the video-on-demand service transferring files to Zedboard for video streaming runs on CPU1. Although ECG network rate is reduced, results are similar. In addition, notice that up to 50% of the total execution time is spent by `easytest` to perform ECG filtering and asynchronous annotation, while server takes ~30% to save the transmitted ECG signal, and `wrsamp` takes ~20% to perform required signal conversions to EC-13 standard. Small variations (< 20ms with rare spikes) are due to file locks at server and animation process. Figure 11.22 shows the same graph for Non-VF 'MemGuardXt' mode (similar to genuine MemGuard). For Non-VF mode, the server cannot meet soft real-time requirements due to guarantee violations; we record ~75K violations for $Q_{min} = 1MB/sec$.

Finally, Figure 11.23 compares the amount of ECG data delivery from each of the two BGW devices to the animator (WAVE application). Notice that while for MG 25/65 configuration, one of the BGW devices has completely stopped due to memory bandwidth starvation; in similar experiments, we have seen that instead both BGWs may lag. However, for MG 75/15 configuration, the server is able to process traffic from both BGW devices in soft real time.

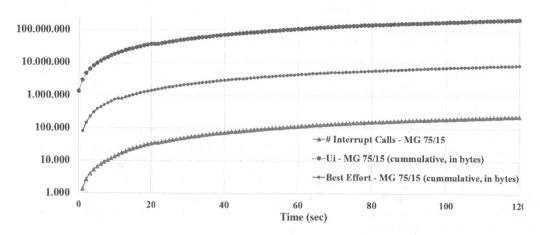

Figure 11.20: Performance of 'MemGuardXt' for MG 75/15 Configuration; the Vertical Axis Units are in Bytes for Total Used Bandwidth (U_i) and Best Effort

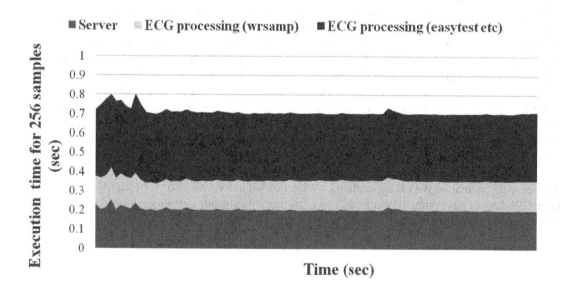

Figure 11.21: Delays at Home Media Gateway for MG 75/15 Script, VF Mode

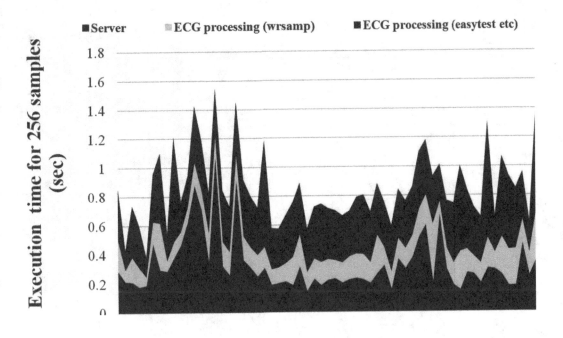

Figure 11.22: Delays at Home Media Gateway for MG 75/15 Script, Non-VF Mode

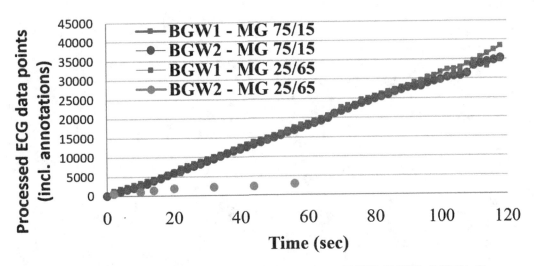

Figure 11.23: Real-Time Performance of ECG Analysis for two BGWs (BGW1, BGW2). Two 'MemGuardXT' Configurations are Compared

References

[1] K. R. Wheeler, D. A. Timuçin, I. X. Twombly, K. F. Goebel, and P. F. Wysocki. Aging aircraft wiring fault detection survey. NASA Ames Research Center, 2007.

[2] Çağatay Tokgöz and Sameh Dardona. Interrogation of electrical connector faults using miniaturized uwb sources. *Radio Science*, 52(1):94–104, 2017.

[3] RTCA DO-178C/EUROCAE ED-12C, Software Considerations in Airborne Systems and Equipment Certification. Standard, Radio Technical Commission for Aeronautics Incorporated (RTCA, Inc.)/European Organisation for Civil Aviation Equipment (EUROCAE), 2011.

[4] ISO. *ISO 26262-1 Road vehicles – Functional safety*. ISO, 2011.

[5] S. Bensalem, K. Goossens, C. M. Kirsch, R. Obermaisser, E. A. Lee, and J. Sifakis. Time-predictable and composable architectures for dependable embedded systems. In *2011 Proceedings of the Ninth ACM International Conference on Embedded Software (EMSOFT)*, pages 351–352, October 2011.

[6] Fred Pollack. Pollack's rule of thumb for microprocessor performance and area. `http://en.wikipedia.org/wiki/Pollacks_Rule`.

[7] R. Obermaisser, Z. Owda, M. Abuteir, H. Ahmadian, and D. Weber. End-to-End Real-Time Communication in Mixed-Criticality Systems Based on Networked Multicore Chips. In *Proceedings of 17th Euromicro Conference on Digital System Design (DSD)*, pages 293–302, 2014.

[8] DO-178B: Software Considerations in Airborne Systems and Equipment Certification. Standard, Radio Technical Commission for Aeronautics Incorporated (RTCA, Inc.)/European Organisation for Civil Aviation Equipment (EUROCAE), December 1992.

[9] International Electrotechnical Commission. IEC 61508-1 Functional Safety of Electrical-/Electronic/Programmable Electronic Safety-Related Systems - Part 1: General Requirements. Standard, IEC, IEC Central Office, 3, rue de Varembé, CH-1211 Geneva 20, Switzerland, April 2010.

[10] ARINC Industry Activities. Aircraft Data Network Part 7 Avionics Full Duplex Switched Ethernet - AFDX Network. Standard A664P7, ARINC, September 2009.

[11] AUTOSAR GbR. *AUTOSAR – Requirements on Gateway, R4.1*, 2013.

[12] Hamidreza Ahmadian, Roman Obermaisser, and Mohammed Abuteir. Time-triggered and rate-constrained on-chip communication in mixed-criticality systems. In *Proceedings of 10th International IEEE Symposium on Embedded Multicore/Many-core Systems-on-Chip (MC-SOC16)*, pages 117–124, 2016.

[13] IEEE standard for local and metropolitan area networks: Media Access Control (MAC) bridges. *IEEE Std 802.1D-2004 (Revision of IEEE Std 802.1D-1998)*, pages 1 –277, September 2004.

[14] M. Abuteir and R. Obermaisser. Scheduling of rate-constrained and time-triggered traffic in multi-cluster ttethernet systems. In *Proceedings of 13th IEEE International Conference on Industrial Informatics (INDIN)*, July 2015.

[15] M. Abuteir, R. Obermaisser, Z. Owda, and T. Moudouthe. Off-chip/on-chip gateway architecture for mixed-criticality systems based on networked multi-core chips. In *Proceedings of IEEE 18th International Conference on Computational Science and Engineering*, October 2015.

[16] Mohammed Abuteir. Architecture design for distributed mixed-criticality systems based on multi-core chips, 2017.

[17] ISO/IEC. ISO/IEC 7498-2:1989. Information processing systems – Open Systems Interconnection – Basic Reference Model – Part 2: Security Architecture. Standard, International Organization for Standardization, Geneva, CH, January 1989.

[18] Roman Obermaisser. Time-triggered communication. In *Industrial Information Technology*, pages 14–1 to 14–16. CRC Press, June 2009.

[19] IEEE. IEEE 1588-2008 Standard for a Precision Clock Synchronization Protocol for Networked Measurement and Control Systems. Technical Report 1588-2008, International Organization for Standardization, 2008.

[20] TTTech. TTEthernet (SAE AS 6802), November 2011.

[21] DREAMS Consortium. DREAMS deliverable D1.2.1: Architectural style of DREAMS. resreport D 1.2.1, DREAMS Consortium, July 2014.

[22] E. Carrascosa, J. Coronel, M. Masmano, P. Balbastre, and A. Crespo. XtratuM hypervisor redesign for LEON4 multicore processor. *SIGBED Rev.*, 11(2):27–31, September 2014.

[23] A. Crespo, M. Masmano, J. Coronel, S. Peiro, P. Balbestre, and J. Simó. Multicore partitioned systems based on hypervisor. In *IFAC World Congress 2014*, Cape Town, South Africa, 2014.

[24] Robert S. Boyer and J. Strother Moore. MJRTY – a fast majority vote algorithm. In Robert S. Boyer and William Pase, editors, *Automated Reasoning*, volume 1 of *Automated Reasoning Series*, pages 105–117. Springer, Netherlands, 1991.

[25] J. Zhang, K. Chen, B. Zuo, R. Ma, Y. Dong, and H. Guan. Performance analysis towards a kvm-based embedded real-time virtualization architecture. *In 5th (ICCIT 10) International Conference on Computer Sciences and Convergence Information Technology*, pages 421–426, 2010.

[26] Aeronautical Radio, Inc. ARINC specification 653P3 — Avionics application software standard interface. Part 3 - Conformity Test Specification. Technical Report A653P3, October 2006.

[27] Aeronautical Radio, Inc. ARINC specification 653P1 — Avionics application software standard interface. Part 1 - Required services. Technical Report A653P1-3, November 2010.

[28] Aeronautical Radio, Inc. ARINC specification 653P2 — Avionics application software standard interface. Part 2 - Extended services. Technical Report A653P2-2, June 2012.

[29] Laura L. Pullum. *Software Fault-Tolerance - Techniques and Implementation*. Artech House Boston, London, September 2001.

[30] Behrooz Parhami. Voting algorithms. *IEEE Transactions on Reliability*, 43(4):617–629, December 1994.

[31] Algirdas Avižienis. The N-version approach to fault-tolerant software. *IEEE Transactions on Software Engineering*, SE-11(12):1491–1501, 1985.

[32] David F. McAllister, Chien-En Sun, and Mladen A. Vouk. Reliability of voting in fault-tolerant software systems for small output-spaces. *IEEE Transactions on Reliability*, 39(5):524–534, December 1990.

[33] Paul R. Lorczak, Alper K. Caglayan, and Dave E. Eckhardt. A theoretical investigation of generalized voters for redundant systems. In *Digest of Papers of the Nineteenth International Symposium on Fault-Tolerant Computing (FTCS)*, pages 444–451, June 1989.

[34] R. B. Broen. New voters for redundant systems. *Journal of Dynamic Systems, Measurement, and Control*, 97(1):41–45, March 1975.

[35] R.E. Lyons and W. Vanderkulk. The use of triple-modular redundancy to improve computer reliability. *IBM Journal of Research and Development*, 6(2):200–209, April 1962.

[36] Robert Hammett. Design by extrapolation: an evaluation of fault-tolerant avionics. In *Proceedings of the 20th Digital Avionics Systems Conference (DASC)*, volume 1, pages 1C5/1–1C5/12 vol.1, October 2001.

[37] Ricky W. Butler. A primer on architectural level fault tolerance. Technical Report NASA/TM-2008-215108, NASA Langley Research Center, Hampton, VA, USA, February 1 2008.

[38] Simon Barner, Alexander Diewald, Jörn Migge, Ali Syed, Gerhard Fohler, Madeleine Faugère, and Daniel Gracia Pérez. DREAMS toolchain: Model-driven engineering of mixed-criticality systems. In *Proceedings of the ACM/IEEE 20th International Conference on Model Driven Engineering Languages and Systems (MODELS '17)*, Austin, TX, USA, September 2017.

[39] Stuart Kent. Model driven engineering. In Michael Butler, Luigia Petre, and Kaisa Sere, editors, *Integrated Formal Methods*, volume 2335 of *LNCS*, pages 286–298. Springer, Berlin Heidelberg, 2002.

[40] Jean Bézivin. In search of a basic principle for model driven engineering. *Upgrade*, 5(2):21–24, April 2004.

[41] Hermann Kopetz. *Real-Time Systems: Design Principles for Distributed Embedded Applications*. Real-Time Systems Series. Springer, New York, Dordrecht, Heidelberg, London, 2nd edition, 2011.

[42] Ed Seidewitz. What models mean. *IEEE Software*, 20(5):26–32, September 2003.

[43] Heiko Kern, Axel Hummel, and Stefan Kühne. Towards a comparative analysis of meta-metamodels. In *Proceedings of the Compilation of the Co-Located Workshops on DSM'11, TMC'11, AGERE!'11, AOOPES'11, NEAT'11, & VMIL'11 (SPLASH '11 Workshops)*, pages 7–12, New York, NY, USA, 2011. ACM.

[44] Object Management Group (OMG). Meta Object Facility (MOF) Core Specification, Version 2.5.1. OMG Document Number formal/16-11-01 (http://www.omg.org/spec/MOF/2.5.1/PDF), 2016.

[45] Object Management Group (OMG). Unified Modeling Language™ (UML®), V2.4.1 - Superstructure specification. OMG Document Number formal/2011-08-06 (http://www.omg.org/spec/UML/2.4.1/Superstructure/PDF), 2011.

[46] Object Management Group (OMG). Unified Modeling Language™ (UML®), V2.4.1 - Infrastructure specification. OMG Document Number formal/2011-08-05 (http://www.omg.org/spec/UML/2.4.1/Infrastructure/PDF), 2011.

[47] Dave Steinberg, Frank Budinsky, Marcelo Paternostro, and Ed Merks. *EMF: Eclipse Modeling Framework*. Addison-Wesley Longman, Amsterdam, 2nd (revised) edition, December 16 2008.

[48] Thomas Stahl, Markus Völter, and Krzysztof Czarnecki. *Model-Driven Software Development: Technology, Engineering, Management*. John Wiley & Sons, 2006.

[49] ISO. *ISO 13849-1: Safety of machinery – Safety-related parts of control systems – Part 1: General principles for design*, November 2006.

[50] Jon Perez, David Gonzalez, Salvador Trujillo, and Anton Trapman. *A safety concept for an IEC 61508 compliant fail-safe wind power mixed-criticality embedded system based on multi-core partitioning*, volume 9111 of *Lecture Notes in Computer Science*. Springer International Publishing, 2015.

[51] IEEE standard for local and metropolitan area networks–media access control (MAC) bridges and virtual bridged local area networks–amendment 17: Priority-based flow control. *IEEE Std 802.1Qbb-2011 (Amendment to IEEE Std 802.1Q-2011 as amended by IEEE Std 802.1Qbe-2011 and IEEE Std 802.1Qbc-2011)*, pages 1–40, September 2011.

[52] R. Obermaisser and P. Peti. A fault hypothesis for integrated architectures. In *2006 International Workshop on Intelligent Solutions in Embedded Systems*, pages 1–18, June 2006.

[53] H. Kopetz. Fault containment and error detection in the time-triggered architecture. In *Proceedings of The Sixth International Symposium on Autonomous Decentralized Systems (ISADS'03)*, ISADS '03, pages 139–, Washington, DC, USA, 2003. IEEE Computer Society.

[54] A. B. Campbell, O. Musseau, V. Ferlet-Cavrois, W. J. Stapor, and P. T. McDonald. Analysis of single event effects at grazing angle. In *RADECS 97. Fourth European Conference on Radiation and its Effects on Components and Systems (Cat. No.97TH8294)*, pages 528–536, September 1997.

[55] B. Pauli, A. Meyna, and P. Heitmann. Reliability of electronic components and control units in motor vehicle applications. VDI Berichte 1415:1009–1024, January 1998.

[56] International Electrotechnical Commission. IEC 61508-1 functional safety of electrical/electronic/programmable electronic safety-related systems. Standard, IEC, April 2010.

[57] Johan Karlsson, Peter Folkesson, Jean Arlat, Yves Y. Crouzet, and Günther Leber. Integration and comparison of three physical fault injection techniques. In *Predictably Dependable Computing Systems, Chapter V: Fault Injection*, pages 309 – 329, December 1999.

[58] Roman Obermaisser and P. Peti. The fault assumptions in distributed integrated architectures. In *SAE Technical Papers*, September 2007.

[59] F. Swiderski and W. Snyder. *Threat Modeling*. Microsoft professional. Microsoft Press, 2004.

[60] Tal Mizrahi. Time synchronization security using IPsec and MACsec. In *2011 IEEE International Symposium on Precision Clock Synchronization for Measurement, Control and Communication*. Institute of Electrical and Electronics Engineers (IEEE), September 2011.

[61] RTCA DO-254/EUROCAE ED-80, Design Assurance Guidance for Airborne Electronic Hardware. Standard, Radio Technical Commission for Aeronautics Incorporated (RTCA, Inc.)/European Organisation for Civil Aviation Equipment (EUROCAE), 2000.

[62] Integrated Modular Avionics (IMA) Development Guidance and Certification Considerations. Standard RTCA DO-297/EUROCAE ED-124, Radio Technical Commission for Aeronautics Incorporated (RTCA, Inc.)/European Organisation for Civil Aviation Equipment (EUROCAE), 2005.

[63] Single European Sky ATM Research (SESAR). https://www.sesarju.eu/. EU Project, 2004-2020.

[64] Single European Sky ATM Research (SESAR). https://www.cleansky.eu/. EU Project, 2008-2024.

[65] UBER Elevate — Fast-Forwarding to a Future of On-Demand Urban Air Transportation. https://www.uber.com/elevate.pdf, October 2016. Whitepaper presenting the Elevate project from UBER.

[66] Dubai completes first flying taxi test flight. https://nypost.com/2017/09/26/dubai-completes-first-flying-taxi-test-flight/. 'New York Post' journal article, Sept 26, 2017.

[67] Dubai starts testing crewless two-person 'flying taxis'. https://www.theverge.com/2017/9/26/16365614/dubai-testing-uncrewed-two-person-flying-taxis-volocopter. 'The Verge' news website article, Sept 26, 2017.

[68] Vahana — The next technological breakthrough in urban air mobility. https://www.airbus-sv.com/projects/1. Airbus Vahana project on urban air mobility (last visited Dec 3, 2017).

[69] Jan Nowotsch and Michael Paulitsch. Leveraging Multi-core Computing Architectures in Avionics. In *Dependable Computing Conference (EDCC), 2012 Ninth European*, pages 132–143. IEEE, 2012.

[70] Jingyi Bin, Sylvain Girbal, Daniel Gracia Pérez, Arnaud Grasset, and Alain Merigot. Studying co-running avionic real-time applications on multi-core COTS architectures. In *Embedded Real Time Software and Systems conference*, 2014.

[71] T. Ungerer, C. Bradatsch, M. Gerdes, F. Kluge, R. Jahr, J. Mische, J. Fernandes, P. G. Zaykov, Z. Petrov, B. Böddeker, S. Kehr, H. Regler, A. Hugl, C. Rochange, H. Ozaktas, H. Cassé, A. Bonenfant, P. Sainrat, I. Broster, N. Lay, D. George, E. Quiñones, M. Panic, J. Abella, F. Cazorla, S. Uhrig, M. Rohde, and A. Pyka. parMERASA – Multi-core Execution of Parallelised Hard Real-Time Applications Supporting Analysability. In *2013 Euromicro Conference on Digital System Design*, pages 363–370, September 2013.

[72] Deep sub-micron microprocessor for spAce rad-Hard appLIcation Asic (DAHLIA). http://dahlia-h2020.eu/. EU Project, 2017-2019.

[73] DEep SubMicron System-on-Chip (SoC) for Harsh Environment applicaTions using EuRopean Technologies (DEMETER). https://demeter-eniac.eu/. EU Project, 2014-2018.

[74] Leonidas Kosmidis, Eduardo Quiñones, Jaume Abella, Tullio Vardanega, Carles Hernandez, Andrea Gianarro, Ian Broster, and Francisco J Cazorla. Fitting processor architectures for measurement-based probabilistic timing analysis. *Microprocessors and Microsystems*, 47:287–302, 2016.

[75] Lui Sha, Marco Caccamo, Renato Mancuso, Jung -eun Kim, Man -ki Yoon, Rodolfo Pellizzoni, Heechul Yun, Russell Kegley, Dennis R. Perlman, Greg Arundale, and Richard M. Bradford. Single Core Equivalent Virtual Machines for Hard Real-time Computing on Multicore Processors. Technical Report at UIUC, 2014.

[76] Xavier Jean. *Hypervisor Control of COTS Multi-Cores Processors in Order to Enforce Determinism for Future Avionics Equipment*. PhD thesis, Telecom ParisTech, 2015.

[77] Sylvain Girbal, Daniel Gracia Pérez, Jimmy Le Rhun, Madeleine Faugère, Claire Pagetti, and Guy Durrieu. A Complete Toolchain for an Interference Free Deployment of Avionic Applications on Multi-Core Systems. In *Digital Avionics Systems Conference*. IEEE, 2015.

[78] Cláudio Maia, Luis Nogueira, Luis Miguel Pinho, and Daniel Gracia Pérez. A Closer Look into the AER Model. In *21st IEEE International Conference on Emerging Technologies & Factory Automation*. IEEE, 2016.

[79] Cláudio Maia, Geoffrey Nelissen, Luis Nogueira, Luis Miguel Pinho, and Daniel Gracia Pérez. Schedulability Analysis for Global Fixed-Priority Scheduling of the 3-Phase Task Model. In *23rd IEEE International Conference on Embedded and Real-Time Computing Systems and Applications (RTCSA 2017)*. IEEE Computer Society, 2017.

[80] R. Tabish, R. Mancuso, S. Wasly, S. S. Phatak, R. Pellizzoni, and M. Caccamo. A Reliable and Predictable Scratchpad-centric OS for Multi-core Embedded Systems. In *2017 IEEE Real-Time and Embedded Technology and Applications Symposium (RTAS)*, pages 377–388, April 2017.

[81] Francisco J. Cazorla, Jaume Abella, Jan Andersson, Tullio Vardanega, Francis Vatrinet, Iain Bate, Ian Broster, Mikel Azkarate-Askasua, Franck Wartel, Liliana Cucu, et al. PROXIMA: Improving measurement-based timing analysis through randomisation and probabilistic analysis. In *Digital System Design (DSD), 2016 Euromicro Conference on*, pages 276–285. IEEE, 2016.

[82] Ankit Agrawal, Gerhard Fohler, Johannes Freitag, Jan Nowotsch, Sascha Uhrig, and Michael Paulitsch. Contention-Aware Dynamic Memory Bandwidth Isolation with Predictability in COTS Multicores: An Avionics Case Study. In *29th Euromicro Conference on Real-Time Systems (ECRTS)*, June 2017.

[83] Sylvain Girbal, Xavier Jean, Jimmy Le Rhun, Daniel Gracia Pérez, and Marc Gatti. Deterministic Platform Software for Hard Real-Time Systems Using Multi-Core COTS. In *IEEE/AIAA 32nd Digital Avionics Systems Conference (DASC)*. IEEE, October 2013.

[84] Stuart Fisher. Whitepaper: Certifying Applications in a Multi-Core Environment. 2014.

[85] H. Yun, G. Yao, R. Pellizzoni, M. Caccamo, and L. Sha. Memguard: Memory bandwidth reservation system for efficient performance isolation in multi-core platforms. In *Real-Time and Embedded Technology and Applications Symposium (RTAS), 2013 IEEE 19th*, pages 55–64, April 2013.

[86] Jingyi Bin, Sylvain Girbal, Daniel Gracia Pérez, and Alain Merigot. Using monitors to predict co-running safety-critical hard real-time benchmark behavior. In *3rd International Conference on Intelligent Technologies and Engineering Systems (ICITES 2014)*, 2014.

[87] Angeliki Kritikakou, Claire Pagetti, Christine Rochange, Matthieu Roy, Madeleine Faugère, Sylvain Girbal, and Daniel Gracia Pérez. Distributed run-time wcet controller for concurrent critical tasks in mixed-critical systems. In *Proceedings of the 22th International Conference on Real-Time and Network Systems (RTNS'14)*, pages 139–148, 2014.

[88] SAFURE - Safety And Security By Design For Interconnected Mixed-Critical Cyber-Physical Systems. `https://safure.eu/`. EU Project, 2015-2018.

[89] J. Windsor, K. Eckstein, P. Mendham, and T. Pareaud. Time and space partitioning security components for spacecraft flight software. In *2011 IEEE/AIAA 30th Digital Avionics Systems Conference*, pages 8A5–1–8A5–14, October 2011.

[90] EURO-MILS: Secure European virtualisation for trustworthy applications in critical domains. `http://www.euromils.eu/index.html`. EU Project, 2012-2016.

[91] SYSGO. Multi-Levels Safe & Secure Solution for Industrial Automation. 2016.

[92] Wind River. Open, Secure Industrial Automation Systems. 2017.

[93] Jon Perez, David Gonzalez, Carlos Fernando Nicolas, Ton Trapman, and Jose Miguel Garate. A safety concept for a wind power mixed-criticality embedded system based on multicore partitioning. In *11th International Symposium - Functional Safety in Industrial Applications (TÜV Rheinland)*, Cologne, Germany, 2014.

[94] M. Hadjem, O. Salem, and F. Nait-Abdesselam. An ECG monitoring system for prediction of cardiac anomalies using wban. In *e-Health Netw. Appl. and Services Conf.*, pages 431–436, 2014.

[95] J. Ko, J. H. Lim, Y. Chen, R. Musvaloiu-E, et al. Medisn: Medical emergency detection in sensor networks. *ACM Trans. on Embedded Comput. Syst.*, 10(1):1–29, 2010.

[96] J. A. Walsh, E. J. Topol, and S. R. Steinhubl. Novel wireless devices for cardiac monitoring. *New Drugs and Technologies*, pages 573–581, 2014.

[97] American Health Organization. `https://www.heart.org/idc/groups/ahamah-public/@wcm/@sop/@smd/documents/downloadable/ucm{\_}491265.pdf`.

[98] Alivecor. `https://www.alivecor.com/`.

[99] L. A. Saxon. Ubiquitous wireless ECG recording: a powerful tool physicians should embrace. *J. Cardiovascular Electrophysiology*, 24(4):480–483, 2013.

[100] BodyGuardian Heart. `http://www.preventicesolutions.com/services/body-guardian-heart.html`.

[101] Life Monitor. `http://www.equivital.co.uk/products/tnr/sense-and-transmit`.

[102] NowCardio. `https://contex-tech.com/medical/nowcardio`.

[103] Physiomem. `http://www.getemed.net/en/telemonitoring/physiomemr-pm-1000`.

[104] S. Gradl, P. Kugler, C. Lohmüller, and B. Eskofier. Real-time ECG monitoring and arrhythmia detection using android-based mobile devices. In *IEEE Engin. Conf.*, pages 2452–2455. Medicine and Biology Society, 2012.

[105] J. J. Oresko, Z. Jin, J. Cheng, S. Huang, et al. A wearable smartphone-based platform for real-time cardiovascular disease detection via electrocardiogram processing. *IEEE Trans. Info. Tech. Biomedicine*, 14(3):734–740, 2010.

[106] T.-H. Yen, C.-Y. Chang, and S.-N. Yu. A portable real-time ECG recognition system based on smartphone. In *IEEE Engin. Medicine and Biology Society*, pages 7262–7265, 2013.

[107] S. Hu, H. Wei, and Y. Chen. A real-time cardiac arrhythmia classification system with wearable sensor network. *Sensors*, 12:12844–12869, 2012.

[108] A. M. Patel, P. K. Gakare, and A. N. Cheeran. Real-time ECG feature extraction and arrhythmia detection on mobile platform. *J. Comp. Appl.*, 44(23):40–45, 2012.

[109] J. Weng, X. M. Guo, L. S. Chen, and Z. H. Yuan et al. Study on real-time monitoring technique for cardiac arrhythmia based on smartphone. *J. Medical and Biological Engineering*, 33(4), 2012.

[110] A. Iglesias, R. Istepanian, and J. G. Moros. Enhanced real-time ECG coder for packetized telecardiology applications. *IEEE Trans. Info Tech. Biomedicine*, 10(2):229–236, 2006.

[111] Vincent Aravantinos, Sebastian Voss, Sabine Teufl, Florian Hölzl, and Bernhard Schätz. AutoFOCUS 3: Tooling concepts for seamless, model-based development of embedded systems. In *Proc. 8th Int. Workshop Model-based Architecting Cyber-Physical and Embedded Systems (ACES-MB)*, pages 19–26, 2015.

[112] Jia Huang, Simon Barner, Andreas Raabe, Christian Buckl, and Alois Knoll. A framework for reliability-aware embedded system design on multiprocessor platforms. *Microprocess. Microsyst.*, 38(6):539–551, March 12 2014.

[113] Simon Barner, Andreas Raabe, Christian Buckl, and Alois Knoll. Beschreibung der Plattformabhängigkeit eingebetteter Applikationen mit Dienstmodellen. In *Tagungsband des Dagstuhl-Workshop MBEES: Modellbasierte Entwicklung eingebetteter Systeme VII*, 2011.

[114] TIMMO-2-USE Consortium. Language syntax, semantics, metamodel V2. D11, 2012.

[115] Simon Barner, Alexander Diewald, Fernando Eizaguirre, Anatoly Vasilevskiy, and Franck Chauvel. Building product-lines of mixed-criticality systems. In *Proceedings of the Forum on Specification and Design Languages (FDL 2016)*, Bremen, Germany, September 2016. IEEE.

[116] fortiss GmbH. AutoFOCUS 3. `http://af3.fortiss.org/`.

[117] OMG. Object constraint language (OCL). `www.omg.org/spec/OCL/2.4/`, 2014.

[118] TIMMO-2-USE. Timing Model - TOols, algorithms, languages, methodology, USE cases. `https://itea3.org/project/timmo-2-use.html`.

[119] TIMMO-2-USE Consortium. Methodology description V2. D13, 2012.

[120] Tim Kelly and Rob Weaver. The Goal Structuring Notation – A safety argument notation. In *Proceedings of Dependable Systems and Networks 2004 Workshop on Assurance Cases*, 2004.

[121] J. Fenn, R. Hawkins, P. Williams, and T. Kelly. Safety case composition using contracts - refinements based on feedback from an industrial case study. In F Redmill and T Anderson, editors, *The Safety of Systems*, number Print ISBN 978-1-84628-805-0, Online ISBN 978-1-84628-806-7. Springer, 2007.

[122] Andreas Svendsen, Xiaorui Zhang, Roy Lind-Tviberg, Franck Fleurey, Øystein Haugen, Birger Møller-Pedersen, and Gøran K. Olsen. Developing a software product line for train control: A case study of cvl. In *Proceedings of the 14th International Conference on Software Product Lines: Going Beyond*, SPLC'10, pages 106–120, Berlin, Heidelberg, 2010. Springer-Verlag.

[123] Don Batory. Feature models, grammars, and propositional formulas. In *Proceedings of the 9th International Conference on Software Product Lines*, SPLC'05, pages 7–20, Berlin, Heidelberg, 2005. Springer-Verlag.

[124] Martin Fagereng Johansen, Øystein Haugen, and Franck Fleurey. Properties of realistic feature models make combinatorial testing of product lines feasible. In *Proceedings of the 14th International Conference on Model Driven Engineering Languages and Systems*, MODELS'11, pages 638–652, Berlin, Heidelberg, 2011. Springer-Verlag.

[125] Martin Fagereng Johansen, Øystein Haugen, and Franck Fleurey. An algorithm for generating t-wise covering arrays from large feature models. In *Proceedings of the 16th International Software Product Line Conference - Volume 1*, SPLC '12, pages 46–55, New York, NY, USA, 2012. ACM.

[126] D. R. Kuhn, D. R. Wallace, and A. M. Gallo. Software fault interactions and implications for software testing. *IEEE Transactions on Software Engineering*, 30(6):418–421, June 2004.

[127] J. M. Jazequel and B. Meyer. Design by contract: the lessons of ariane. *Computer*, 30(1):129–130, January 1997.

[128] John H. Miller and Scott E. Page. *Complex Adaptive Systems: An Introduction to Computational Models of Social Life*. Princeton University Pres, 2007.

[129] B. J. Garvin and M. B. Cohen. Feature interaction faults revisited: An exploratory study. In *2011 IEEE 22nd International Symposium on Software Reliability Engineering*, pages 90–99, November 2011.

[130] Michaela Steffens, Sebastian Oster, Malte Lochau, and Thomas Fogdal. Industrial evaluation of pairwise spl testing with moso-polite. In *Proceedings of the Sixth International Workshop on Variability Modeling of Software-Intensive Systems*, VaMoS '12, pages 55–62, New York, NY, USA, 2012. ACM.

[131] Matej Črepinšek, Shih-Hsi Liu, and Marjan Mernik. Exploration and Exploitation in Evolutionary Algorithms: A Survey. *ACM Comput. Surv.*, 45(3):35:1–35:33, July 2013.

[132] International Electrotechnical Commission. IEC 61508-2 Functional Safety of Electrical/Electronic/Programmable Electronic Safety-Related Systems - Part 2: Requirements for Electrical/Electronic/Programmable Electronic safety-related systems. Standard, IEC, IEC Central Office, 3, rue de Varembé, CH-1211 Geneva 20, Switzerland, April 2010.

[133] International Electrotechnical Commission. IEC 61508-6 Functional Safety of Electrical/Electronic/Programmable Electronic Safety-Related Systems - Part 6: Guidelines on the application of IEC 61508-2 and IEC 61508-3. Standard, IEC, IEC Central Office, 3, rue de Varembé, CH-1211 Geneva 20, Switzerland, April 2010.

[134] Alexander Diewald, Sebastian Voss, and Simon Barner. A lightweight design space exploration and optimization language. In *Proceedings of the 19th International Workshop on Software and Compilers for Embedded Systems*, SCOPES '16, pages 190–193, New York, NY, USA, 2016. ACM.

[135] Jens Meinicke, Thomas Thüm, Reimar Schröter, Fabian Benduhn, and Gunter Saake. An overview on analysis tools for software product lines. In *Proceedings of the 18th International Software Product Line Conference: Companion Volume for Workshops, Demonstrations and Tools-Volume 2*, pages 94–101. ACM, 2014.

[136] Jihyun Lee, Sungwon Kang, and Danhyung Lee. A survey on software product line testing. In *Proceedings of the 16th International Software Product Line Conference-Volume 1*, pages 31–40. ACM, 2012.

[137] Martin Fagereng Johansen, Øystein Haugen, and Franck Fleurey. A survey of empirics of strategies for software product line testing. In *Proceedings of IEEE Fourth International Conference on Software Testing, Verification and Validation Workshops (ICSTW)*, pages 266–269. IEEE, 2011.

[138] Martin Fagereng Johansen. *Testing Product Lines of Industrial Size: Advancements in Combinatorial Interaction Testing*. PhD thesis, University of Oslo, 2013.

[139] Harald Cichos, Sebastian Oster, Malte Lochau, and Andy Schürr. Model-based coverage-driven test suite generation for software product lines. In *International Conference on Model Driven Engineering Languages and Systems*, pages 425–439. Springer, 2011.

[140] D. Richard Kuhn, Dolores R. Wallace, and Albert M. Gallo. Software fault interactions and implications for software testing. *IEEE Transactions on Software Engineering*, 30(6):418–421, 2004.

[141] Anatoly Vasilevskiy, Øystein Haugen, Franck Chauvel, Martin Fagereng Johansen, and Daisuke Shimbara. The BVR tool bundle to support product line engineering. In *Proceedings of the 19th International Conference on Software Product Line*, pages 380–384. ACM, 2015.

[142] Øystein Haugen and Ommund Øgård. BVR–better variability results. In *International Conference on System Analysis and Modeling*, pages 1–15. Springer, 2014.

[143] Øystein Haugen, Birger Møller-Pedersen, Jon Oldevik, Gøran K Olsen, and Andreas Svendsen. Adding standardized variability to domain specific languages. In *Software Product Line Conference, 2008. SPLC'08. 12th International*, pages 139–148. IEEE, 2008.

[144] Object Management Group (OMG). OMG Systems Modeling Language™, Version 1.5. OMG Document Number formal/2017-05-01 (http://www.omg.org/spec/SysML/1.5/PDF), 2017.

[145] Anatoly Vasilevskiy, Franck Chauvel, and Øystein Haugen. Toward robust product realisation in software product lines. In *Proceedings of the 20th International Systems and Software Product Line Conference*, pages 184–193. ACM, 2016.

[146] Yi Li, Aws Albarghouthi, Zachary Kincaid, Arie Gurfinkel, and Marsha Chechik. Symbolic optimization with smt solvers. In *ACM SIGPLAN Notices*, volume 49, pages 607–618. ACM, 2014.

[147] Aldeida Aleti, Barbora Buhnova, Lars Grunske, Anne Koziolek, and Indika Meedeniya. Software Architecture Optimization Methods: A Systematic Literature Review. *IEEE Transactions on Software Engineering*, 39(5):658–683, May 2013.

[148] S. Kugele, G. Pucea, R. Popa, L. Dieudonné, and H. Eckardt. On the deployment problem of embedded systems. In *2015 ACM/IEEE International Conference on Formal Methods and Models for Codesign (MEMOCODE)*, pages 158–167, Sept 2015.

[149] Eike Martin Thaden. *Semi-Automatic Optimization of Hardware Architectures in Embedded Systems.* Dissertation, Universität Oldenburg, 2013.

[150] Joachim Keinert, Martin Streubuhr, Thomas Schlichter, Joachim Falk, Jens Gladigau, Christian Haubelt, Jurgen Teich, and Michael Meredith. SystemCoDesigner - an automatic ESL synthesis approach by design space exploration and behavioral synthesis for streaming applications. *ACM Transactions on Design Automation of Electronic Systems*, 14(1):1–23, January 2009.

[151] M. Thompson, H. Nikolov, T. Stefanov, A. D. Pimentel, C. Erbas, S. Polstra, and E. F. Deprettere. A framework for rapid system-level exploration, synthesis, and programming of multimedia mp-socs. In *2007 5th IEEE/ACM/IFIP International Conference on Hardware/-Software Codesign and System Synthesis (CODES+ISSS)*, pages 9–14, Sept 2007.

[152] Martin Lukasiewycz, Michael Glaß, Felix Reimann, and Jürgen Teich. Opt4J: A modular framework for meta-heuristic optimization. In *Proceedings of the 13th Annual Conference on Genetic and Evolutionary Computation*, GECCO '11, pages 1723–1730, New York, NY, USA, 2011. ACM.

[153] Martin Lukasiewycz, Sebastian Steinhorst, Florian Sagstetter, Wanli Chang, Peter Waszecki, Matthias Kauer, and Samarjit Chakraborty. Cyber-Physical Systems Design for Electric Vehicles. In *2012 15th Euromicro Conference on Digital System Design*, pages 477–484. IEEE, September 2012.

[154] Y. Papadopoulos, M. Walker, M.-O. Reiser, M. Weber, D. Chen, M. Törngren, David Servat, A. Abele, F. Stappert, H. Lonn, L. Berntsson, Rolf Johansson, F. Tagliabo, S. Torchiaro, and Anders Sandberg. Automatic allocation of safety integrity levels. In *Proceedings of the 1st Workshop on Critical Automotive Applications: Robustness & Safety*, CARS '10, pages 7–10, New York, NY, USA, 2010. ACM.

[155] Thilo Streichert, Michael Glaß, Christian Haubelt, and Jürgen Teich. Design space exploration of reliable networked embedded systems. *Journal of Systems Architecture*, 53(10):751–763, October 2007.

[156] Á. Hegedüs, Á. Horváth, I. Ráth, and D. Varró. A model-driven framework for guided design space exploration. In *2011 26th IEEE/ACM International Conference on Automated Software Engineering (ASE 2011)*, pages 173–182, Nov 2011.

[157] Ken Vanherpen, Joachim Denil, Paul De Meulenaere, and Hans Vangheluwe. Design-space exploration in model driven engineering. *ACM/IEEE 17th International Conference on Model Driven Engineering Languages and Systems*, 2014.

[158] Jia Huang, Andreas Raabe, Kai Huang, Christian Buckl, and Alois Knoll. A framework for reliability-aware design exploration on MPSoC based systems. *Design Automation for Embedded Systems*, 16(4):189–220, 2012.

[159] Marc Boyer, Hugo Daigmorte, Nicolas Navet, and Jörn Migge. Performance impact of the interactions between time-triggered and rate-constrained transmissions in ttethernet. In *Proceedings of the 8th European Congress on Embedded Real Time Software and Systems*, ERTSS '16, 2016.

[160] Theodore P. Baker and Alan C. Shaw. The cyclic executive model and Ada. In *Proceedings of the 9th IEEE Real-Time Systems Symposium (RTSS '88), December 6-8, 1988, Huntsville, Alabama, USA*, pages 120–129, 1988.

[161] Juan Zamorano, Alejandro Alonso, and Juan Antonio de la Puente. Building safety-critical real-time systems with reusable cyclic executives. *Control Engineering Practice*, 5(7):999 – 1005, 1997.

[162] R. I. Davis and A. Burns. Hierarchical fixed priority pre-emptive scheduling. In *IEEE Real-Time Systems Symposium*, 2005.

[163] Irina Iulia Lupu, Pierre Courbin, Laurent George, and Joël Goossens. Multi-criteria evaluation of partitioning schemes for real-time systems. In *Proceedings of 15th IEEE International Conference on Emerging Technologies and Factory Automation, ETFA 2010, September 13-16, 2010, Bilbao, Spain*, pages 1–8, 2010.

[164] Theodore P. Baker. A stack-based resource allocation policy for realtime processes. In *Proceedings of the Real-Time Systems Symposium - 1990, Lake Buena Vista, Florida, USA, December 1990*, pages 191–200, 1990.

[165] Z. Shi and A. Burns. Real-time communication analysis for on-chip networks with worm-hole switching. In *Second ACM/IEEE International Symposium on Networks-on-Chip (nocs 2008)*, pages 161–170, April 2008.

[166] Wilfried Steiner. An evaluation of SMT-based schedule synthesis for time-triggered multi-hop networks. In *Proceedings of the 31st Real-Time Systems Symposium (RTSS'10)*, pages 375–384. IEEE, 2010.

[167] Silviu S. Craciunas and Ramon Serna Oliver. Combined task- and network-level scheduling for distributed time-triggered systems. *Real-Time Systems*, 52(2):161–200, 2016.

[168] Hermann Kopetz, A. Ademaj, P. Grillinger, and K. Steinhammer. The time-triggered Ethernet (TTE) design. In *Proc. ISORC*. IEEE, 2005.

[169] Wilfried Steiner and Bruno Dutertre. Automated formal verification of the TTEthernet synchronization quality. In *NASA Formal Methods*, volume 6617 of *Lecture Notes in Computer Science*. Springer, 2011.

[170] Hermann Kopetz. *Real-Time Systems: Design Principles for Distributed Embedded Applications*. Kluwer Academic Publishers, 1997.

[171] Licong Zhang, D. Goswami, R. Schneider, and S. Chakraborty. Task- and network-level schedule co-synthesis of Ethernet-based time-triggered systems. In *Proc. ASP-DAC*. IEEE Computer Society, 2014.

[172] J. Forget, E. Grolleau, C. Pagetti, and P. Richard. Dynamic priority scheduling of periodic tasks with extended precedences. In *Proc. ETFA*. IEEE Computer Society, 2011.

[173] R. Zurawski. *Industrial Communication Technology Handbook, Second Edition*. Industrial Information Technology. Taylor & Francis, 2014.

[174] Wilfried Steiner, Günther Bauer, Brendan Hall, and Michael Paulitsch. TTEthernet: Time-Triggered Ethernet. In Roman Obermaisser, editor, *Time-Triggered Communication*. CRC Press, August 2011.

[175] Clark Barrett, Roberto Sebastiani, Sanjit Seshia, and Cesare Tinelli. Satisfiability modulo theories. In *Handbook of Satisfiability*, volume 185. IOS Press, 2009.

[176] Roberto Sebastiani. Lazy satisfiability modulo theories. *JSAT*, 3(3-4):141–224, 2007.

[177] Leonardo Moura and Nikolaj Bjørner. Satisfiability modulo theories: An appetizer. In *Formal Methods: Foundations and Applications*, volume 5902, pages 23–36. Springer Berlin Heidelberg, 2009.

[178] Leonardo De Moura and Nikolaj Bjørner. Satisfiability modulo theories: Introduction and applications. *Commun. ACM*, 54(9):69–77, 2011.

[179] Inc. Gurobi Optimization. Gurobi optimizer reference manual, version 6.0, 2014.

[180] Nikolaj Bjørner and Anh-Dung Phan. νz - maximal satisfaction with Z3. In *Proc. SCSS*. EasyChair, 2014.

[181] Nikolaj Bjørner, Anh-Dung Phan, and Lars Fleckenstein. νz - an optimizing SMT solver. In *Proc. TACAS*. Springer, 2015.

[182] Roberto Sebastiani and Patrick Trentin. OptiMathSAT: A Tool for Optimization Modulo Theories. In *Proc. CAV*, volume 9206 of *LNCS*. Springer, 2015.

[183] S.P. Bradley, A.C. Hax, and T.L. Magnanti. *Applied Mathematical Programming*. Addison-Wesley, 1977.

[184] J. Bisschop. *Aimms Optimization Modeling*. Paragon Decision Technology, 2006.

[185] Computer Science Laboratory – SRI International. The Yices SMT Solver. `http://yices.csl.sri.com/`.

[186] Leonardo De Moura and Nikolaj Bjørner. Z3: An efficient SMT solver. In *Proc. TACAS*. Springer-Verlag, 2008.

[187] Gerhard Fohler. *Flexibility in Statically Scheduled Real-Time Systems*. PhD thesis, TNF, Wien, Österreich, April 1994.

[188] H. Kopetz, R. Nossal, R. Hexel, A. Krueger, D. Millinger, R. Pallierer, C. Temple, and M. Krug. Mode handling in the Time-Triggered Architecture. *Control Engineering Practice*, 1998.

[189] Jorge Real, Sergio Saez, and Alfons Crespo. Combining time-triggered plans with priority scheduled task sets. In *Reliable Software Technologies - Ada-Europe 2016 - 21st Ada-Europe International Conference on Reliable Software Technologies, Pisa, Italy, June 13-17, 2016, Proceedings*, pages 195–212, 2016.

[190] Patrick Graydon and Iain Bate. Safety assurance driven problem formulation for mixed-criticality scheduling. 2013.

[191] H. Kopetz, Gerhard Fohler, G. Grünsteidl, H. Kantz, G. Pospischil, P. Puschner, J. Reisinger, R. Schlatterbeck, W. Schütz, A. Vrchoticky, and R. Zainlinger. The programmers view of mars. In *PRTS92*, pages 223–226, Phoenix, Arizona, USA, December 1992.

[192] H. Kopetz. Sparse time versus dense time in distributed real-time systems. In *Proceedings of the 12th International Conference on Distributed Computing Systems*, pages 460–467, June 1992.

[193] Steve Vestal. Preemptive scheduling of multi-criticality systems with varying degrees of execution time assurance. In *IEEE Conference on Real-Time Systems Symposium (RTSS)*, 2007.

[194] Alexandre Esper, Geoffrey Nelissen, Vincent Nélis, and Eduardo Tovar. How realistic is the mixed-criticality real-time system model? In *International Conference on Real-Time and Networks Systems (RTNS), 2015*.

[195] Jens Theis. *Certification-Cognizant Mixed-Criticality Scheduling in Time-Triggered Systems*. PhD thesis, March 2015.

[196] Georgia Giannopoulou, Nikolay Stoimenov, Pengcheng Huang, and Lothar Thiele. Mapping Mixed-Criticality Applications on Multi-Core Architectures. In *Design, Automation & Test in Europe Conference & Exhibition (DATE), Hot-Topic Session on Predictable Multicore Computing*, pages 1–6, Dresden, Germany, March 2014. IEEE.

[197] Aeronautical Radio, Inc. Avionics application software standard interface. Technical Report ARINC653, November 2010.

[198] A. Crespo, I. Ripoll, M. Masmano, P. Arberet, and J.J. Metge. XtratuM an open source hypervisor for TSP embedded systems in aerospace. *Data Systems In Aerospace (DASIA), 2009*.

[199] Björn B. Brandenburg. *Scheduling and Locking in Multiprocessor Real-Time Operating Systems*. PhD thesis, The University of North Carolina at Chapel Hill, 2011.

[200] Adam Lackorzyński, Alexander Warg, Marcus Völp, and Hermann Härtig. Flattening hierarchical scheduling. In *ACM International Conference on Embedded Software (EMSOFT), 2012*.

[201] Ali Abbas Jaffari Syed, Gerhard Fohler, and Daniel Gracia Pérez. Online admission of non-preemptive aperiodic mixed-critical tasks in hierarchic schedules. In *The 23rd IEEE International Conference on Embedded and Real-Time Computing Systems and Applications (RTCSA)*, August 2017.

[202] J. P. Lehoczky and S. Ramos-Thuel. An optimal algorithm for scheduling soft-aperiodic tasks in fixed-priority preemptive systems. In *IEEE Conference on Real-Time Systems Symposium (RTSS), 1992*.

[203] S. R. Thuel and J. P. Lehoczky. Algorithms for scheduling hard aperiodic tasks in fixed-priority systems using slack stealing. In *IEEE Conference on Real-Time Systems Symposium (RTSS), 1994*.

[204] Too-Seng Tia, Jane W. S. Liu, and Mallikarjun Shankar. Algorithms and optimality of scheduling soft aperiodic requests in fixed-priority preemptive systems. *Real-Time Systems*, 1996.

[205] Damir Isović and Gerhard Fohler. Efficient Scheduling of Sporadic, Aperiodic, and Periodic Tasks with Complex Constraints. In *IEEE Conference on Real-time Systems Symposium (RTSS), 2000*.

[206] Stefan Schorr. *Adaptive Real-Time Scheduling and Resource Management on Multicore Architectures*. PhD thesis, March 2015.

[207] Kevin Jeffay, Donald F. Stanat, and Charles U. Martel. On non-preemptive scheduling of period and sporadic tasks. In *IEEE Conference on Real-time Systems Symposium (RTSS), 1991*.

[208] Joseph Y.-T. Leung and M.L. Merrill. A note on preemptive scheduling of periodic, real-time tasks. *Information Processing Letters*, 1980.

[209] Enrico Bini and Giorgio C. Buttazzo. Measuring the performance of schedulability tests. *Real-Time Systems*, 30(1-2):129–154, 2005.

[210] T. Koller, G. Gala, D. Gracia Pérez, C. Ruland, and G. Fohler. Dreams: Secure communication between resource management components in networked multi-core systems. In *2016 IEEE Conference on Open Systems (ICOS)*, pages 99–104, Oct 2016.

[211] Eclipse Foundation. Acceleo. https://www.eclipse.org/acceleo/.

[212] OMG. Meta-object-facility model-to-text transformation language (MOFM2T). http://www.omg.org/spec/MOFM2T/1.0/, 2008.

[213] R.P. Goldberg. Survey of virtual machine research. *IEEE Computer Magazine*, 7(6):34–45, 1974.

[214] John Rushby. Design and verification of secure systems. In *ACM Operating Systems Review*, volume 15, 5, pages 12–21, 1981.

[215] Airlines Electronic Engineering Committee. Avionics Application Software Standard Interface. Technical Report ARINC-653, 1996.

[216] Fent Innovative Software Solutions. XtratuM. http://www.fentiss.com/en/products/xtratum.html, 2014.

[217] Miguel Masmano, Ismael Ripoll, Alfons Crespo, Jean-Jacques Metge, and Paul Arberet. XtratuM: An open source hypervisor for TSP embedded systems in aerospace. In *DASIA 2009. Data Systems In Aerospace.*, Istanbul, May 2009.

[218] GMV. AIR. https://www.gmv.com/en/Products/air/, 2014.

[219] Anthony Velte and Toby Velte. *Microsoft Virtualization with Hyper-V*. McGraw-Hill, Inc., New York, NY, USA, 1 edition, 2010.

[220] XEN. XEN. https://www.xenproject.org/, 2017.

[221] Christoffer Dall and Jason Nieh. KVM/ARM: Experiences building the Linux ARM hypervisor, 2013.

[222] SYSGO. PikeOS hypervisor. http://www.sysgo.com/products/pikeos-rtos-and-virtualization-concept/, 2015.

[223] A. Crespo, I. Ripoll, and M. Masmano. Partitioned embedded architecture based on hypervisor: The XtratuM approach. In *European Dependable Computing Conference (EDCC)*, pages 67–72, 2010.

[224] Salvador Trujillo, Alfons Crespo, Alejandro Alonso, and Jon Perez. MultiPARTES: Multicore partitioning and virtualization for easing the certification of mixed-criticality systems. *Microprocessors and Microsystems*, 38(8, Part B):921 – 932, 2014.

[225] Miguel Masmano, Alfons Crespo, and Javier Coronel. XtratuM hypervisor: User manual. Technical Report 14-035-03.005.sum.02, Fent Innovative Software Solutions, S.L., January 2016.

[226] Miguel Masmano, Javier O. Coronel, and Alfons Crespo. XtratuM hypervisor: Reference manual. Technical Report 14-035-03.006.sum.02, Fent Innovative Software Solutions, S.L., 2016.

[227] Miguel Masmano, Alfons Crespo, and Javier Coronel. XtratuM hypervisor: User manual annex XML schema specification. Technical Report 14-035-03.005.sum.02, Fent Innovative Software Solutions, S.L., January 2016.

[228] Fei Liu, Lanfang Ren, and Hongtao Bai. Secure-turtles: Building a secure execution environment for guest VMs on turtles system. *Journal of Computers*, 9(3):741–749, 2014.

[229] Linux kernel paravirt ops documentation. `http://lxr.free-electrons.com/source/Documentation/virtual/paravirtops.txt`. Accessed: 2017-08-27.

[230] S. Stabellini. Xen ARM/ARM64 config paravirt patch series. `http://lists.xen.org/archives/html/xen-devel/2014-01/msg00851.html`. Accessed: 2017-08-27.

[231] Linux process scheduler core code file. `http://lxr.free-electrons.com/source/kernel/sched/core.c`. Accessed: 2017-08-27.

[232] Alexander Spyridakis, Daniel Raho, and Jérémy Fanguéde. Virtual-BFQ: A coordinated scheduler to minimize storage latency and improve application responsiveness in virtualized systems. *International Journal on Advances in Software*, 7(3):642–652, 2014.

[233] Jérémy Fanguéde, Alexander Spyridakis, and Daniel Raho. Towards coordinated task scheduling in virtualized systems. In *The Ninth International Conference on Advanced Engineering Computing and Applications in Sciences (ADVCOMP)*, 2015.

[234] Heechul Yun, Gang Yao, Rodolfo Pellizzoni, Marco Caccamo, and Lui Sha. Memory bandwidth management for efficient performance isolation in multi-core platforms. In *IEEE Transactions on Computers*, pages 562 – 576. IEEE, 2016.

[235] N. Dagieu, A. Spyridakis, and D. Raho. Memguard: A memory bandwith management in mixed criticality virtualized systems memguard KVM scheduling. In *10th International Conference on Mobile Ubiquitous Computing, Systems, Services and Technologies (UBICOMM)*, 2016.

[236] Heechul Yun. Improving real-time performance on multicore platforms using memguard.

[237] Engin Ipek, Onur Mutlu, and José F. Martinez. Self-optimizing memory controllers: A reinforcement learning approach. In *Proceedings of 35th International Symposium on Computer Architecture, ISCA '08*. IEEE, 2008.

[238] Pierre Lucas, Kevin Chappuis, Michele Paolino, Nicolas Dagieu, and Daniel Raho. VOSYSmonitor, a low latency monitor layer for mixed-criticality systems on ARMv8-A. In *Proceedings of 29th Euromicro Conference on Real-Time Systems (ECRTS)*, 2017.

[239] ARM Ltd. Trustzone. https://developer.arm.com/technologies/trustzone.

[240] Richard Grisenthwaite. Armv8 technology preview. In *IEEE Conference*, 2011.

[241] Malcolm S Mollison, Jeremy P Erickson, James H Anderson, Sanjoy K Baruah, and John A Scoredos. Mixed-criticality real-time scheduling for multicore systems. In *Proceedings of IEEE 10th International Conference on Computer and Information Technology (CIT)*, pages 1864–1871. IEEE, 2010.

[242] ARM Ltd. *ARM Cortex -A Series*, March 2015. Programmer's Guide for ARMv8-A.

[243] ARM Ltd. *Juno ARM Development Platform SoC*, r1p0 edition, June 2013.

[244] Renesas. *R-Car Series, 3rd Generation User's Manual: Hardware*, February 2016.

[245] NVIDIA. *NVIDIA Tegra X1 Mobile Processor Technical Reference Manual*, November 2015.

[246] ARM Ltd. *SMC Calling Convention*, June 2013.

[247] ARM Ltd. *Power State Coordination Interface*, August 2012.

[248] ARM Ltd. *ARM Architecture Reference Manual*, January 2016. ARMv8, for ARMv8-A architecture profile.

[249] High Integrity Systems Ltd. SAFERTOS safety certified RTOS. https://www.highintegritysystems.com/safertos/.

[250] ARM Ltd. ARM compiler 6. https://developer.arm.com/products/software-development-tools/compilers/arm-compiler-6.

[251] Andrew Hopkins. The functional safety imperative in automotive design. Standard, ARM Ltd, September 2016.

[252] ARM Ltd. ARM compiler safety package. https://developer.arm.com/products/software-development-tools/compilers/arm-compiler/safety.

[253] Rusty Russell. Ubuntu manpage: hackbench - scheduler benchmark/stress test.

[254] A. Radulescu, J. Dielissen, S. G. Pestana, O. P. Gangwal, et al. An efficient on-chip NI offering guaranteed services, shared-memory abstraction, and flexible network configuration. *IEEE Trans. CAD of Integr. Circ. and Syst.*, 24(1):4–17, 2005.

[255] D. Lo, L. Cheng, R. Govindaraju, P. Ranganathan, et al. Heracles: improving resource efficiency at scale. In *Int. Symp. Computer Architecture*, pages 450–462, 2015.

[256] Marco Paolieri, Eduardo Quiñones, Francisco J. Cazorla, and Mateo Valero. An analyzable memory controller for hard real-time CMPs. *IEEE Embedded Systems Letters*, 1(4):86–90, 2009.

[257] T. Kwon, S. Lee, and J. Rho. Scheduling algorithm for real-time burst traffic using dynamic weighted round-robin. In *IEEE Int. Symp. Circ. and Syst.*, pages 506–509, 1998.

[258] R. Obermaisser, C. E. Salloum, B. Huber, and H. Kopetz. The time-triggered system-on-a-chip architecture. In *Industrial Electronics, 2008. ISIE 2008. IEEE International Symposium on*, pages 1941–1947, June 2008.

[259] W. Wolf, A. A. Jerraya, and G. Martin. Multiprocessor system-on-chip (mpsoc) technology. *IEEE Transactions on Computer-Aided Design of Integrated Circuits and Systems*, 27(10):1701–1713, October 2008.

[260] G. Tsamis, S. Kavvadias, A. Papagrigoriou, M. D. Grammatikakis, et al. Efficient bandwidth regulation at memory controller for mixed criticality applications. In *Reconfigurable SoC*, pages 1–8, 2016.

[261] B. Akesson and K. Goossens. Architectures and modeling of predictable memory controllers for improved system integration. In *Design, Automation Test in Europe Conference Exhibition (DATE), 2011*, pages 1–6, March 2011.

[262] Kyle J. Nesbit, Nidhi Aggarwal, James Laudon, and James E. Smith. Fair queuing memory systems. In *Proceedings of the 39th Annual IEEE/ACM International Symposium on Microarchitecture*, MICRO 39, pages 208–222, Washington, DC, USA, 2006. IEEE Computer Society.

[263] Y. Li, B. Akesson, and K. Goossens. Dynamic command scheduling for real-time memory controllers. In *26th Euromicro Conference on Real-Time Systems (ECRTS)*, pages 3–14, July 2014.

[264] Keith I. Farkas, Paul Chow, Norman P. Jouppi, and Zvonko Vranesic. Memory-system design considerations for dynamically-scheduled processors. In *Proceedings of the 24th Annual International Symposium on Computer Architecture*, ISCA '97, pages 133–143, New York, NY, USA, 1997. ACM.

[265] HSoC SystemC Virtual Platform. https://sourceforge.net/projects/hsoc/, 2013.

[266] François Pellegrini and Jean Roman. Scotch: A software package for static mapping by dual recursive bipartitioning of process and architecture graphs. In *Proceedings of the International Conference and Exhibition on High-Performance Computing and Networking*, HPCN Europe 1996, pages 493–498, London, UK, 1996. Springer-Verlag.

[267] Pier Stanislao Paolucci, Francesca Lo Cicero, Alessandro Lonardo, Mersia Perra, Davide Rossetti, Carlo Sidore, Piero Vicini, Marcello Coppola, Luigi Raffo, Gianni Mereu, Francesca Palumbo, Luca Fanucci, Sergio Saponara, and Francesco Vitullo. Introduction to the Tiled HW Architecture of SHAPES. In *Design, Automation and Test in Europe (DATE)*, April 2007.

[268] Lenovo. Optimizing memory performance of Lenovo servers based on Intel Xeon E7v3 processor. Technical Report lp0048, 2016.

[269] Arteris NoC. http://www.arteris.com.

[270] Sonics NoC. https://www.semiwiki.com/forum/content/4742-sonics-new-noc.html.

[271] Miltos D. Grammatikakis, Kyprianos Papadimitriou, Polydoros Petrakis, and Antonis Papagrigoriou et al. Security in MPSoCs: A NoC firewall and an evaluation framework. *IEEE Transactions on Computer-Aided Design of Integrated Circuits and Systems (TCAD)*, pages 1344–1357, August 2015.

[272] M.D. Grammatikakis, K. Papadimitriou, P. Petrakis, M. Coppola, and M. Soulie. Address interleaving in low cost NoCs. In *Reconfigurable SoC*, pages 1–8, 2016.

[273] John Rushby. *Partitioning in avionics architectures: Requirements, mechanisms, and assurance*, volume NASA/CR-1999-209347 of *NASA contractor report*. 1999.

[274] N.D. Enright Jerger and L.-S. Peh. *On-chip Networks*. Morgan & Claypool Publishers, 2009.

[275] Hermann Kopetz and Günther Bauer. *The Time-Triggered Architecture*. 2003.

[276] W. Steiner. Synthesis of static communication schedules for mixed-criticality systems. In *Proc. of 14th IEEE Int. Symposium on Object / Component / Service-Oriented Real-Time Distributed Computing*, pages 11–18, 2011.

[277] Gerald J. Popek and Robert P. Goldberg. Formal requirements for virtualizable third generation architectures. *Commun. ACM*, 17(7):412–421, July 1974.

[278] H. Kopetz and J. Reisinger. The non-blocking write protocol NBW: A solution to a real-time synchronization problem. In *1993 Proceedings Real-Time Systems Symposium*, pages 131–137, Dec 1993.

[279] Edited by: Roman Obermaisser. *Time-Triggered Communication*. Embedded Systems. CRC Press, USA, 2012.

[280] Precision clock synchronization protocol for networked measurement and control systems, September 2004.

[281] Hamidreza Ahmadian and Roman Obermaisser. Temporal partitioning in mixed-criticality NoCs using timely blocking. In *Proceedings of 11th International IEEE Symposium on Embedded Multicore/Many-core Systems-on-Chip (MCSOC17)*. 2017.

[282] W. J. Dally. Virtual-channel flow control. *IEEE Transactions on Parallel and Distributed Systems*, 3(2):194–205, March 1992.

[283] Hamidreza Ahmadian and Roman Obermaisser. A configurable simulation model for mixed-criticality multi-processor systems-on-chips. In *4th International Conference on Knowledge-Based Engineering and Innovation (KBEI-2017) Dec. 22nd, 2017 - Iran University of Science and Technology – Tehran, Iran*. IEEE, 2017.

[284] James W. Dally and Brian Towles. *Principles and Practices of Interconnection Networks*. Morgan Kaufmann Publishers, San Francisco, CA, 2003.

[285] W. Steiner and G. Bauer. Mixed-criticality networks for adaptive systems. In *29th Digital Avionics Systems Conference*, pages 5.A.3–1–5.A.3–10, October 2010.

[286] W. Steiner, G. Bauer, B. Hall, M. Paulitsch, and S. Varadarajan. TTEthernet dataflow concept. In *2009 Eighth IEEE International Symposium on Network Computing and Applications*, pages 319–322, July 2009.

[287] M. Abuteir and R. Obermaisser. Simulation Environment for Time-Triggered Ethernet. In *Proceedings of 11th IEEE International Conference on Industrial Informatics (INDIN)*, pages 642–648, July 2013.

[288] SAE. *SAE AS6802: Time-Triggered Ethernet*. SAE, November 2016.

[289] IEEE. IEEE 802.1D-2004 Standard for Local and Metropolitan Area Networks-Media Access Control (MAC) Bridges. Technical Report 802.1D-2004, International Organization for Standardization, 2004.

[290] IEC. IEC 61158 Digital data communications for measurement and control - Fieldbus for use in industrial control systems, June 2003.

[291] EtherCAT. *ETG.2100 EtherCAT Network Information (ENI) Specification*. EtherCAT, September 2015.

[292] IEC. *IEC 61784-3 Industrial communication networks - Profiles - Part 3: Functional safety fieldbuses - General rules and profile definitions*. IEC, June 2010.

[293] IEC. *IEC 62280 - Railway applications - Communication, signalling and processing systems*. IEC, October 2002.

[294] M. C. Magro, P. Pinceti, and L. Rocca. Can we use IEC 61850 for safety related functions? In *IEEE 16th International Conference on Environment and Electrical Engineering (EEEIC)*, pages 1–6, 2016.

[295] IEEE. IEEE 802.1AE-2006 Standard for Local and Metropolitan Area Networks-Media Access Control (MAC) Security. Technical Report 802.1AE-2006, International Organization for Standardization, 2006.

[296] Thomas Koller and Donatus Weber. *Security Services for Mixed-Criticality Systems Based on Networked Multi-core Chips*, pages 210–221. Springer International Publishing, Cham, 2016.

[297] IEEE. IEEE 802.1X-2010 Standard for Local and metropolitan area networks–Port-Based Network Access Control. Technical Report 802.1X-2010, International Organization for Standardization, 2010.

[298] Morris Dworkin. Recommendation for block cipher modes of operation: Galois/Counter Mode (GCM) and GMAC. Technical Report SP 800-38D, National Institute of Standards and Technology (NIST), 2007.

[299] NIST. Advanced Encryption Standard (AES). Technical Report FIPS-197, National Institute of Standards and Technology (NIST), 2001. http://csrc.nist.gov/publications/fips/fips197/fips-197.pdf.

[300] ISO/IEC. ISO/IEC 19772:2009. Information Technology – Security Techniques – Authenticated Encryption Mechanisms. Standard, International Organization for Standardization, Geneva, CH, February 2009.

[301] ISO/IEC. ISO/IEC 29192-1:2012. Information technology – Security techniques – Lightweight cryptography – Part 1: General. Standard, International Organization for Standardization, Geneva, CH, June 2012.

[302] ISO/IEC. ISO/IEC 29192-2:2012. Information technology – Security techniques – Lightweight cryptography – Part 2: Block ciphers. Standard, International Organization for Standardization, Geneva, CH, January 2012.

[303] Phillip Rogaway. OCB: Background. Website, 2015. Last visited on 15/03/2017.

[304] Horst Feistel. Cryptography and computer privacy. *Scientific American*, 228(5):15–23, May 1973.

[305] Y. Nir and A. Langley. ChaCha20 and Poly1305 for IETF Protocols. Technical Report RFC7539, RFC Editor, 2015.

[306] A. Langley and W. Chang. ChaCha20-Poly1305 Cipher Suites for Transport Layer Security (TLS). Technical Report RFC7905, RFC Editor, 2016.

[307] Radio Technical Commission for Aeronautics (RTCA) and EURopean Organisation for Civil Aviation Equipment (EUROCAE). DO-297: Software, electronic, integrated modular avionics (IMA) development guidance and certification considerations.

[308] DREAMS consortium. DREAMS final integration. http://www.uni-siegen.de/dreams/publications/2016-12-20-newsletter.pdf, 2016.

[309] Gautam Gala, Thomas Koller, Daniel Gracia Pérez, Gerhard Fohler, and Christoph Ruland. Timing analysis of secure communication between resource managers in dreams. In *1st Workshop on Security and Dependability of Critical Embedded Real-Time Systems in conjunction with RTSS 16*. CERTS 2016, IEEE Real-Time Systems Symposium, November 2016.

[310] H. Kopetz. The complexity challenge in embedded system design. In *11th IEEE International Symposium on Object Oriented Real-Time Distributed Computing (ISORC)*, pages 3–12, 2008.

[311] J. Perez, D. Gonzalez, C. F. Nicolas, T. Trapman, and J. M. Garate. A safety certification strategy for IEC-61508 compliant industrial mixed-criticality systems based on multicore partitioning. In *17th Euromicro Conference on Digital System Design (DSD)*, pages 394–400, August 2014.

[312] Darren Buttle. Real-time in the prime-time - ECRTS keynote talk. Report, ETAS GmbH, 2012.

[313] Jürgen Leohold and Christian Schmidt. Communication requirements for automotive systems. In *5th IEEE Workshop in Factory Communication Systems (WCFS)*, 2004.

[314] ERRAC. Joint Strategy for European Rail Research. Report, The European Rail Research Advisory Council, 2001.

[315] Hubert Kirrmann and Pierre A. Zuber. The IEC/IEEE communication network. *IEEE Micro*, vol. 21, no. 2:81–92, 2001.

[316] F. Corbier, L. Kislin, and E. Forgeau. How train transportation design challenges can be addressed with simulation based virtual prototyping for distributed systems. In *3rd European Congress ERTS - Embedded Real-Time Software*, 2006.

[317] Ondrej Kotaba, Jan Nowotsch, Michael Paulitsch, Stefan M. Petters, and Henrik Theilingx. Multicore in real-time systems - temporal isolation challenges due to shared resources. In *Workshop on Industry-Driven Approaches for Cost-Effective Certification of Safety-Critical, Mixed-Criticality Systems (WICERT)*, 2013.

[318] Risto Nevalainen, Oscar Slotosch, Dragos Truscan, Uwe Kremer, and Vicky Wong. Impact of multicore platforms in hardware and software certification. In *Workshop on Industry-Driven Approaches for Cost-effective Certification of Safety-Critical, Mixed-Criticality Systems (WICERT)*, 2013.

[319] J. Schneider, Michael Bohn, and Robert Röbger. Migration of automotive real-time software to multicore systems: First steps towards an automated solution. In *22nd EUROMICRO Conference on Real-Time Systems*, 2010.

[320] R. Fuchsen. How to address certification for multi-core based IMA platforms: Current status and potential solutions. In *IEEE/AIAA 29th Digital Avionics Systems Conference (DASC)*, 2010.

[321] Christian El Salloum, Martin Elshuber, Oliver Hoftberger, Haris Isakovic, and Armin Wasicek. The ACROSS MPSoC – a new generation of multi-core processors designed for safety-critical embedded systems. In *Digital System Design (DSD), 2012 15th Euromicro Conference on*, pages 105–113, 2012.

[322] J. Abella, F. J. Cazorla, E. Quinones, A. Grasset, S. Yehia, P. Bonnot, D. Gizopoulos, R. Mariani, and G. Bernat. Towards improved survivability in safety-critical systems. In *IEEE 17th International On-Line Testing Symposium (IOLTS)*, pages 240–245, 2011.

[323] Asier Larrucea. *Development and Certification of Dependable Mixed-Criticality Embedded Systems*. Thesis, Universität Siegen, 2017.

[324] Carlos Fernando Nicolas Ramirez. *Re-use of Tests and Arguments for Assessing Dependable Mixed-criticality Systems*. Thesis, Mondragon Unibertsitatea, 2017.

[325] Peio Onaindia. *Design patterns for mix-criticality applications*. Master, University of Kaiserslautern, 2017.

[326] Generic embedded system platform (GENESYS). `https://cordis.europa.eu/project/rcn/85392_en.html`.

[327] Roman Obermaisser and Hermann Kopetz. *GENESYS: A candidate for an ARTEMIS Cross-Domain reference architecture for embedded systems*. Number 202. Südwestdeutscher Verlag für Hochschulschriften (SVH) Aktiengesellschaft & Co. KG, 2009. FP7 ARTEMIS.

[328] Trusted computing engineering for resource constrained embedded systems application (TERESA). `https://cordis.europa.eu/project/rcn/93271_en.html`.

[329] Brahim Hamid, Jacob Geisel, Adel Ziani, Jean-Michel Bruel, and Jon Perez. Model-driven engineering for trusted embedded systems based on security and dependability patterns. In Ferhat Khendek, Maria Toeroe, Abdelouahed Gherbi, and Rick Reed, editors, *SDL 2013: Model-Driven Dependability Engineering*, pages 72–90, Berlin, Heidelberg, 2013. Springer Berlin Heidelberg.

[330] Brahim Hamid and Jon Perez. Supporting pattern-based dependability engineering via model-driven development: Approach, tool-support and empirical validation. *Journal of Systems and Software*, 122:239 – 273, 2016.

[331] Multi-cores partitioning for trusted embedded systems (MultiPARTES). `https://cordis.europa.eu/project/rcn/99802_en.html`.

[332] Probabilistic real-time control of mixed-criticality multicore and manycore systems (PROXIMA). `https://cordis.europa.eu/project/rcn/109947_en.html`.

[333] I. Agirre, M. Azkarate-Askasua, A. Larrucea, J. Perez, T. Vardanega, and F. J. Cazorla. A safety concept for a railway mixed-criticality embedded system based on multicore partitioning. In *2015 IEEE International Conference on Computer and Information Technology; Ubiquitous Computing and Communications; Dependable, Autonomic and Secure Computing; Pervasive Intelligence and Computing*, pages 1780–1787, October 2015.

[334] Irune Agirre, Mikel Azkarate-askasua, Asier Larrucea, Jon Perez, Tullio Vardanega, and Francisco J. Cazorla. Automotive safety concept definition for mixed-criticality integration on a COTS multicore. In Amund Skavhaug, Jérémie Guiochet, Erwin Schoitsch, and Friedemann Bitsch, editors, *Computer Safety, Reliability, and Security*, pages 273–285. Springer International Publishing, 2016.

[335] Wikipedia. Product liability. `https://en.wikipedia.org/wiki/Product_liability`, August 2017.

[336] Council of the European Union European Parliament. *Council Directive 85/374/EEC of 25 July 1985 on the approximation of the laws, regulations and administrative provisions of the Member States concerning liability for defective products*. European Parliament, Council of the European Union, The Member States, July 1985.

[337] Verkehr und Technologie Bayerisches Staatsministerium für Wirtschaft, Infrastruktur. *Pflichten der Wirtschaftsakteure - Kurzinformationen*. Bayerisches Staatsministerium für Wirtschaft, Infrastruktur, Verkehr und Technologie, Bayerisches Staatsministerium für Wirtschaft, Infrastruktur, Verkehr und Technologie, Prinzregentenstraße 28, 80538 München, October 2012.

[338] Council of the European Union European Parliament. *Regulation (EC) No 765/2008 of the European Parliament and of the Council of 9 July 2008 setting out the requirements for accreditation and market surveillance relating to the marketing of products and repealing*

Regulation (EEC) No 339/93. European Parliament, Council of the European Union, The Member States, July 2008.

[339] Council of the European Union European Parliament. *Regulation (EC) No 764/2008 of the European Parliament and of the Council of 9 July 2008 laying down procedures relating to the application of certain national technical rules to products lawfully marketed in another Member State and repealing Decision No 3052/95/EC.* European Parliament, Council of the European Union, The Member States, July 2008.

[340] Council of the European Union European Parliament. *Directive 2001/95/EC of the European Parliament and of the Council of 3 December 2001 on general product safety.* European Parliament, Council of the European Union, The Member States, December 2001.

[341] Council of the European Union European Parliament. *Decision No 768/2008/EC of the European Parliament and of the Council of 9 July 2008 on a common framework for the marketing of products, and repealing Council Decision 93/465/EEC.* European Parliament, Council of the European Union, The Member States, July 2008.

[342] European Commission. New legislative framework. https://ec.europa.eu/growth/single-market/goods/new-legislative-framework_en, August 2017.

[343] Council of the European Union European Parliament. *Directive 2006/42/EC of the European Parliament and of the Council of 17 May 2006 on machinery, and amending Directive 95/16/EC.* European Parliament, Council of the European Union, The Member States, May 2006.

[344] International Organization for Standardization. *Conformity assessment - Vocabulary and general principles.* ISO International Organization for Standardization, Geneva, Switzerland, 1st edition, 2004.

[345] International Organization for Standardization. *Conformity assessment - General requirements for accreditation bodies accrediting conformity assessment bodies.* ISO International Organization for Standardization, Geneva, Switzerland, 1st edition, April 2004.

[346] European Committee for Standardization. *EN ISO/IEC 17020, Conformity assessment - Requirements for the operation of various types of bodies performing inspection.* CEN European Committee for Standardization, Brussels, Belgium, 2nd edition, March 2012.

[347] European Committee for Standardization. *EN ISO/IEC 17025, General requirements for the competence of testing and calibration laboratories.* CEN European Committee for Standardization, Brussels, Belgium, 2nd edition, May 2005.

[348] European Committee for Standardization. *EN ISO/IEC 17065, Conformity assessments - Requirements for bodies certifying products, processes and services.* CEN European Committee for Standardization, Brussels, Belgium, 1st edition, September 2012.

[349] Wikipedia. Ce marking. https://en.wikipedia.org/wiki/CE_marking?oldformat=true, July 2017.

[350] European Commission. Harmonised standards. http://ec.europa.eu/growth/single-market/european-standards/harmonised-standards/, August 2017.

[351] Wikipedia. European committee for standardization. https://en.wikipedia.org/wiki/European_Committee_for_Standardization, July 2017.

[352] European Commission. Legislations. `http://ec.europa.eu/growth/tools-databases/nando/index.cfm?fuseaction=directive.main`, July 2017.

[353] International Electrotechnical Commission. IEC 61508-3 Functional Safety of Electrical/-Electronic/Programmable Electronic Safety-Related Systems - Part 3: Software requirements. Standard, IEC, IEC Central Office, 3, rue de Varembé, CH-1211 Geneva 20, Switzerland, April 2010.

[354] Adelard. Adelard safety case development manual (ASCAD), 1998.

[355] OCM. Structured assurance case metamodels (SACM), 2016.

[356] Asier Larrucea, Jon Perez, Irune Agirre, Vicent Brocal, and Roman Obermaisser. A modular safety case for an IEC 61508 compliant generic hypervisor. In *18th Euromicro Conference on Digital Systems Design (DSD 2015)*, pages 571–574, August 2015.

[357] Asier Larrucea, Jon Perez, and Roman Obermaisser. A modular safety case for an IEC 61508 compliant COTS multi-core device. In *The 13th International Conference on Dependable, Autonomic and Secure Computing (DASC)*, page 8, October 2015.

[358] Asier Larrucea, Jon Perez, Carlos F. Nicolas, Hamidreza Ahmadian, and Roman Obermaisser. A realistic approach to a network-on-chip cross-domain pattern. In *19th Euromicro Conference on Digital System Design (DSD 2016)*, pages 396–403, October 2016.

[359] Asier Larrucea, Imanol Martinez, Roman Obermaisser, Jon Perez, and Carlos F. Nicolas. Modular development of dependable mixed-criticality embedded systems. In *20th Euromicro Conference on Digital System Design (DSD 2017)*, pages 419–426, August 2017.

[360] IEC. *IEC 61784-3-12 Industrial communication networks - Profiles - Part 3-12: Functional safety fieldbuses - Additional specifications for CPF 12*. IEC, June 2010.

[361] EtherCAT. EtherCAT slave implementation guide. Techreport, EtherCAT, January 2012.

[362] EtherCAT. EtherCAT – Press Release, April 2013.

[363] Scott Henninger and Victor Corrêa. Software pattern communities: Current practices and challenges. In *Proceedings of the 14th Conference on Pattern Languages of Programs*, PLOP '07, pages 14:1–14:19, New York, NY, USA, 2007. ACM.

[364] Christopher Preschern, Nermin Kajtazovic, Andrea Höller, and Christian Kreiner. Pattern-based safety development methods: Overview and comparison. In *Proceedings of the 19th European Conference on Pattern Languages of Programs*, EuroPLoP '14, pages 28:1–28:20, New York, NY, USA, 2014. ACM.

[365] Asier Larrucea, Imanol Martinez, Vicent Brocal, Hamidreza Ahmadian, Salvador Peiró, Roman Obermaisser, and Jon Perez. Reusable generic design patterns for mixed-criticality systems based on DREAMS. *Microprocessors and Microsystems*, 54:35 – 46, 2017.

[366] Wind River. Wind River hypervisor. http://www.windriver.com/products/hypervisor/, 2015.

[367] QNX. QNX hypervisor. http://www.qnx.com/products/hypervisor/, 2015.

[368] Woo-Cheol Kwon and Li-Shiuan Peh. A universal ordered NoC design platform for shared-memory MPSoC. In *Proceedings of the IEEE/ACM International Conference on Computer-Aided Design*, ICCAD '15, pages 697–704, 2015.

[369] Sai Prashanth Muralidhara, Lavanya Subramanian, Onur Mutlu, Mahmut Kandemir, and Thomas Moscibroda. Reducing memory interference in multicore systems via application-aware memory channel partitioning. In *Proceedings of the 44th Annual IEEE/ACM International Symposium on Microarchitecture*, MICRO-44, pages 374–385, New York, NY, USA, 2011. ACM.

[370] Freescale Semiconductor. P4080 development system user's guide. Report, Freescale Semiconductor, August 2010.

[371] Jan Gustafsson, Adam Betts, Andreas Ermedahl, and Björn Lisper. The Mälardalen WCET benchmarks - past, present and future. In Björn Lisper, editor, *10th International Workshop on Worst-Case Execution Time Analysis (WCET 2010)*, volume 15, pages 136–146. Schloss Dagstuhl, 2010.

[372] Yoongu Kim, Dongsu Han, Onur Mutlu, and Mor Harchol-Balter. Atlas: A scalable and high-performance scheduling algorithm for multiple memory controllers. In *2010 International Conference on High Performance Computing Systems (HPCS)*, page 12, 2010.

[373] Hardik Shah, Kai Huang, and Alois Knoll. Timing anomalies in multi-core architectures due to the interference on the shared resources. In *2014 19th Asia and South Pacific Design Automation Conference (ASP-DAC)*, pages 708–713, 2014.

[374] Asier Larrucea, Imanol Martinez, Hamidreza Ahmadian, Roman Obermaisser, Vicent Brocal, Salvador Peiró, and Jon Perez. DREAMS: Cross-domain mixed-criticality patterns. In *Mixed-Criticality Workshop on Real Time System Symposium (RTSS)*, page 8, 2016.

[375] XILINX. Zynq-7000 All Programmable SoC: Technical Reference Manual. Report, XILINX, September 2014.

[376] R. Das, R. Ausavarungnirun, O. Mutlu, A. Kumar, and M. Azimi. Application-to-core mapping policies to reduce memory system interference in multi-core systems. In *2013 IEEE 19th International Symposium on High Performance Computer Architecture (HPCA)*, pages 107–118, February 2013.

[377] Freescale Semiconductor. Chip errata for the i.MX 6Dual/6Quad, 2013.

[378] Kees Goossens, John Dielissen, and Adrei Radulescu. AEthereal network on chip: concepts, architectures and implementations. *IEEE Design Test of Computers*, 22:414–421, September 2005.

[379] Heinz Gall. Certificate - functional safety management structure - IEC-61508 (no.: 968/FSM 138.00/11). Report, IKERLAN, 2011.

[380] DREAMS Consortium. DREAMS deliverable D1.3.1: Description of development process with model transformations. Report, DREAMS Consortium, 2014.

[381] L. A. Zadeh. Outline of a new approach to the analysis of complex systems and decision processes. *IEEE Transactions on Systems, Man, and Cybernetics*, SMC-3(1):28–44, Jan 1973.

[382] Sven Apel and Christian Kaestner. An overview of feature-oriented software development. *Journal of Object Technology*, 8(5):49–84, 2009.

[383] K. Kang, S. Cohen, J. Hess, W. Novak, and A. Peterson. Feature-oriented domain analysis (FODA) feasibility study. Technical Report CMU/SEI-90-TR-21, Software Engineering Institute, Carnegie Mellon University, November 1990.

[384] P. Schobbens, P. Heymans, and J. C. Trigaux. Feature diagrams: A survey and a formal semantics. In *Requirements Engineering, 14th IEEE International Conference*, pages 139–148, 2006.

[385] Klaus Pohl, Guenter Boeckle, and Frank J. van der Linden. *Software Product Line Engineering: Foundations, Principles and Techniques*. Springer-Verlag New York, Inc., 2005.

[386] Clemens A. Szyperski. *Component Software - Beyond Object-Oriented Programming*. Addison-Wesley-Longman, 1998.

[387] Bertrand Meyer. Applying "design by contract". *Computer*, 25(10):40–51, 1992.

[388] C. A. R. Hoare. An axiomatic basis for computer programming. *Commun. ACM*, 12(10):576–580, 1969.

[389] Antoine Beugnard, Jean-Marc Jézéquel, Noël Plouzeau, and Damien Watkins. Making components contract aware. *Computer*, 32(7):38–45, July 1999.

[390] Albert Benveniste, Benoit Caillaud, Dejan Nickovic, Roberto Passerone, Jean-Baptiste Raclet, Philipp Reinkemeier, Alberto Sangiovanni-Vincentelli, Werner Damm, Thomas Henzinger, and Kim G. Larsen. Contracts for system design. Research Report HAL-00757488, INRIA, 2012.

[391] C. A. R. Hoare. Communicating sequential processes. *Commun. ACM*, 21(8):666–677, 1978.

[392] Martn Abadi and Leslie Lamport. Conjoining specifications. *ACM Trans. Program. Lang. Syst.*, 17(3):507–535, 1995.

[393] Martn Abadi and Leslie Lamport. Composing specifications. *ACM Trans. Program. Lang. Syst.*, 15(1):73–132, 1993.

[394] Manfred Broy and Ketil Stølen. *Specification and development of interactive systems: focus on streams, interfaces, and refinement*. Springer-Verlag New York, Inc., 2001.

[395] Marius Bozga, Mohamad Jaber, Nikolaos Maris, and Joseph Sifakis. Modeling dynamic architectures using Dy-BIP. In Thomas Gschwind, Flavio De Paoli, Volker Gruhn, and Matthias Book, editors, *Software Composition*, volume 7306 of *Lecture Notes in Computer Science*, book section 1, pages 1–16. Springer Berlin Heidelberg, 2012.

[396] Robert J. Allen. *A Formal Approach to Software Architecture*. PhD thesis, Carnegie Mellon University, 1997.

[397] Modelica Association. Modelica® - A Unified Object-Oriented Language for Systems Modeling, Language Specification, Version 3.4. https://modelica.org/documents/ModelicaSpec34.pdf, 2017.

[398] E. Althammer, E. Schoitsch, G. Sonneck, H. Eriksson, and J. Vinter. Modular certification support – the decos concept of generic safety cases. In *6th IEEE International Conference on Industrial Informatics*, pages 258–263, 2008.

[399] Andrew Kornecki and Janusz Zalewski. Certification of software for real-time safety-critical systems: state of the art. *Innovations in Systems and Software Engineering*, 5(2):149–161, 2009.

[400] Frank Dordowsky, Richard Bridges, and Holger Tschope. Implementing a software product line for a complex avionics system. In Eduardo Santana de Almeida, Tomoji Kishi, Christa Schwanninger, Isabel John, and Klaus Schmid, editors, *Software Product Lines - 15th International Conference, SPLC 2011, Munich, Germany*, pages 241–250. IEEE, 2011.

[401] Stuart Hutchesson and John McDermid. Development of high-integrity software product lines using model transformation. In Erwin Schoitsch, editor, *Computer Safety, Reliability, and Security*, volume 6351, pages 389–401. Springer Berlin Heidelberg, 2010.

[402] Ibrahim Mustafa Habli. *Model-based assurance of safety-critical product lines.* Thesis, University of York, Department of Computer Science, 2009.

[403] Rosana T. Vaccare Braga, Onofre Trindade Junior, Kalinka R. L. J. Castelo Branco, and Jaejoon Lee. Incorporating certification in feature modeling of an unmanned aerial vehicle product line. In Eduardo Santana de Almeida, Christa Schwanninger, and David Benavides, editors, *16th International Software Product Line Conference, SPLC '12, Salvador, Brazil - September 2-7, 2012, Volume 1*, pages 249–258. ACM, 2012.

[404] P. Conmy and I. Bate. Assuring safety for component based software engineering. In *2014 IEEE 15th International Symposium on High-Assurance Systems Engineering*, pages 121–128, Jan 2014.

[405] Carlos-F. Nicolas, Fernando Eizaguirre, Asier Larrucea, Simon Barner, Franck Chauvel, Goiuria Sagardui, and Jon Perez. GSN support of mixed-criticality systems certification. In Stefano Tonetta, Erwin Schoitsch, and Friedemann Bitsch, editors, *Computer Safety, Reliability, and Security*, pages 157–172. Springer International Publishing, 2017.

[406] Luca P. Carloni, Fernando De Bernardinis, Claudio Pinello, Sangiovanni-Vicentelli, Alberto L., and Marco Sgroi. *Platform-Based Design for Embedded Systems*, volume 11. Mobile Networks and Applications, August 2006.

[407] A. Ruiz, H. Espinoza, S. Nair, K. Attwood, and P. Conmy. Compositional certification conceptual framework. Report Tech. Report D5.3, OPENCOSS project, 2013.

[408] Assurance and certification of cps (AMASS). http://www.amass-ecsel.eu/.

[409] Open platform for evolutionary certification of safety-critical systems (OPENCOSS). http://www.opencoss-project.eu/.

[410] OPENCOSS Consortium. OPENCOSS deliverable D5.3: Compositional certification conceptual framework. Report, 2013.

[411] A. Larrucea, I. Agirre, C. F. Nicolas, J. Perez, M. Azkarate-Askasua, and T. Trapman. Temporal independence validation of an IEC-61508 compliant mixed-criticality system based on multicore partitioning. In *Forum on Specification and Design Languages (FDL)*, pages 1–8, 2015.

[412] DREAMS Consortium. DREAMS deliverable D2.3.1: XtratuM support of enhanced hypervisor layer services. description and interfaces. Report, DREAMS Consortium, 2015.

[413] DREAMS Consortium. DREAMS deliverable D1.4.1: Meta-models for application and platform. Report, DREAMS Consortium, 2015.

[414] Gerhard Fohler, Gautam Gala, Daniel Gracia Perez, and Claire Pagetti. Evaluation of DREAMS resource management solutions on a mixed-critical demonstrator. In *9th European Congress on Embedded Real Time Software and Systems (ERTS)*, 2017.

[415] Soft Real Time ECG Analysis and Visualization. https://physionet.org/works.

[416] G. Tsamis, M. D. Grammatikakis, A. Papagrigoriou, P. Petrakis, V. Piperaki, A. Mouzakitis, and M. Coppola. Soft real-time smartphone ECG processing. In *12th IEEE International Symposium on Industrial Embedded Systems (SIES)*, pages 1–4, 2017.

[417] J. D.-Ferrer, D. Sánchez, and G.R.-Torrell. Anonymization of nominal data based on semantic marginality. *Info Sciences*, 242:35–48, 2013.

[418] J. Porquet, A. Greiner, and C. Schwarz. NoC-MPU: A secure architecture for flexible co-hosting on shared memory MPSoCs. In *Design Autom. Test in Europe*, pages 591–594, 2011.

[419] A. Wiggins, S. Winwood, H. Tuch, and G. Heiser. Legba: Fast hardware support for fine-grained protection. In *Asia-Pac. Conf. Adv. Comput. Syst. Arch.*, pages 320–336, 2003.

[420] L. Fiorin, G. Palermo, S. Lukovic, V. Catalano, and C. Silvano. Secure memory accesses on networks-on-chip. *IEEE Transactions on Computers*, 57(9):1216–1229, 2008.

[421] WFDB and OSEA. https://www.physionet.org/physiotools/wfdb.shtml.

[422] P. S. Hamilton, S. Patrick, and W. J. Tompkins. Quantitative investigation of QRS detection rules using the MIT/BIH arrhythmia database. *IEEE Transactions on Biomedical Engineering*, 12:1157–1165, 1986.

Index

Accreditation, 408
Adaptability in mixed-criticality systems, 378
Adaptation strategies, 221
    mode change, 232
Address interleaving, 330
    evaluation, 332
Avionics domain
    ARINC, 338
    challenges, 80, 81
    cybersecurity, 82
    DASL, 80
    federated architecture, 81
    IMA, 81
    mixed-criticality, 82
    partitioning, 80

Bandwidth management schemes, 318
Base variability resolution, 169

Communication paradigms, 333
    best-effort, 333
    rate-constrained, 333
    time-triggered, 333
Core failure management, 387
Core interface, 335
Cryptographic algorithms, 371
    ChaCha20-Poly1305, 371, 372
    CLEFIA-OCB, 371

Deadline overrun management, 388
Design space exploration, 155
    multi-objective, 170
    variability, 166
Dispatcher, 337
DRAM bandwidth management, 324
DREAMS abstraction layer, 53, 378
DREAMS execution architecture, 267
    application layer, 271
    hardware layer, 267
    runtime layer, 270
    virtualization layer, 268
DREAMS safety certification, 404
    functional safety management, 425

mixed-criticality pattern, 69, 414
    modular, 69
    modular safety case, 409
    product line, 428, 437
    strategy, 67, 405
DREAMS security
    application level, 363, 369
    cluster level, 363, 365
DREAMS services
    communication, 17, 19, 26–28, 318, 342
    core, 12
    execution, 46
    global time, 29
    optional, 15
    resource management, 40, 377, 378
    security, 59, 363, 365, 369
    virtualization, 48, 268, 273, 294

EtherCAT, 354
Execution services, 46

Failure modes, 72
Fault assumptions, 70
Fault containment regions, 70
Fault management, 458
Flow-control, 340
Functional safety management, 425

Gateway, 23
Generic methodology pattern, 117
Global reconfiguration graph, 395
Global time, 12, 29
    off-chip, 36
    on-chip, 32
Goal structuring notation, 128, 409, 435, 440,
    450
GRec, 222
GRM, 14, 41, 394
    implementation, 394
Guarding windows, 338

Health-care domain, 85
    challenges, 86
    demonstrator, 86, 459

Hypervisor, 262, 334, 414
Hypervisor Linux-KVM, 294
    'memguard', 299
    scheduling, 295
    security, 304
Hypervisor XtratuM, 273
    access to devices, 282
    configuration, 285
    deployment, 292
    health monitor, 281
    partitions, 277
    scheduling, 278
    services, 284

I-NoC, 332
I/O memory management unit, 27
Independent kernel-level monitoring, 322
Integration of TT and ET messages, 338
Interrupt virtualization, 50

Job shifting, 242

LRM, 15, 42, 387
LRS, 15, 43, 384
    general approach, 384
    implementation, 385

MemGuard, 318, 319
    architecture, 325
    evaluation, 325
    hardware, 323
    NoC-based evaluation, 327
MemGuardXt, 318
Message exchange format, 369
Meta-NoC, 332
Metamodel, 91
    deployment, 134
    hypervisor, 109
    NoC, 104
    node, 100
    processor, 105
    schedule, 138
    tile, 103
    timing, 114
    virtual link, 135
MINT, 338
Mixed-criticality, 1
Mixed-criticality pattern, 69, 414
    hypervisor, 414
    multi-core device, 418
    network, 423
Model-driven engineering, 64

Modeling viewpoints, 89
    architectural, 92
    configuration, 90
    deployment, 90
    logical, 89
    safety, 90
    technical, 89, 95
    timing, 89
    variability, 90
Modular safety case, 129, 409
MON, 15, 43, 379
    core failure, 380
    deadline overrun, 381
    quality of service, 382

Namespace, 10
NetGuard extension, 322

Off-chip, 19, 342
    network interface, 19
    router, 21
    security services, 360
On-chip
    network interface, 334
    router, 334
On-chip ports, 335
    architecture, 335
    configuration unit, 336
    data area, 335
    direction, 335
    event ports, 336
    state ports, 336
    status unit, 336
Order channel, 399

Partition, 47, 277
    multi-core, 52
    real-time system, 52
    states, 53
    system, 52
Partitioning, 13
Partitioning kernel, 47, 268
Platform, 9
Platform-independent model, 67
Platform-specific model, 67
Port, 17
    queuing, 52
    sampling, 52
Priorities for ET messages, 339
Priority unit, 337
Processor virtualization, 50

Product line, 166, 428
    safety certification, 130, 428, 435
    safety method, 437

QoS and deadline overrun managers, 393
QoS improvements, 392
QoS management, 390

Reconfiguration and adaptation, 379
Recovery, 222
Relaying of BE messages, 338
Resource allocation, 143
Resource management, 40
RM communication, 397

Safety
    architecture, 182
    case, 123
    constraints and rules checker, 128
    functions, 121
    management tool, 120
    manual, 122
    model, 120
Safety certification, 67, 406, 409
    compliant item, 69, 126, 407, 409
    modular, 69
Safety communication layer, 354, 410
Safety-critical demonstrator, 448, 453
Satisfiability modulo theory, 218
Scheduling domains, 185
    off-chip communication, 191
    on-chip communication, 191
    task, 190
Secure time synchronization, 367
Security, 59
    RM communication, 400
    services, 29, 360
Shared memory, 26
Shuffling, 339
Software analysis, 167
Spatial partitioning, 333
STNoC, 330
    address interleaving, 331
    evaluation of address interleaving, 332
    routing, 331
    simulation in Gem5, 332
Synchronization of schedulers, 337
System model, 9

Technology independence, 16
Temporal partitioning, 333
Threats, 73

Tile, 333
Time-triggered communication schedule, 337
Timely blocking, 338
    clean-up slot, 339
    white slot, 339
Timing chain, 115
Timing constraint, 116
Timing decomposition, 184
Timing event, 114
Transmission of RC messages, 338
Transmission of TT messages, 337
TTEthernet, 342

Update channel, 398

V model, 119
Variability, 149
    exploration, 151
    mixed criticality, 152
    technical, 159
Virtual channel, 339
Virtual link, 11, 136
Virtualization, 48, 262
    bare-metal, 264, 273
    full virtualization, 265
    hosted, 264, 294
    para-virtualization, 266
Virtualization layer, 48
Voting, 60

Waistline, 11
Wind-power domain, 83
    challenges, 84
    demonstrator, 448
    mixed-criticality, 85